# Lecture Notes in Electrical Engineering

## Volume 381

*About this Series*

"Lecture Notes in Electrical Engineering (LNEE)" is a book series which reports the latest research and developments in Electrical Engineering, namely:

- Communication, Networks, and Information Theory
- Computer Engineering
- Signal, Image, Speech and Information Processing
- Circuits and Systems
- Bioengineering

LNEE publishes authored monographs and contributed volumes which present cutting edge research information as well as new perspectives on classical fields, while maintaining Springer's high standards of academic excellence. Also considered for publication are lecture materials, proceedings, and other related materials of exceptionally high quality and interest. The subject matter should be original and timely, reporting the latest research and developments in all areas of electrical engineering.

The audience for the books in LNEE consists of advanced level students, researchers, and industry professionals working at the forefront of their fields. Much like Springer's other Lecture Notes series, LNEE will be distributed through Springer's print and electronic publishing channels.

More information about this series at http://www.springer.com/series/7818

Ahmed El Oualkadi · Fethi Choubani
Ali El Moussati
Editors

# Proceedings of the Mediterranean Conference on Information & Communication Technologies 2015

## MedCT 2015 Volume 2

 Springer

*Editors*
Ahmed El Oualkadi
National School of Applied Sciences
  of Tangier
Abdelmalek Essaadi University
Tangier
Morocco

Ali El Moussati
National School of Applied Sciences, Oujda
Mohammed Premier University
Oujda
Morocco

Fethi Choubani
Technopark el Ghazala
Sup'Com
Ghazala
Tunisia

ISSN 1876-1100                    ISSN 1876-1119   (electronic)
Lecture Notes in Electrical Engineering
ISBN 978-3-319-80773-7           ISBN 978-3-319-30298-0   (eBook)
DOI 10.1007/978-3-319-30298-0

Printed on acid-free paper

This Springer imprint is published by Springer Nature
The registered company is Springer International Publishing AG Switzerland

# Preface

*The Mediterranean Conference on Information & Communication Technologies* (MedICT 2015) was held at the wonderful Moroccan Blue Pearl city of Saidia, Morocco during May 7–9, 2015. MedICT is a Mediterranean premier networking forum for leading researchers in the highly active fields of information and communication technologies. The general theme of this edition is Information and communication technologies for sustainable development of Mediterranean countries.

MedICT provides an excellent international forum to the researchers and practitioners from both academia as well as industry to meet and share cutting-edge development. The conference has also a special focus on enabling technologies for societal challenges, and seeks to address multidisciplinary challenges in information and communication technologies such as health, demographic change, well-being, security, and sustainability issues.

The proceedings publish high-quality papers which are closely related to the various theories, as well as emerging and practical applications of particular interest to the ICT community.

This volume provides a compact yet broad view of recent developments in the data, systems, services and education, and covers exciting research areas in the field including control systems, software engineering, data mining and big data, ICT for education and support activities, networking, cloud computing and security, ICT-based services and applications, mobile agent systems, software engineering, data mining and big data, online experimentation and artificial intelligence in education, networking, cloud computing and security, ICT-based education and services, ICT challenges and applications, advances in ICT modeling and design, and ICT developments and recent progress.

Ahmed El Oualkadi
Fethi Choubani
Ali El Moussati

# Contents

**Part VII  Online Experimentation and Artificial Intelligence in
Education**

**Part VIII  Networking, Cloud Computing and Security**

# Part I
# Control Systems

# A Novel Velocity-Separation Difference Model for Traffic Flow Theory

Hajar Lazar, Khadija Rhoulami and Moulay Driss Rahmani

**Abstract** The main objectives of paper is understanding the traffic theory, discussed the extended optimal velocity models proposed in many research and we proposed to modify VSDM model by incorporating between simple optimal velocity function and inverse of time to collision which its used as a safely indicator in braking state. The simulation results shows that MVSDM react better than VSDM in order to brake in a short time with a safely distance.

**Keywords** Velocity-separation difference model · Traffic flow · Microscopic simulation · Car following model · Optimal velocity function

## 1 Introduction

Traffic flow problem started some 40 years ago, when Lighthill and Whitham [1] presented a model based on the analogy of vehicles in traffic flow. Then, mathematical description of traffic flow has been an interest subject of research and debate for traffic engineers. This has resulted in a broad scope of models describing different aspects of traffic flow operations, either by considering the time-space behavior of individual drivers under the influence of vehicles in their proximity [2]. For modeling traffic flow problem, there are two classes of models: Macroscopic, which is concerned with average behavior, such as traffic density, average speed and module area [3]. Microscopic models attempt to model the motion of individual vehicles within a system. They are typically functions of position, velocity, and acceleration. This type of models is created using ordinary differential equations, with each vehicle having its own equation. Because the behavior of these models is usually dictated by a lead vehicle, they are termed "car-following" models. Car-following models, which describe the processes by which drivers follow each

H. Lazar (✉) · K. Rhoulami · M.D. Rahmani
Faculty of Sciences, Rabat GSCM-LRIT Laboratry Associate Unit to CNRST (URAC 29),
Mohammed V University, B.P. 1014 Rabat, Morocco
e-mail: hajar.lazar@gmail.com

© Springer International Publishing Switzerland 2016
A. El Oualkadi et al. (eds.), *Proceedings of the Mediterranean Conference on Information & Communication Technologies 2015*, Lecture Notes in Electrical Engineering 381, DOI 10.1007/978-3-319-30298-0_1

3

other in the traffic stream [4]. Car-following itself forms one of the main processes in all microscopic simulation models as well as in modern traffic flow theory, which attempts to understand the interplay between phenomena at the individual driver level and global behavior. The most famous one is the Car-Following model particularly optimal velocity model [5], where the driver adjusts his or her acceleration according to the conditions in front. In that model the vehicle position is treated as a continuous function and each vehicle is governed by an ordinary differential equation that depends on speed and distance of the car in front. In this work, we discussed and gives a detailed analysis of extended optimal velocity models OVM, FVDM, and VSDM. Then we proposed to modify the VSDM model by introduces a new optimal velocity function using the safely indicator inverse of time to collision which takes into account the both of spacing and the relative speed.

The rest of the paper is organized as follows: Sect. 2 gives a short discussion of previous car-following models particularly optimal velocity model. Section 3 presents the modified velocity-separation difference model (modified VSDM in short). Simulation results are presented in Sect. 4.

## 2 Optimal Velocity (OV) Model and Some Extended Models

The optimal velocity (OV) model is one of the traffic flow models, which has been put forward by to describe some realistic features [6]. OV model may be one of the most representative car-following models for its simplicity and good performance. Generally the microscopic model with continuous variables is expressed by the equation of motion:

$$\ddot{X}_n(t) = k * F \begin{pmatrix} \ldots, \Delta X_{n+1}, \Delta X_n, \Delta X_{n-1}, \ldots \\ \ldots, \dot{X}_{n+1}, \dot{X}_n, \dot{X}_{n-1}, \ldots \\ \ldots, X_{n+1}, X_n, X_{n-1}, \ldots \end{pmatrix} \tag{1}$$

where $X_n$, $\dot{X}_n$, $S(t) = \Delta X_n = X_{n+1} - X_n$, and coefficient k are position, velocity, headway of the n-th car, and sensitivity respectively. Equation (1) simply states that the acceleration of the n-th car is decided by the motion of the n-th car and the surrounding traffic, i.e. the headway, velocity, and position of the preceding car and those of the car that follows and so on. The car-following models are based on the idea that the dominant part of stimulus comes from the preceding car [7]. Along this idea, the equation of motion (1) can be simplified. In this paper, we presented three models that are constructed in different viewpoints with respect to the behavior of a driver: How a driver controls his car.

## 2.1 Model 1: Optimal Velocity Model Proposed by [6]

In the OV model, the acceleration of the n-th vehicle at time t is determined by the difference between the actual velocity, $\dot{X}_n$, and an optimal velocity $V_{opt}(S(t))$, which depends on the headway S(t) to the car in the front. The equation of motion is:

$$\ddot{X}(t) = k * [V_{opt}((S(t)) - \dot{X}_n(t)]$$  (2)

The optimal velocity function $V_{opt}(S(t))$ is a monotonically increasing function and has an upper bound. The purpose of this paper is to evaluate and clarify the dynamical changes of traffic states using optimal velocity function in traffic system. The choice of the optimal velocity function is crucial to achieve the goal of research. Here we adopt the optimal velocity function calibrated by using actual measurement data as follows [8]:

$$V_{opt}(S(t)) = V_1 + V_2 \tanh[C_1(S(t) - l) - C_2]$$  (3)

## 2.2 Model 2: Full Velocity Difference Model Proposed by [9]

Simulation results show that the OV model produces too high acceleration and unrealistic deceleration, and crash phenomenon may be appeared. The authors [8] proposed Generalized Force Model (GFM) to avoid these problems. Based in this theory, the velocity difference is appended to the OV model when the velocity of leading car is smaller than the following [10]. In 2001, [9] conducted a detailed analysis of GFM and found out that it cannot predict correctly the delay time of car motion when using GFM. They thought it was due to that GFM does not take the effect of positive $\Delta\dot{X}$ on traffic dynamics into account, and then they suggested a full velocity difference model (FVDM) which includes the positive $\Delta\dot{X}$ factor and obtained a more systematic model as follows:

$$\ddot{X}(t) = k * [V_{opt}((S(t)) - \dot{X}_n(t)] + \lambda\Delta\dot{X}$$  (4)

## 2.3 Model 3: Velocity-Separation Difference Model (VSDM) Proposed by [5]

In 2006, Li Zhi-Peng et al. [5] conduct that the second term in the right side of FVDM of (4) makes no allowance of the effect of the inter car spacing independently of the relative velocity. However, they proposed to separate between cars because it plays an important role in traffic dynamics. They justified that the acceleration controlled by a driver is different when he responds to two different stimuli which have the same relative velocity but different headway. According to the above analysis, they modified the FVDM by taking the separation between cars into account. Since this model takes velocity difference dependent on separation into account, they call it a velocity-difference-separation model (VDSM). The dynamic equation as follows:

$$\ddot{X}(t) = k * [V_{opt}((S(t)) - \dot{X}_n(t)]$$
$$+ \lambda\Theta(\Delta\dot{X})\Delta\dot{X}(1 + \tanh[C_1(S(t) - l) - C_2])^3 \qquad (5)$$
$$+ \lambda\Theta(-\Delta\dot{X})\Delta\dot{X}(1 - \tanh[C_1(S(t) - l) - C_2])^3$$

where $\Theta(.)$ is step function.

## 3 Proposed Model

In this paper, we proposed to modify VSDM that takes velocity difference dependent on separation into account using a new optimal velocity function (OVF). The selection of the OVF thus depends on a users choice. However, this choice cannot be completely arbitrary since the OVF must satisfy several analytical conditions to describe the observed relation between spacing and velocity. The OVF should be a continuous non-negative function defined for $S(t) > 0$ and must be a monotone function [11]. In this work, we using a new OVF introducing a weighting factor that depends on the ratio of the relative speed to spacing proposed by [12], that is the opposite of the inverse of time to collision (TTC). In Ref. [12], the authors used this new optimal velocity function to modify OV model. In our case, we used this new one to modify VSDM model to make it more reactive on braking state. The optimal velocity function (3) is thus changed to $V_{opt}^{new}(S, \dot{S})$ given by:

$$V_{opt}^{new} = V_{opt}(S) * W(S, \dot{S}) \qquad (6)$$

where the weighting factor is:

$$W(S, \dot{S}) = [A(1 + \tanh B(\frac{\dot{S}}{S} + C)] \qquad (7)$$

We extended VSDM model by incorporating the new OVF that combined between the simple OVF and the concept of inverse time to collision to get a novel model that called a modified VSDM model is expressed by the equation of motion:

$$\ddot{X}(t) = k * [V_{opt}^{new}((S(t), \dot{S}(t)) - \dot{X}_n(t)]$$
$$+ \lambda\Theta(\Delta\dot{X})\Delta\dot{X}(1 + \tanh[C_1(S(t) - l) - C_2])^3 \tag{8}$$
$$+ \lambda\Theta(-\Delta\dot{X})\Delta\dot{X}(1 - \tanh[C_1(S(t) - l) - C_2])^3$$

We proposed in Fig. 1 a modified VSDM car-following model that uses the new optimal velocity function introducing the weighting factor. However, the weighting factor introduces the inverse of time to collision as a very important safely indicator using in braking state.

In order to do numerical analysis, we rewrite the second-order ODE of modified VSDM (8) as a system of first-order ODEs. Let $x_1 = X$ and $x_2 = \dot{X}$ are position and velocity respectively. Let $x = \begin{pmatrix} x_1 \\ x_2 \end{pmatrix}$ the vector depending on position and velocity of each car and $f\begin{pmatrix} x_1 \\ x_2 \end{pmatrix} = \begin{pmatrix} x_2 \\ g(x_1, x_2, p) \end{pmatrix}$ the function to resolve of first-order ODEs. Where p = (t, x, k, $v_0$, l, A, $C_1$, $C_2$, $V_1$, $V_2$, C, B, $\lambda$) parameters used in a simulation.

$$g(x_1, x_2, p) = \alpha_0 * x_2 + \alpha_1 \tag{9}$$

where

$$\alpha_0 = -k \quad \text{and} \quad \alpha_1 = k * V_{opt}^{new}$$
$$+ \lambda\Theta(\Delta\dot{X})\Delta\dot{X}(1 + \tanh[C_1(S - l) - C_2])^3 \tag{10}$$
$$+ \lambda\Theta(-\Delta\dot{X})\Delta\dot{X}(1 - \tanh[C_1(S - l) - C_2])^3$$

## 4   Simulation

In these simulations we used ODE45 in Matlab which uses Runge-Kutta methods to solve systems of ordinary differential equations to find the position and velocity of each car. The parameters values used in (3) are $V_1 = 6.75\,\text{m/s}$, $V_2 = 7.91\,\text{m/s}$, and $C_1 = 0.13$, $C_2 = 1.57$. The parameters values used in (7) are $B = 5\,\text{s}$, $C = 0.4\,\text{m}^{-1}$, and $A = 1/2$. The sensitivities coefficients are $k = 0.85$, $\lambda = 0.4$.

Figure 2 simulate the motion of car using VSDM and MVSDM model. Figure 3 illustrates the time evolution of the speed of both VSDM and a modified VSDM vehicles. The simulation result shows that the vehicle velocity maximum of modified VSDM reach 8 m/s, however, the vehicle velocity maximum of VSDM reach 14 m/s. We deducted that the maximum of speed of a modified VSDM under the maximum of speed that related to VSDM model. MVSDM velocity begins to

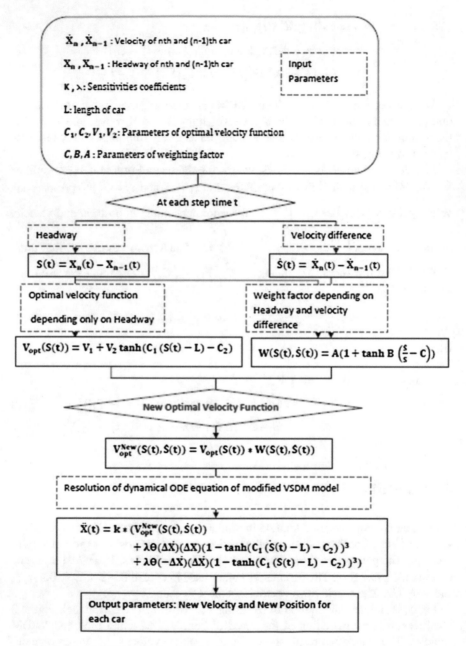

**Fig. 1** Modified VSDM Car-following algorithm

decrease before the VSDM vehicle velocity reaches its maximum. This due to introduction of weighting factor which used the inverse of time to collision which reduces the optimal velocity and make it the model more reactive in braking state.

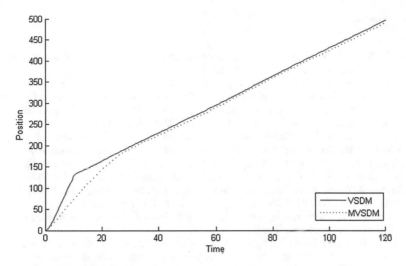

**Fig 2** Position of motion of car using VSDM and MVSDM models

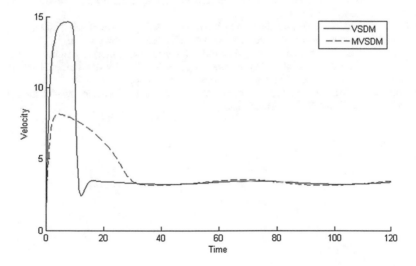

**Fig. 3** Velocity of car using VSDM model and MVSDM models

## 5 Conclusion

Microscopic traffic flow model described a traffic flow in a realistic manner. In this paper we describe a VSDM model used a optimal velocity function and we compare them with a modified VSDM model that introduce the new optimal velocity function based on weighting factor which depending on spacing and relative speed for control a braking situations to avoid a collision.

# References

1. Lighthill, M. J., Whitham, G. B.: On kinematic waves. II. A theory of traffic flow on long crowded roads. Proc. R. Soc. Lond. (1955)
2. Serge, P., Piet,. B: State-of-the-art of vehicular traffic flow modeling. Special Issue on Road Traffic Modeling and Control of the Journal of Systems and Control Engineering (1999)
3. Sadeq, J.: New Models for Crowd Dynamics and Control. Faculty of the Virginia Polytechnic Institute and State University. Thesis, Blacksburg, Virginia (2006)
4. Mark, B., Mike, M.: Car-Following: A Historical Review. Department of Civil and Environmental Engineering, Transportation Research Group, University of Southampton (2000)
5. Li, Z., Gong, X., Liu,Y.: A Velocity-Separation Difference Model for Car-Following Theory and Simulation Tests. Department of Control Science and Technology, IEEE (2006)
6. Bando, M., Haseb, K., Nakayama, A., Shibata, A., Sugiyama, Y.: Dynamical Model of Traffic Congestion and Numerical Simulation, vol. 51. Aichi University (1995)
7. Nakayama, A., Haseb, K., Sugiyama, Y.: Optimal Velocity Model and Its Applications. Gifu Keizai University, Ohgaki (2003)
8. Li, Z., Yi, Y.: Study on Traffic States and Jamming Transitions for Two Lane Highway Including a Bus by Using a Model with Calibrated Optimal Velocity Function. The Key Laboratory of Embedded System and Service Computing supported by Ministry of Education, Tongji University, Shanghai, China (2014)
9. Rui, J., Qingsong, W., Zuojin, Z.: Full velocity difference model for a car-following theory. Phys. Rev. E 017–101 (2001)
10. Liu, L., Zhang, N.: A New Car Following Model Considering the Acceleration of Leading Car School of Economics and Management. Beihang University, Beijing (2010)
11. Milan, B., Elen, T.: Optimal velocity functions for car-following models. Journal of Zhejiang University-Science A (Applied Physics and Engineering), Faculty of Maritime Studies and Transport, University of Ljubljana, 6320 Portoro, Slovenia (2012)
12. Mammar, S., Mammar, S., Haj Salem, H.: A Modified Optimal Velocity Model for Vehicle Following. Marne-la-Vallee, France (2005)

# Artificial Intelligence (AI) Contribution to GIS in Optimal Positioning of Hydrophone Sensors Using Genetic Algorithm (Case Study: Water Network, Casablanca, Morocco)

Hatim Lechgar, Abdelouahed Mallouk, Mohamed El Imame Malaainine, Tarik Nahhal and Hassane Rhinane

**Abstract** All over the world, potable water networks suffer from substantial leaking which may occur in places that are hard to locate. In addition to economic loss, there is potential hazard of epidemics. Thus, real time detection and hence fixing water leaks is essential for an efficient distribution network. Hydrophone sensors come in handy thanks to their ability to detect leaks. Nevertheless, a comprehensive covering of the network is impossible financially. This leads to searching for an arrangement of sensors to have an optimal coverage of distribution network. Linear search for optimal solutions in big networks breaks into an explosion of calculations which is unpractical in terms of resources allocation and time execution. We present how artificial intelligence (AI), and especially genetic algorithms (GA), offer both efficient and fast ways of figuring out the best disposition of sensors for an optimal coverage of the water network.

**Keywords** GIS · Artificial intelligence · Network's sensors · Network coverage · Genetic algorithm · SLOTS

H. Lechgar (✉) · A. Mallouk · M.E.I. Malaainine · T. Nahhal · H. Rhinane
Faculté Des Sciences Ain Chock, Université Hassan II-Casablanca, Casablanca, Morocco
e-mail: h.lechgar@gmail.com

A. Mallouk
e-mail: wahid.mallouk@gmail.com

M.E.I. Malaainine
e-mail: mohamed.malaainine@gmail.com

T. Nahhal
e-mail: t.nahhal@fsac.ac.ma

H. Rhinane
e-mail: h.rhinane@gmail.com

© Springer International Publishing Switzerland 2016
A. El Oualkadi et al. (eds.), *Proceedings of the Mediterranean Conference on Information & Communication Technologies 2015*, Lecture Notes in Electrical Engineering 381, DOI 10.1007/978-3-319-30298-0_2

# 1  Introduction

GIS problems are often subject to what is known as curse of dimensionality, which means that the state space grows rapidly when the number of parameters increases. However, the use of intelligent algorithms reduces considerably the size of the state space and helps to quickly find the optimal configurations.

In this chapter we will study the case of optimization of hydrophone sensors placement in a water network. Hydrophone sensors are devices that could listen to noise along a certain distance in water pipe.

There are several approaches of optimizing placement that have been used, especially in the problem of the Battle of the Water Sensor Networks (BWSN) [1], which studies the optimization of placement of sensors that detects a contamination in a water network. Other studies related to flaws detection apply other methods using pressure drop sensors [2]. Nevertheless pressure is not quite trustworthy unless to detect minor bursts, due to daily fluctuation and variation which enables spotting only major flaws.

We will show in this chapter the application of the methods used in BWSN, in the positioning of Hydrophone sensors namely greedy, and slots algorithms, and use their results as inputs to an elaborated Genetic algorithm.

## 1.1  Problem Formulation

Hydrophone sensors are devices placed on valves of water network and can listen to noise along a specific distance in a pipe, and hence allow the detection of an abnormal high noise; which is correlated with "bursts or leaks in pipes" [3].

An exhaustive coverage of the network is financially impossible, so water distribution companies tend to place sensors only on areas of recurrent burst. However, using sequential search techniques to maximize the coverage of a network or at least achieve a high percentage, bursts rapidly into what is known as combinatorial explosion. For instance the computing of all possible configurations (state space) of 20 sensors in a small subset of water network that contains 130 valves (possible placements) would take 3000 years even with a computer that does 1 trillion operations per second.

$$\binom{130}{20} / (10^{12} * 3600 * 24 * 365) \cong 3000\,\text{years}) \tag{1}$$

It is worth noting also that even with a complete coverage of valves there may be still some dark areas of network that could not possibly be listened to. So we had to restrict our study to the subset of water network that is listenable.

We formulated the problem as follows:

- Let **NetW** be a network of water with **n** valves.
- Let **m** be the number of sensors.
- Let **TargPrc** be the target percentage of covered pipes.
- Find the minimum number of sensor **m** and their optimal placement on **n** candidate valves in order to reach at least **TargPrc** in **NetW**.

By solving this problem we guarantee that we can detect leaks across the wanted percentage of water network with **m** sensors.

## 1.2  Related Work

There are several works in similar problems, especially regarding BWNS [4]. Other problems based on calculating pressure drop to detect leaks, use structural analysis and integer optimization [2]. However acoustic based detection is more reliable thanks to its better sensitivity to abnormal noise. Casillas et al. (2013) continue in the same optic of using integer optimization and enhanced the performance with genetic algorithm [5]. Nevertheless there are other techniques that may be less optimal but more straightforward and efficient. For instance SLOTS is the best non evolutionary algorithm that solves the BWNS [6]. Genetic algorithm comes in handy because it can optimize far more initial suboptimal solutions. So we can use outputs from a greedy algorithm or SLOTS and optimize them afterward in an evolutionary way.

## 1.3  Overview of Our Method

We proposed to apply at first greedy algorithm and SLOTS algorithm to compare their results and efficiency then opt for the two best results to incorporate in a genetic algorithm.

## 2  Data Preparation

### 2.1  Subset of Listenable Network

We restricted the study domain to the reachable pipes. The reachable pipes are pipes that could be listened to while having a sensor on each valve in the network.

## 2.2 Cutting to Elementary Segments

Processing segments in their variety of length could slow down computation because that would imply for example cutting a pipe to the distance it is heard by a sensor, and hence write in the spatial database 2 new segments. To avoid this trap, we split all segments into equi-length segments of size **d**, so we did not have to bother about the geometry or length of each segment at once. This resulted in a bigger segment table, but the speed tradeoff was worthwhile.

## 2.3 Equivalent Homogenous Network

A hydrophone sensor come with a manufacturer note on how long it can listen across a specific materiel. For example the sensor could listen up to a distance 'a' along a steel pipe, while it could listen to a different distance 'b' along concrete one. That adds a complexity to our problem, because computations would require checking the type of each pipe to figure out how much length it could be listened to.

To reduce that complexity, we transformed the real length $LP_{real}$ of each pipe to an equivalent length $LP_{eq}$ as follows:

- Let **SeP** be the sensor sensitivity to the materiel of pipe **P**.
- Let $Se_{min}$ be the lowest sensor sensitivity to all materiels.
- And **KP = SeP/Se$_{min}$**

So:

$$LP_{eq} = KP * LP_{real}. \tag{2}$$

## 2.4 Adjacency Tables

Spatial data are inherently well formatted and indexed in spatial databases.

However a massive querying in problems like the one we are tackling could make the computation process inefficient.

That is why we opted for the use of the representation of our network as an attribute table rather than attacking spatial entities directly. A common representation of this table is adjacency table, in which we correspond to each valve, all correspondent network (elementary) segments and to each segment its adjacent segments. These generated tables were then indexed, and we could acknowledge the evident enhancement in queries speed.

## 2.5   Preparation of Initial Heuristic Table

A heuristic is a function that estimates at each state, in a search problem space, the proximity to the goal state.

In order to implement AI search techniques in our problem, we built a table of heuristics in which each valve (a possible sensor position) is matched to the sum of all lengths of the pipes that can detect.

Each time a segment is matched to a sensor (that is a valve position) it gets subtracted from all other positions that could reach it.

This heuristic is used while searching for the next state in graph search both in greedy and SLOTS.

# 3   Methodological Approach

The couple {m,TargPrc} presents a trade-off. That is we don't know what the maximum achievable **TargPrc** with m sensors is. So in order not to complexify the problem, we applied for each number **m** of sensors both Greedy algorithm and SLOTS to compare their results. Afterwards we chose two outputs of better-off method to plug in Genetic algorithm as inputs.

At the end, it is the job of the decision-maker to determine which couple is affordable.

## 3.1   Greedy Algorithm

We applied at first the greedy algorithm. The greedy algorithm uses mainly the heuristic we built. It is simple but can be very suboptimal because it does not question at any step its previous choices.

Greedy algorithm is as follows:

- Let **N** be the number of sensors to be placed and **Sj** the sensor at location (**j**).
- The greedy Algorithm proceeds in **N** iterations by choosing each time a **Sj** that enables the sensor to listen to the maximum length of network (our heuristic), and adds its location to the set of solution.

## 3.2   SLOTS Algorithm

SLOTS stands for Sensors local optimal transformation system by Dorini et al. (2010) [7]. It was developed in response to the BWNS and it is as follows:

- Begins with an arbitrary set of **m** sensor locations.
- One sensor moves while other sensors are held fixed, performance changes are measured each time.
- The sensor is placed in the location corresponding to the best improvement in performance.
- The algorithm terminates when there is no improvement from loop to loop.

## 3.3  Genetic Algorithm

Genetic algorithm (GA) is a search technique that mimics the process of natural selection [8].

To prevent the algorithm from converging on suboptimal local minima or maxima, an initial population of 'parent' solutions is first generated (Huang et al. 2006) [9].

That is why to process GA for **m** sensors we applied SLOTS to **m** sensor with 2 arbitrary initial locations, and use their respective results as the initial individual in GA.

- Let **Confg** be a set of bits such as its cardinal is equal to the number of valves in the network. **Confg** stands for configuration.
- Let **bi** be the bit that corresponds to the valve number **i** and **nbSns** the number of sensors.

So:

$$\mathbf{Confg} = \{\mathbf{b1}, \mathbf{b2}, \ldots, \mathbf{bn}\}. \tag{3}$$

$$\mathbf{bi} = \begin{cases} \mathbf{0} & \text{if no sensor is placed on it.} \\ \mathbf{1} & \text{if a sensor is placed on it.} \end{cases} \tag{4}$$

$$\sum_{1}^{n} bi = \mathbf{nbSns} \tag{5}$$

Each state in the search space of our problem is consequently represented by a different configuration.

The Genetic algorithm goes as follows:

- 2 initial random propositions, said individuals, are picked.
- The configuration, or in GA terminology chromosome, of each individual is **Confg**.
- A utility function called 'fitness function' calculates for each individual its total coverage.

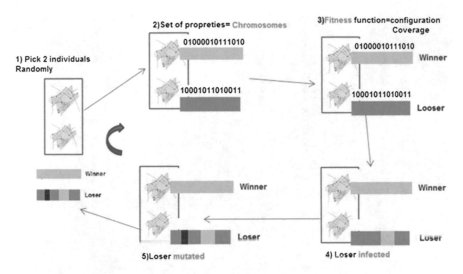

**Fig. 1** GA is a search technique that mimics the competition and mutations in evolution theory. The process goes on again and again until hitting the maximum allocated time

- The configuration that gets the best coverage is the winner while the other one is declared looser.
- The looser configuration gets infected randomly by 10 % of the winner chromosomes.
- The looser get 5 % of its chromosome changed or muted.
- And the computations go on again until hitting the maximum allocated time (Fig. 1).

## 4 Case Study: Sensing a Water Network in Casablanca

### 4.1 Study Area and Settings

We chose for test, a medium sized subset of Casablanca water network with 130 valves and 13 km of pipes. We run firstly the greedy algorithm then 2 SLOTS with different initial configurations. The outputs of SLOTS were plugged in as inputs in GA (Fig. 2).

### 4.2 Results

As expected, for each number **m** of sensors, the SLOTS algorithm outperformed the greedy algorithm while the GA outran both of them.

Nevertheless, to achieve high coverage percentage we had to allocate a fair amount of time (2 h) for every **m** sensors case (Fig. 3).

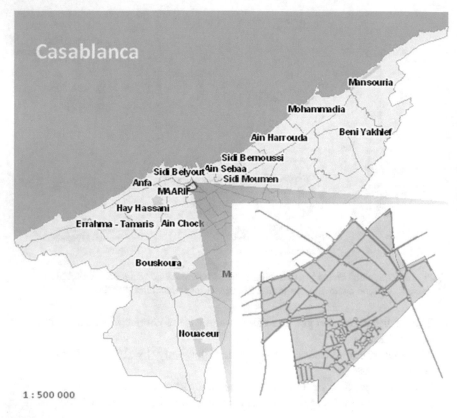

**Fig. 2** A subset of Casablanca water network, extracted from GIS database, with 130 valves and 13 km of pipes

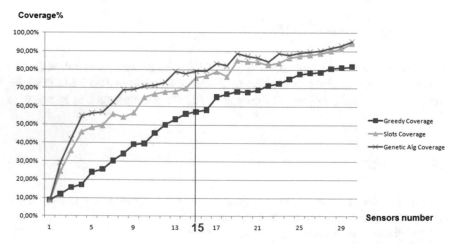

**Fig. 3** The SLOTS curve represents the best results of the 2 SLOTS results. With 15 sensors we achieved 57 % coverage with greedy, 76 % with SLOTS and 80 % with GA

# 5   Conclusion and Further Thoughts

AI search techniques enable a fast and efficient way to optimize placement of spatial entities in a network. Furthermore, Genetic algorithm offers to optimize already optimized solutions.

While multithreading would be a good alternative to process simultaneously computations for different numbers of sensors, more sophisticated heuristics could however accelerate both SLOTS and GA.

# References

1. Ostfeld, A., et al.: The battle of the water sensor networks (BWSN): a design challenge for engineers and algorithms. J. Watei Resour. Plann. Manag. **134**(6), 556–568 (2008)
2. Sarrate, R., Nejjari, F., Rosich, A.: Sensor placement for fault diagnosis performance maximization in distribution networks. In: 20th Mediterranean Conference on Control and Automation (MED), 2012. IEEE (2012)
3. Hunaidi, O., et al.: Detecting leaks. J. AWWA **92**(2), 82–94 (2000)
4. Thompson, K.E., et al.: Optimal macro-location methods for sensor placement in urban water systems. Technology **37**(1), 205–213 (2005)
5. Casillas, M.V., et al.: Optimal sensor placement for leak location in water distribution networks using genetic algorithms. Sensors **13**(11), 14984–15005 (2013)
6. Thompson, K.E., et al.: Optimal macro-location methods for sensor placement in urban water systems. Technology **37**(1), 72–75 (2005)
7. Dorini, G., et al.: SLOTS: effective algorithm for sensor placement in water distribution systems. J. Water Resour. Plann. Manag. **136**(6), 620–628 (2010)
8. Mitchell, R.J., Chambers, B., Anderson, A.P.: Array pattern control in the complex plane optimised by a genetic algorithm. In: Tenth International Conference on Antennas and Propagation (Conf. Publ. No. 436). vol. 1. IET (1997)
9. Huang, J.J., McBean, E.A., James, W.: Multi-objective optimization for monitoring sensor placement in water distribution systems. In: 8th Annual Symposium on Water Distribution Systems Analysis. Environmental and Water Resources Institute of ASCE (EWRI of ASCE), New York (2006)

# Direct Torque Control of Induction Machine Using Fuzzy Logic MRAS Speed Estimator Implemented for Electric Vehicle

Abderrahmane Ouchatti, Ahmed Abbou, Mohammed Akherraz
and Abderrahim Taouni

Abstract This paper discusses a sensorless speed control of an induction machine, used for traction application in electric vehicle. The controller allows operating the machine in all four quadrants. The implemented controller is based on classical Direct Torque and Flux Control (DTC) technique, due to its simplicity and good performance. The speed rotor and stator flux-linkage components are estimated by using a model reference adaptive system (MRAS). All regulators are based on conventional fuzzy-logic. The performance obtained by variable-speed drive system are presented first by simulation results, and secondly by the experimental results based on a dSPACE DS1104 platform.

Keywords Induction motor · Direct torque and flux control (DTC) · Fuzzy logic controller (FLC) · Electric vehicle · Model reference adaptive systems (MRAS)

## 1 Introduction

The induction machine is widely used as motor in industrial applications for its high reliability, relatively low cost and modest maintenance requirements. However, the problem is that control of the induction machine is more complex since its dynamic and highly nonlinear model, and most state variables are not measurable. Scalar-controlled drives have been widely used in industry due to their easier

A. Ouchatti (✉) · A. Abbou · M. Akherraz · A. Taouni
Mohammadia School of Engineers, University of Mohammed V Agdal,
Rabat, Morocco
e-mail: ouchatti_a@yahoo.fr

A. Abbou
e-mail: abbou@emi.ac.ma

M. Akherraz
e-mail: akherraz@emi.ac.ma

A. Taouni
e-mail: taouni40@hotmail.com

© Springer International Publishing Switzerland 2016
A. El Oualkadi et al. (eds.), *Proceedings of the Mediterranean Conference
on Information & Communication Technologies 2015*, Lecture Notes
in Electrical Engineering 381, DOI 10.1007/978-3-319-30298-0_3

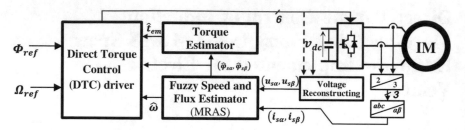

**Fig. 1** Proposed control scheme

implementation. However, the superior performances (demanded in many recent applications) of vector-control, have diminished the use of scalar control.

In the literature there are many papers which discuss and propose various vector drives. Only a few papers discuss implementations and mainly concentrate on simulations. To this end, induction machine control was treated according to different approaches. These include simple linear techniques such as Field Oriented Control (FOC), Direct Torque Control (DTC) and more involved nonlinear techniques like input-output linearization, Backstepping [1, 2], passivity, sliding mode [3, 4] etc. Most of the proposed controllers use speed sensor, for economical and maintenance reasons, the sensorless control constitutes one of the main preoccupations of industry, and so there are several techniques to estimate the rotor speed [5, 6, 7].

The purpose of this paper is sensorless speed control of an induction machine using a classical DTC technique associated with fuzzy rotor speed estimator based on Fuzzy MRAS method. The inertia of the rotor and the load torque are considered unknown parameters. In this paper it will be shown that the control performs well with Fuzzy Logic. Figure 1 shows the block diagram of the proposed control design.

As mentioned above the IM controller's objective is to motorize the electric vehicle, the power supply is composed of a main storage battery (e.g. battery pack lead acid) and an ultra-capacitor bank.

This paper is organized as following: induction motor model and fuzzy speed controller based on classical DTC are presented in the second section. Section 3 treats the fuzzy technique used to estimate the rotor speed. The Sect. 4 is devoted to simulation and experimental results.

## 2 Direct Torque Control (DTC) of Induction Machine

In direct-torque-controlled (DTC) induction motor drive, supplied by a voltage source inverter, it is possible to control directly the stator flux linkage and the electromagnetic torque by the selection of optimum inverter switching modes (see Sect. 2.2).

## 2.1 Induction Machine Model

The induction motor is described by the following model in the $(\alpha, \beta)$ axis (where magnetic saturation effect is neglected)

$$\dot{\omega} = \frac{1}{J}\left(t_{em} - t_l\right) \tag{1}$$

$$t_{em} = p\left(\varphi_{s\alpha}i_{s\beta} - \varphi_{s\beta}i_{s\alpha}\right) \tag{2}$$

$$\dot{\varphi}_{s\alpha} = u_{s\alpha} - R_s i_{s\alpha} + \omega_s \varphi_{s\beta} \tag{3}$$

$$\dot{\varphi}_{s\beta} = u_{s\beta} - R_s i_{s\beta} - \omega_s \varphi_{s\alpha} \tag{4}$$

$$\dot{\varphi}_{r\alpha} = -\frac{1}{T_r}\varphi_{r\alpha} + \frac{L_m}{T_r}i_{s\alpha} \mid \omega_{slp}\varphi_{r\beta} \tag{5}$$

$$\dot{\varphi}_{r\beta} = -\frac{1}{T_r}\varphi_{r\beta} + \frac{L_m}{T_r}i_{s\beta} - \omega_{slp}\varphi_{r\alpha} \tag{6}$$

$$\omega_{slp} = \omega_s - p\omega \tag{7}$$

$$T_r = \frac{L_r}{R_r} \tag{8}$$

where $u_s$ is the terminal voltage of the machine, $i_s$ is the stator current, $\varphi_s$ is the stator flux linkage, $t_{em}$ is the electromagnetic torque, $\omega_s$ is the synchronous angular speed of the stator flux vector, $\omega_{slp}$ is the angular slip frequency, $R_s$ is the stator winding resistance, $L_m$ is the machine magnetizing self-inductance, $T_r$ is the rotor time-constant and $p$ is the number of pole-pairs.

## 2.2 Stator-Flux-Based DTC Induction Machine Drive

Direct torque controller consist calculating an estimate of the stator flux and electromagnetic torque based on the measured stator current and stator voltage. The simplified schematic of stator-flux-based DTC induction motor drive with voltage source inverter is shown in Fig. 2. It consists of two-level stator flux hysteresis comparator, and three-level torque hysteresis comparator.

After determining the stator flux and electromagnetic torque errors, the adequate switching-vector is selected using an optimal voltage switching look-up table. In DTC induction machine drive, a decoupled control of torque and flux can be achieved by two independent control loops. However, the estimation of stator voltage when the machine is operating at low speed introduces error in flux

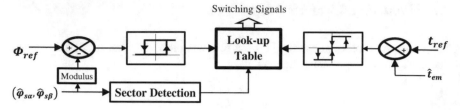

**Fig. 2** Stator-flux-based DTC induction machine drive

**Fig. 3** Sectors and required
switching-voltage space
vector

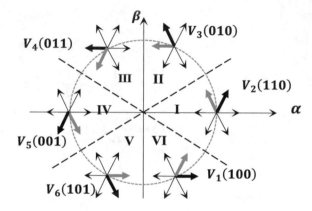

estimation which also affects the estimation of torque and speed in case of sensorless drive [8, 9].

There are eight states whose two of them have null voltage. The six states are illustrated in Fig. 3. It shows that the stator flux plane is divided into six sectors where each one has a set of voltage vectors. Therefore, the stator flux is controllable according to actual sector of flux vector [7]. As shown In Fig. 3, the bold arrow increases the stator flux, while light arrow decreases it. Similarly, it can be seen that, if an increase of the torque is required, then the torque is controlled by applying voltage vectors that advance the flux-linkage space vector in the direction of rotation, and if a decrease of torque is required, voltage vectors are applied which oppose the direction of the torque. If zero torque is required then zero switching vector $V_0$ or $V_7$ is applied. According to previous inverter state, the zero switching vector ($V_0$ or $V_7$) to be applied is chosen so as to switch one only arm of inverter.

## 2.3 Flux and Torque Estimator

The electromagnetic torque can be estimated easily by using Eq. (2). However, we must firstly have the stator flux-linkage components.

It was not convenient to use flux sensors; that is why the stator flux-linkage components must be estimated. In the DTC induction machine drive the stator flux-linkage has a double interest: First, stator flux-linkage components are required for the estimation of the electromagnetic torque as mentioned above. Secondly they are also required in the optimum switching vector selection table discussed in the previous section. It should be noted that, in general, the stator flux follows the applied stator voltage vector during an interval of time (dt) according to Eq. (9).

$$\varphi_s = \begin{pmatrix} \varphi_{s\alpha} \\ \varphi_{s\beta} \end{pmatrix} = \begin{pmatrix} \int (u_{s\alpha} - R_s i_{s\alpha})dt \\ \int (u_{s\beta} - R_s i_{s\beta})dt \end{pmatrix} \tag{9}$$

It should be noted that the performance of the DTC drive using Eq. (9) depend greatly on the accuracy of the estimated stator flux components, and these depend on the accuracy of the measured voltages and currents, and also of stator resistance. This method qualified as an open-loop flux linkage estimator is simple and work well excepting at very low frequency [7].

## 2.4 Rotor Speed Controller

In order to control the rotor speed, the reference torque $t_{ref}$ is changed as following:

- If rotor speed is less than the reference speed, the torque must be increased;
- If the speed exceeds its reference the torque must be reduced.

The rotor speed is obtained by using a fuzzy MRAS estimator. The following section provides a description of the fuzzy controller.

## 3 Fuzzy MRAS Speed Estimator

The proposed scheme of the MRAS-based rotor speed observer is shown in Fig. 4. It should be noted that a linear state observer for the rotor flux can then be derived as follows by considering the mechanical time-constant is much greater than the electrical time-constants [7].

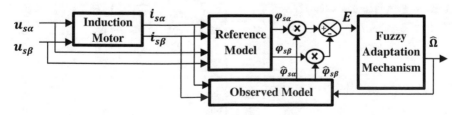

**Fig. 4** Structure of the fuzzy MRAS speed estimator

The reference model is a model that doesn't depend on the rotation speed; it allows calculating the components of rotor flux (in the stationary reference frame) from stator voltage equations:

$$\frac{d\Phi_{r\alpha}}{dt} = \frac{1}{a}\left(u_{s\alpha} - R_s i_{s\alpha} - \frac{1}{\delta}\frac{di_{s\alpha}}{dt}\right); \frac{d\Phi_{r\beta}}{dt} = \frac{1}{a}\left(u_{s\beta} - R_s i_{s\beta} - \frac{1}{\delta}\frac{di_{s\beta}}{dt}\right) \qquad (10)$$

The observer model uses the speed of rotation in its equations and permits to estimate the components of rotor flux and thus:

$$\frac{d\hat{\Phi}_{r\alpha}}{dt} = -k\Phi_{r\alpha} - p\hat{\Omega}\Phi_{r\beta} + kL_m i_{s\alpha}; \frac{d\hat{\Phi}_{r\beta}}{dt} = -k\Phi_{r\beta} + p\hat{\Omega}\Phi_{r\alpha} + kL_m i_{s\beta} \qquad (11)$$

where the subscript ^ denotes the estimated value of the specified signal. Figure 4 shows the schematic of the rotor speed estimator. The adaptation mechanism is derived by using Popov's criterion of hyperstability. This results in a stable and quick response system [7], where the differences between the state-variables of the reference model and adaptive model (state errors) are manipulated into a speed tuning signal (E), which is then an input into a fuzzy-type of controller (shown in Fig. 5), which outputs the estimated rotor speed.

kE, kdE, kdU are gains called "scale factor". They can change the sensitivity of the controller without changing its structure. The fuzzy controller is composed of three blocks: Fuzzification, rule bases, and Defuzzification. The fuzzy subsets are as following: NB (Negative Big), NM (Negative Medium), NS (Negative Small), Z (Zero), PS (Positive Small), PM (Positive Medium), PB (Positive Big). As shown in Table 1, there are 7 fuzzy subsets for each variable, which gives 49 possible rules.

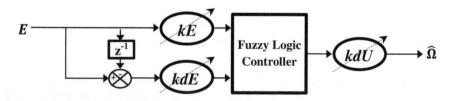

**Fig. 5** Fuzzy logic controller elements

| dE/E | NB | NM | NS | Z | PS | PM | PB |
|------|----|----|----|----|----|----|----|
| PB | Z | PS | PM | PB | PB | PB | PB |
| PM | NS | Z | PS | PM | PB | PB | PB |
| PS | NM | NS | Z | PS | PM | PB | PB |
| Z | NB | NM | NS | Z | PS | PM | PB |
| NS | NB | NB | NM | NS | Z | PS | PM |
| NM | NB | NB | NB | NM | NS | Z | PS |
| NB | NB | NB | NB | NB | NM | NS | Z |

**Table 1** Fuzzy rule base with 49 rules

# 4 Experimental and Simulation Result

The experimental setup arrangement is based on dSPACE DS1104 board (based on a Texas Instruments TMS320F240 DSP) and a SEMIKRON inverter switching at 10 kHz (noted by VSI). The load torque of the induction motor is controlled by the DC machine (see Fig. 6).

The induction machine parameters are mentioned in Table 2. The simulation and experimental results are shown in Fig. 7.

The applied speed reference profile is shown in Fig. 7b. During start-up the load torque is maintained at 10 Nm; at t = 1 s, the load torque value changes to −10 Nm.

We can see a fast speed response during start-up and during fast changes of speed reference or load torque. The stator currents and stator flux are approximately sinusoidal. The driver has the capability of four-quadrant operation (it should be noted that between 0.75 and 0.83 s, (see Fig. 7e): the induction machine operates in the regenerating mode and the kinematic energy stored in the system inertia is converted into electrical energy).

**Fig. 6** Experimental setup test

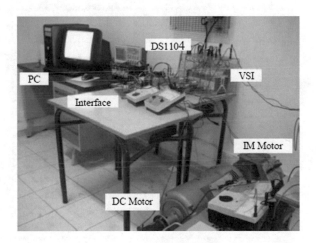

**Table 2** Induction machine parameters

| Parameter | Symbol | Rating value |
|---|---|---|
| Rated power | $Pn$ | 1.5 kW |
| stator voltage/frequency | $U/f$ | 380 V/50 Hz |
| Number of pole-pairs/rated speed | $P/n$ | 2/1460 rpm |
| Resistance (stator/rotor) | $Rs/Rr$ | 5.7/3.4 Ω |
| Inductance (stator/rotor) | $Ls/Lr$ | 0.23 H/0.22 H |
| Mutual inductance | $Lm$ | 0.21 H |
| Polar moment of inertia | $J$ | 0.18 kg m$^2$ |

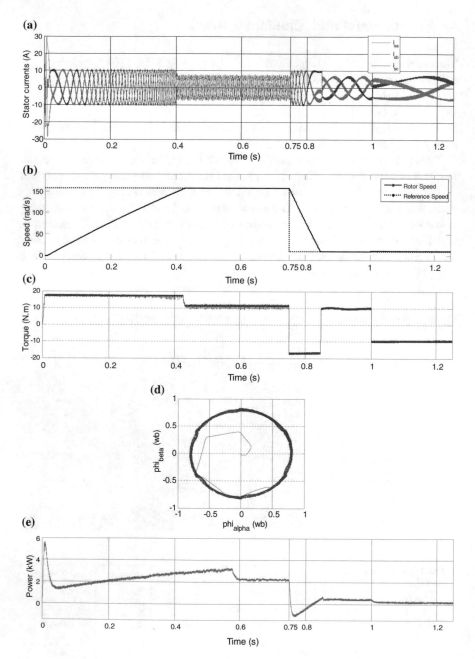

**Fig. 7** Simulation and experimental results: **a** Stator currents; **b** Rotor speed; **c** Estimated electromagnetic torque; **d** Stator flux trajectory; **e** Instantaneous power supply

# 5  Conclusion

In this paper we have presented the study of the classical direct torque and flux controller of induction machine. The main advantages of the DTC are:

- Direct control of flux and torque (by the selection of optimum inverter switching vectors);
- Very quick response;
- Indirect control of stator currents and voltages (current regulators not required);
- Absence of coordinate transformations (which are required in most of the vector-controlled drive implementations);

From simulation results, it can be shown that the DTC combined with fuzzy logic speed controller ensures robust start and operation in the zero-speed region. And the adaptive control can be achieved by stator resistance estimation.

Our future research work would take into account the variation of the rotor resistance and the implementation of this control strategy using a 16-bit Fixed-point DSP.

# References

1. Abbou, A., Mahmoudi, H.: Design of a new sensorless control of induction motor using backstepping approach. IREE 3(1), 166–173 (2008)
2. Ouchatti, A., Abbou, A., Akherraz, M., Taouni, A.: Induction motor controller using fuzzy mras and backstepping approach. IREE 9(3), 511–518 (2014)
3. Ortega, R., Loria, A., Nicklasson, P.J., Sira-Ramirez, H.: Passivity-based Control of Euler-Lagrange Systems: Mechanical Electrical and Electromechanical Applications, in Communications and Control Engineering. Springer, Berlin (1998)
4. Dybkowski, M., Orlowska-Kowalska, T., Tarchata, G.: Sensorless traction drive system with sliding mode and MRAS estimators using direct torque control. Automatika 329–336 (2013)
5. Abu-Rub, H., Stando, D., Kazmierkowski, M.P.: Simple speed sensorless DTC-SVM scheme for induction motor drives. Tech. Sci. 61(2), 301–307 (2013)
6. Marino, R., Tomei, P.: Nonlinear Control Design Adaptive and Robust. Prentice-Hall (1995)
7. Vas, P.: Sensorless Vector and Direct Torque Control. Oxford University Press (1998)
8. Bimal Bose, K.: Modern Power Electronics and AC Drives. Prentice Hall PTR (2002)
9. Raymond, D.: DTC-Theoretical Etude and Simulation, Renesas—Application Note (2009)

# Control Synthesis for General Positive 2D Models

M. Bolajraf

**Abstract** This paper is about control and synthesis for general positive 2D models. We present a method for designing a state feedback controller by using Linear programing (LP) conditions to address the stability problem for general positive 2D models. Necessary and sufficient conditions for stability of positive systems has been provided.

**Keywords** Control and synthesis · Positive 2D models · Linear programing

## 1 Introduction

During the last decade, many researchers have proposed and analyzed linear state-space models for two-dimensional (2D) systems. The most popular models were presented by Roesser [1], Fornasini-Marchesini [2] and Kurek [3]. The models have been also extended to positive 2D models [4–8] and multidimensional systems [9].

The stability of 2D positive systems described by Roesser model and synthesis of state-feedback controllers has been considered in [10], some other results of sufficient conditions for asymptotic stability and stabilization problem for 2D discrete linear systems can be found in the literature [11] and some recent results [5–8, 12]. In this paper, we analyze the stability of positive 2D general time-invariant state-space model [3, 7] to obtain necessary and sufficient condition for its stability, then we have to propose a simple numerical method based on a previous approach for 1D positive systems initiated in [13–16] where some synthesis problems are solved in term of linear programing (LP), based on this approach we proved necessary and sufficient conditions for the stability for positive 2D general state-space model expressed in terms of LP.

M. Bolajraf (✉)
PMMAT Lab, Faculty of Science Ain Chock, University Hassan II,
Casablanca, Morocco
e-mail: mohamed.bolajraf@gmail.com

© Springer International Publishing Switzerland 2016
A. El Oualkadi et al. (eds.), *Proceedings of the Mediterranean Conference on Information & Communication Technologies 2015*, Lecture Notes in Electrical Engineering 381, DOI 10.1007/978-3-319-30298-0_4

This paper is organized as follows. In Sect. 2 basic definition and preliminary results are given. In Sect. 3 necessary and sufficient conditions concerning the asymptotic stability of general 2D positive system are presented. Section 4 studies the synthesis problem. Finally, Sect. 5 gives some numerical example to illustrate the proposed results.

The following notations will be used. $\mathbb{Z}_+$ denotes the set of nonnegative integers, $\mathbb{R}^{n \times m}$ is the set of $n \times m$ real matrices and $\mathbb{R}_+^{n \times m}$ denotes the set of real $n \times m$ matrices with nonnegative entries. $A^T$ denotes the transpose of the real matrix $A$. $\rho(M)$ denotes the spectral radius of a matrix $M \in \mathbb{R}^{n \times n}$ and is defined as: $\rho(M) = \max(|\lambda_1|, \ldots, |\lambda_n|)$. $\mathbf{vec}(E)$ denotes the vector formed by the columns of a given matrix $E$. $\mathbf{diag}(v)$ denotes a diagonal matrix with diagonal components are the elements of $v$. $1_n$ is a vector of $n$ elements equal to 1. $0_{n \times p}$ is a matrix of dimension $n \times p$ where all its elements equal to 0 and finally, the $n \times n$ identity matrix will be denoted by $I_n$.

## 2   Problem Formulation

Consider the general 2D system [2, 3] described by the following

$$
x_{k+1,t+1} = A \begin{bmatrix} x_{k,t} \\ x_{k+1,t} \\ x_{k,t+1} \end{bmatrix} + B \begin{bmatrix} u_{k,t} \\ u_{k+1,t} \\ u_{k,t+1} \end{bmatrix} \tag{1}
$$

where $x(t, k) \in \mathbb{R}^n$ is the state vector at the point $(k, t)$, $u \equiv [u(k, t)$ $u(k+1, t) \, u(k, t+1)]^T \in \mathbb{R}^{3m}$ is the input, $A = [A_0 \, A_1 \, A_2] \in \mathbb{R}^{n \times 3n}$ and $B = [B_0 \, B_1 \, B_2] \in \mathbb{R}^{n \times 3m}$ are known matrices.

The boundary conditions for (1) are defined by

$$
\begin{cases} x_{k,0} = x_{k0}, \, \forall k \in \mathbb{Z}_+, \\ x_{0,t} = x_{0t}, \, \forall t \in \mathbb{Z}_+, \end{cases} \tag{2}
$$

where $x_{k0}$ and $x_{0t}$ are given sequences of vectors.

The main problem considered in this paper is to design a control law in such way that the resulting 2D system is positive and asymptotically stable. In other words, the main problem reduces to look a state feedback control $u = K [x(k, t) \, x(k+1, t) \, ; x(k, t+1)]^T$, where

$$
K = \begin{bmatrix} K_0 & 0 & 0 \\ 0 & K_1 & 0 \\ 0 & 0 & K_2 \end{bmatrix}, \tag{3}
$$

leading to the closed-loop system defined by

$$x_{k+1,t+1} = [A_0 + B_0 K_0 \quad A_1 + B_1 K_1 \quad A_2 + B_2 K_2] \begin{bmatrix} x_{k,t} \\ x_{k+1,t} \\ x_{k,t+1} \end{bmatrix} \tag{4}$$

**Definition 1** The free system (1) ($u \equiv 0$) is called a positive 2D system if the corresponding trajectory is nonnegative for any nonnegative boundary conditions (2).

**Definition 2** A real matrix $M$ is called a nonnegative matrix ($M \in \mathbb{R}_+^{n \times q}$) if all its elements are nonnegative $m_{ij} \geq 0$, $i = 1, \ldots, n$, $j = 1, \ldots, q$.

**Proposition 1** *The free system* (1) ($u \equiv 0$) *is positive if and only if* $A \in \mathbb{R}_+^{n \times 3n}$.

In the following we start necessary and sufficient conditions for the asymptotic stability of system (1) with $u \equiv 0$,

**Definition 3** [5] A 2D positive system described by (1) is called asymptotically stable if the free state evolution corresponding to any set of nonnegative boundary conditions (2) asymptotically tends to zero, i.e.,

$$\lim_{k,t \to \infty} x_{k,t} = 0.$$

**Lemma 1** [6, 7] *Let* $(A_0, A_1, A_2)$ *be a triple of* $n \times n$ *nonnegative matrices. The* $(A_0, A_1, A_2)$ *is asymptotically stable if and only if* $\rho(A_0 + A_1 + A_2) < 1$.

By using the Proposition 1 and Lemma 1, we have the following necessary and sufficient condition for the closed-loop system (4) to be positive and asymptotically stable.

$$\begin{cases} \text{Positivity}: \ [A_0 + B_0 K_0 \ A_1 + B_1 K_1 \ A_2 + B_2 K_2] \geq 0, \\ \text{Stability}: \ \rho(\sum_{i=0}^{2}(A_i + B_i K_i) - I) < 1. \end{cases} \tag{5}$$

## 3 Main Results

This section contains the main results, that will be extended in the next sections to other kind of problems.

Note that, the asymptotically stability of the 2D system (1) with free input is equivalent to the asymptotic stability of the following 1D discrete time system

$$x(k+1) = (A_0 + A_1 + A_2)x(k) \tag{6}$$

**Lemma 2** [10] *System* (6) *is positive if and only if all the components of the matrices* $A_0$, $A_1$ *and* $A_2$ *are nonnegative.*

**Lemma 3** [10] *Assume that system* (6) *is positive. The following statements are equivalent*

(i)   *System* (6) *is asymptotically stable.*
(ii)  $\rho(A_0 + A_1 + A_2) < 1.$
(iii) *There exist a vector* $\lambda \in \mathbb{R}^n$ *such that*

$$(A_0 + A_1 + A_2 - I_n)\lambda < 0, \ \lambda > 0. \tag{7}$$

Now, we present a necessary and sufficient condition to the asymptotic stability of the 2D positive system (1) with free input $u = 0$.

**Theorem 1** *Assume that system* (1) *with free input* $(u = 0)$ *is positive. Then the following statements are equivalent*

(i)   *The 2D system* (1) *with free input is asymptotically stable.*
(ii)  *System* (6) *is asymptotically stable.*
(iii) $\rho(A_0 + A_1 + A_2) < 1.$
(iv)  *There exist a vector* $\lambda \in \mathbb{R}^n$ *such that*

$$(A_0 + A_1 + A_2 - I)\lambda < 0, \ \lambda > 0. \tag{8}$$

*Proof* The equivalence (i) $\Leftrightarrow$ (iii) results from Lemma 1, (ii) $\Leftrightarrow$ (iii) $\Leftrightarrow$ (iv) results from Lemma 3 and then the proof will be complete by show (i) $\Leftrightarrow$ (iv).

(i) $\Rightarrow$ (iv) By using Lemma 1 (i) is equivalent to (iii), on other hand (iii) is equivalent to (iv).

Reciprocally, (iv) $\Rightarrow$ (i) it suffices to follow the same argument by using Lemma 3 and Lemma 1.                                                                    $\square$

## 4   Controller Synthesis

In this section, necessary and sufficient condition for the asymptotic stability of positive 2D system described by (1) are presented for a state feedback control of the form

$$\begin{bmatrix} u_{k,t} \\ u_{k+1,t} \\ u_{k,t+1} \end{bmatrix} = \begin{bmatrix} K_0 & 0 & 0 \\ 0 & K_1 & 0 \\ 0 & 0 & K_2 \end{bmatrix} \begin{bmatrix} x_{k,t} \\ x_{k+1,t} \\ x_{k,t+1} \end{bmatrix},$$

then, the resulting closed-loop system is described by (4) and the following main result can be presented.

**Theorem 2** *For any given nonnegative initial boundary conditions* (2), *the closed-loop system* (4) *is positive and asymptotically stable if and only if the following LP problem in the variables* $\lambda \in \mathbb{R}^n$ *and* $Z_i \in \mathbb{R}^{m \times n}$, $i = 0, 1, 2$ *is feasible*

$$\begin{cases} (\sum_{i=0}^{2} A_i - I)\lambda + (\sum_{i=0}^{2} B_i Z_i) 1_n < 0, \\ \lambda > 0, \\ A_i \mathbf{diag}(\lambda) + B_i Z_i \geq 0, \quad i = 0, 1, 2. \end{cases} \tag{9}$$

*Moreover, the gain matrices $K_i$, $i = 0, 1, 2$ are computed as*

$$K_i = Z_i \mathbf{diag}(\lambda)^{-1}, \quad i = 0, 1, 2. \tag{10}$$

*Proof* Assume that condition (9) are satisfied and define the matrices $K_i = Z_i$ $\mathbf{diag}(\lambda)^{-1}, i = 0, 1, 2$. Now, it is easy to see that $[A_0 + B_0 K_0 \ A_1 + B_1 K_1 A_2 + B_2 K_2]$ is nonnegative matrix. Effectively, from the last inequalities in condition (9) we have for $i = 0, 1, 2$.

$A_i \mathbf{diag}(\lambda) + B_i Z_i \geq 0 \Rightarrow A_i + B_i Z_i \mathbf{diag}(\lambda)^{-1} - (A_i + B_i K_i) \geq 0$ (because $\lambda > 0$).

Next, to show the asymptotic stability, we define the matrices $K_i = Z_i \mathbf{diag}(\lambda)^{-1}$, $i = 0, 1, 2. \Rightarrow Z_i = K_i \mathbf{diag}(\lambda)^{-1}$, then, the first inequality in condition (9) leads to $(\sum_{i=0}^{2} A_i + \sum_{i=0}^{2} B_i K - I)\lambda < 0$. Since $\lambda > 0$ and the matrix $[A_0 + B_0 K_0 \quad A_1 + B_1 K_1 \quad A_2 + B_2 K_2]$ is nonnegative, then by using Theorem 1, we conclude that the 2D system described by the closed-loop system (4) is positive and asymptotically stable.

The rest of the proof follows the same line of arguments and then is omitted.

In the following, a positive feedback control can be handled by using a similar LP approach.

**Theorem 3** *The following statements are equivalent*

(i) *There exist a positive feedback control such that the closed-loop system (4) is positive and asymptotically stable for any nonnegative initial boundary condition (2).*

(ii) *There exists a matrix $K$ of the form (3) such that $K \geq 0$ and the closed-loop system (4) is positive and asymptotically stable for any nonnegative initial boundary condition (2).*

(iii) *The following LP problem in the variables $\lambda \in \mathbb{R}^n$ and $Z_i \in \mathbb{R}^{m \times n}$, $i = 0, 1, 2$ is feasible*

$$\begin{cases} (\sum_{i=0}^{2} A_i - I)\lambda + (\sum_{i=0}^{2} B_i Z_i) 1_n < 0, \\ \lambda > 0, \\ Z_i \geq 0, \quad i = 0, 1, 2, \\ A_i \mathbf{diag}(\lambda) + B_i Z_i \geq 0, \quad i = 0, 1, 2. \end{cases} \tag{11}$$

*Moreover, the gain matrices $K_i$, $i = 0, 1, 2$ are computed as*

$$K_i = Z_i \, \mathbf{diag}(\lambda)^{-1}, \quad i = 0, 1, 2.$$

*Proof* This proof follows the same line of arguments of Theorem 2.   □

*Remark 1* By taking $A_0 = 0_n$ and $B = 0_{n \times m}$, the system described by (1) is the well-known Fornasini-Marchesini Second model [2] and it can be stabilizing by using Theorem 2 for a state feedback control or Theorem 3 for a positive state feedback control.

*Remark 2* A negative state feedback control can be considered, by just imposing $Z_i \leq 0$, $i = 0, 1, 2$, instead $Z_i \geq 0$, $i = 0, 1, 2$, in the LP problem (11).

*Remark 3* The LP problem (9) and (11) can be rewritten in standard form that is

$$
\begin{bmatrix}
\sum_{i=0}^{2} A_i - I & 1_n^T \otimes B_0 & 1_n^T \otimes B_1 & 1_n^T \otimes B_2 \\
-I & 0_{n \times p} & 0_{n \times p} & 0_{n \times p}
\end{bmatrix} w < 0,
$$

$$
\begin{bmatrix}
-\sum_{i=1}^{n} e_i k_i^T \otimes A_0^T e_i & -I \otimes B_0 & 0 & 0 \\
-\sum_{i=1}^{n} e_i k_i^T \otimes A_1^T e_i & 0 & -I \otimes B_1 & 0 \\
-\sum_{i=1}^{n} e_i k_i^T \otimes A_2^T e_i & 0 & 0 & -I \otimes B_2
\end{bmatrix} w \leq 0,
\tag{12}
$$

where $w = [\lambda \quad \mathbf{vec}(Z_0) \quad \mathbf{vec}(Z_1) \quad \mathbf{vec}(Z_2)]^T$, $e_1, \ldots, e_n$ is the canonical basis of $\mathbb{R}^n$.

## 5   Numerical Example

### 5.1   Stabilization Example

Consider the following 2D system

$$
A_0 = \begin{bmatrix} -1.5 & 0.1 & 0 \\ 0.2 & 0.5 & 0.3 \\ 0 & 0 & 0 \end{bmatrix}, A_1 = \begin{bmatrix} 0 & 0.3 & 0 \\ 0.1 & 0 & 0 \\ 0 & 0.4 & 0 \end{bmatrix}, A_2 = \begin{bmatrix} 0 & 0 & 0 \\ 0.5 & 0 & 0.3 \\ 1.3 & 0.7 & 0.1 \end{bmatrix},
$$

$$
B_0 = \begin{bmatrix} 1 & 0.5 \\ 0 & 0 \\ 0 & 0 \end{bmatrix}, B_1 = \begin{bmatrix} 0 & 0 \\ 0.8 & 0.2 \\ 0.1 & 0 \end{bmatrix}, B_2 = \begin{bmatrix} 0.3 & 0 \\ 0 & 0 \\ 0.2 & 0.4 \end{bmatrix}.
$$

The objective is to calculate a state feedback controller that stabilizes the system and makes the closed-loop system positive for any boundary condition initial of the form (2). For this, the condition of Theorem 2 must be satisfied. The gain of a stabilizing control can be calculate by solving the LP problem (12), by using Matlab, we find a gain matrix $K$ of the form (3) where

$$K_0 = \begin{bmatrix} 1.2618 & -0.0759 & 0.0107 \\ 0.5145 & -0.0372 & 0.0054 \end{bmatrix}, \ K_1 = \begin{bmatrix} 0.1324 & -0.2227 & 0.1291 \\ -0.9761 & 0.9062 & -0.4728 \end{bmatrix},$$

$$K_2 = \begin{bmatrix} 0.0561 & 0.0182 & 0.0453 \\ -3.2626 & -1.7441 & -0.2232 \end{bmatrix}.$$

The corresponding closed-loop system matrices are given by

$$A_0 + B_0 K_0 = \begin{bmatrix} 0.0190 & 0.0055 & 0.0134 \\ 0.2 & 0.5 & 0.3000 \\ 0 & 0 & 0 \end{bmatrix}, \ A_1 + B_1 K_1 = \begin{bmatrix} 0 & 0.3 & 0 \\ 0.0107 & 0.0031 & 0.0087 \\ 0.0132 & 0.3777 & 0.0129 \end{bmatrix},$$

$$A_2 + B_2 K_2 = \begin{bmatrix} 0.0168 & 0.0055 & 0.0136 \\ 0.5 & 0 & 0.3 \\ 0.0062 & 0.0060 & 0.0198 \end{bmatrix}.$$

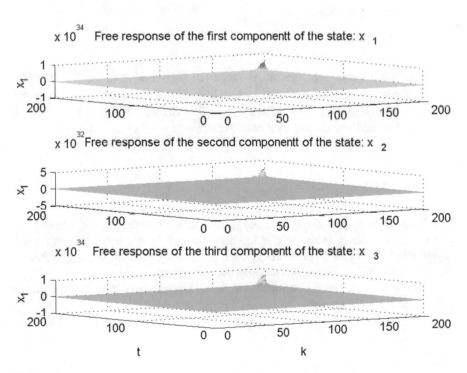

**Fig. 1** System free response

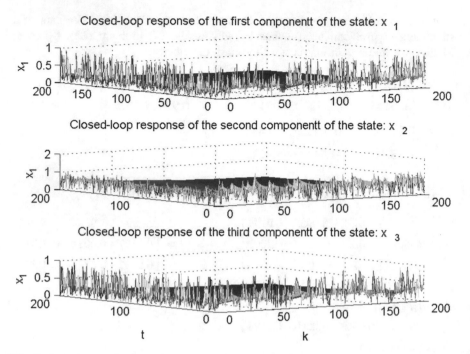

**Fig. 2** System closed-loop response

Then, the closed-loop system is positive $(A_i + B_i K_i \geq 0, \ i = 0, 1, 2)$, and asymptotically stable (the matrix $\sum_{i=0}^{2} A_i + B_i K_i$ has all the eigenvalues inside the unit circle $\lambda_0 = 0.9907$, $\lambda_1 = 0.01$ and $\lambda_2 = -0.4290$). Finally, Figs. 1 and 2 show the open-loop and closed-loop response respectively.

## 6 Conclusion

In this paper, we have considered the design of a state feedback controller for general 2D models. It is shown that the necessary and sufficient condition for the asymptotic stability of positive 2D system can be expressed in terms of LP conditions, as a consequence one can easily compute them.

An extension of these considerations for 2D positive continuous-time linear systems is an open problem.

# References

1. Roesser, R.P.: A discrete state-space model for linear image processing. IEEE Trans. Autom. Control **AC-25**(1), 1–10 (1975)
2. Fornasini, E., Marchesini, G.: State-space realization theory of two-dimentional filters. IEEE Trans. Autom. Control **21**(4), 484–492 (1976)
3. Kurek, J.: The general state-space model for a two-dimensional linear digital system. IEEE Trans. Autom. Control, **AC-30**(6), 600–602 (1985)
4. Alfidi, M., Hmamed, A., Tadeo, F.: Linear programming approach for 2-d stabilization and positivity. In: Third Multidisciplinary International symposium on Positive Systems: Theory and Applications, Valencia, Spain (2009)
5. Kaczorek, T.: Positive 1D and 2D Systems. Springer (2001)
6. Kaczorek, T.: Asymptotic stability of positive 2d linear systems with delays. Bull. Pol. Acad. Sci. Tech. Sci. **57**(2) (2009)
7. Twardy, M.: An lmi approach to checking stability of 2d positive systems. Bull. Pol. Acad. Sci. Tech. Sci. **55**(4) (2007)
8. Valcher, M.E.: On the interval stability and asymptotic behavior of 2d positive systems. IEEE Trans. Circ. Syst. **44**(7), 602–613 (1997)
9. Tzafestas, S.G., Fimenides, T.G.: Exact model-matching control of three-dimensional systems using state and output feedback. J. Syst. Sci. **13**, 1171–1187 (1982)
10. Hmamed, A., Rami, M.A., Alfidi, A.: Control synthesis for positive 2d systems described by roesser model. In: IEEE Conference on Decision and control, Cancun Mexico, 2008
11. Yaz, E.: On state-feedback stabilization of two-dimensional digital systems. IEEE Trans. Circ. Syst. **CAS-32**, 1069–1070 (1985)
12. Kaczorek, T., Busowic, M.: Minimal realization problem for positive multivariable linear systems with delay. Int. J. Appl. Math. Comput. Sci. (2014)
13. Helmke, U., Ait Rami, M., Tadeo, F.: Positive observation problem for linear time-lag positive systems. In: IFAC Symposium on System, Structure and Control Faz do Iguassu, Brasil, 2007
14. Ait Rami,M., Tadeo, F.: Controller synthesis for linear systems to impose positiveness in closed-loop states. In: Proceedings of the IFAC World Congress, Prague (2005)
15. Ait Rami, M., Tadeo, F.: Linear programming approach to impose positiveness in closed-loop and estimated states. In: Proceedings of the International Symposium on Mathematical Theory of Networks and Systems, Kyoto, Japan (2006)
16. Ait Rami, M., Tadeo, F.: Positive observation for discrete positive systems. In: Conference on Decision and Control, San Diego (2006)

# VHDL Design and FPGA Implementation of the PWM Space Vector of an AC Machine Powered by a Voltage Inverter

Elhabib Lotfi, Mustapha Elharoussi and Elhassane Abdelmounim

**Abstract** This paper presents a VHDL design and FPGA implementation of the space-vector pulse-width modulation (SVPWM) strategy. This design is made in a way that each block of the architecture is described on a separate entity. The global block is represented using the above entities as components. The proposed architecture for SVPWM is composed of four blocks. After the implantation of each block, we note that the resources consumed by the global entity, knowing that the same circuit is used for the different blocks namely the Stratix II device EP2S15F484C3, are 1185 ALUTs. The execution time of our architecture is two clock cycles.

**Keywords** SVPWM · FPGA · Vector control · VHDL · Asynchronous machine

## 1 Introduction

The control of AC machines with a voltage inverter frequently uses pulse width modulation techniques to control power switches. The pulse width modulation techniques are multiple. The choice of one of them depends on the type of control that is applied to the machine [1], on the inverter modulation frequency and on the harmonic constraints set by the user [2]. The modulation can be made in various approaches, particularly by comparing the reference to a triangular function [3] or applying the currently used space vector pulse width modulation (SVPWM). The principle of this technique is based on the selection of the sequence and the

E. Lotfi (✉) · M. Elharoussi (✉) · E. Abdelmounim (✉)
ASTI Laboratory, University Hassan 1ER, 26000 Settat, Morocco
e-mail: lothamid@gmail.com

M. Elharoussi
e-mail: m.elharoussi@gmail.com

E. Abdelmounim
e-mail: hassan.abdelmounim@hotmail.fr

© Springer International Publishing Switzerland 2016
A. El Oualkadi et al. (eds.), *Proceedings of the Mediterranean Conference on Information & Communication Technologies 2015*, Lecture Notes in Electrical Engineering 381, DOI 10.1007/978-3-319-30298-0_5

calculation of the conduction or the extinction time. In this work, we present the design of the various blocks of the vector modulation [2]. The proposed architecture is implemented on a FPGA circuit. We have used the VHDL language to describe our architecture and Quartus II tool for logic synthesis.

## 2 Principle of Vector PWM

A three-phase inverter with two voltage levels, has six switching cells, giving eight possible switching configurations. These eight switching configurations (denoted $V_0$–$V_7$) can be expressed in the reference ($\alpha\beta$) by 8 tensions vectors, among which two are null and others are equi-distributed every 60° [4].

## 3 Conception of SVPWM Blocks

### 3.1 Determination of the Reference Voltages $V_\alpha$, $V_\beta$

This block is used to project the three-phase-voltages in the reference ($\alpha\beta$) by performing the transformation of Concordia [2, 5].

### 3.2 Determination of the Sectors

The determination of the sector is by comparing $V_\alpha$ and $V_\beta$ tensions as shown in Table 1.

Once the sector number is determined, it is necessary to apply adjacent vectors to the sector concerned respectively during the period $T_1$ and $T_2$ [6].

**Table 1** Identifying sector

| Sectors | $V_\alpha > 0$ | $V_\alpha > \sqrt{3}V_\beta$ | $V_\alpha > -\sqrt{3}V_\beta$ |
|---------|----------------|------------------------------|-------------------------------|
| I       | 1              | 0                            | 1                             |
| II      | 0              | 0                            | 1                             |
| III     | 0              | 0                            | 0                             |
| IV      | 0              | 1                            | 0                             |
| V       | 0              | 0                            | 0                             |
| VI      | 1              | 1                            | 0                             |

**Table 2** Calculation of the duration $T_1$ and $T_2$ for each sector

| N° sector | i = 1 | i = 2 | i = 3 | i = 4 | i = 5 | i = 6 |
|---|---|---|---|---|---|---|
| The duration of vectors | $T_1 = X$ $T_2 = Z$ | $T_1 = Y$ $T_2 = -X$ | $T_1 = Z$ $T_2 = -Y$ | $T_1 = -X$ $T_2 = -Z$ | $T_1 = -Y$ $T_2 = X$ | $T_1 = -Z$ $T_2 = Y$ |

### 3.3 Calculation of the Variables

We define intermediate variables X, Y, Z

$$X = \sqrt{(2/3)}\, V_\alpha \cdot Tmod/E \quad Y = \left(\frac{V_\alpha}{\sqrt{6}} + \frac{V_\beta}{\sqrt{2}}\right)\frac{Tmod}{E}$$

$$Z = \left(\frac{-V_\alpha}{\sqrt{6}} + \frac{V_\beta}{\sqrt{2}}\right)\frac{Tmod}{E} \tag{1}$$

### 3.4 Calculation of the Duration $T_1$ and $T_2$ for Each Sector

See Table 2.

### 3.5 Generation of the Series of Pulses Ta, Tb and Tc

For each inverter arm, it is necessary to define the timing that defines the time during which the middle point of one arm is E/2 or −E/2 within a switching period of the inverter. There are different PWM implementation strategies ensuring the achievement of the desired tension. To reduce the harmonics it is preferable to generate voltages centered on the modulation period of the inverter.

## 4 Design and Implementation of Our Architecture of the Space Vector Modulation

The proposed architecture for SVPWM is composed of four main parts. The first is the Concordia matrix. The second is the scan_sector block. The third block calculates the duration $T_1\_T_2$ adjacent to each sector of vectors and the last block is generation_ pulse which generates the pulses to be applied to the various arms of the inverter.

Several studies on SVPWM command have already been carried out both at the university or industrial sectors [2]. The FPGA used for implantation is the Altera's Stratix 2 EP2S15F484C3.

## 4.1 Determination of the Reference Voltages $V_\alpha$, $V_\beta$

This block is used to project the three-phase voltages in the reference (αβ) by performing the transformation of Concordia (Fig. 1).

## 4.2 Determination of Sectors

The determination of the sector is by comparing two voltages $V_\alpha$ and $V_\beta$ (Fig. 2).

## 4.3 Calculation of $T_1$ and $T_2$ for Each Sector

The inputs of this block are the values of $V_\alpha$, $V_\beta$, Tmod and sector number provided by the previous block. This block calculates the times $T_1$ and $T_2$ during which the two vectors adjacent to the sector are applied respectively (Fig. 3).

## 4.4 Pulse-Generation

See Fig. 4.

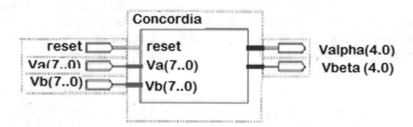

**Fig. 1** RTL schematic of the Concordia matrix

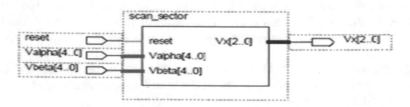

**Fig. 2** RTL block diagram of scan_sector

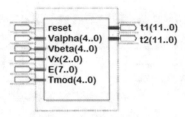

**Fig. 3** RTL schematic of $T_1-T_2$ Duration

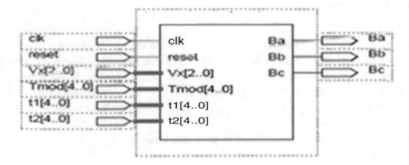

**Fig. 4** RTL schematic of the pulse-generation unit

## 4.5 Validation of the Proposed Architecture

The proposed architecture for SVPWM is presented by Fig. 5. It is based on the selection of the sequence and the calculation of the conduction or the extinction time. The global block is represented using the above entities as components.

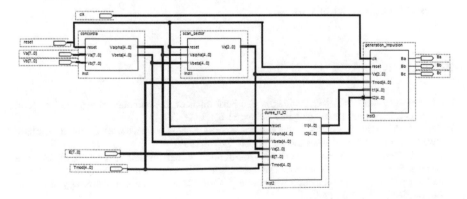

**Fig. 5** RTL schematic of the global entity unit

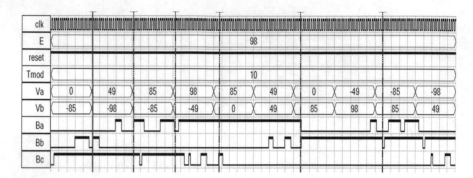

**Fig. 6** Simulation result

This block generates the pulses applied to the arm of the inverter according to $T_1$ and $T_2$ values by having a PWM centered. It is clear that the sector 2 for example the timing of the pulses $B_a$, $B_b$ and $B_c$ match the desired results: $B_a = 1$ during $T_1$, $B_b = 0$ during $T_2$ and $B_c$ is always equal to 1, i.e. We apply vector $V_2 = (010)$ during $T_1$ and vector $V_3 = (110)$ for $T_2$. The vectors $V_2$ and $V_3$ are adjacent vectors in sector 2 (Fig. 6).

Our proposed architecture responds after two clock cycles with a complexity of 1185 LUTs.

# 5  Conclusion

The approach of the VHDL design and FPGA implementation of the algorithm modeling the SVPWM proposed in this paper has permitted to achieve high performance when reducing complexity. Indeed, the proposed architecture responds after two clock cycles with a complexity of 1185 LUTs.

The vector control of an AC machine of this architecture allows exploitation of the results proposed in this work.

# References

1. Chaikhy, H., Khafallah, M.: Evaluation des performances des commandes vectorielles de la machine à induction. Revue de génie industriel **6**, 23–32 (2011)
2. Tzou, Y.-Y., Hsu, H.-J.: FPGA realization of space-vector pwm control IC for three-phase PWM inverters. IEEE Trans. Power Electron. **12**(6) (1997)
3. Rafa, S., Zeroug, H.: Simulation sur Matlab/Simulink et implémentation sur DSP/FPGA de la commande vectorielle de la machine synchrone à aimants permanents (PMSM) alimenté par un onduleur de tension à Modulation vectorielle (SVM). Thesis Doctoral in University of Houari Boumedienne, Algérie (2009)

4. Mailloux, J.-G.: Prototypage rapide de la commande vectorielle sur FPGA à l'aide des outils simulink—system generator. Thesis Doctoral in University of Québec MARS (2008)
5. Akin, O.: The use of FPGA in field-oriented control of an induction machine. Turk. J. Electr. Eng. Comput. Sci. **18**(6) (2010) TUBITAK
6. Mohamed Wissem Naouar: Commande numerique à base de composants FPGA d'une machine synchrone. Thesis of Doctoral in University of Cergy Pontoise (2007)

# Optimized Neural Network Sliding Mode Control for Two Links Robot Using PSO Technique

Siham Massou, El-mahjoub Boufounas and Ismail Boumhidi

**Abstract** This work presents the neural network combined with the sliding mode control (NNSMC) to design a robust controller for the two-links robot system. Sliding mode control (SMC) is well known for its robustness and efficiency to deal with a wide range of control problems with nonlinear dynamics. However, for complex nonlinear systems, the uncertainties are large and produce higher amplitude of chattering due to the higher switching gain. In order to reduce this gain, neural network (NN) is used to estimate the uncertain parts of the system plant with on-line training using backpropagation (BP) algorithm. The learning rate is one of the parameters of BP algorithm which have a significant influence on results. Particle swarm optimization (PSO) algorithm with global search capabilities is used in this study to optimize this parameter in order to improve the network performance in term of the speed of convergence. The performance of the proposed approach is investigated in simulations and the control action used did not exhibit any chattering behavior.

**Keywords** Robot manipulators · Neural network · Sliding mode control · Particle swarm optimization

## 1 Introduction

The motion control design for robot manipulators attracted considerable attention, it's a complex nonlinear system and its dynamic parameters are crucial to estimate accurately. Moreover, it is almost impossible to reach exact dynamic models as the system is described by a nominal model with large uncertainties to name a few: payload parameter, internal friction, and external disturbance. To deal with the

S. Massou (✉) · E. Boufounas · I. Boumhidi
LESSI Laboratory, Department of Physics, Faculty of Sciences,
Sidi Mohammed Ben Abdellah University, Fez, Morocco
e-mail: siham.massou@gmail.com

© Springer International Publishing Switzerland 2016
A. El Oualkadi et al. (eds.), *Proceedings of the Mediterranean Conference on Information & Communication Technologies 2015*, Lecture Notes in Electrical Engineering 381, DOI 10.1007/978-3-319-30298-0_6

parameters uncertainties, several methods have been proposed; introducing the neural network based controls [1–4] and the Sliding Mode Control (SMC) [5, 6].

SMC is one of the most important approaches to handle systems with uncertainties, nonlinearities, and bounded external disturbance. However, there is undesirable chattering in the control effort and bounds on the uncertainties are required in the design of the SMC. It is well known that the main advantage of using the boundary layer solution [5, 7] is eliminating this chattering problem. This method can resolve the problem for only systems with small uncertainties. In case of large uncertainties, a neural network structure is proposed to estimate the unknown parts of the two-links robot model, so that the system uncertainties can be kept small and hence enable a lower switching gain to be used. The neural network weights are trained on-line using the backpropagation algorithm (BP) [8]. The proposed control consists of the predicted equivalent control added to the robust control term, where the neural network estimated function is incorporated in the equivalent control component. The learning rate is one of the parameters of BP algorithm which have a significant influence on results; practically, its value is chosen usually between 0.1 and 1 [9]. Learning rate which is too small or too large may not be favourable for convergence. In order to solve this problem, we proceed to use particle swarm optimization (PSO) algorithm with global search capabilities to optimize this parameter in order to improve the training speed.

This study is organized as follows. The next section presents the proposed optimal neural network sliding mode control. In Sect. 3, simulation results are provided to demonstrate the robust control performance of the proposed approach. Finally, in Sect. 4 a concluding remark is given.

## 2 Optimal Neural Network Sliding Mode Control Design

### 2.1 Controller Design

Consider the dynamic model of the two links robot written in the state space as follows:

$$
\begin{cases}
\dot{x}_1 = x_2 \\
\dot{x}_2 = x_3 \\
\dot{x}_3 = h_{1n}(\underline{x}, \underline{u}) + \xi_1(\underline{x}, t) \\
\dot{x}_4 = x_5 \\
\dot{x}_5 = x_6 \\
\dot{x}_6 = h_{2n}(\underline{x}, \underline{u}) + \xi_2(\underline{x}, t)
\end{cases} \tag{1}
$$

where $h_{1n}(\underline{x}, \underline{u})$ and $h_{2n}(\underline{x}, \underline{u})$ are the nominal representations of the system, and the unknown parts $\xi_1(\underline{x}, t)$ and $\xi_2(\underline{x}, t)$. $\underline{u} = [u_1 \quad u_2]^T$ and

$\underline{x} = [x_1 \quad x_2 \quad x_3 \quad x_4 \quad x_5 \quad x_6]^T$. are respectively the vector inputs and the outputs of the system. More physical representation details and the lyapunov function are given in [4].

The robot manipulator control law is given as [4]:

$$\underline{u} = g_n^{-1}(\underline{x}) \left( -\left(f_n(\underline{x}) + \hat{\xi}(\underline{x},t)\right) + \begin{pmatrix} \dot{x}_{3d} \\ \dot{x}_{6d} \end{pmatrix} - \gamma \ddot{e} - \beta \dot{e} - k sat(S) \right) \qquad (2)$$

## 2.2 Neural Network Representation

In this paper, we consider a NN with two layers of adjustable weights. $\underline{x}$ is the state input variables and the output variables are: $y_1 = \hat{\xi}_1(\underline{x},t)$ and $y_2 = \hat{\xi}_2(\underline{x},t)$ $y_k = W_k^T \sigma(W_j^T \underline{x})$ $k = 1,2$. Where $\sigma(.)$ represents the hidden-layer activation function considered as a sigmoid function given by: $\sigma(s) = \frac{1}{1+e^{-s}}$ $W_k = [W_{k1} W_{k2} \ldots W_{kN}]^T$ and $W_j = [W_{j1} W_{j2} \ldots W_{jN}]^T$ are respectively interconnection weights between the hidden and the output layers and between the input and the hidden layers. The actual output $y_{dk}(\underline{x})$ (desired output which is the difference between the actual and nominal functions) is:

$$y_{dk}(\underline{x}) = y_k(\underline{x}) + \varepsilon(\underline{x}) \qquad (3)$$

where $\varepsilon(\underline{x})$ is the NN approximation error.

The network weights are adjusted during the online implementation. The method used is based on the gradient descent method (GD). The essence of the GD consists of iteratively adjusting the weights in the direction opposite to the gradient of $E$, so as to reduce the discrepancy according to:

$$\frac{\partial W_{kj}}{\partial t} = -\eta_k \frac{\partial E}{\partial W_{kj}} \qquad (4)$$

where $\eta_k > 0$ is the usual learning rate. The gradient terms $\frac{\partial E}{\partial w_{kj}}$ can be derived using the backpropagation algorithm [8]. The cost function $E$ is defined is the error index and the least square error criterion is often chosen as follows: $E = \frac{1}{2} \sum_{k=1}^{2} \varepsilon_k^2$

## 2.3 PSO-Based Training Algorithm

Particle swarm optimization (PSO) is a type of derivative free evolutionary search algorithms resulted by an intelligent and well organized interaction between individual members in a group of birds or fish for example. This algorithm was planned to simulate the positioning and dynamic movements in biological swarms as they look for sources of food or keep away from adversaries [10].

In PSO, $m$ particles fly through an $n$-dimensional search space. For each particle $i$, there are two vectors: the velocity vector $V_i = (v_{i1}, v_{i2}, \ldots, v_{in})$ and the position vector $X_i = (x_{i1}, x_{i2}, \ldots, x_{in})$. Similar to bird socking and fishes schooling, the particles are updated according to their previous best position $P_i = (p_{i1}, p_{i2}, \ldots, p_{in})$ and the whole swarm's previous best position $P_g = (p_{g1}, p_{g2}, \ldots, p_{gn})$. This means that particle $i$ adjust its velocity $V_i$ and position $X_i$ in each generation according to the equations bellow [11]:

$$v_{id}(t+1) = v_{id}(t) + c_1 \times rand()_1 \times (p_{id} - x_{id}) + c_2 \times rand()_2 \times (p_{gd} - x_{id}) \quad (5)$$

$$x_{id}(t+1) = x_{id}(t) + v_{id}(t+1) \quad (6)$$

where $d = 1, 2, \ldots, n$; $c_1, c_2$ are the acceleration coefficients with positive values; $rand()_1$, $rand()_2$ are random numbers between 0 and 1. The new velocity and position for each particle are calculated using the Eqs. (5) and (6) based on its velocity $v_{id}(t)$, best position $P_{id}$ and the swarm's best position $P_{gd}$.

In order to calculate the optimized parameter of learning rate $\eta_k$ given in Eq. (4), the PSO is used *off-line* to minimize the neural network prediction error.

We define the quadratic errors $e_{rq}$ as: $e_{rq}(t_i) = \frac{1}{2} \sum_{k=1}^{2} \varepsilon_k^2(t_i)$

The objective function $f$ to be minimized is chosen as the norm of the quadratic error: $f = norm(E_{rq})$ with $E_{rq}$ the vector that contains all errors $e_{rq}(t_i)$.

The PSO algorithm test the search space using $m$ particles according to (5) and (6). Each particle $i$ moves in search space and stores its best position $p_{id}$ ($\eta_k$), then, it compares all positions to finally take out the chosen $\eta_{k-optimum}$ that give the minimum value of the objective function $f$.

## 3   Simulation Results

In this section, we test the proposed control approach on a two links robot described by the model (1). The control objective is to maintain the system to track the desired angle trajectory:$x_{1d} = (\pi/3)\cos(t)$ and $x_{4d} = \pi/2 + (\pi/3)\sin(t)$.

The masses are considered to be $m_1 = 0.6$ and $m_2 = 0.4$. The considered uncertainties are a vector random noise with the magnitude equal to unity.

$$E = \begin{pmatrix} 5 & 0 \\ 0 & 5 \end{pmatrix}, \quad B = \begin{pmatrix} 10 & 0 \\ 0 & 10 \end{pmatrix} \text{ and } J = \begin{pmatrix} 100 & 0 \\ 0 & 100 \end{pmatrix}$$

The switching functions coefficients are defined as: $\gamma_{11} = \gamma_{22} = \beta_{11} = \beta_{22} = 4$. A swarm population of 20 particles is used in this paper (Table 1).

From Figs. 1 and 3, it can be seen that the tracking performance is obtained without any oscillatory behaviour even in the presence of large uncertainties. The corresponding control current signals are given in Figs. 2 and 4.

**Table 1** The global optimum particle of learning rate $\eta_k$

| Iteration number | Variation of $\eta_k$ | Iteration number | Variation of $\eta_k$ |
|---|---|---|---|
| 1 | 0.4232 | 21 | 0.1124 |
| 2 | 0.6444 | 22 | 0.1210 |
| 3 | 0.8449 | 23 | 0.1191 |
| 4 | 0.5226 | 24 | 0.1075 |
| 5 | 0.6521 | 25 | 0.1024 |
| 6 | 0.6984 | 26 | 0.1008 |
| 7 | 0.5981 | 27 | 0.1015 |
| 8 | 0.6954 | 28 | 0.1085 |
| 9 | 0.3215 | 29 | 0.1112 |
| 10 | 0.1201 | 30 | 0.1102 |
| 11 | 0.0544 | 31 | 0.1094 |
| 12 | 0.1410 | 32 | 0.1084 |
| 13 | 0.1727 | 33 | 0.1081 |
| 14 | 0.1710 | 34 | **0.1082** |
| 15 | 0.1654 | 35 | **0.1082** |
| 16 | 0.1591 | 36 | **0.1082** |
| 17 | 0.1053 | 37 | **0.1082** |
| 18 | 0.1722 | 38 | **0.1082** |
| 19 | 0.1834 | 39 | **0.1082** |
| 20 | 0.1949 | 40 | **0.1082** |

**Fig. 1** Angle response $x_1$ and desired trajectory $x_{1d}$

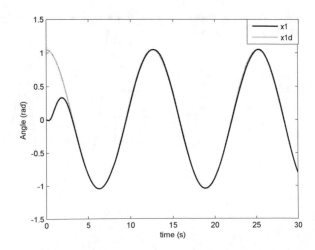

**Fig. 2** Control $u_1$ (input current of join actuator 1)

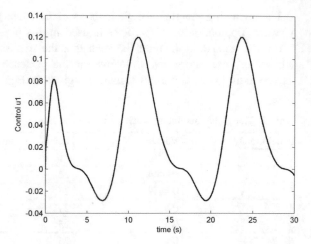

**Fig. 3** Angle response $x_4$ and desired trajectory $x_{4d}$

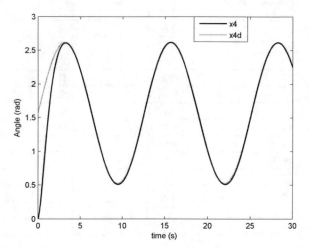

**Fig. 4** Control $u_2$ (input current of join actuator 2)

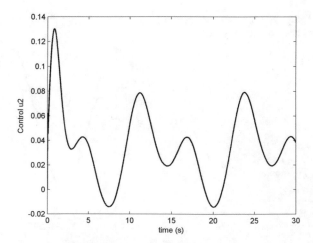

# 4 Conclusion

This paper addressed the robust optimal reference tracking problem for two-links robot manipulators. The designed method is a combination of traditional sliding mode control approach and neural network. The later is employed to approximate the unknown nonlinear model function with online adaptation of parameters via BP learning algorithm. This provides a better description of the plant, and hence enables a lower switching gain to be used despite the presence of large uncertainties. The particle swarm optimization (PSO) algorithm is used to optimize the learning rate of BP algorithm in order to get faster convergence. The simulation results have shown a good performance of the proposed method to track the reference without any oscillatory behavior. Further studies would focus on more effective optimization methods for the gain of the sliding additive control.

# References

1. Patino, H.D., Carelli, R., Kuchen, B.R.: Neural networks for advanced control of robot manipulators. IEEE Trans. Neural Networks **13**, 343–354 (2002)
2. Hussain, M.A., Ho, P.Y.: Adaptive sliding mode control with neural network based hybrid models. J. Process Control **14**, 157–176 (2004)
3. Liu, P.X., Zuo, M.J., Meng, M.Q.H.: Using neural network function approximation for optimal design of continuous-state parallel-series systems. Comput. Oper. Res. **30**, 339–352 (2003)
4. Sefreti, S., Boumhidi, J., Naoual, R., Boumhidi, I.: Adaptive neural network sliding mode control for electrically-driven robot manipulators. Control Eng. Appl. Inform. **14**, 27–32 (2012)
5. Slotine, J.J.: Sliding controller design for non-linear systems. Int. J. Control **40**, 421–434 (1984)
6. Utkin, V.I.: Sliding modes in control optimization, Springer (1992)
7. Slotine, J.J., Sastry, S.S.: Tracking control of nonlinear systems using sliding surfaces with applications to robot manipulators. Int. J. Control **39**, 465–492 (1983)
8. Rumelhart, D.E., Hinton, G.E., Williams, RJ.: Learning internal representations by error propagation. In: Parallel Distributed Processing, vol. 1. Cambridge, MIT Press (1986)
9. Fu, L.M.: Neural Networks in Computer Intelligence. McGraw-Hill, New York (1995)
10. Eberhar, R.C., Kennedy, J.: A new optimizer using particle swarm theory. In: Proceedings of the Sixth International Symposium on Micro-Machine and Human Science, pp 39–43 (1995)
11. Cavuslua, M.A., Karakuzub, C., Karakayac, F.: Neural identification of dynamic systems on FPGA with improved PSO learning. Appl. Soft Comput. **12**, 2707–2718 (2012)

# Advanced MPPT Controller Based on P&O Algorithm with Variable Step Size and Acceleration Mechanism for Solar Photovoltaic System

**Sanae Dahbi, Abdelhak Aziz, Naima Benazzi, Mohamed Elhafyani and Nourddine Benahmed**

**Abstract** This paper presents an improvement of maximum power point tracking (MPPT) for photovoltaic system (PV) by the use of an algorithm based on the classical method of perturbation and observation (P&O). The proposed algorithm is built around the assembly of two separate principles: stability of the system at the operating point of the maximum power point and the precision which is translated by total elimination of the oscillations around the point. The improvement to this algorithm implements:

- A variable step size of duty cycle and consequently an acceleration to catch up operating point: rapidity of the system.
- To locate exactly the point of operation and to maintain it as much as the period of sunlight is constant. This has the effect of eliminating oscillations around the maximum power point (MPP) and pointing at this point.

**Keywords** MPPT · P&O · Variable step size · Acceleration · PV · MPP · Rapidity · Stability

S. Dahbi · A. Aziz (✉) · N. Benazzi (✉) · M. Elhafyani (✉) · N. Benahmed (✉)
Laboratory of Electrical Engineering and Maintenance, Higher School of Technology, Oujda, Morocco
e-mail: aziz.abdelhak67@gmail.com

N. Benazzi
e-mail: benazzin@gmail.com

M. Elhafyani
e-mail: elhafyani77@gmail.com

N. Benahmed
e-mail: nourddine324@hotmail.com

S. Dahbi
e-mail: dahbisanae@hotmail.fr

© Springer International Publishing Switzerland 2016
A. El Oualkadi et al. (eds.), *Proceedings of the Mediterranean Conference on Information & Communication Technologies 2015*, Lecture Notes in Electrical Engineering 381, DOI 10.1007/978-3-319-30298-0_7

# 1 Introduction

Known by its ubiquity, abundance, and the respect of sustainable development criteria, Photovoltaic's solar energy (PV) is now considered as one of the most promising energy resources among all other renewable sources. However, the PV system presents two major problems: low efficiency of conversion in produced electrical energy, and its intermittent character (the output characteristic of a PV system is not linear and varies according to the ambient temperature and the levels of solar irradiance). Therefore a MPPT technique is necessary to obtain the maximum power from photovoltaic system.

In fact, the algorithm perturb and observe (P&O) continues to be the most widely used and studied in the MPPT method for PV [1–3], due to its simplicity of application, high reliability and efficiency tracking. In spite of these advantages, the method (P&O) is still handicapped by the fact that its control algorithm based on fixed size iteration concerning the duty cycle, which presents a common obstacle to overcome: The compromise between the dynamic performance and oscillations state around the MPP [4]. Many authors have proposed different improvements of the basic P&O algorithm as in [5–7], in order to optimize the effectiveness of P&O method and globally the efficiency of PV supply system. This paper will focus on one hand, to the maximization and stability of the output power PV system around the MPP for all levels of solar illumination by using variable step size. In other hand, the acceleration of the search for optimal point by introducing a new factor K which is responsible to speed up the time of response and reach the MPP.

In the next, we present and analyze the electrical models of a photovoltaic panel and a step-down converter. Then we describe profoundly the proposed algorithm MPPT based on the principle of variable step and acceleration mechanism. We present simulation results to evaluate this new control algorithm and some conclusions are drawn.

# 2 Photovoltaic System Modeling

Figure 1 shows an implementation of a PV system compound of a photovoltaic panel Mutsibuchi-180 type which can deliver in the standard test conditions (1000 W/m$^2$ and a temperature of 25 °C) a power of 174 W, buck converter power (step-down) already sized and designed to work at a frequency of 100 kHz and autonomous resistive load [8].

## 2.1 Photovoltaic Module

A photovoltaic panel is a set of solar cells that convert rays or photons of the sun directly in electric energy. A solar cell is generally represented by a current source

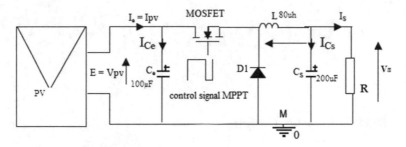

**Fig. 1** Diagram of the PV system

**Fig. 2** Equivalent circuit of a solar cell

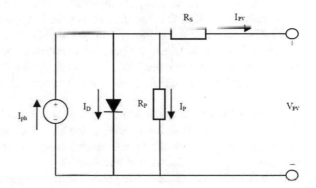

connected in parallel with a diode, a series resistance and a shunt resistance (Fig. 2). Therefore, $I_{PV}$-$V_{pv}$ characteristic equation of a solar cell is given by [9]:

$$I_{PV} = I_{PH} - I_S \left[ \exp \cdot \left( \frac{q \cdot (V_{PV} + R_S \cdot I_{PV})}{k \cdot T \cdot A} \right) - 1 \right] - \frac{(V_{PV} + R_S \cdot I_{PV})}{R_P} \quad (1)$$

where $I_{PV}$: photovoltaic generated current (A), $V_{PV}$: photovoltaic generated voltage (V), $I_{PH}$: light-generated current (photo-current) (A), $I_s$: saturated diode current (A), q: Unsigned electron charge (C), A: ideal factor (varies between 1.2 and 5), k: Boltzmann's constant (J/K), Tc: absolute cell temperature (K), Rs: series resistance (Ω), Rp: shunt resistance (Ω).

As a PV cell produces only low power at low voltage ($\sim 0.5$ V), solar cells must be connected in series-parallel configuration to provide a sufficiently high output power, and a sufficient voltage to meet the needs of any application. Therefore, the equivalent circuit of the PV module arranged in series cell Ns and parallel cell Np is mainly based on the equations of a solar cell [9]:

$$I_{PV} = N_P \cdot I_{PH} - I_S \cdot N_P \cdot \left[ \exp \cdot \left( \frac{q \cdot (V_{PV} + R_S \cdot I_{PV})}{k \cdot T \cdot A} \right) - 1 \right] - \frac{\left( \left( \frac{N_P}{N_V} \right) \cdot V_{PV} + R_S \cdot I_{PV} \right)}{R_P}$$

$$(2)$$

## 2.2 Buck Converter Design

The characteristic I (V) whom presents by the PV is not linear; the output power varies according to the solar irradiances, temperature and nature of the connected load. Under these conditions, photovoltaic systems should interpose an adapting stage which is a DC/DC converter to reach an optimum operation.

This is realized in our design by interposing between the PV and the load a buck converter in order to adjust the point of maximum operation; so this block (converter) receives the signal modulated in width of pulse (PWM) from the algorithmic unit and regulates the operating voltage and current of the PV panel in the optimal values.

Figure 3 shows the step-down buck converter model used for the regulation of the voltage in the simulation in the Matlab/Simulink environment. The buck converter contains a switch, diode and passive components: an inductor L, both input and output capacitor Ce, Cs and a resistor R.

The step-down buck converter components are already dimensioned [10] and adapted to operate at a frequency of 100 kHz by the following relationships:

**Fig. 3** Buck converter model in Matlab/simulink

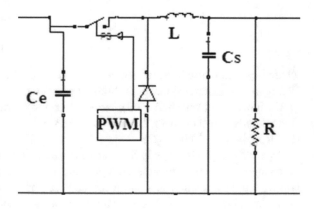

## 3   Perturb and Observe (P&O) Algorithm

Perturb and observe method P&O remains one of the most used in PV systems among all strategies MPPT methods. The Fig. 4 shows the P&O flowchart [11] which is based on the calculation of the PV array output power and the power change by sensing both the PV current and voltage. The tracker operates periodically by comparing the actual value of the power with the previous value to determine the change (incrementing or decrementing) on the solar array voltage or current (depending on the control strategy). In general, the P&O algorithm uses a fixed step size, which causes low precision and slow dynamic response. In fact, if a given perturbation leads to an increase (decrease) in the output power of the PV, then the subsequent perturbation is generated in the same (opposite) direction [12, 13]. When the MPP is reached, the system then oscillates around the MPPT.

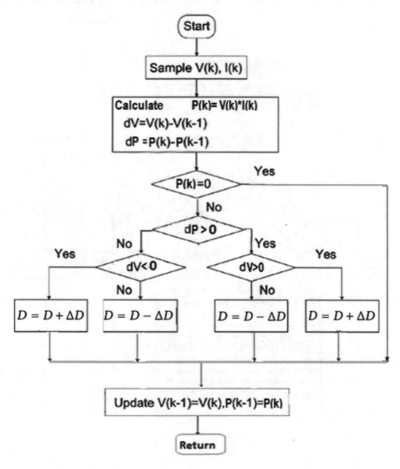

**Fig. 4** Flowchart of traditional P&O

In order to minimize the oscillation, the perturbation step size should be reduced. However, a smaller step size slows down the MPPT.

## 4 Improved P&O Algorithm

In this chapter, the P&O MPPT algorithm proposed is developed to finding a simple effective way to help the system converge to the optimal point MPP by an acceleration mechanism (rapidity) and earning at the same time precision (avoid oscillation) without losing stability. Note that these two parameters (accuracy/stability) obey the uncertainty principle in conventional systems; improving a one results to degradation of the other and therefore to a decrease in overall system performance.

The flowchart of the P&O MPPT algorithm with variable and accelerated step size is shown in Fig. 5 wherein the step size is automatically adjusted according to

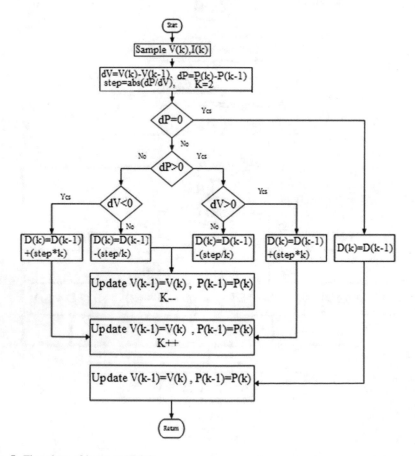

**Fig. 5** Flowchart of improved P&O

the PV array operating point. If a radical change in solar radiation occurs, the step size is set again according to the new position of the operating point. Indeed, the proposed MPPT controller will change the duty cycle of buck converter according to the law expressed by the following relationship:

$$D_k = D_{k-1} \pm \Delta P / \Delta V \tag{3}$$

If the operating point is far from the MPP, algorithm increases the step size, if not it decreases. In the case where the power reaches the MPP, the value of the duty cycle does not receive any modification.

To speed up the time response and reach the MPP, we introduce another factor K which allows:

- The acceleration of the search for optimal point by adjusting the number of iteration (and therefore the duty cycle), when it is far from the point MPP.
- Deceleration when is near to MPP.

This factor allows the rapidity convergence of the system to the point of maximum power.

## 5 Simulation Results Evaluation

To verify the performance of the proposed algorithm based on the P&O method with variable step size and acceleration mechanism of duty cycle, a model of a photovoltaic system in Matlab/Simulink was already developed and simulated. The PV module contains 50 cells in series and can deliver in the standard test conditions a power of 174 W, a current of 8.3 A under optimal voltage 24 V. The modeling of physical components PV is made in the Simscape software and the modeling of numerical part is performed by the S-Function CMEX tool using the language C programming. To compare the performance of the proposed P&O method with the common fixed step size (traditional) P&O MPPT method, the simulations are configured under exactly the same conditions.

The Figs. 6 and 7 represent the simulation of different parameters (power, current, voltage and solar irradiance).

Compared with the P&O MPPT with fixed step size, the new P&O MPPT with Variable Step Size and Acceleration Mechanism of duty cycle shows a good dynamic performance. As indicated in the results of simulations performed, the oscillations around the maximum power point are now completely reduced (almost deleted) and the response time to reach the MPP is less than 5 ms against 12 ms in MPPT with fixed step size. These results prove the efficiency and reliability of this new algorithm and its rapidity to reach the MPP.

**Fig. 6** PV array output
(power, current, voltage, and
solar irradiance) of fixed step
size P&O MPPT

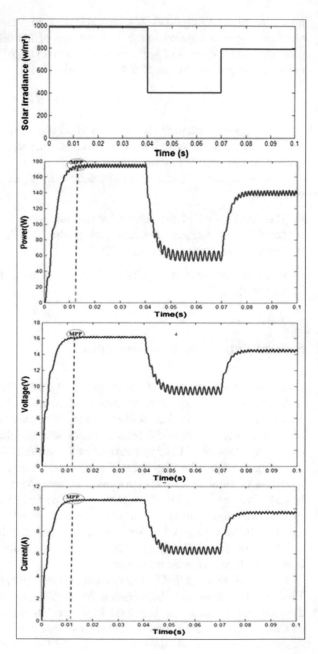

**Fig. 7** PV array output (power, current, voltage, and solar irradiance) of proposed P&O MPPT

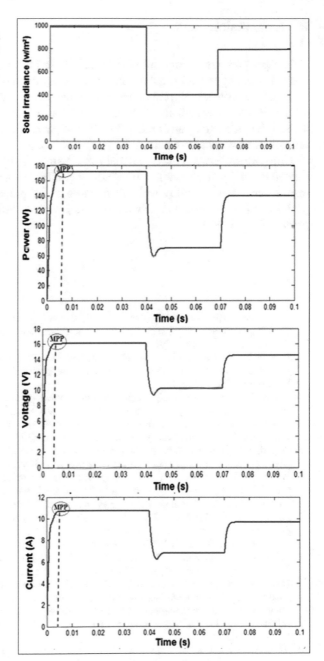

## 6 Conclusion

This paper presents an improvement of maximum power point tracking (MPPT) of a PV system by using an algorithm based on the method of perturb and observe (P&O). The proposed algorithm consists on one hand, a variable step-size and also an acceleration mechanism of the duty cycle in order to reduce the oscillations around the maximum power point (MPP), and get a fast response and precise MPP regardless of weather conditions.

Comparison between the traditional P&O method and proposed P&O MPPT was implemented by a simulation in Matlab-simulink environment. The simulation results verify the feasibility and effectiveness of the proposed method to solve the problems of conventional P&O with fixed step size and enhance the performance of photovoltaic systems.

## References

1. Abdelsalam, A.K., Massoud, A.M., Ahmed, S., Enjeti, P.N.: High performance adaptive perturb and observe MPPT technique for photovoltaic-based microgrids. IEEE Trans. Power Electron. **26**, 1010–1021 (2011)
2. Petrone, G., Spagnuolo, G., Vitelli, M.: A multivariable perturb and observe maximum power point tracking technique applied to a single-stage photovoltaic inverter. IEEE Trans. Ind. Electron. **58**, 76–84 (2011)
3. Boico, F., Lehman, B.: Multiple-input maximum power point algorithm for solar panels with reduced sensing circuitry for portable applications. Sol. Energy **86**, 463–475 (2012)
4. Esram, T., Chapman, P.L.: Comparison of photovoltaic array maximum power point tracking techniques. IEEE Trans. Energy Convers. **22**, 439–449 (2007)
5. Piegari, L., Rizzo, R.: Adaptive perturb and observe algorithm for photovoltaic maximum power point tracking. J. Renew. Power Gener. **4**, 317–328 (2010)
6. Wang, H., Jianhui Su., Nayar, C., Peng, Z.: Adaptive maximum power point tracker in photovoltaic grid-connected system. In: 2010 2nd IEEE International Symposium on Power Electronics for Distributed Generation Systems, pp. 374–377. Hefei, China (2010)
7. Aashoor, F.A.O., Robinson, F.V.P.A: Variable step size perturb and observe algorithm for photovoltaic maximum power point tracking. In: 47th International Universities Power Engineering Conference (UPEC), pp. 1–6. IEEE Trans, London (2012)
8. Dahbi, S., Aziz, A., Benazzi, N., Zahboune, H.: Toward a new method to improving hydrogen production by an adaptive photovoltaic system. In: The 2nd International Renewable and Sustainable Energy Conference, IREC. Ouarzazate, Morocco (2014)
9. Maammeura, H., Hamidatb, A., Loukarfia, L.: A numerical resolution of the current-voltage equation for a real photovoltaic cell. In: TerraGreen 13 International Conference 2013— Advancements in Renewable Energy and Clean Environment. J. Energy Procedia. **36**, 1212–1221 (2013)
10. Aziz, A.: Propriétés Electriques des Composants Electroniques Minéraux et Organiques, Conception et Modélisation d'une Chaîne Photovoltaïque pour une Meilleure Exploitation de l'Energie Solaire. Thesis, Toulouse III-Paul Sabatier University (2006)
11. Setti, M., Tanouti, J., Aziz, A., Zdravko, K., El mamoun, A.: Efficient modeling of photovoltaic systems using C MEX S-function under Matlab-simulink environment. J. Environ. Sci. Eng. **5**, 857–865 (2011)

12. Harrag, H., Messalti, S.: Variable step size modified P&O MPPT algorithm using GA-based hybrid offline/online PID controller. J. Renew. Sustain. Energy Rev. **49**, 1247–1260 (2015)
13. Al-Diab, A., Sourkouni, C.: Variable step size P&O MPPT algorithm for PV systems. In: 12th International Conference on Optimization of Electrical and Electronic Equipment, pp. 1097–1102. IEEE Trans, Basov (2010)

# Online Local Path Planning for Mobile Robot Navigate in Unknown Indoor Environment

Mohamed Emharraf, Mohammed Saber, Mohammed Rahmoun
and Mostafa Azizi

**Abstract** In this paper, the problem of path planning is studied for the case of a mobile robot moving in a priori unknown indoor environment, with static obstacles. There is often a need to replan paths online based on information extracted from the surroundings. The environment is modeled as a grid-map form of environment. The online replanning problem is solved using a new approach for $A^*$ algorithm, called $OA^*$; Online $A^*$ allowing the mobile robot navigate through obstacles in unknown environment and find the shortest feasible path from an initial position to a target position by avoiding the obstacles. Simulation results show the applicability and the feasibility of the approach.

**Keywords** Online path planning · Unknown environment · Indoor environment · Mobile robot

## 1 Introduction

Navigation in a priori unknown environments has a wide spectrum of applications in advanced robotics. Traditionally, this problem has been addressed either by having the robot build a map of the environment [1] (what can be seen from actual

M. Emharraf (✉) · M. Saber · M. Rahmoun
Laboratory Electronics, Computer and Image Systems,
National School of Applied Sciences, Oujda, Morocco
e-mail: m.emharraf@gmail.com

M. Saber
e-mail: mosaber@gmail.com

M. Rahmoun
e-mail: moha1rahmoun@gmail.com

M. Azizi
Laboratory Mathmatiques appliques, traitement du signal et informatique,
First Mohammed University, Oujda, Morocco
e-mail: azizi.mos@gmail.com
URL: http://www.ump.ma

© Springer International Publishing Switzerland 2016
A. El Oualkadi et al. (eds.), *Proceedings of the Mediterranean Conference
on Information & Communication Technologies 2015*, Lecture Notes
in Electrical Engineering 381, DOI 10.1007/978-3-319-30298-0_8

position) before planning the path, or by applying a deterministic algorithms that are able to cope with unknown environments [2].

Path planning is an important task in mobile robot intelligent control which should be performed efficiently. Planning a path means generate a collision free path in an environment with obstacles and optimize it [3, 4]. The environment map may be imprecise, vast, dynamical or non-structured [5]. In such environment, path planning depends on the robot sensory information about the environment, which might be associated with imprecision and uncertainty. Thus, to have a suitable path planning in such environment, the control system must be adaptive in nature of work area. If the environment is known and static, then the path generate in advance, it called off-line algorithm [6]. The planning is online [6], if it is capable of producing a new path in response to environmental changes. The path planning is the art of deciding which route to take for navigation under environment criterion. The idea presented in this paper is a new approach of $A^*$ algorithm called Online $A^*$. In the order to update the A* algorithm for the unknown environment, which allow an autonomous mobile robot navigate in static unknown indoor environment based on sensors information about the environment and there position. Without any need such external information or a priori information about the environment. It can help the advantages of $A^*$ in path determination and eliminating many of the drawbacks.

The rest of the paper is outlined as follows. Section 2 gives a description of path planning method. The new algorithm for the path planning is described in detail in Sect. 3. Section 4 provides the result through simulation and Sect. 5 concludes the paper.

## 2 Path Planning

Basic assumptions used in this approach:

- The robot has a short sensing range compared to the size of the environment.
- Robot senses radially from its position. Some obstacles stop the sensing in there directions.
- Robot knows its current position (coordinates, orientation, using local localization (dead reckoning).

### 2.1 Environment Presentation

For modeling the system, the local work area must be discretized. Discretizing the environment excludes lots of solutions. In this paper the terrain is described by nodes formed on squares/rectangles of equal size. Then one assigns a cost for a transition between two squares. Another straight forward approach, used by [7],

**Fig. 1 a** 4-neighbors,
**b** 8-neighbors

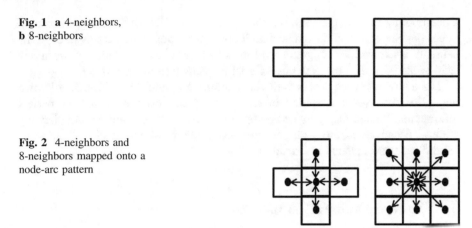

**Fig. 2** 4-neighbors and
8-neighbors mapped onto a
node-arc pattern

is described by nodes connected by arcs. Each arc has a specified cost for moving along it.

A similar description of the environment is to first place nodes in a grid pattern, then assign costs for the transition between two nodes (squares). Two obvious representations are shown in Fig. 1. The two patterns in Fig. 1 are, as can be seen in Fig. 2, almost equivalent to the approach using nodes and arcs squares in a pattern. The squares structure (map grid) allows more planning options and makes complete presentation of the environment. For a map grid, to be an interesting approach. Must have a sensing range longer than the length to the longest considerable neighbor.

## 2.2 The Classic $A^*$ Algorithm

The $A^*$ operates [8] essentially the same as Dijkstras algorithm except that it guides its search towards the most promising states, potentially saving a significant amount of computation. A $A^*$ plans a path from an initial start state to a goal state. To plan the path, the algorithms store an estimate g(s) of the path cost from the initial state to each state s. Initially, g(s) = $\infty$ for all states s. The algorithm begins by updating the path cost of the start state to be zero, then places this state onto a priority queue known as the OPEN list. Each element s in this queue is ordered according to the sum of its current path cost from the start, g(s), and a heuristic estimate of its path cost to the goal, h(s, goal). The state with the minimum such sum is at the front of the priority queue. The heuristic h(s, sgoal) typically underestimates the cost of the optimal path from s to goal and is used to focus the search.

The algorithm then pops the states at the front of the queue and updates the cost of all states reachable from this state through a direct edge: if the cost of state s, g(s), plus the cost of the edge between s and a neighboring state s0, c(s, s0), is less than the current cost of state s0, then the cost of s0 is set to this new, lower value. If the cost of a neighboring state s0 changes, it is placed on the OPEN list. The algorithm continues popping states off the queue until it pops off the goal state. At this stage, if the heuristic is admissible, i.e. guaranteed to not overestimate the path cost from any state to the goal, then the path cost of goal is guaranteed to be optimal.

The above approach works well for planning an initial path through a known map. However, when operating in real world scenarios, robots haven't perfect information. Rather, they may be equipped with incomplete or inaccurate planning graphs. In such cases, any path generated using the initial map may turn out to be invalid as it gets updated information.

## 3   The New Online $A^*$ Approach

The general idea of $OA^*$ algorithm is that a path is planned based on what is known right now. When the robot gets new information it considers updating the path. Gradual learning about the surroundings results in better plans. The information about the environment is translated to nodes costs. If the terrain is completely unknown the nodes can be initially assigned not affected by obstacles costs.

### 3.1   Operating Principle

The $OA^*$ algorithm determinate the path to the goal in unknown a prior environment by following the next steps:

Step 1:   Based on the present knowledge a path from the current node to the goal node is planned using an optimal solution (direct path) Fig. 3a.

Step 2:   The robot follows the optimal path and explorer the environment in same time. If they reach the goal, the robot stops. If an obstacle detected in the path to the goal, the algorithm searches a new path to the goal step 3.

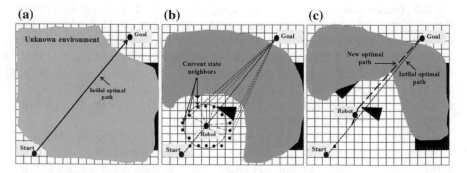

**Fig. 3** Online OA star steps

Step 3:  determinate all neighbors of the current state Sc in a circle with center Sc and radius the distance between Sc and the detected obstacle as Fig. 3b.

Step 4:  compute the cost function for each neighbors determinate in step 3, using the cost function described in Sect. 3.2, and store the result in a queue (OPEN list) where the state (neighbor) with minimal cost are placed in the top.

Step 5:  robot move to the state with minimal cost, return to step 1 Fig. 3c.

## 3.2  The Cost Function

The OA* method use the cost function to determinate the current optimal path to the goal. The cost function can be defined as below:

$$F(n) = G(n) + H(n) + T(n) + O(n) + M(n)$$

where

G(n)   is the generation cost of the node i.e. from current node to node.

H(n)   is heuristic cost of the node; here we consider the Euclidean distance between the goal and n node.

T(n)   is the cost of trajectory node i.e. how often robot visit the node, the T(n) value given by mapping system [9].

O(n)   present the occupancy of node i.e. low values for free or unknown node and very high value for occupied nodes, the O(n) value computed by mapping system [9].

M(n)   is the cost of node neighbors i.e. low value if all neighbors are free nodes, and become high with number of occupied nodes neighbors.

The next Fig. 4 presents the parameter of cost function.

## 4  Simulation Results

In this section, we give examples to illustrate the new path planning algorithm. The working space is of 36 × 36 grids, and all obstacles in the working space are described such that the mobile robot can be viewed as a point in a grid.

The path planning algorithm is programmed in LABVIEW platform on a computer. We use two matrices, one for the map occupancy O, and the other for map trajectory T [9]. The value of the element of the matrix O represents the occupancy of nodes (free, occupied, unknown); for example, if the n node is an obstacle, then the value of the element of matrix O(n) is infinite (in a program,

**Fig. 4** Cost function parameters

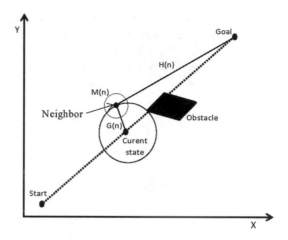

assigned 100); if the m node is a free space, then the value of the element of matrix O(m) is 0. The value of the element of matrix O that represents mobile robot start or goal is 1, and the value of the element of matrix T that expresses the mobile robots trajectory grid is increment by 1 each time robot pass through the node (the values of these elements of the matrices were 0 before mobile robot passes).

Example 1: A mobile robot is passing through a simple environment with static obstacles, as shown in Fig. 5a. According to the path planning algorithm, the mobile robot will use initial path planning until an obstacle passes. Then, the mobile robot will program new path based on OA* algorithm. Figure 5b gives the path planning result. Where the red square is the position of the goal, the black squares are the positions of the obstacles, and the blue squares show the path planning. The algorithm find an optimal path to the goal through the unknown environment.

**(a)** **(b)**

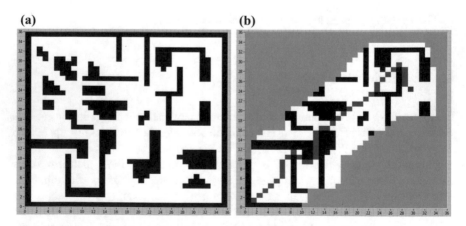

**Fig. 5** Simple environment path planning

(a)                                    (b)

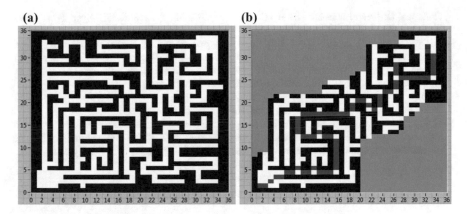

**Fig. 6** Maze path planning

Example 2: for improving the feasibility of the algorithm, we use the robot in a small maze Fig. 6a, the result of simulation improve the feasibility of the algorithm in complexes environments Fig. 6b.

## 5 Conclusion

This paper presented an algorithm for path planning in an unknown environment based on the metric map. An improvement method for mobile robots path planning was explored. The simulation results have shown that the algorithm is efficient. The path planning algorithm has the following features:

(1) Easy to program
(2) Applicability to different static environments
(3) Convergence

## References

1. Meryer, J.A., Filliat, D.: Map-based navigation in mobile robots: II. A review of map-learning and path planning strategies. Cogn. Syst. Res. **4**, 283–317 (2003)
2. Ferguson, D., et al.: A guide to heuristic-based path planning. In: Proceedings of the Workshop on Planning Under Uncertainty, for Automated Planning and Scheduling (ICAPS-05), Monterey, CA (2005)
3. Howard, A., Matari, M.J., Sukhatme, G.S.: An incremental self-deployment algorithm for mobile sensor networks, autonomous robots. Intell. Embedd. Syst. **13**(2), pp. 113–126 (2002)
4. Florczyk, S.: Robot Vision Video-based Indoor Exploration with Autonomous and Mobile Robots. WILEY-VCH Verlag GmbH & Co. KGaA, Weinheim (2005)

5. Janglova, D: Neural networks in mobile robot motion. Int. J. Adv. Robot. Syst. **1**(1), 15–22 (2004). ISSN 1729-8806
6. Al Marzouqi, M., Jarvis, R.A.: Robotic covert path planning: a survey. In: 2011 IEEE Conference on Robotics, Automation and Mechatronics (RAM), pp. 77, 82, 17–19 Sept 2011
7. Ersson, T., Hu, X.: Path planning and navigation of mobile robots in unknown environments. In: Proceedings of the IEEE International Conference on Intelligent Robots and Systems (IROS) (2001)
8. Bellman, R.: Dynamic Programming. Princeton University Press (1957)
9. Emharraf, M., Rahmoun, M., Saber, M., Azizi, M.: Mobile robot unknown indoor environment exploration using self-localization and grid map building. In: 2014 9th International Conference on Intelligent Systems: Theories and Applications (SITA-14), pp. 1, 5, 7–8 May 2014

# Part II
# Software Engineering, Data Mining and Big Data

# Process Mining: On the Fly Process Discovery

**Souhail Boushaba, Mohammed Issam Kabbaj, Zohra Bakkoury
and Said Mohamed Matais**

**Abstract** Process mining is a set of techniques helping enterprises to avoid process modeling, which is time consuming, and error prone task. The goal of such techniques is to extract the process as it has been executed. However, the increase of data production in event logs of process aware information systems makes it necessary to mine the processes in real time. For this purpose, it is necessary to define new approaches for process discovery analyzing data on the fly. This paper presents a new process discovery approach aiming to extract data on the fly by discovering the set of blocks composing the process.

**Keywords** Process mining · Process discovery · On the fly process discovery

## 1 Introduction

Process discovery aims to extract a business process model as it has been executed [1]. Using such techniques permits to avoid process modeling, which is time consuming, and error prone task.

Many algorithms in literature are developed in order to extract process models from event log (α-algorithm [1], automated block discovery algorithm [2], genetic process mining [3]).

S. Boushaba (✉) · M.I. Kabbaj · Z. Bakkoury · S.M. Matais
AMIPS Research Group, Ecole Mohammadia d'Ingénieurs,
Mohammed V University, Rabat, Morocco
e-mail: souhailboushaba@research.emi.ac.ma

M.I. Kabbaj
e-mail: Kabbaj@emi.ac.ma

Z. Bakkoury
e-mail: bakkoury@emi.ac.ma

S.M. Matais
e-mail: saidmatais@reseach.emi.ac.ma

© Springer International Publishing Switzerland 2016
A. El Oualkadi et al. (eds.), *Proceedings of the Mediterranean Conference
on Information & Communication Technologies 2015*, Lecture Notes
in Electrical Engineering 381, DOI 10.1007/978-3-319-30298-0_9

As the mining algorithms extract one process model from "complete" process event log, it is necessary to wait for all the data to be stored before analyzing it.

The aim of this paper is to use our block discovery method [2] to discover process models on the fly by detecting the set of blocks composing an initial event log and changing them when new traces are extracted. A block is defined as a set of tasks having the same behavior with respect to other tasks.

The remainder of this paper is organized as follows. In Sect. 2, we present related works. In Sect. 3, we present our preliminaries. We introduce on the fly process discovery concepts Sects. 4 and 5. The methodology of our discovering processes is presented in Sect. 6. An illustrative example in Sect. 7. Finally, Sect. 8 concludes the paper.

## 2  Related Works

The present paper introduces a new approach of process discovery. In this area many algorithms can be used to extract workflow nets from event logs. Among them the α-algorithm presented in [1] which extracts SWF nets (structured work-flow nets) used to represent a large set of business process models. However, it has its limits when it comes to dealing with noise, discovering short loops and some specific categories of Petri Nets (invisible tasks and non-free choice constraints). An attempt was made in [4] to fix the noise problem by using some metrics expressed in literature. In [5] authors dealt with the problem of non-free choice constraint.

Our approach presented in [2, 6] discovers the set of blocks composing the process model from workflow log. The discovery is done by building a matrix presentation of the process and using filters to automate the extraction of the necessary information.

The previously mentioned methods suppose the existence of a complete event log. That is why, in order to make a proper mining, it is necessary to wait for data to be completely stored (i.e. the execution of all process cases). In this paper, we aim to mine the process using only available data and changing the discovered blocks using the new generated data. This challenge have been already mentioned in the process mining manifesto [7], that's why Burattin et al. [8] have proposed an adaptation of the heuristic process mining algorithm [4] to the so called event streams. Unfortunately, the heuristic miner itself is a slow process as it uses the same concepts of direct succession detection.

# 3 Preliminaries

## 3.1 Indirect Succession Operator

Let W be a workflow log over the set of activities T and let two activities a, b $\in$ T (for more information about the indirect succession principal the reader is referred to [2, 6]):

> a $\ggg_w$ b (Indirect succession) if and only if there exists a trace $\sigma = t_1 t_2 t_3 \ldots t_n$ and i, j $\in \{1, \ldots, n\}$ such that $\sigma \in$ W, $t_i = a$, $t_j = b$ and i < j we also define :
> The causality relation:          a $\rightarrow_w$ b if and only if a $\ggg_w$ b and b $\ggg_w$ a;
> Absence succession:              a $\not\equiv_w$ b if and only if a $\ggg_w$ b and b $\ggg_w$ a;
> The parallelism relation:        a $\|\|_w$ b if and only if a $\ggg_w$ b and b $\ggg_w$ a;

## 3.2 Characteristic Matrix

Based on the indirect succession operator, we build a characteristic matrix as follows:

Let L be a workflow log, $(A_i)_{1 \le i \le n}$ the set of activities (n is the number of tasks composing the log), the matrix $(C_{i,j})_{\substack{1 \le i \le n \\ 1 \le j \le n}}$ is the binary matrix where:

- $C_{i,j} = 1$ if $A_i \ggg_L A_j$;
- Else $C_{i,j} = 0$.

## 3.3 Block Formalism

The block concept is a subject that we introduced in [2]. Our main idea is to extract the set of blocks composing the process model from process event log. Note that a block is a set of tasks having the same behavior with respect to other tasks outside it.

To automate the detection of blocks we created a combinatorial logic operator formalized as follows:

Let $(A_i)_{1 \leq i \leq n}$ be a set of tasks having the characteristic matrix $(Cij)_{\substack{1 \leq i \leq n \\ 1 \leq j \leq n}}$

the logical similarity operator calculus is defined as:

$$
\overline{\oplus_{i=1}^{n} A_i} = \begin{cases} 1, & \sum_{i=1}^{n} Cij(mod\, n) = 0 \\ 0, & \sum_{i=1}^{n} Cij\,(mod\, n) > 0 \end{cases}
$$

## 3.4   Detection Filters

### 3.4.1   Pattern Detection

We adopt three design patterns allowed by the petri net notation [2]. We apply the logical similarity operator to each characteristic matrix design patterns. The following tables show the results (Tables 1, 2 and 3):

**Results:**

- In each table, tasks B and C are in block (see the blue boxes)
- The logical similarity operator gives different result in every pattern which proves its ability to detect patterns.

### 3.4.2   Detecting First and Last Set of Tasks

Let L be a workflow log and let $(C_{i,j})_{\substack{1 \leq i \leq n \\ 1 \leq j \leq n}}$ its corresponding characteristic matrix:

**Theorem 1** *Task Ai is the first task iff the sum of its corresponding column equals min (SumColumn).*

*Task Ai is the last task iff the sum of its corresponding row equals min (SumRow).*

**Table 1** Succession pattern detection

| >>> | A | | C | D | $\overline{B \oplus C}$ |
|---|---|---|---|---|---|
| A | 0 | 1 | 1 | 1 | 1 |
| B | 0 | 0 | 1 | 1 | 0 |
| C | 0 | 0 | 0 | 1 | 1 |
| D | 0 | 0 | 0 | 0 | 1 |
| $\overline{B \oplus C}$ | 1 | 1 | 0 | 1 | |

**Table 2** Parallel pattern detection

| >>> | A | B | C | D | $\overline{B \oplus C}$ |
|---|---|---|---|---|---|
| A | 0 | 1 | 1 | 1 | 1 |
| B | 0 | 0 | 1 | 1 | 0 |
| C | 0 | 1 | 0 | 1 | 0 |
| D | 0 | 0 | 0 | 0 | 1 |
| $\overline{B \oplus C}$ | 1 | 0 | 0 | 1 | |

**Table 3** Xor pattern detection

| >>> | A | B | C | D | $\overline{B \oplus C}$ |
|---|---|---|---|---|---|
| A | 0 | 1 | 1 | 1 | 1 |
| B | 0 | 0 | 0 | 1 | 1 |
| C | 0 | 0 | 0 | 1 | 1 |
| D | 0 | 0 | 0 | 0 | 1 |
| $\overline{B \oplus C}$ | 1 | 1 | 1 | 1 | |

# 4 On the Fly Process Discovery: The Fundamentals

Our goal is to discover blocks on the fly. In the first subsection, we introduce a new concept of partial event log. We give a matrix composition and then we show some rules used to upgrade the discovered blocks.

A partial log is a set of traces obtained by the execution of the process model. To formalize the concept we consider a workflow log L.

**Definition 1** $L_i$ is a partial log of L iff: $L_i = \{\theta_j \in L / 1 \leq j < n$, where $n <$ card $(L)\}$, $\theta_j$ is process case

The goal of our mining processes on the fly is to extract the process behavior for each partial log separately and composing the results to find the new process model representation. The main issue is to know if the appearance of a new partial event log changes the existing patterns or not. To compose partial logs, we define the following three operators: $\sum_1^n L_i = \{\theta_j / \exists i \in [1, n], \theta_j \in L_i\}$ (composition), $\bigcap_1^n L_i = \{A_k / \exists i \in [1, n], \exists j \in [1, \text{card}(L_i)], \theta_j \in L_i \text{ and } A_k \in \theta_j\}$ (intersection) and $\bigcup_1^n L_i = \{A_k / \forall i \in [1, n], \forall j \in [1, \text{card}(L_i)], \theta_j \in L_i \text{ and } A_k \in \theta_j\}$ (union).

## 5    On the Fly Process Discovery: Data Extraction

### 5.1    Detection of the Transformed Relation Between Activities

To extract processes on the fly, it is necessary to detect changes generated by the occurrence of new succession relations. For this purpose, we redefine the logical similarity to take as input a set of partial logs matrices as:

$$\overline{\oplus_{k=1}^n C_{ij}} = \begin{cases} 1, & \sum\limits_{k=1}^m C_{k;ij}(mod\ m) = 0 \\ \\ 0, & \sum\limits_{j=1}^m C_{k;ij}(mod\ m) > 0 \end{cases} \quad where\ m = \text{card}(\bigcup_1^n L_i)$$

- The value 1 in resulting calculus indicates that the existing relation between a couple of activities $A_i$ and $A_j$ did not change.
- However, the value 0 indicates the occurrence of changes.

So to detect the changed relations, we apply our logical similarity operator (as defined before) to sub-characteristics matrices (containing only common activities) of the characteristic matrix corresponding to each partial log.

### 5.2    Transformation Directives of the Common Activities

We mean by transformation directives the set of rules allowing changing the process model on the fly. We note that a pattern may change if and only if 0 values appear in its corresponding row and column in the logical similarity calculus.

**Lemma 1** *Let's consider A and B two activities in partial logs $L_i$ and $L_j$ with 0 occurs in the logical similarity calculus*

$$\text{If } A \twoheadrightarrow_{L_i} B \text{ and } B \twoheadrightarrow_{L_j} A \rightarrow A \;|||_{L_i + L_j} B$$

$$\text{If } A \twoheadrightarrow_{L_i} B \text{ and } A \;|||_{L_j} B \rightarrow A \;|||_{L_i + L_j} B$$

$$\text{If } A \twoheadrightarrow_{L_i} B \text{ and } A \not\equiv_{L_j} B \rightarrow A \twoheadrightarrow_{L_i + L_j} B$$

$$\text{If } A \not\equiv_{L_i} B \text{ and } A \;|||_{L_j} B \rightarrow A \;|||_{L_i + L_j} B$$

## 5.3 Add New Activities

As we are able to detect changed blocks, not common activities must be added. For this purpose, we add each task to the characteristic matrix of the composed partial log $\left(\sum_1^n L_i\right)$ such that the new cells must be added with a 0 value (due to the absence of relationship). However, the newest activity saves the same value with common activities $\bigcap L_i$. Using the previous example, $B \in L_1$ but $B \notin L_2$ and $E \in L_2$ but $E \notin L_1$ so $B \not\equiv_{L_1 + L_2} E$. To compose L1 and L2, we need to introduce in the first matrix the activity E.

## 6 On the Fly Process Discovering: The Methodology

Let's $L_{t0}$ be the first partial event log, to discover the set of blocks we execute the following steps:

- First step: compute the characteristic matrix, and the sum of rows ($S_{row}$) and columns $S_{column}$ for each activity;
- Second step: select tasks having the same value of $S = S_{row} + S_{column}$ (if multiple sets select those having the maximum is row's sum);
- Third step: for the selected tasks, we apply the logical similarity operator;
- Fourth step: if the selected tasks are in a bloc, we try to detect its type:

  - First, we look for Xor pattern;
  - Second, we look for the parallel pattern;
  - Third, we assume that it is a succession pattern;

- Fifth step: create a frame of discovered Petri Net;
- Sixth step: we repeat all steps for the unselected tasks.

By the extraction of new partial event log, we execute the following steps:

- First step: detect common activities,
- Second step: compute sub matrices for common activities,
- Third step: compute the composition of sub matrices,
- Fourth step: apply transformation rules to upgrade the patterns,
- Fifth step: add the new activities to the characteristic matrix.

# 7 Illustrative Example

Let's consider for example the workflow log L = [(A, B, C, D), (A, E, D), (A, C, B, D)].

We suppose for example that in a first time we obtained the log: $L_1$ = [(A, B, C, D)].

- We compute the characteristic matrix corresponding to the log $L_1$ (see Table 4)
- Task A is the first task (sum of column equals to 0)
- Task D is the last task (sum of row equals to 0)
- The characteristic matrix shown in Table 4 represents a succession pattern (see Table 3)
- The process model corresponding to the partial event log L1is as follows:

Now, we assume that we found a new partial event log $L_2$: $L_2$ = [(A, E, D)]

- First: we detect common tasks: {A, D}
- Second: we compute the characteristic sub-matrices for common activities (Tables 5 and 6)

**Table 4** Characteristic matrix corresponding to $L_1$

| ≫ | A | B | C | D | $S_{row}$ |
|---|---|---|---|---|---|
| A | 0 | 1 | 1 | 1 | 3 |
| B | 0 | 0 | 1 | 1 | 2 |
| C | 0 | 0 | 0 | 1 | 1 |
| D | 0 | 0 | 0 | 0 | 0 |
| $S_{column}$ | 0 | 1 | 2 | 3 | |

**Table 5** Characteristic sub-matrix of $L_1$

| ≫ | A | D |
|---|---|---|
| A | 0 | 1 |
| D | 0 | 0 |

**Table 6** Characteristic
sub-matrix of $L_2$

| ⋙ | A | D |
|---|---|---|
| A | 0 | 1 |
| D | 0 | 0 |

**Table 7** Matrix composition
calculus

| $\overline{C_1 \oplus C_2}$ | A | D |
|---|---|---|
| A | 1 | 1 |
| D | 1 | 1 |

**Fig. 1** Petri net corresponding to partial event log $L_1$

**Table 8** Global characteristic
matrix of $L_1 + L_2$

| ⋙ | A | B | C | D | E |
|---|---|---|---|---|---|
| A | 0 | 1 | 1 | 1 | 1 |
| B | 0 | 0 | 1 | 1 | 0 |
| C | 0 | 1 | 0 | 1 | 0 |
| D | 0 | 0 | 0 | 0 | 0 |
| E | 0 | 0 | 0 | 1 | 0 |

- Third: we compute the composition of sub-matrices (see Table 7), the set of 1 values means that the relation between A and D by the appearance of the partial event log $L_2$
- Fourth: no changes to make.
- Fifth: we add the unique uncommon task {E} to the petri net (Fig. 1) as an Xor pattern (see the characteristic matrix Table 8):

Again, we suppose that we found a new partial event log L3: L3 = [(A, C, B, D)]

- First: we detect common tasks: {A, B, C, D}
- Second: we compute the characteristic sub-matrices for common activities (see Tables 9 and 10),
- Third: we compute the composition of sub-matrices (see Table 11), the occurrence of 0 values (red cells) means that the relation between B and C changed.
- Fourth: we apply transformation rules presented in (Sect. 5.2) to the block B, C as follows:

  - The block B, C is a **succession block** B→$_{L1+L2}$C (green box Table 9)
  - The block B,C is **succession block** C→$_{L_3}$B; (green box Table 10)
  - Then B, C is a parallel block (application of transformation rules)

- Fifth: no uncommon tasks to add, the petri net presented in Fig. 2 becomes as shown in Fig. 3:

**Table 9** The characteristic sub-matrix of $L_1 + L_2$

|   | A | B | C | D |
|---|---|---|---|---|
| A | 0 | 1 | 1 | 1 |
| B | 0 | 0 | 1 | 1 |
| C | 0 | 0 | 0 | 1 |
| D | 0 | 0 | 0 | 0 |

**Table 10** The characteristic sub-matrix of $L_3$

| ≫ | A | B | C | D |
|---|---|---|---|---|
| A | 0 | 1 | 1 | 1 |
| B | 0 | 0 | 0 | 1 |
| C | 0 | 1 | 0 | 1 |
| D | 0 | 0 | 0 | 0 |

**Table 11** Matrix composition calculus

|   | A | B | C | D |
|---|---|---|---|---|
| A | 1 | 1 | 1 | 1 |
| B | 1 | 1 | 0 | 1 |
| C | 1 | 0 | 1 | 1 |
| D | 1 | 1 | 1 | 1 |

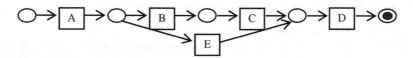

**Fig. 2** Petri net corresponding to partial event log $L_1 + L_2$

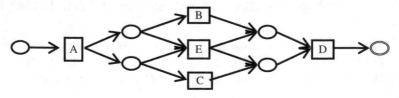

**Fig. 3** Process model corresponding to $L = L_1 + L_2 + L_3$

# 8 Conclusion and Perspectives

In this paper, we discuss the issue of discovering process models on the fly from event logs. The proposed approach, which is based on a matrix calculus, allows discovering a petri net from a partial event log and making changes by the arrival of new set of activities without waiting for all the data to be stored. The paper introduces also the concept of partial event log, which can be mined as a complete event log. In addition, we introduce the concept of matrix composition to detect changed patterns. Besides, to mine process model on the fly, our method starts by detecting blocks of a partial event log, detects the changed blocks using matrix composition. Finally, connects each new block as an Xor pattern. As a perspective of this research, we intend creating new mechanisms to handle loop patterns and automating the procedure by implementing the algorithm in our process discovery package presented in [2].

# References

1. van der Aalst, W.M.P., Weijters, A.J.M.M., Maruster, L.: Workflow mining: discovering process models from event logs. IEEE Trans. Knowl. Data Eng. **16**(9), 1128–1142 (2004) (IEEE Transactions)
2. Boushaba, S., Kabbaj, M.I., Bakkoury, Z.: Process discovery: automated approach block discovery, evavluation of novel approaches in software engineering (ENASE) 2013
3. De Medeiros, A.K.A., Weijters, A.J.M.M., Van Der Aalst, W.M.P.: Using Genetic Algorithms to Mine Process Models: Representation, Operators and Results. Eindhoven University of Technology, Eindhoven (2004)
4. Weijters, A.J.M.M., Ribeiro, J.T.S.: Flexible Heuristics Miner (FHM). BETA Working Paper Series, WP 334, Eindhoven University of Technology, Eindhoven (2010)
5. Wen, L., van der Aalst, W.M.P., Wang, J., Sun, J.: Mining process models with non-free-choice constructs. Data Min. Knowl. Disc. **15**, 145–180 (2007)
6. Boushaba, S., Kabbaj, M.I., Bakkoury, Z.: Process mining: matrix representation for block discovery. In: Intelligent Systems: Theories and Applications (SITA), IEEE (2013)
7. IEEE Task Force on Process Mining: Process Mining Manifesto. In: BPM Workshops. LNBIP, vol. 99, pp. 169–194. Springer (2012)
8. Burattin, A., Sperduti, A., van der Aalst, W.M.P.: Control-flow Discovery from Event Streams IEEE Congress on Evolutionary Computation, IEEE (2014)

# A New Approach Based on PCA and CE-SVM for Hepatitis Diagnosis

Naoual Elaboudi and Laila Benhlima

**Abstract** In this study, we propose a new approach that aims improving the medical diagnosis of hepatitis disease via Machine Learning techniques. The proposed solution consists of two stages In the first one, Principal Component Analysis (PCA) is used to reduce the dimension of features vector. In the second one, classification process is implemented using the remaining features through Support Vector Machine (SVM) method which its parameters are determined using Cross Entropy Optimisation (CEO) that represents an efficient stochastic optimization tool. We have performed experiments on UCI datasets with the combination of SVM with CE. Classification accuracy obtained with the proposed approach is very promising with regard to the existing classification methods for hepatitis disease diagnosing.

**Keywords** Machine learning · Support Vector Machine · Hepatitis diagnosis · Principal Component Analysis · Cross Entropy Optimization

## 1 Introduction

More than 1.4 million people die annually in the world as a consequence of hepatitis which is an inflammation of the liver [10]. It is most commonly caused by a viral infection. Therefore more efforts should be deployed to reduce such high mortality. In the fight against hepatitis, enhancing diagnosis is a key success factor since that

---

The original version of this chapter was revised: The author name has been changed to "Naoual Elaboudi" instead of "Naoual EL Aboudi". The erratum to this chapter is available at DOI 10.1007/978-3-319-30298-0_79

---

N. Elaboudi (✉) · L. Benhlima
Ecole Mohammadia d'ingénieurs, Mohammed V University, Rabat, Morocco
e-mail: nawal.elaboudi@gmail.com

L. Benhlima
e-mail: benhlima@emi.ac.ma

© Springer International Publishing Switzerland 2016
A. El Oualkadi et al. (eds.), *Proceedings of the Mediterranean Conference on Information & Communication Technologies 2015*, Lecture Notes in Electrical Engineering 381, DOI 10.1007/978-3-319-30298-0_10

harmful illness could be treated efficiently if early detected. In this context, many algorithms of Machine Learning have been used as automatic diagnosis systems in medical applications, in particular for hepatitis with linear discriminant analysis and FSM without rotations algorithms used in [5, 16]. In the field of medical diagnosis, pursuing the best performance is a critical issue. Indeed, to enhance the accuracy of the Hepatitis Diagnosis, we reduce the features space dimensionality of hepatitis dataset, then we combine stochastic optimization with Machine Learning algorithms, in particular SVM algorithm. This later is one of the most well-known algorithms in medical diagnosis due to its high predictive power [8, 11], although it is very sensitive to the setting of its hyper-parameters and kernel parameters that influence to a great degree the quality of the resulting model. Thus, the purpose of this study is to propose an approach to improve the medical diagnosis of hepatitis disease by reducing the number of features included initially in the hepatitis dataset through (PCA) in one hand, then classify the resulting dataset by SVM which its parameters are optimizing through cross entropy optimization (CEO) and finally we provide a comparative study between our approach and existing methods designed for diagnosing hepatitis. The rest of the paper is organized as follows. In Sect. 2, we introduce the related works and then we give the background of our study in Sect. 3. In Sect. 4 we detail the new approach that we propose for hepatitis diagnosis. We present the experimental results in Sect. 5. Finally, Sect. 6 discusses research opportunities for future works.

## 2   Related Works

Many researchers have proposed several Machine Learning ML algorithms to design medical diagnosis systems. In this context, Neural Network (NN) methods were presented by Raj Anand in [1] to diagnose Type II Diabetes. Termurtas proposed several methods in [17] to diagnose thyroid. SVM algorithm based on Genetic Algorithm (GA-SVM) was proposed in [7] by Huang and Wang in the frame of diagnosis to several medical datasets. Their solution proved to be computationally time consuming due to low speed convergence [13]. In the field of hepatitis diagnosis problem, which is studied in this paper, Multilayer perceptron, RBF and GRNN algorithms were applied in [11], FS-fuzzy-AIRS and CSFNN algorithms were used in [12, 11] showing limited results in terms of the fulfilled accuracy. In addition, Sartakhti presented an SVM algorithm tuned by simulated annealing to reach better accuracy than previous work, since it enables to use best SVM parameters automatically [15]. In this paper, we propose a new approach based on PCA for feature reduction and SVM with stochastic optimization methods that is CEO for classification process so as to achieve high accuracy, compared to existing approaches for hepatitis diagnosis, within affordable computational time.

## 3 Background

### 3.1 Support Vector Machine

SVM was first introduced by Vapnik and colleagues in 1995 [4], and has been used for several ML problems such as bioinformatics and pattern recognition. It is a supervised learning algorithm used in classification and regression problems that implements the following idea: construct an optimal separating hyperplane that maximizes the margin between the two classes, so that the risk of misclassifying entries of the test dataset can be minimized.

Given dataset $M = \{(x_1, y_1)...(x_n, y_n)\}$ SVM classifies linear separable data by solving the following optimization problem:

$$\underset{b,w}{\text{minimize}} \quad \frac{1}{2}\|w\|^2$$
$$\text{s.t} \qquad y_i(w.x_i + b) \succeq 1 \quad \forall i = 1,\ldots,m. \tag{1}$$

- $x_i$ m-dimensional is an instance of data
- w is m-dimensional weight vector, the perpendicular vector characterising the separate hyperplan
- The offset parameter b scalar which allows to increase the margin

To separate non-linear data which is the case of hepatitis dataset, SVM efficiently performs, and the optimization problem becomes as follows:

$$\underset{b,w,\xi}{\text{minimize}} \quad \frac{1}{2}\|\mathbf{w}\|^2 + C\sum_{i=1}^{m}\xi_i$$
$$\text{s.t} \qquad y_i(w.x_i + b) \succeq 1 - \xi_i \quad \forall i = 1,\ldots,m \tag{2}$$
$$\qquad \xi_i \succeq 0 \qquad\qquad\qquad \forall i = 1,\ldots,m$$

- $\xi_i$ are non-negative slack variables
- C penalty parameter which is particularly important for non-separable training data creates a soft margin that permits some misclassifications, such as it allows some training points on the wrong side of the hyperplane [18] for details.

To solve the optimization problem shown in (2), SVM uses what is called the kernel trick as polynomial, gaussian and radial basis functions. They implicitly map their inputs into high-dimensional feature spaces. In our study, we focus on the radial basis function kernel for its good performance and wide application. One of the important issues in SVM is how to set the best kernel parameters as this setting influences highly the performance of the algorithm. Hence, the parameters that shall be optimized are penalty parameter C and the kernel function parameters such as $\gamma$ for the radial basis function kernel RBF [18]. In SVM, these parameters have to be manually tuned seeking a good accuracy, such an option can be time consuming.

In the next section, we present our solution that consists on using a stochastic optimization method for automatically tuning the C and $\gamma$ parameters.

## 4  Our Approach for Hepatitis Diagnosis

Recall that our objective is to classify hepatitis patients into two classes. The first class corresponds to patients who will react to the treatment and the second one includes other patients. For this purpose we use PCA for feature reduction then SVM for classification based on the radial basis function which needs tuning C and $\gamma$ parameters. In order to automatically perform this tuning, we propose to use CEO. In this section, we describe each of these methods and then, we explain steps of our implementation.

### 4.1  Principal Component Analysis

Since we seek to retain only the most significant features before feeding the classification algorithm, we adopt in our approach Principal Component Analysis (PCA) [6] as a feature reduction method. PCA is known to be one of the popular methods allowing dimensionality reduction. Given a set of data on n dimensions, PCA proceeds by analysing data with the objective of finding patterns that reduce the dimensions of the dataset with minimal loss of information. The PCA approach is summarized by the following steps:

- Take the whole dataset consisting of d-dimensional samples ignoring the class labels
- Compute the d-dimensional mean vector (the means for every dimension of the whole dataset)
- Compute the scatter matrix (alternatively, the covariance matrix) of the whole data set
- Obtain the Eigenvectors and Eigenvalues from the covariance matrix
- Sort the eigenvectors by decreasing eigenvalues and choose k eigenvectors with the largest eigenvalues
- Construct the projection matrix W from the selected k eigenvectors.
- Transform the original dataset X via W to obtain a k-dimensional feature subspace Y This can be summarized by the mathematical equation: $y = W^T \times x$ (where x is a d $\times$ 1 dimensional vector representing one sample, and y is the transformed k $\times$ 1 dimensional sample in the new subspace.)

## 4.2 Cross Entropy Based Optimization of SVM Parameters

In classification phase, the features set obtained at the end of the features reduction stage is given to SVM classifier which its hyperparameters have to be tuned through CEO. The cross-entropy (CE) method was proposed by Rubinstein in [14] as an adaptive importance sampling procedure for the estimation of rare-event probabilities. Subsequent work by Rubinstein has shown that many optimization problems can be translated into a rare-event estimation problem. Since its beginnings, the CE method has been successfully applied to many optimization problems [3, 9].

Thus, we will apply the CE method to tune SVM parameters $(C, \gamma)$ regarding medical diagnosis specially for hepatitis diagnosis, which has never been done before. We will gradually change the sampling distribution of the random search to get the optimal C and $\gamma$. We chose the normal distribution in our study since it gives satisfactory results. Our CEO SVM is an iterative method that consists in repeating the next two steps until the variations of samples $(C, \gamma)$ become smaller than a predefined threshold:

1. The representation and generation of a sample using two normal distributions, the first one provides C values and the second one gives $\gamma$ values.
2. The updating of the sampling distribution parameters based on the best pairs of previous sample.

We note that normal distribution is characterized by $\mu$ which represents the mean of sample and $\sigma$ that provides the variance of the sample. Our method has several important initializing parameters:

1. N = the size of population.
2. Initializing parameters of the two normal distributions.
3. A parameter which tunes the size of the best fitness set in step 2, i.e. the set of best pairs $(C, \gamma)$ according to the SVM accuracy criterion.

## 5 Experimental Results

Prior to comparing proposed methods with existing ones, we describe the Hepatitis dataset attributes that we used for conducting our experiments and we make it ready for simulation by preprocessing data before scaling them. Then, we use k-cross validation method to assess the performance of our solutions. This assessment consists in dividing the whole data into k mutually exclusive and approximately equal size subsets. The classification algorithm is trained and tested k times. In each case, one of the folds is taken as test data and the remaining folds are added to form training data. Thereby, k different test results exist for each training-test configuration. Finally, the average of these results is computed to determine the test accuracy of the algorithm. In our study we chose 10 cross validation.

## 5.1  Hepatitis Disease Dataset

The source of hepatitis data set is the UCI machine learning repository [2]. The data source uses 155 samples with two class problems: die with 32 cases (20.6 %) and live with 123 cases (79.4 %). The dataset consists of 19 attributes, 13 binary and 6 numerical attributes. The attributes description is shown in Table 1. The result of removing missing values is 13 samples in life class and 67 samples in the die class. Since SVM use vector as entries, we have to preprocess the data in order to transform them into vector form.

## 5.2  Preprocessing and Scaling Data

Before conducting our experiments, the dataset which includes 13 categorical attributes and 6 numerical ones must be represented as a vector by converting each categorical attribute into numerical data. The following step consists of scaling the numerical data in order to avoid attributes in greater numeric ranges dominating those in smaller numeric ranges by using a linear transformation described in (3)

**Table 1** Hepatitis attributes

| Attributes | Values |
|---|---|
| Age | 10, 20, 30, 40, 50, 60, 70, 80 |
| Sex | Male, Female |
| Steroid | No, Yes |
| Antivirals | No, Yes |
| Fatigue | No, Yes |
| Malaise | No, Yes |
| Anorexia | No, Yes |
| Liver Big | No, Yes |
| Liver Firm | No, Yes |
| Spleen Palpable | No, Yes |
| Spiders | No, Yes |
| Ascites | No, Yes |
| Varices | No, Yes |
| Bilirubin | 0.39, 0.80, 1.20, 2.00, 3.00, 4.00 |
| Alk Phosphate | 33, 80, 120, 160, 200, 250 |
| Sgot | 13, 100, 200, 300, 400, 500 |
| Albumin | 2.1, 3.0, 3.8, 4.5, 5.0, 6.0 |
| Protime | 10, 20, 30, 40, 50, 60, 70, 80, 90 |
| Histology | No, Yes |

where X is the original data, $X_{min}$ and $X_{max}$ are the minimum and the maximum of X, and $X_{normalized}$ is the normalized data.

$$X_{normalized} = \frac{X - X_{min}}{X_{max} - X_{min}} \qquad (3)$$

## 5.3  Performance Evaluation Metrics

A confusion matrix is a specific table layout as shown in Fig. 1, that allows visualization of the performance of an algorithm, typically with supervised learning, it is usually called a matching matrix in which the columns denote the actual cases and the rows denote the predicted cases, the accuracy of model obtained from training set (model fitness) and testing set (prediction accuracy) are calculated.

- True positive (TP): If an input is determined as die with an optic nerve diagnosed by the automatic clinicians.
- True negative (TN): If an input is determined as live, which was labeled as live by the automatic clinicians.
- False positive (FP): If an input is determined as die, which was labeled as live by the automatic clinicians.
- False negative (FN): If an input is determined as live with an optic nerve diagnosed by the automatic clinicians.

## 5.4  Simulation Results

In the scope of this experiment, the dimension of the hepatitis dataset used in our study corresponds to 19 features. It was reduced to 10 features by PCA method and then fed as input to our proposed classification method CEO-SVM which its sample was set to have the size N = 100. The computations were implemented on a PC under the Ubuntu operating system having a Intel Core 2 Duo at 2.00 GHz processor and 4 GB of RAM. PCA-CEO-SVM classification accuracy for hepatitis

| Predicted | Actual | |
|---|---|---|
| | Negative | Positive |
| Negative | TN | FN |
| Positive | FP | TP |

Fig. 1  Confusing matrix

**Table 2** Simulation results

| Algorithms | Accuracy | Runtime (s) |
|---|---|---|
| PCA-CE-SVM | 97.2 | 90 |
| SA-SVM | 96 | 250 |
| LDA | 86.4 | – |
| FSM without rotations | 88.4 | – |
| Multilayer Perceptron (MLP) | 74.3 | – |
| PCA-LSSVM | 95 | – |
| LFDA-SVM | 96.77 | – |

disease dataset is compared in Table 2 to other classification methods. According to the obtained results, it is obvious that our proposed approach performs better than previously existing algorithms.

## 6 Conclusion and Future Works

In this study, a new approach PCA-CEO-SVM was provided for hepatitis diagnosis. Cross validation technique was applied to evaluate the classification accuracy of our proposed method. According to the results shown in Table 2, PCA-CE-SVM diagnosis approach for hepatitis diagnosis performs better than existent proposed methods in terms of classification accuracy. Therefore, this approach can be very helpful to the physicians in taking accurate decisions for their patients treatment. In future studies on the diagnosis of hepatitis diseases, we are working on different learning methods to keep enhancing the accuracy of the prediction. Finally, we plan to apply the proposed method on diverse dataset provided by medical care institutes.

## References

1. Anand, R., Pratap, V., Kirar, S., Burse, K.: Data pre-processing and neural network algorithms for diagnosis of type ii diabetes: a survey. Int. J. Eng. Adv. Technol. **2**(1), 49–52 (2012)
2. Cheung, N.: Machine learning techniques for medical analysis. Master's thesis, University of Queensland (2001)
3. Cohen, I., Golany, B., Shtub, A.: Resource allocation in stochastic, finite-capacity, multi-project systems through the cross entropy methodology. J. Sched. **10**(3), 181–193 (2007)
4. Cortes, C., Vapnik, V.: Support-vector networks. Mach. Learn. **20**(3), 273–297 (1995)
5. Duch, W., Adamczak, R., Grabczewski, K.: Optimization of logical rules derived by neural procedures. In: International Joint Conference on Neural Networks, 1999. IJCNN'99, vol. 1, pp. 669–674 (1999)
6. Duygu Calisir, E.D.: A new intelligent hepatitis diagnosis system: Pcalssvm. Expert Syst. Appl. **38**(8), 10705–10708 (2011)
7. Huang, C.L., Wang, C.J.: A ga-based feature selection and parameters optimizationfor support vector machines. Expert Syst. Appl. **31**(2), 231–240 (2006)

8. Jiang, Z., Yamauchi, K., Yoshioka, K., Aoki, K., Kuroyanagi, S., Iwata, A., Yang, J., Wang, K.: Support vector machine-based feature selection for classification of liver fibrosis grade in chronic hepatitis c. J. Med. Syst. **30**(5), 389–394 (2006)
9. Kothari, R.P., Kroese, D.P.: Optimal generation expansion planning via the cross-entropy method. In: Proceedings of the 2009 Winter Simulation Conference (WSC), pp. 1482–1491 (2009)
10. Lee, W.M.: Hepatitis b virus infection. N. Engl. J. Med. **337**(24), 1733–1745 (1997)
11. Ozyilmaz, L., Yildirim, T.: Artificial neural networks for diagnosis of hepatitis disease. In: Proceedings of the International Joint Conference on Neural Networks, 2003, vol. 1, pp. 586–589 (2003)
12. Polat, K., Gunes, S.: A hybrid approach to medical decision support systems: combining feature selection, fuzzy weighted pre-processing and airs. Comput. Methods Prog. Biomed. **88**(2), 164–174 (2007)
13. Ren, Y., Bai, G.: Determination of optimal svm parameters by using ga/pso. J. Comput. 5(8) (2010)
14. Rubinstein, R.Y.: Optimization of computer simulation models with rare events. Eur. J. Oper. Res. **99**, 89–112 (1996)
15. Sartakhti, J.S., Zangooei, M.H., Mozafari, K.: Hepatitis disease diagnosis using a novel hybrid method based on support vector machine and simulated annealing (svm-sa). Comput. Methods Programs Biomed. **108**(2), 570–579 (2012)
16. Ster, B., Dobnikar, A.: Neural network in medical diagnosis: comparison with other methods. In: Proceedings of the International Conference EANN'96, pp. 427–430 (1996)
17. Temurtas, F.: A comparative study on thyroid disease diagnosis using neural networks. Expert Syst. Appl. **36**(1), 944–949 (2009)
18. Zhang, X.L., Chen, X.F., He, Z.J.: An aco-based algorithm for parameter optimization of support vector machines. Expert Syst. Appl. **37**(9), 6618–6628 (2010)

# Optimized Approach for Dynamic Adaptation of Process Models

Tarik Chaghrouchni, Mohammed Issam Kabbaj
and Zohra Bakkoury

**Abstract** Process model deviations when occurring can be handled by adjusting the model and considering some properties that ensure that all constraints are satisfied. However, sometimes we cannot replace the whole missing fragment and wait the full execution of the sub-activities as structured in the initial order. For example, during software development process, Coding can start while missing the Design activity, hence deviation is raised. We have to put in place the adjusted fragments to ensure that coding will be done correctly; In other words, producing a Design-Model artifact before completing the coding will remediate to the deviation and minimize the risk. In this article, we will suggest a new approach that considers criticality and dependency of the sub-activities pertaining to the missing fragment and handles the deviation smartly and quickly.

**Keywords** PSEE · Deviation · Dynamic adaptation · Process model · Fragment library · Optimizing · Optional · Criticality · Execution order

## 1 Introduction

Process-centred Software Engineering Environments (PSEEs) are giving the option to monitor the process execution: Process activities order and artifacts production and consuming; Hence possibility to raise deviation when detected. They must accept a permanent evolution of process models, tolerate and manage inconsistencies and deviations.

T. Chaghrouchni (✉) · M.I. Kabbaj · Z. Bakkoury
AMIPS Research Group, Ecole Mohammadia d'Ingénieurs,
Mohammed V University, Rabat, Morocco
e-mail: tarik.chaghrouchni@research.emi.ac.ma

M.I. Kabbaj
e-mail: Kabbaj@emi.ac.ma

Z. Bakkoury
e-mail: bakkoury@emi.ac.ma

© Springer International Publishing Switzerland 2016          101
A. El Oualkadi et al. (eds.), *Proceedings of the Mediterranean Conference on Information & Communication Technologies 2015*, Lecture Notes in Electrical Engineering 381, DOI 10.1007/978-3-319-30298-0_11

The problem of process evolution is well known in software process community but only few works have addressed it. The existing approaches do not respond to real-world situations that require context-dependent handling of process evolution.

The objectives of the work presented in this paper are to support smartly and quickly the late process enactment evolution by managing deviations occurring during the enactment time using the notion of optimized dynamic adaptation.

In other words, the system suggested will analyze the observed model, retrieve the mandatory activities and artifacts, optimize execution of new activities by ordering them and defer the optional activities. The new dynamic adaptation model should guide smartly the change management, ignore the obsolete activities and provide the activity-based model that handles the deviation.

This paper is organized as follows: Sect. 2 presents related works conducted around software process deviation problems. Section 3 deals with problems and concepts of process deviation and process enactment evolution and outlines the approach we propose; Sect. 4 illustrates the new approach through a simplistic example. Section 5 concludes the paper and gives some perspectives.

## 2    Related Works

Process deviation problem is one of the issues that have not been enough covered despite being well known in software process community. Process models enactment must be flexible enough to allow process changes to face unexpected states, undergo changes and refinements to increase their ability to deal with requirements and expectations. This requirement reflects the nature of a creative activity such as software development, where consistency is the exception, not the rule.

Deviations are defined as operations executed that violate process constraints, activities order and process artifacts.

The process deviation problem in SPADE [1] was treated by providing features for changing process models. This approach may be effective to cope with major deviations from process models that are expected to occur again in the future, but the approach is unsuitable for situations that require minor or temporary deviations.

Other approach based on SENTINEL [2] tolerates deviations while critical requirements of processes are not violated. But it makes no difference between deviations according to their origin, and has no consistency handling policies.

Marcos Almeida et al. [3] treated the problem of process deviation handling and given an excellent approach to handle: Early/Late deviation detection, risk assessment, correction guidance. The solution was based on proposing sequences of actions from process model that will correct the deviation: Once deviation is detected, the set of missing activities will be executed in a suggested order, the model will change his activity flow but its components will remain the same.

The previous chapter [4] which is considered as a continuation of some previous works [5, 6] done around deviation management systems tackled the problem using a new approach based on dynamic handling of the deviation: Put in place some other systems that could better enact the process by implementing a Fragment library.

# 3 Dynamic Adaptation of Process Model

The main goal of this paper is to define a smart approach handle dynamically the deviation once it is qualified as major one, this means we will be required to change the process model: The enacting process model will be modified accordingly.

As seen in a previous paper [4]; once a deviation is raised, a new fragment is put in place to update the process model, taking in consideration that some properties: This fragment (WorkDefinition) must validate the constraints of the missing WorkDefinition. The Figure below illustrates the proposed approach (Fig. 1).

However, the existence of a deviation is often synonymous with a delay in the execution duration of the process or with a new procedure with uncontrolled execution duration. Hence, this approach will not optimize the execution time as the new fragment will only replace the old one without taking in consideration if the activities can be reported or are obsolete.

In real situation, when an activity is missing and the following is being executed, we must handle quickly the deviation and concentrate on the activities that will provide the artifacts needed in order to minimize the risk and ensure the quality of the output.

The running activity can reach a level of execution maturity, if we choose to replace the missing fragment with new one verifying all the constraints, this level may increase considerably, if we ignore the obsolete fragments in the missing

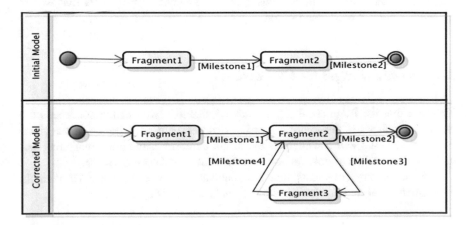

**Fig. 1** Dynamic adaptation of process model

activity and concentrate only on the mandatory ones with an optimized order, it may increase slightly and the risk will be minimized with a better quality. Following this logic, we will set some new rules considering the criticality of the activities, the order that they will take in case of deviation and their execution optionality. Thus, the new fragment should be optimized in terms of runtime.

To materialize this, we will three attributes to the WorkDefinition class; those attributes value will be calculated automatically based on the history of the process model execution and can be updated as well:

- Mandatory (True/False): If value = True, then the elementary fragment (WorkDefinition) must be executed during the enactment of the new fragment; If value = False, it can be skipped (will become optional).
- OptionalToExecute (True/False): If value = True, then the WorkDefinition cannot be ignored, it need to be executed before the full process model takes end. Else, it can be ignored once for all.
- Mandatory (True/False): This attribute will define if the activity needs to be re-executed or not. The sub activities consuming the input of the parent activity will have the value true as well as the ones producing the output used by the following activity.

## 3.1 Fragment Library

In addition to the attributes that we will add to the WorkDefinition class, additional information will be considered in the Fragment library, it will express the order of the elementary activities during the re-work, and this order will be validated by the process model engineer as per the experience from the past executions. Each fragment in the library will be constituted of elementary fragments that will be ranked based on a predefined order.

## 3.2 Process Fragment Conception

Knowing that the fragment to be considered should verify all the conditions that would be realized if the missing activity was executed as expected in the represented model. We will consider the execution duration of the elementary fragment under each activity and take in consideration the mandatory attribute.

The elementary fragments that are mandatory should be executed with the established order and stocked in the Fragment library (Fig. 2).

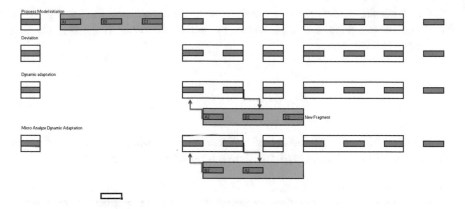

**Fig. 2** Optimized approach for dynamic adaptation

## 3.3 Logical Formalization of Process Model

In the below, we will consider that

$$Fragmenti = WorkDefinition\ i= Wi$$

$$Fragmenti.j= SubWorki.j= SWi.j$$

We consider that deviation occurs since WorkDefinitioni was not executed; hence, we will use the dynamic process model adaptation and replace the missing activity with a new fragment that will take some specific properties as explained in the paper [4].

To formalize this representation, we will consider the same structure "M" which is the union of the set of formulas that describe structure and behavior of an enacting process model; According to the conditions that the new process fragment will verify, there are some new first-order predicate rules that will be added to M:

goal(WorkDefinitioni)→goal(NewWorkDefiniton)
deviation("launch"    ,WorkDefinition(i+1))∧    (precedence(WorkDefinitioni)    ∨
validated(NewWorkDefinitoni))→precedence(WorkDefinition(i+1))
output(output(WorkDefinitioni),NewWorkDefinition)
inOutput(inOutput(WorkDefinitioni),NewWorkDefinition)
input(Output(WorkDefinitio(i+1)),NewWorkDefinition)
input(inOutput(WorkDefinition(i+1)),NewWorkDefinition)
deviation("launch"    ,WorkDefinition(i+1))∧    validated(NewWorkDefiniton)    →
inOutput((output(WorkDefinition(i+1)),WorkDefinition(i+1))

Behaviour of an enacting process model is described by state-machine diagram.

For example, activatable, enactable and enacting are the states of a WorkDefinition which are represented by activatable(Wi), enactable(Wi) and enacting(Wi).

## 3.4 Rules

An artefact which is Input/Output of a WorkDefinition which is a composition of elementary sub-works is as well an artefact Input/Output of a sub-work.

Wi.subWorks -> notEmpty() implies
Wi.output -> forAll(p | Wi.subWorks -> select(SW | SW.output -> exists(p)) -> size() = 1)
Wi= SWj v ... v SW(j+m)

The approach will be expressed as (Fig. 3).

Following the approach adopted, WNi is the WorkDefinition that will replace Wi.

WNOi represents the new approach fragment that will be considered in the new process model taking in consideration the notion of mandatory, optionality and the order established in the Fragment library. The most optimized fragment will replace better the missing activity with better quality and less risk.

WNi = SWi1 v SWi2 v …v SWin   implies
WNOi = Union (SWi where  SWi.mandatory= True)
WNO will be the optimized fragment  implies
WNO.duration = min (WNOi.duration)

The SWi will contain the value True in optionalToExecute, it will be stored as per defined order and be executed once the activity subject of deviation is completed.

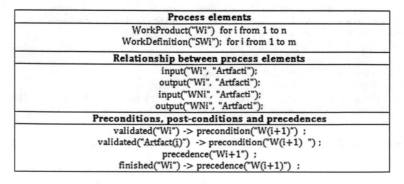

| Process elements |
|---|
| WorkProduct("Wi") for i from 1 to n<br>WorkDefinition("SWi"); for i from 1 to m |
| **Relationship between process elements** |
| input("Wi", "Artfacti");<br>output("Wi", "Artfacti");<br>input("WNi", "Artfacti");<br>output("WNi", "Artfacti"); |
| **Preconditions, post-conditions and precedences** |
| validated("Wi") -> precondition("W(i+1)") ;<br>validated("Artfact(i)") -> precondition("W(i+1) ");<br>precedence("Wi+1") ;<br>finished("Wi") -> precedence("W(i+1)") ; |

**Fig. 3** Rules

# 4 Case Study

We will take as a case study the below model that will represent a software development process: (Fig. 4).

We will consider that Design Activity is not executed and a deviation is occurring.

We will implement a new WorkDefinition: ReDesign tol replace Design Activity.

The dynamic adaptation will ensure the below constraints:

goal(Design)→goal(ReDesign)
deviation("launch",Design)∧ (precedence(Design) ∨ validated(ReDesign))→ precedence(ReDesign)
output(output(Design),ReDesign)
inOutput(InOutput(Design), ReDesign)
input(Output(Coding),ReDesign)
input(inOutput(Coding), ReDesign)
deviation("launch" ,Coding)∧ validated(ReDesign)→inOutput((output(Coding), Coding)

The approach will produce new fragments that will replace the missing ones:

Re-Design= Design = N_UseCase v N_DesignModel v N_Documentation

where N_UseCase, N_DesignModel and N_Documentation are the sub works of the activity Re-Design (the fragment that will replace the Design activity).

Since we have only one fragment to replace the missing one, there will be no need to take in consideration the duration of each fragment as per the proposed approach.

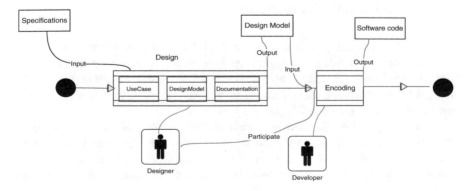

**Fig. 4** Case study model

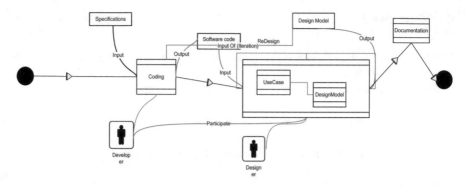

**Fig. 5** Optimized approach for dynamic adaptation

Mandatory attribute will have the values:

True for UseCase: Mandatory to adjust and validate the Design Model.
True for DesignModel: Input of the Coding activity(Output of the Design Activity)
False for Documentation: Elementary activity that is producing documentation

Since N_UseCase and N_DesignModel are considered as mandatory, they will be executed in the same order (predefined); Documentation will be skipped or let till the end of the execution. As per the rules we will consider (value contained under optionalToExecute), we will consider its value as true hence Documentation activity will be executed after the main activity is completed.

The optimized method will verify the below:

$$O\_Design = UseCase \text{ v } DesignModel$$

The new process model will be as following: (Fig. 5).

## 5   Conclusion and Perspectives

Software development processes are always subject to dynamic evolution. The previous works presented some approaches to support process enactment evolution. The presented approach makes two process models coexist: a pre-set process model that guides development, and an observed process model. A first-order logic representation is used for detecting process deviations.

In addition to this, the new approach considers that process models guides humans by optimizing the execution duration and minimizing the risk; humans must always remain the masters of decision, and be allowed to decide whether to adopt dynamic adaptation process models whenever they need to face unexpected situations.

This makes the new approach very innovative, with improved quality, less risky and radically different from the approaches proposed so far.

# References

1. Bandinelli, S., et al.: SPADE: an environment for software process analysis, design and enactment. In: Finkelstein, A., et al. (eds.) Software Process Modeling and Technology, pp. 223–244. Wiley, London (1994)
2. Cugola, G.: Tolerating deviations in process support systems via flexible enactment of process models. J. IEEE Trans. Soft. Eng. **24**(11), 982–1001 (1998)
3. Almeida da Silva, M.A., Bendraou, R., Robin, J., Blanc, X.: Flexible deviation handling during software process enactment. In: Enterprise Distributed Object Computing Conference Workshops (EDOCW), 2011 15th IEEE International, pp. 34, 41, Aug. 29 2011–Sept. 2 (2011)
4. Chaghrouchni, T., Kabbaj, M.I., Bakkoury, Z.: Towards dynamic adaptation of the software process (IEEE). In: The ninth International Conference on Intelligent Systems: Theories and Applications (SITA14), INPT, Rabat, Morocco (2014)
5. Kabbaj, I.M.: Deviation Management in software process enactment Gestion des déviations dans la mise en oeuvre des procédés logiciel. Université de Toulouse (2009)
6. Kabbaj, M., Lbath, R., Coulette, B.: A deviation management system for handling software process enactment evolution. In: Wang, Q., Pfahl, D., Raffo, D.M. (eds.) Proceedings of the Software Process, 2008 International Conference on Making Globally Distributed Software Development a Success Story (ICSP'08), pp. 186–197. Springer, Berlin, Heidelberg (2008)
7. Lbath, R., Coulette, B., et al.: A multi-Agent Approach to a SPEM-based modeling and enactment of software development processes. In: 7th International Conference on Software Engineering and Knowledge Engineering (SEKE), Taipei, Taiwan, pp. 241–246 (2005)

# Clustering Problem with 0–1 Quadratic Programming

**Khalid Haddouch, Ahmad El Allaoui, Abdelhafid Messaoudi,
Karim El Moutaouakil and El Wardani Dadi**

**Abstract** The most unsupervised methods of classification suffer from several performance problems, especially the class number, the initialization start points and the solution quality. In this work, we propose a new approach to estimate the class number and to select a set of centers that represent, fiddly, a set of given data. Our key idea consists to express the clustering problem as a bivalent quadratic optimization problem with linear constraints. The proposed model is based on three criterions: the number of centers, the density data and the dispersion of the chosen centers. To validate our proposed approach, we use a genetic algorithm to solve the mathematical model. Experimental results applied on IRIS Data, show that the proposed solution selects an adequate centers and leads to a reasonable class number.

**Keywords** Bivalent quadratic optimization problem · Clustering problem · Data IRIS · Genetic algorithm

K. Haddouch (✉) · A.E. Allaoui · A. Messaoudi · K.E. Moutaouakil · E.W. Dadi
National School of Applied Sciences of Al Hoceïma,
University Mohammed First, Oujda, Morocco
e-mail: haddouchk@yahoo.fr

A.E. Allaoui
e-mail: hmad666@gmail.com

A. Messaoudi
e-mail: messaoudi1968abdelhafid@gmail.com

K.E. Moutaouakil
e-mail: yassirkarimimane@gmail.com

E.W. Dadi
e-mail: wrd.dadi@gmail.com

© Springer International Publishing Switzerland 2016
A. El Oualkadi et al. (eds.), *Proceedings of the Mediterranean Conference on Information & Communication Technologies 2015*, Lecture Notes in Electrical Engineering 381, DOI 10.1007/978-3-319-30298-0_12

# 1    Introduction

The classification problem consists of partitioning a set of data into clusters (classes, groups, subsets, …) [1]. A cluster is described by considering the internal homogeneity and the external separation; in this sense, the elements of the same class should be similar to each other, while those belonging to different classes should be not.

The classification problem has been applied in a wide variety of fields, especially engineering, computer science, life and medical science, social science, and economy [2]. Several methods are proposed to solve the classification problem such as K-means, ISODATA, SVM, SOM, trees method and the Bayesian method [1–4]. Most of them require a priori specifying the number of classes $k$. Indeed, the quality of resulting clusters is, largely, dependents on the estimated $k$. An algorithm can, always, generate a division, no matter whether the structure exists or not.

Determining the number of clusters in a data set is a frequent problem in data clustering and is a distinct issue from the process of solving the clustering problem [5]. This works aims to propose a new approach to estimate the number of classes $k$ and to select a set of centers that represent fiddly a given data set. Our solution consists to express the clustering problem as a bivalent quadratic optimization problem with linear constraints. The proposed model is based on three criterions: the number of centers, density data and dispersion of the chosen centers. For it performance, the genetic algorithm is used in order to solve this model. In this context, we have proposed the classical mutation and crossover operators; the selection function is nothing but the objective function of the proposed model.

This paper is organized as follow: in Sect. 2, we present the 0–1 mathematical programming model for the unsupervised classification problem. The genetic algorithm for solving the proposed model is presented in Sect. 3. While in Sect. 5, the performances of this new method are evaluated by some experimental results. The last section concludes this work.

# 2    The 0–1 Mathematical Programming Model for the Unsupervised Classification Problem

Let $D = \{d_1,...,d_n\} \subset IR^m$ be the set of data, where $n$ and $m$ are integer numbers. The Unsupervised Classification Problem (UCP) looks for a set $S$, called set of centers, of an optimal size that represents fiddly the set $D$. In this part, we express the UCP problem as a bivalent quadratic problem with linear constraints, such that S is a subset of $D$. To this end, we define some necessary concepts. From a computational point of view, the UCP is one of the most difficult optimization problems; it is an $NP$-hard problem and the existence of efficient heuristic approaches for the general case cannot be assured.

## 2.1 The Sample Density

Let $\sigma$ be a non negative number, for each sample $d_i$, we introduce the set $A_i = \{d_j/ \|d_i - d_j\| \leq \sigma\}$. The *density parameter* of $d_i$ is given by $\alpha_i = |A_i|/n$, where $|A_i|$ is the size of the set $A_i$. In this sense, an isolated sample is the one that has a low density; an interior point is the one that has a large density. In this context, the Fig. 1 represents the IRIS Data for $\sigma = 0.4$ and 0.086 as density threshold.

Noted that, the right centers of the IRIS Data are $c1 = (1.464, 0.244)$, $c2 = (4.242, 1.336)$, $c3 = (5.57, 2.016)$. Basing on the samples density, the problem *UCP* looks for one set $S$ of center among $D$ such that:

- The elements of $S$ should have a large dispersion,
- The $|density(S)\text{-}density(D)|$ is small as possible,
- The size of the center set $S$ is small as possible.

## 2.2 The Decision Variable

Let $S$ be a centers set among $D$; for each sample $d_i$, we introduce the binary variable $x_i$ that equals to 1 if the sample $d_i$ is into $S$, 0 else. The vector decision is denoted by $x = (x_1, ..., x_n)^t$.

**Fig. 1** The isolated point, represented by the character 'i', and the interior ones, represented by the character 'o', of the IRIS data

## 2.3  The Decision Dispersion

Let $\beta_{ii} = \|d_i - d_j\|$ be the distance between the samples $d_i$ and $d_j$. The deviation of the selected samples around the sample $d_i$ is given by $x_i \sum_{j=1}^{n} \beta_{ij} x_j$. In this sense, for some vector decision $x = (x_1, ..., x_n)^t$, we measure the dispersion of this decision by the quantity:

$$f_1(x) = \sum_{i,j=1}^{n} x_i \beta_{ij} x_j \tag{1}$$

*Remark 1* There exist several methods to measure this dispersion; such as interquartile range and standard deviation. These criterions of dispersion measure geometrically de dispersion of all data in order to specify automatically the center which represents these data. This type of dispersion is chosen for three resents: Is less sensitive to extreme data, Represent fiddly all data and is easy to implement it.

## 2.4  The Decision Density

For a decision vector $x = (x_1, ..., x_n)^t$, the total density of the selected centers is given by:

$$f_2(x) = \sum_{i=1}^{n} \alpha_i x_i \tag{2}$$

## 2.5  The Centers Set Size

For a decision vector $x = (x_1, ..., x_n)^t$, the centers number is calculated by:

$$f_3(x) = \sum_{i=1}^{n} x_i \tag{3}$$

Basing on the criterions $f_1$, $f_2$ and $f_3$, we can define several bivalent quadratic problems with linear constraints. In our case, we fix some thresholds for the decision density, and then we look for one decision that maximizes the centers dispersion and minimizes the number of centers:

$$(P_\sigma) : \begin{cases} Max\left( \sum_{i,j=1}^{n} x_i \beta_{ij} x_j - \lambda \sum_{i=1}^{n} x_i \right) \\ SC : \\ \sum_{i=1}^{n} \alpha_i x_i \geq td \\ x_i \in \{0,1\}, \quad i = 1, \ldots, n \end{cases} \qquad (4)$$

where $\lambda$ is a penalty parameter. This penalty term is must be chosen in order to equilibrate the compromise between the dispersion and the number of centers terms.

In the next section, we will use the genetic algorithm to solve the problem $P_\sigma$.

# 3   Genetic Algorithm for Solving the Proposed Model

Since the proposed model is NP-complete, we use the genetic algorithm as optimization tool [6]. In this regard, we define our own coding, fitness function, selection mechanisms, crossover and mutation operators.

## 3.1   Coding

As it is known, the UCP problem is a 0–1 quadratic programming. Then, it is natural to use the binary codes in order to produce our population; see Fig. 2. The size of each individual is equals to $n$ (number of data).

## 3.2   Fitness Function

In order to evolve good solutions and to implement natural selection, the notion of fitness, which evaluates the solution, is used. In this case, the fitness function is no think but the objective function of our model:

$$f(x) = \sum_{i=1}^{n} \sum_{j=1}^{n} x_i \beta_{ij} x_j - \lambda \sum_{i=1}^{n} x_i \qquad (5)$$

**Fig. 2** Example of the encoding individuals

| 0 | 1 | 0 | 0 | 0 | 1 | 0 | 0 | 1 |
|---|---|---|---|---|---|---|---|---|

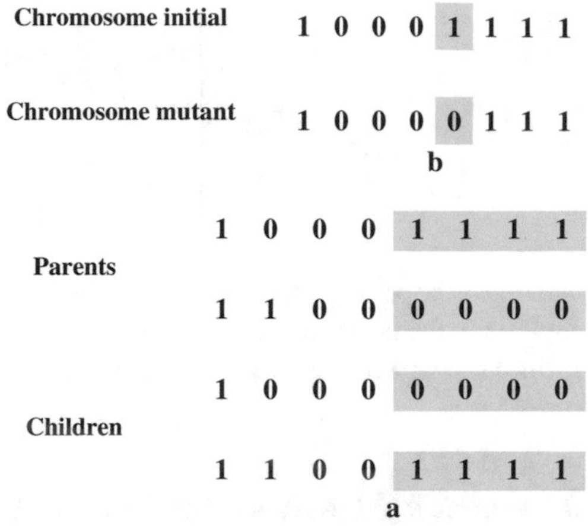

Fig. 3 Crossover and
mutation operators

Chromosome initial    1  0  0  0  1  1  1  1

Chromosome mutant    1  0  0  0  0  1  1  1
                                    b

Parents              1  0  0  0  1  1  1  1

                     1  1  0  0  0  0  0  0

Children             1  0  0  0  0  0  0  0

                     1  1  0  0  1  1  1  1
                                    a

## 3.3  Mechanism of Selection Genetic Operators

At each step, a new population is created by applying the genetic operators: selection, crossover and mutation [7].

At the level of selection, the main idea is to prefer better solutions over worse ones. In this work, the type of the selection used is roulette-wheel selection. This type of selection guarantees a good luck to the good individual in order to be selected in the future generation, and a bad luck to the unsuitable individual targeted to be vanished in the future generation.

Crossover consists in building two new chromosomes from two old ones referred to as the parents, Fig. 3a. Mutation realizes the inversion of one or several genes in a chromosome, Fig. 3b.

After several experiments, the population size is fixed in function of the number $n$ of variables. So, the size of population is equal to the number of data set $n$. In general, the probability of applying crossover operator is equal to 0.6 and the probability of applying mutation operator is equal to 0.02. The stopping criterion is based on a maximal number of iterations and/or when the performance function doesn't change.

## 4  Experimentation

To evaluate the performance of our method, the parameters $\alpha$ and $\beta$ are calculated from data iris. Several runs have been conducted for different values of the density parameter. To measure the performance of used algorithms, we used the following measures:

**Table 1** Results for different values of the density parameter

| σ | Decision | | Class 1 | | | | Class 2 | | | | Class 3 | | | |
|---|---|---|---|---|---|---|---|---|---|---|---|---|---|---|
| | MCN | MTE | MCN | MInP | MIsP | MEC1 | MCN | MInP | MIsP | MEC2 | MCN | MInP | MIsP | MEC3 |
| 0.4 | 7 | 12.234 | 2 | 2 | 0 | 0.280 | 3 | 3 | 0 | 0.853 | 3 | 3 | 1 | 0.874 |
| 0.36 | 7 | 12.461 | 3 | 2 | 1 | 0.568 | 2 | 1 | 1 | 0.695 | 2 | 1 | 1 | 0.738 |
| 0.32 | 6 | 12.461 | 2 | 1 | 1 | 0.568 | 2 | 1 | 1 | 0.695 | 2 | 1 | 1 | 0.738 |
| 0.28 | 5 | 10.371 | 2 | 0 | 0 | 0.459 | 2 | 2 | 0 | 0.782 | 2 | 1 | 1 | 0.899 |
| 0.24 | 5 | 10.371 | 2 | 0 | 0 | 0.459 | 2 | 2 | 0 | 0.782 | 2 | 1 | 1 | 0.899 |
| 0.2 | 7 | 12.528 | 2 | 1 | 1 | 0.459 | 3 | 2 | 1 | 0.781 | 3 | 2 | 1 | 0.899 |
| 0.18 | 7 | 12.121 | 2 | 2 | 0 | 0.459 | 3 | 2 | 1 | 0.781 | 2 | 0 | 2 | 0.899 |
| 0.16 | 6 | 11.802 | 2 | 2 | 0 | 0.459 | 2 | 1 | 1 | 0.781 | 2 | 1 | 1 | 0.899 |
| 0.12 | 7 | 10.886 | 1 | 1 | 0 | 0.240 | 4 | 2 | 2 | 3.826 | 2 | 1 | 1 | 0.474 |
| 0.11 | 7 | 10.886 | 1 | 1 | 0 | 0.240 | 4 | 2 | 2 | 3.86 | 2 | 1 | 1 | 0.474 |

- MCN: Mean Class Number,
- MTE: Mean Total Error,
- MInP: Mean Interior Point,
- MIsP: Mean Isolated Point,
- MEC$i$: Mean Error of the Class $i$.

The parameter $\sigma$ is randomly chosen from the interval *[0.4,0.11]*; see the Table 1. The density parameter was fixed near to 70 %.

As shown in the Table 1, the mean number of selected centers using our approach is 6. The selected centers are equitably reparted between the three classes of the Iris Data; see the Figs. 4 and 5. In fact, our method assigns almost two centers to each class; see Table 1.

As shown in the Fig. 6, for $\sigma > 0.15$, the errors MEC1, MEC2 and MEC3 become almost constant. In this sense, the density has a low impact on the decision errors. Then, we can deduce that our method is consistent.

The most selected data are interiors samples; this means all the selected samples are representative ones. It should be noted that we can use the Euclidian distance to group the collected centers to obtain the right centers of the data under study [8].

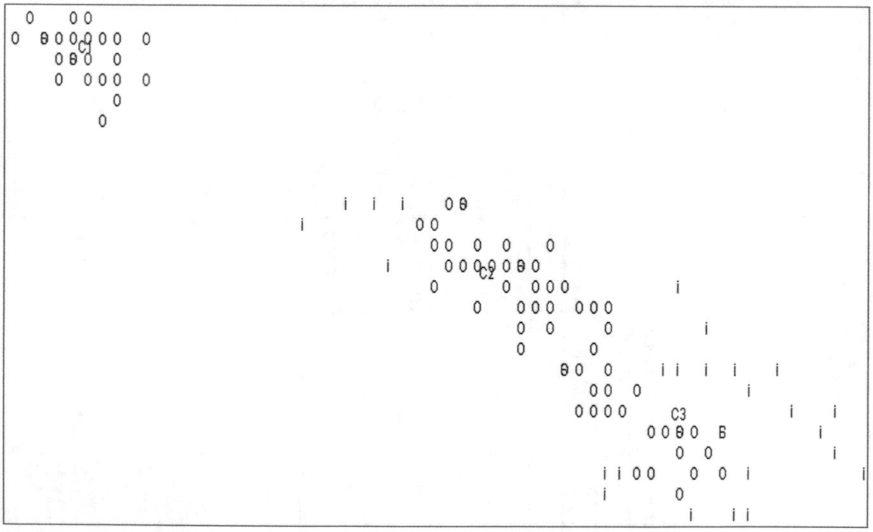

**Fig. 4** The selected centers, for $\sigma = 0.4$, are represented by the character 's'

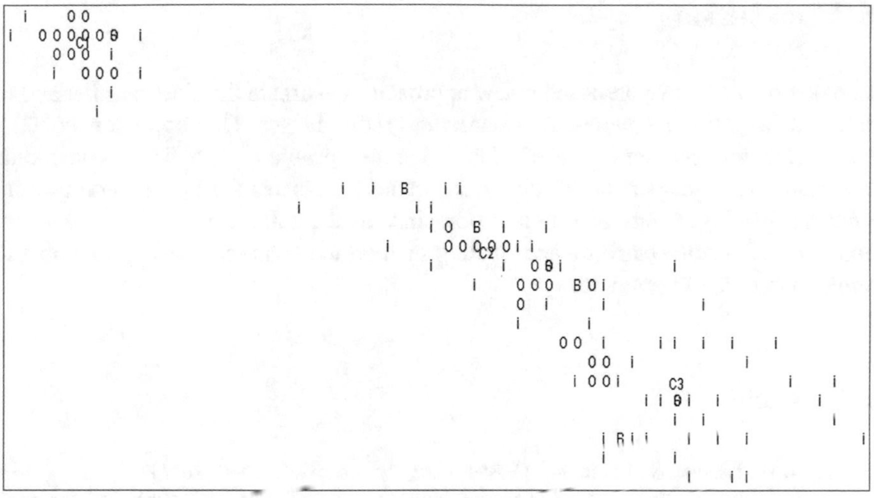

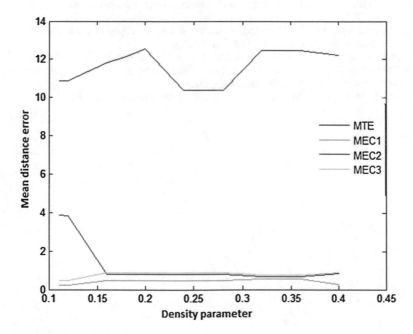

**Fig. 5** The selected centers for $\sigma = 0.12$ are represented by the character 's'

**Fig. 6** Mean errors density versus density parameter $\sigma$

## 5 Conclusion

In this work, we have proposed a new approach to estimate the class number and to select a set of centers that represent fiddly the D set. Our approach consists expressing this problem as a bivalent quadratic problem with linear constraints ensuring a large dispersion of the chosen centers, respects the information percent imposed by the experts of different domains. In the future, we will combine our approach with some performance labeling method to construct a new system able to solve classification problem.

## References

1. Everitt, B., Landau, S., Leese, M.: Cluster Analysis. Arnold, London (2001)
2. Xu, R., Wunsch, D.: Survey of clustering algorithms neural networks. IEEE Trans. Neural Netw. **16**(3), 645–678 (2005)
3. Hall, L.Q., Özyurt, I.B., Bezdek, J.C.: Clustering with a genetically optimized approach. IEEE Trans. Evol. Comput. **3**(2), 103–110 (1999)
4. Ball, G., Hall, D.: A clustering technique for summarizing multivariate data. Behav. Sci. **12**, 153–155 (1967)
5. Abascal, F., Valencia, A.: Clustering of proximal sequence space for the identification of protein families. Bioinformatics **18**, 908–921 (2002)
6. Holland, J.: Adaptation in Natural an Artificial Systems, 2nd edn. Press, M.I.T (1992)
7. Goldberg, D.E.: Design of Innovation: Lessons From and for Competent Genetic Algorithms. Kluwer Academic Publishers, Boston, MA (2002)
8. Jain A.K., Dubes, R.C.: Algorithms for Clustering Data. Prentice Hall (1988)

# Part III
# ICT for Education and Support Activities

# A New Architecture Based Business Intelligence for the Assessment and Management of E-Learning 2.0 Processes

**Rhizlane Seltani, Noura Aknin, Souad Amjad and Kamal Eddine El Kadiri**

**Abstract** After the integration of the web 2.0 in online education, the learner becomes more involved in the process and the platform offers new features stemming from the principles of web 2.0. Nevertheless, monitoring is becoming an increasingly difficult task. This paper presents our approach, which uses the functions of business intelligence in a relevant way to assess different sides of the e-learning system based on the generation of a variety of statistics on one hand, and to fill the need in terms of pertinent researches and analyzes which complement those which exist already to allow managers to make better decisions, on the other hand. Our system will not just control the platform, but also the entire process, in order to improve the decision-making task, by providing better assessment and management of online training processes and therefore, increasing their efficiency and performance.

**Keywords** Web 2.0 · Business intelligence · E-learning 2.0 · Assessment · Management

R. Seltani (✉) · N. Aknin · S. Amjad
Information Technology and Modeling Systems Research Unit, Faculty of Science, Abdelmalek Essaadi University, 93030 Tetuan, Morocco
e-mail: sel.rhizlane@gmail.com

N. Aknin
e-mail: aknin@ieee.org

S. Amjad
e-mail: amjad_souad@yahoo.fr

R. Seltani · N. Aknin · S. Amjad · K.E. El Kadiri
Computer Science, Operational Research and Applied Statistics Laboratory, Faculty of Science, Abdelmalek Essaadi University, 93030 Tetuan, Morocco
e-mail: elkadiri@uae.ma

© Springer International Publishing Switzerland 2016
A. El Oualkadi et al. (eds.), *Proceedings of the Mediterranean Conference on Information & Communication Technologies 2015*, Lecture Notes in Electrical Engineering 381, DOI 10.1007/978-3-319-30298-0_13

# 1 Introduction

Today, e-learning environments become a usual source of training. E-learning refers to the gathering of network and computer technologies to ensure the task of education [1, 2]. Despite all the benefits that the e-learning platform offers and like any system, it must undergo continuous evaluation to assess its performance. The monitoring and the assessment are becoming more difficult after the appearance of e-learning 2.0.

We propose in this paper, a new approach based on business intelligence to evaluate e-learning processes by generating a decision support system. It will improve and enrich analyzes in order to better evaluate and assess e-learning processes, by adopting new concepts, interconnected in a way to get relevant and innovative performance indicators and thus, giving a solid support to managers in order to make their decisions. Our decision-making system is based on the concept of the data warehouse which is an analytical database used as the basis of a decision support system to ensure intuitive access to information that will be manipulated to make decisions [3]. Our approach will also give the ability to involve analyzes of other e-learning platforms, if there are needs, so the model will be extended depending on the number of these platforms. This will allow us in addition to controlling the platform, to compare its various statistics with those of its similar systems using the same analytical criteria to identify areas of improvement.

# 2 Literature Review and Research Issues

## 2.1 Evaluation of E-Learning Environments

Evaluation of e-learning platforms becomes increasingly a rich subject of research. Several studies have addressed this issue according to the educational side dealing with the evaluation of the usability, the external quality of the system to test the characteristics that influence user satisfaction about e-learning systems, and usually researches are focused on the pedagogical analysis of e-learning systems [4–6]. Nevertheless, some reservations last and problems persist in terms of the relevance and the significance of analyzes, and the time required for them.

## 2.2 Business Intelligence Based Data Warehousing for E-Learning Environments

Business intelligence is a field of huge importance to all business professionals. It is based on emerging techniques and technologies, and generates significant business impacts, but it needs to be managed well [7], in order to give the organizations the ability to make decisions, based on multiple functions and rules [8]. Our approach is based on one of its essential aspects which is data warehousing.

Data warehousing is one of the important topics in the computing industry, it is more than a product, it is a process for collecting and managing data to get an entire view of a component or all of a business [9]. With the integration of the collaborative nature of web 2.0 in e-learning environments, the benefits are increasingly important, but the control goes to other more complicated and rich horizons because the tools of evaluation in these platforms do not cover all aspects that must be monitored [10, 11]. Opting for data warehousing in the field of e-learning is due to several reasons: analyzes in the e-learning environments are not sufficient for the moment, the increasing number of users of e-learning platforms makes the extraction of valuable information a very difficult task, and the need for performance analysis tools [12]. Some studies have addressed this use of data warehousing [12, 13] to calculate for example the number of: resource views, read messages, written messages, as well as details about the participation of learners and the time spent by them viewing multimedia courseware. However, the presence of several and unresolved problems was noted, such as the need to integrate new data sources, instructors cannot always ensure the success of the learning process because they do not have all the tools with which to supervise and evaluate the performance of learners in their virtual courses, the lack of tools to determine and assess the performance of learners in all types of activities. Also, it was found that most of these works are based on introspective systems which only contain the usual information, so the results are typical and focused on the control of the user activity and not the entire process.

# 3 Proposed Approach

## 3.1 An Architecture Based Data Warehousing and Sentiment Analysis for E-Learning 2.0 Environments Assessment

Our proposed architecture, solves some of persistent problems about the significance of analyzes in the field of e-learning, by designing an approach based on a data warehouse following a multidimensional fact constellation schema. Dimensional modeling offers several advantages such as: the simplicity due to the reduced number of tables, the use of significant business descriptors, which makes it less likely that mistakes will happen, database optimizers will process these simple schemas more proficiently with fewer joins, data is easier to navigate and to comprehend, and these models are gracefully extensible to accommodate changes [14]. We choose the multidimensional fact constellation schema, to ensure a better design of the data warehouse. Our approach will address the problem of limited relevance of decisions which are based on the introspective evaluation of e-learning

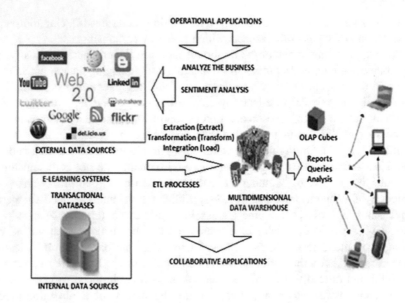

**Fig. 1** The general outline of the different parts of our global architecture

systems by integrating data from web 2.0 in our data warehouse using sentiment analysis [15]. What people think, constitutes a significant piece of information. That is why we choose to rely on it, to enrich our system, by studying and analyzing opinions and attitudes of people. The general outline of our system is shown in the Fig. 1.

## 3.2 Design of the Multidimensional Model of the Data Warehouse

A decision-making environment is said ideal, if it is possible to easily navigate from star to star and from constellation to constellation. Thing which is respected by our approach, which makes our decision-making system an important tool for evaluating e-learning processes. It is based on a multidimensional data warehouse according to the fact constellation model, which is characterized by a strong and remarkable architecture. In addition to typical data necessary for the management and evaluation of e-learning environments, we incorporate innovative and significant data to better assess and handle these systems.

The implementation of our approach is represented by the architecture shown in Figs. 2 and 3. It consists of two main parts that we can easily connect to each other through Time, Date and common dimensions.

The first part is largely introspective and based on the internal information of the system. It generates analyzes of it as shown in the Fig. 2.

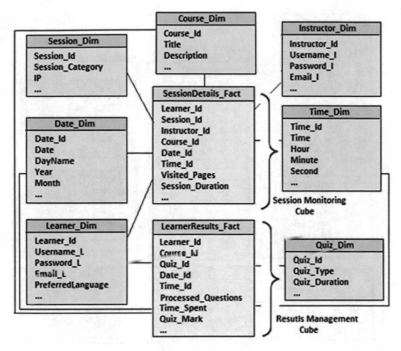

**Fig. 2** The introspective division of our multidimensional data warehouse for an e-learning 2.0 environment assessment

Consequently, several statistics can be generated about for example:

- Sessions according to the learner, instructor, time, course…
- Results by learner, quiz, date, course…

The description of some key terms of this part of the architecture is given by:

- Visited_Pages: Number of visited pages during a session.
- Session_Duration: Time between the beginning and the end of the session.
- Processed_Questions: Number of treated questions during the quiz.
- Time_Spent: Duration of the quiz.
- Quiz_Mark: Assessment for the quiz.

The second part as shown in the Fig. 3 is filled from web 2.0 data and involves techniques of opinion analysis. Thus, we can take into consideration the preferences and opinions of learners about everything related to the e-learning system in question. These opinions are extracted from different sources: forums, blogs, discussion boards, social networks. So, we have the possibility to generate several new performance indicators. Our system allows us to engender reports of statistics about for example:

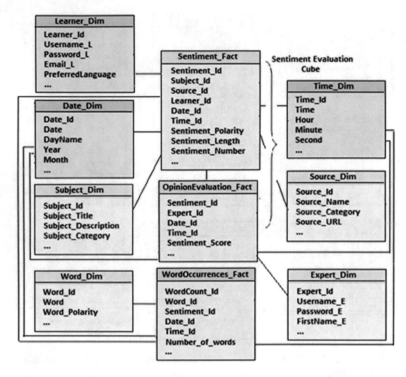

**Fig. 3** The external division of our multidimensional data warehouse for an e-learning 2.0 environment assessment

- Sentiments according to the subject, learner, source, date...
- Opinion's Evaluation by sentiment, expert, time...
- Words used in sentiment comments according to the sentiment, date, word...

The description of some key terms of this part of the architecture is given by:

- Sentiment_Polarity: the polarity of an opinion (positive or negative).
- Sentiment_Length: the opinion length (long or short message).
- Sentiment_Number: the calculated accumulative number of opinions.
- Sentiment_Score: the evaluation of the opinion.
- Number_of_words: the number of occurrences of the word in sentiments.

The global architecture is based on eleven dimensions which correspond to the axes of analysis, connected to five fact tables which contain the observable data (facts) that we have about an aspect and we want to study depending on various axes of analysis (dimensions). Each fact table represents a business process [14]. Then, our system produces five business processes.

We integrate the notion of Expert into the system (Expert Dimension), to better assess opinions of learners and consequently improve the procedure of treatment.

On the other hand, a scalable approach using Time and Date dimensions can be adopted by combining information about multiple processes or e-learning platforms to generate statistics for a comparison in accordance with standard criteria. So, we can connect our data warehouse to other data warehouses related to other e-learning environments by linking the systems firstly by Date and Time dimensions.

The extraction and the analysis mechanism of sentiments is based on a process that exploits the opinions included in user comments using a classification based on supervised and unsupervised methods. Thus, many questions regarding the control of an e-learning environment will be answered, we cite for example:

- How many opinions have we by source? By learner? By subject?
- What are the different opinions of the learner X about the subject Y?
- What are the most used words by learners to express their opinions?
- How many pages the learner X visited during the session Y?
- How long the learner X spent in the session Y, consulting the course Z?
- What is the platform's component which is the most criticized by learners?
- What are the subjects the most discussed by learners?
- What is the most active period of the use of the platform?
- What are the results of the learner X in relation to the quiz Y?
- What is the time spent by the learner X in the quiz Y?

OLAP allows querying data quickly and effectively. We integrate the use of OLAP cubes to minimize the time required for queries execution, and to better structure analyzes.

Some examples of OLAP cubes in our system are shown in Figs. 2 and 3: session monitoring cube, sentiment evaluation cube, and results management cube which is shown in the Fig. 4.

An example of a simple query model outcome from the implementation of the Results Management Cube is shown in the Table 1.

**Fig. 4** The architecture of one of our system cubes: results management cube

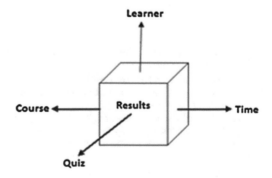

**Table 1** A query representation from the results management cube (without including the time dimension)

| Course = "Mathematics" | | | | |
|---|---|---|---|---|
| Learner | Quiz 1 | Quiz 2 | Quiz 3 | ⋯ |
| L1 | 16,50 | 13,45 | 15,90 | ⋯ |
| L2 | 12,25 | 11,75 | 10,50 | ⋯ |
| L3 | 16,80 | 14,68 | 15,20 | ⋯ |
| L4 | 19,00 | 17,30 | 18,33 | ⋯ |
| ⋯ | ⋯ | ⋯ | ⋯ | ⋯ |

## 4   Conclusion and Future Works

This paper gives a vision of a detailed approach that ensures the control of the e-learning environment, especially in its new form: the e-learning 2.0 which gives rise to several new features and services, but with a possibility to control more difficult.

Our approach is based on the gathering of diverse concepts such as data warehousing, data analytics, and sentiment analysis, which generates a decision support system producing innovative and relevant performance indicators. Our data warehouse supports special types of analyzes, to elaborate different reports based on different queries about a large amounts of data, incorporating in addition to internal and common data, other information that will improve analysis, such as profiles and backgrounds of users, also information from external sources like web 2.0 sources. So, we have more complete and meaningful analysis.

A data warehouse is easily expandable like we said before, so our architecture is open to adaptation and extension by adding new dimensions. Therefore, we can also adapt, in the future, our solution to the needs or to other fields. We also intend to work more on this solution seeking the best way to process data from web 2.0.

## References

1. Horton, W.: E-learning by Design. Wiley, New Jersey (2011)
2. Clark, R.C., Mayer, R.E.: E-learning and the Science of Instruction: Proven Guidelines for Consumers and Designers of Multimedia Learning. Wiley, New Jersey (2011)
3. Poe, V., Brobst, S., Klauer, P.: Building a data warehouse for decision support. Prentice-Hall, Inc (1997)
4. Costabile, M.F., De Marsico, M., Lanzilotti, R., Plantamura, V.L., Roselli, T.: On the usability evaluation of e-learning applications. In: 38th Annual Hawaii International Conference on System Sciences, pp. 6b–6b. IEEE (2005)
5. Padayachee, I., Kotze, P., van Der Merwe, A.: ISO 9126 external systems quality characteristics, sub-characteristics and domain specific criteria for evaluating e-learning systems. In: The Southern African Computer Lecturers' Association. University of Pretoria, South Africa (2010)
6. Britain, S., Liber, O.: A framework for pedagogical evaluation of virtual learning environments. Research report (2004)

7. Sabherwal, R., Becerra-Fernandez, I.: Business Intelligence. Wiley, New Jersey (2010)
8. Loshin, D.: Business Intelligence: The Savvy Manager's Guide. Newnes (2012)
9. Gardner, S.R.: Building the data warehouse. Commun. ACM **41**(9), 53 (1998)
10. Zorrilla, M.E., Marín, D., Álvarez, E.: Towards virtual course evaluation using web intelligence. In: Moreno Díaz, R., Pichler, F., Quesada Arencibia, A. (eds.) Computer Aided Systems Theory–EUROCAST 2007. LNCS, vol. 4739, pp. 392–399. Springer Heidelberg (2007)
11. Mazza, R., Dimitrova, V.: Visualising student tracking data to support instructors in web-based distance education. In: 13th international World Wide Web conference on Alternate Track Papers & Posters, pp. 154–161. ACM (2004)
12. Nebic, Z., Mahnič, V.: Data warehouse for an e-learning platform. In: 14th WSEAS International Conference on Computers, pp. 415–420. World Scientific and Engineering Academy and Society (2010)
13. Falakmasir, M.H., Habibi, J., Moaven, S., Abolhassani, H.: Business intelligence in e-learning (case study on the iran university of science and technology dataset). In: 2nd International Conference on Software Engineering and Data Mining, pp. 473–477. IEEE (2010)
14. Kimball, R., Ross, M.: The Data Warehouse Toolkit: The Complete Guide to Dimensional Modeling. Wiley, New Jersey (2011)
15. Pang, B., Lee, L.: Opinion mining and sentiment analysis. Found. Trends Inform. Retrieval **2**, 1–135 (2008)

# Hospital Information Systems Management: Towards a Comprehensive Maturity Model

João Vidal de Carvalho, Álvaro Rocha
and José Braga de Vasconcelos

**Abstract** In the present paper we put forth a preliminary research aimed at the development of an encompassing maturity model for the management of hospital information systems. The development of this model is justified to the extent that current maturity models in the field of hospital information systems management are still in an early development stage, and especially because they are poorly detailed, do not provide tools to determine the maturity stage nor structure the characteristics of maturity stages according to different influencing factors.

**Keywords** Stages of growth · Maturity models · Hospital information systems · Strategy · Management

## 1 Introduction

Health care institutions and governmental organizations are starting to understand that the reasons underlying a certain inadequacy in the management of health processes directly relates to infrastructural limitations and their inefficient management [16, 55]. Hospital Information Systems (HIS) managers usually contemplate the errors that occurred in these organizations and wonder what could have been done to avoid them. They conclude that these errors are usually a natural growth and maturation symptom of organizations, and are often the result of the

J.V. de Carvalho
ISCAP, Instituto Politécnico do Porto, Porto, Portugal
e-mail: j.vidal.carvalho@gmail.com

Á. Rocha (✉)
Departamento de Engenharia Informática, Universidade de Coimbra,
Coimbra, Portugal
e-mail: amrocha@dei.uc.pt

J.B. de Vasconcelos
Universidade Atlântica, Oeiras, Portugal
e-mail: jose.braga.vasconcelos@gmail.com

© Springer International Publishing Switzerland 2016
A. El Oualkadi et al. (eds.), *Proceedings of the Mediterranean Conference on Information & Communication Technologies 2015*, Lecture Notes in Electrical Engineering 381, DOI 10.1007/978-3-319-30298-0_14

development that brought the organization to its current maturity stage [52]. This phenomenon of change fits the principles behind the growth stages theory and in the current context surrounding Information Systems (IS) of health organizations.

Based on this presupposition which highlights the relevance of Maturity Models in the HIS fields, this research work is intended at the development of a maturity model that is especially adapted to the needs of Hospital Information Systems Management. To develop this new model, we carried out a preliminary study on IS Maturity Models and respective specificities. Based on this preliminary review of the State of the Art, concerning these two types of maturity models, we define a research methodology to propose and validate the new maturity model.

Besides this section, this article is organized in fourth more. Accordingly, the second section proposes a preliminary systematization of the state of the art concerning maturity models in IS Management and HIS Management. The third section defines the problem underlying this research work. The fourth section presents the questions and objectives of the research. And, finally, the fifth section offers some final remarks.

## 2   Preliminary Review of Literature

Information Systems Management (ISM) is the activity responsible for the tasks of an organization pertaining to Information Management, Information Systems and the adoption of Communication and Information Technologies (CIT) [1]. The maturity of this activity is a key factor for the success of organizations, to the extent that an IS is fundamental for their survival, competitive edge and success. In this context, there are several instruments that help the ISM achieve an enhanced maturity, namely the so called Maturity Models. Indeed, maturity models provide organization managers an important way for the identification of the maturity stage of an IS in order to plan and implement actions that will allow them to move towards an enhanced maturity stage, and thus achieve the proposed goals [50, 51]. Maturity Models can be perceived as conceptual models, comprised by discreet stages that are used to identify "anticipated, typical, logical or desired evolution paths towards maturity" [4]. We observe that these models have been used in multiple areas to describe a wide variety of phenomena [9, 12, 28, 29, 35].

Maturity Models are sustained by the principle that people, organizations, functional areas, processes, etc., evolve, towards an enhanced maturity and following a development or growth process, which covers a number of different stages [50, 51]. That is, Maturity Models are based on the theory of cyclic stages of growth, where the changes observed in an IS over the course of time occur in a sequential and predictable mode, covering a certain number of cumulative and hierarchically sequential stages, which can be described and linked to a specific level of maturity [4, 43, 50, 52]. In the same sense, Caralli and Knight [10] argue that maturity models provide organizations with a tool to address their problems and challenges in a structured way, offering both a reference point to evaluate their capabilities and a guide to improve them.

Over the last four decades several maturity models have been proposed, with differences as to the number of stages, influencing factors and intervention areas. Each one of these factors identifies the characteristics that typify the focus of each maturity stage, that is, these factors work as reference descriptors or variables to characterize each stage and provide the necessary criteria to achieve a specific maturity stage [4].

## 2.1  Evolution of Maturity Models in IS Management

Richard Nolan is considered the mentor of the IS maturity approach. Indeed, after studying/researching the use of IS in the biggest US organizations, Nolan proposed a maturity model that initially included 4 stages [43]. Later, with a view to improve his first proposal, Nolan included two additional stages to the initial model [44]. In this second version, Nolan suggests that organizations start slowly in the Initiation stage, followed by a rapid spread in the use of ITs during the Contagion stage. Subsequently, the need for Control emerges, and this stage is followed by the Integration of different technological solutions. Data Management allows for development without increasing IS related costs and, finally, constant growth promotes the achievement of Maturity.

Although this approach to the maturity models developed by Nolan, has been recognized as significantly ground breaking, it also raised a lot of debate and controversy within the scientific community. Several researchers have published studies that, on the one hand, validated and, on the other hand, expanded the model proposed by Nolan. Indeed, resulting from the researches in this field several researchers have proposed new models [e.g.: 14, 17, 24, 27, 34].

Amongst these new models proposed after the initial approach developed by Nolan, the most widely accepted, detailed and comprehensive is the Revised Model of Galliers and Sutherland [9, 50, 51]. This model provides an improved perspective of how an organization plans, develops, uses and organizes an IS and offers suggestions towards an enhanced maturity stage. This model involves six stages of maturity and assumes that an organization can occupy different maturity stages in any given moment and be conditioned by influencing factors. Moreover, it presents the characteristics of the stages aligned with modern network organizations and offers a data collection tool to evaluate maturity [50, 51].

More recently, after the model proposed by Galliers and Sutherland [17], other models have been resealed (e.g.: [3, 26, 28, 39]), including a new Nolan model with nine stages of maturity [49], developed as an answer to the technological evolution in the IS field and its management. As to the field of IS Management, another solid example of a Maturity Model is the model developed by de Khandelwal and Ferguson [26], proposing nine stages of maturity and combining stages theory with

Critical Success Factors. Notwithstanding, the model proposed by Galliers and Sutherland [17] is still perceived as the most complete and updated in IS management [52].

Additionally, these maturity models are still being used and implemented in multiple types of organizations and to different areas inside them. Mutafelija and Stromberg [38] refer that the concept of maturity has been applied to more than 150 areas inside IS. In fact, there are several examples of maturity models focused on different organization and IS areas, namely the maturity model for Intranet implementation, by Damsgaard and Scheepers [11]; the maturity model for ERP systems by Holland and Light [22]; and the CMMI maturity model for the software development process [56]. We can also mention maturity models for fields such as Software Management [3], Business Management [30], Project Management [8, 25], Project Portfolio and Program Management [37], Information Management [56], IS/ICT Management [46], e-Business [14, 15, 18, 31], e-Learning [32], Knowledge Management [5, 33], BPM—Business Process Management [53], Enterprise Architecture [13, 40], etc.

## 2.2  Maturity Models for HIS Management

Health related organizations, and more specifically Hospital IS Management organizations, are also increasingly adopting maturity models. This use is connected to a growing provision of health care services based on electronic systems, supported by enhanced computer capacity and an increased ability to seize and share knowledge in a digital format. It is widely agreed that ISs offer significant opportunities for health care providers and health provision in general, as well as access to information required by users [52].

In this context, some maturity models emerged, namely the Quintegra Maturity Model for electronic Healthcare [55], proposed as a model that goes beyond the limits of an organization, incorporating every service linked to the medical process applied to each health care provider in each maturity stage. Another example of a maturity model in the health field is the HIMSS Maturity Model for Electronic Medical Record, which identifies different maturity stages in the Electronic Medical Record (EMR) of hospitals [19]. IDC (Health Industry Insights) has also developed a maturity model which describes the five stages of development in Hospital SIs. This maturity model has been used all over the world by IDC, both to evaluate the maturity of IS in hospitals and to compare maturity average differences between regions and countries in different continents [23]. To these models we can add the Maturity Model for Electronic Patient Record, directed to the system that manages every patient related information, that is, a system that manages the EPR (Electronic Patient Record) [47] and the maturity model for PACS by Wetering and Batenburg [57].

National health services from several countries have also started to develop and adopt Maturity Models. That is the case of the model created by the National E-health

Transition Authority of Australia (NEHTA) [41] and baptized Interoperability Maturity Model (IMM). This model focuses on interoperability associated with technical, informational and organizational capabilities of the different players involved in health care services. Another example concerns the Maturity Model of the NHS Infrastructure Maturity Model (NIMM) [46]. This is a maturity evaluation model that helps NHS organizations carry out an objective self-assessment in terms of technological infrastructures.

# 3 The Problem

Health care institutions and governmental organizations are starting to understand that the reasons underlying a certain inadequacy in the management of health processes directly relates to infrastructural limitations and their inefficient management [49, 55]. An analysis to the current health context clearly reveals the weight of the technological transition problem [55]. Moreover, operational information technologies have increased in complexity to answer the demands of the sector. This increase in complexity, in its turn, led to the integration of several new enterprise integration systems, processes and approaches, and the emergence of new companies providing services in this field. Consequently, many underdeveloped products and services are being consumed by HISs undergoing a process of change and demanding, more than ever, a degree of performance and effectiveness that will answer their needs. In this scenario, several questions that require a convincing answer emerge:

- How can we know if we are doing a good job when managing these changes and monitoring their progress on an ongoing basis?
- How can we manage the interactions of systems and processes in constant evolution?
- How can we manage the impact of low interoperability, security, reliability, efficiency and effectiveness processes?
- How can we evaluate the impact of current clinical and hospital software applications in the maturity development of their respective IS?

We observe that the benefits brought by modern technology to the health field, and supported by better methods and tools, cannot be obtained via undisciplined and chaotic processes [20, 21]. That is why we believe that IS Management in health organizations must be carried out based on maturity models.

Several maturity models have been proposed in the course of time, for personal evolution purposes, for the general evolution of organizations and for the evolution of the IS Management task in particular. The differences in these models lie specifically in the number of stages, evolution variables and focus areas [35, 52]. Each of these models identifies certain characteristics that typify the target of different growth or maturity stages and are implemented in different organizations.

Where health related organizations are concerned, several maturity models are also proposed. Notwithstanding the specificities of these models that distinguish them from the models of other areas, these are still in an early development stage [35, 52]. In the research that was carried out, we observed that the models pertaining to the health field are poorly detailed, do not provide maturity measuring tools and do not structure the characteristics of maturity stages according to influencing factors. This reality offers an opportunity for the development of new maturity models focused on IS management in the health field that are capable of filling the previously identified gaps. Within the universe of the maturity models that we know, we believe that the model of maturity stages reviewed by Galliers and Sutherland [17] can serve as an inspiration and reference, both to define influencing factors and to develop a measuring instrument for Hospital IS maturity.

Additionally, the very concept of Maturity Model is not devoid of criticism. For instance, Pfeffer and Sutton [46] argue that the purpose of maturity models is to identify a gap that can be filled with subsequent improvement measures. However, most of these models fail to describe how to effectively carry out these actions, as the closing of such gaps can be extremely difficult to illustrate. The strongest point of criticism, where maturity models are involved, is their weak theoretical basis [7]. Most of these models are based on "best practices" or "success factors" connected with processes from organizations that have shown favorable results. Therefore, although these practices are compatible with the maturity model, they provide no guarantee as to the success of the organization. There is no agreement surrounding the "real path" that will ensure a positive result [36]. According to deBruin and Rosemann [12] the reasons for the, sometimes, ambiguous results obtained with the maturity models stem from the insufficient testing of the models in terms of validity, reliability and generalization, as well as from the lack of documents addressing the design and development process behind this type of model. For this reason, it is fundamental to describe the work underlying the development of a maturity model based on an approach sustained by DSR (Design Science Research) principles.

## 4   Research Question and Objectives

After describing the problem we elaborate the following research question:

- What is the best model and respective stages of maturity to be applied in HIS Management?

  From this research question we can pose more specific questions:

- What influencing factors, associated with the stages of maturity, are perceived as being fundamental in the health care field by stakeholders?
- How can these factors be determined, quantified and integrated in the context of HIS maturity stages?

To answer these questions we defined the following objectives:

- Through a systematic literature review, the identification of the main maturity models adopted in IS and HIS Management and characteristics of their different stages;
- The identification and characterization of a set of influencing factors that can be used in different stages of maturity of HISs;
- The proposal of a conceptual model that will allow us to put into context, classify and describe the factors that influence the maturity stages of the HIS;
- The validation of the proposed conceptual model and methodology through interviews with hospitals' CIOs.
- The development of an automatic tool that will allow us to identify the stage of maturity of a given HIS and the influencing factors that must be improved in order to achieve an enhanced stage of maturity.

## 5 Final Remarks

In the present paper we put forth the initial stage of a research aimed at the development of a comprehensive maturity model for hospital IS management, justified by several existing limitations in current maturity models in the health field.

A future work will involve systematic reviews of the available literature concerning maturity models for information systems management in general and hospital information systems management in particular and interviews with hospitals' CIOs, which will allow us to identify a set of potential influencing factors and maturity characteristics that should be considered during the development of the maturity model, which will be validated and subsequently we will develop an automatic tool that will allow us to identify the stage of maturity of a given HIS and the path that must be pursued towards a growing maturity.

## References

1. Amaral, L., Varajão, J.: Planeamento de Sistemas de Informação, 4th Ed. FCA (2007)
2. April, A., Abran, A., Dumke, R.: Assessment of software maintenance capability: a model and its architecture. In: Proceedings of the 8th European Conference on Software Maintenance and Reengineering (CSMR2004) (pp. 243–248), Los Alamitos CA: IEEE Computer Society Press (2004)
3. Auer, T.: Beyond IS implemention: a skill-based aproach to IS use. In: Paper presented at the 3rd European Conference on Information Systems, Athens, Greece (1995)
4. Becker, J., Knackstedt, R., Pöppelbuß, J.: Developing maturity models for IT management. Business Inform. Syst. Eng. 1(3), 213–222 (2009)
5. Berztiss, A.T.: Capability maturity for knowledge management. In: DEXA Workshop, IEEE Computer Society, 162–166 (2002)

6. Bhidé, Amar V.: The Origin and Evolution of New Businesses. Oxford University Press, New York, NY, USA (2000)
7. Biberoglu, E., Haddad, H.: A survey of industrial experiences with CMM and the teaching of CMM practices. J. Comput. Sci. Coll. **18**(2) (2002)
8. Brookes, N., Clark, R.: Using Maturity Models to Improve Project Management Practice. In: POMS 20th Annual Conference, May 1–4. Orlando Florida USA (2009)
9. Burn, J.M.A.: Revolutionary staged growth model of information systems planning. In: Proceedings of the International Conference on Information Systems, Vancouver, British Columbia, Canada, pp. 395–406 (1994)
10. Caralli, R., Knight, M.: Maturity Models 101: A Primer for Applying Maturity Models to Smart Grid Security, Resilience, and Interoperability. Carnegie Mellon University, Software Engineering Institute (2012)
11. Damsgaard, J., Scheepers, R.: Managing the crises in intranet implementation: a stage model. Inform. Syst. J. **10**(2), 131–149 (2000). doi:10.1046/j.1365-75.2000.00076.x
12. deBruin, T., Rosemann, M.: Understanding the main phases of developing a maturity assessment model. In: Proceedings of the 16th Australasian Conference on Information Systems, Sydney, Australia (2005)
13. Doc, IT Architecture Capability Maturity Model, Department of Commerce, USA Government Introduction (2003)
14. Earl, M.J.: Management Strategies for Information Technologies, Upper Saddle River. Prentice Hall, NJ (1989)
15. Earl, M.J.: Evolving the EBusiness. Bus. Strateg. Rev. **11**(2), 33–38 (2000)
16. Freixo, J., Rocha, Á.: Arquitetura de Informação de Suporte à Gestão da Qualidade em Unidades Hospitalares. RISTI—Revista Ibérica de Sistemas e Tecnologias de Informação **14**, 1–18 (2014). doi:10.17013/risti.14.1-18
17. Galliers, R.D., Sutherland, A.R.: Information systems management and strategy formulation: the 'stages of growth' model revisited. J. Inform. Syst. **1**(2), 89–114 (1991). doi:10.1111/j.1365-2575.1991.tb00030.x
18. Gardler, R., Mehandjiev, N.: Supporting component-based software evolution. In: Aksit, M., Mezini, M., Unland, R. (eds.) Objects, Components, Architectures, Services, and Applications for a Networked World, Series: Lecture Notes in Computer Science, 2591 (pp. 103–120), Springer (2003)
19. Garets, D., Davis, M.: Electronic Medical Records versus Electronic Health Records: Yes, there is a difference, Chicago. HIMSS Analytics, IL (2006)
20. Gonçalves, J., Rocha, Á.: A decision support system for quality of life in head and neck oncology patients. Head Neck Oncol. **4**(3), 1–9 (2012). doi:10.1186/1758-3284-4-3
21. Gonçalves, J., Silveira, A., Rocha, Á.: A platform to study the quality of life in oncology patients. Int. J. Inf. Syst. Change Manage. **5**(3), 209–220 (2011). doi:10.1504/IJISCM.2011.044501
22. Holland, C., Light, B.: A stage maturity model for enterprise resource planning systems. Data Base Adv. Inform. Syst. **32**(2), 34–45 (2001)
23. Holland, M., Piai, S., Dunbrack, L.A.: Healthcare IT Maturity Model: Western European Hospitals—The Leading Countries (Tech. Rep. No. HI210231), Framingham, MA: IDC Health Insights (2008)
24. Huff, Sidney L., Munro, Malcolm C., Martin, Barbara H.: Growth stages of end-user computing. Commun. ACM **31**(5), 542–550 (1988)
25. Kerzner, H.: Using the Project Management Maturity Model: Strategic Planning for Project Management, 2nd edn. Wiley, New York (2005)
26. Khandelwal, V., Ferguson, J.: Critical success factors (CSFs) and the growth of IT in selected geographic regions. In: Proceedings of 32nd Hawaii International Conference on Systems Sciences (HICSS-32), USA (1999)
27. King, J., Kraemer, K.: Evolution and organizational information systems: an assessment of Nolan's stage model. Commun. ACM **27**(5), 466–475 (1984)

28. King, W.R., Teo, T.S.H.: Integration between business planning and information systems planning: validating a stage hypothesis. Decis. Sci. **28**(2) (1997)
29. Kohlegger, M., Maier, R., Thalmann, S.: Understanding maturity models results of a structured content analysis. In: Proceedings of I-KNOW and I-SEMANTIC 09. Graz, Austria, University of Innsbruck, School of Management, Information Systems, Austria (2009)
30. Levin G., Nutt, H.: Achieving Excellence in Business Development: The Business Development Capability Maturity Model (2005)
31. Ludescher, G., Usrey, M.: Towards an ECMM (E-Commerce Maturity Model). In: Proceedings of the First International Research Conference on Organizational Excellence in the Third Millennium, Colorado State University, Estes Park (2000)
32. Marshall, S.: E-Learning Maturity Model (2007). http://www.utdc.vuw.ac.nz/research/emm/ (Retrieved in Set/2014)
33. Maybury, M.T.: Knowledge Management at the MITRE Corp (2002) http://www.mitre.org
34. McKenney, James L., McFarlan, Franklin W.: The information archipelago—maps and bridges. Harvard Bus. Rev. **60**(5), 109–119 (1982)
35. Mettler, T.: A Design Science Research Perspective on Maturity Models in Information Systems. University of St. Gallen, St. Gallen (2009)
36. Montoya-Weiss, M.M., Calantone, R.J.: Determinants of new product perfomance: a review and meta-analysis. J. Prod Innovation Manage. **11**(5) (1994)
37. Murray, A.. Capability Maturity Models—Using P3M3 to Improve Performance. vol. 2; Issue 0616–01-12 (2006). Available: www.outperform.co.uk (Retrieved in Set/2014)
38. Mutafelija, B., Stromberg, H.: Systematic Process Improvement Using ISO 9001:2000 and CMMI. Artech House, Boston (2003)
39. Mutsaers, E., Zee, H., Giertz, H.: The Evolution of Information Technology. Inform. Manage. Comput. Secur. **6**(3), 115–126 (1998)
40. Nascio, NASCIO Enterprise Architecture Maturity Model, Version 1.3, National Association of State Chief Information Officers, December (2003)
41. NEHTA, NEHTA Interoperability Maturity Model. 2007 cd. Level 25, 56 Pitt Street, Sydney, NSW, 2000, Australia.: National EHealth Transition Authority Ltd. (2007)
42. NHS, National Infrastructure Maturity Model (Online) (2011), Available: http://www. connectingforhealth.nhs.uk/systemsandservices/nimm (Retrieved in Set/2014)
43. Nolan, R.: Managing de computer resource: a stage hypotesis. Commun. ACM **16**(7), 399–405 (1973)
44. Nolan, R., Managing the crisis in data processing. Harvard Bus. Rev. **57**(2) (1979)
45. Nolan, R., Koot, W.: Nolan stages theory today: a framework for senior and IT management to manage information technology. Holland Manage. Rev. **31** (1992)
46. Pfeffer, J., Sutton, R.: Knowing "what" to do is not enough: turning knowledge into action. Calif. Manag. Rev. **42**(1), 83–108 (1999)
47. Priestman, W., ICT Strategy 2007-2011 for The Royal Liverpool and Broadgreen University Hospitals NHS Trust. Trust Board Meeting 6th November (2007)
48. Renken, J.: Developing an IS/ICT management capability maturity framework. In: Research Conference of the South African Institute for Computer Scientists and Information Technologists (SAICSIT), Stellenbosch, pp. 53–62 (2004)
49. Rocha, Á., Rocha, B.: Adopting nursing health record standards. Inform. Health Soc. Care **39** (1), 1–14 (2014). doi:10.3109/17538157.2013.827200
50. Rocha, Á., Vasconcelos, J.: Os Modelos de Maturidade na Gestão de Sistemas de Informação. Revista da Faculdade de Ciência e Tecnologia da Universidade Fernando Pessoa **1**, 93–107 (2004)
51. Rocha, Á.: Maturidade da Função Sistema de Informação: Teoria de Estádios, Modelos e Avaliação. Universidade Fernando Pessoa, Porto, Portugal (2002)
52. Rocha, Á.: Evolution of information systems and technologies maturity in healthcare. Int. J. Healthcare Inform. Syst. Inform. (IJHISI) **6**(2), 28–36 (2011). doi:10.4018/jhisi.2011040103
53. Rosemann, M., deBruin, T., Business Process Management Maturity—A Model for Progression. In: Proceedings of the 13th ECIS, May, Regensburg (2005)

54. SEI Software Engineering Institute, CMMI® for Development, Version 1.3, Improving processes for developing better products and services, (Tech. Rep. No. CMU/SEI-2010-TR-033), Carnegie Mellon University (2010)
55. Sharma, B.: Electronic Healthcare Maturity Model (eHMM), India: Quintegra (2008)
56. Venkatesh, V., et al.: User acceptance of information technology: toward a unified view. MIS Q. **27**(3), 425–478 (2003)
57. Wetering, R., Batenburg, R.: A PACS maturity model: A systematic meta-analytic review on maturation and evolvability of PACS in the hospital enterprise. Int. J. Med. Informa. **78**, 127–140 (2009). doi:10.1016/j.ijmedinf.2008.06.010

# A Mathematical Ontology for a Pertinent Research of Didactic Exercises

Imane Lmati, Habib Benlahmar and Naceur Achtaich

**Abstract** This paper presents a semantic annotation of mathematics exercises based on the canonical form of variables or formulas used in the text of exercise. A MathML representation of mathematical formulas has been exploited for the extraction of canonical forms. An extension of the educational ontology Math-Bridge is used for semantic annotation. The ontology consists of all the themes and sub-themes of algebra hierarchically structured by generalization relationships. Each themes of ontology contain useful new attributes for annotation.

**Keywords** Exercise · Evaluation · Ontology · Mathematic

## 1 Introduction

The term "semantic annotation" refers to the activity of fixing the interpretation of a document by associating a formal and explicit semantics [1]. It leads immediately to a multitude of practices as: the comments of the reviewers, the indexing affixed by librarians…

Semantic annotation is one of the best known process in the field of search for knowledge. It was studied by several scientific works dedicated to the extraction of

I. Lmati (✉) · N. Achtaich
Analysis, Modeling and Simulation Laboratory,
University Hassan II, Casablanca, Morocco
e-mail: lmati2010@gmail.com

N. Achtaich
e-mail: nachtaich@gmail.com

H. Benlahmar
Laboratory of Information Technology and Modeling,
University Hassan II, Casablanca, Morocco
e-mail: h.benlahmer@gmail.com

© Springer International Publishing Switzerland 2016
A. El Oualkadi et al. (eds.), *Proceedings of the Mediterranean Conference on Information & Communication Technologies 2015*, Lecture Notes in Electrical Engineering 381, DOI 10.1007/978-3-319-30298-0_15

143

knowledge from logical content such as mathematics. For example, we find the Mias project [2] which operates semantic annotations in the conception and architecture of its system for the recovery of mathematical knowledge. The system adds to mathematics texts (including mathematical formulas) additional representations concerning semantic information (formulas developed as text, canonical text …). The system is dedicated to research applications that use the library DML (Digital Libraries mathematics). It uses techniques of Natural Language Processing NLP and MathML ("Mathematical Markup Language") representation.

Another case study was conducted by Kristianto [3]. The approach allows the annotation of scientific articles in XML format for research mathematical formulas represented by MathML. Although these formulas can be indexed and searched by their XML tree structures, they usually do not have enough information to semantic interpretation. The approach provides an annotation model to connect mathematical formulas to descriptions in natural language based text that surrounds it.

The project [4] also studied the semantic annotation; he introduced a new Framework for adding semantics in e-learning system. The proposed approach is based on RDFa [5] and MathML for collaborative annotation of the content of the e-learning and also on ontology to categorize the content of e-learning. The annotation of the Framework adds great value to meet the semantic queries (for example, SPARQL [6]) to retrieve the information requested or desired by a user.

The exercises were also treated in the field of indexing and annotation. The project [7] indexes the geometry exercises by the properties and theorems that serve for their resolution thereby facilitate their research. Indexing is performed using automatic theorem prover Argo, it generates rules (in relation to the themes of ontology of geometry theorems) from the properties that have been provided to it.

## 2   Problematic

Most systems have used semantic annotation for information search. They have annotated the mathematical content by text only (natural language).

The idea is to use logical expressions in the annotation process to facilitate research especially for pedagogical exercises. For the mathematical text, the annotation formalism can be difficult caused by the mix of textual expression and logical relationships. To overcome this problem, we use the ontology Math-Bridge [8] for the textual part and MathML representation for the logic part.

In the following paragraph, we introduce an extension of the educational ontology Math-Bridge [8] useful for annotation. Then we present the semantic annotation algorithm and we conclude with some perspectives.

# 3 Extension of Ontology Math-Bridge

Math-Bridge European project [8] is financed by the european program eContent Plus and project partner institutions. The purpose is to provide a broad base of customized courses in mathematics data, computerized in an online platform. The target group is students in first or second year of post-baccalaureate training, having mathematics in their courses.

During the preparations didactic project, all mathematical themes were organized hierarchically in the form of ontology of concepts relevant to the target group. See for example Fig. 1 for concepts in algebra.

The organization of mathematical concepts in such a tree structure is not obvious.

Some branches of mathematics such as the theory of categories do not appear, because in the beginning of the university, they are not taught in any european country, other mathematical concepts are relevant in one country but not in another.

In our study, we thought to extend the ontology by other attributes and concepts useful for annotation of mathematical exercises. For the topic of digital functions, each polynomial function has a degree and each degree has a canonical form and name.

Let:

$$F(x) = 3x^2 + 2x + 1. \tag{1}$$

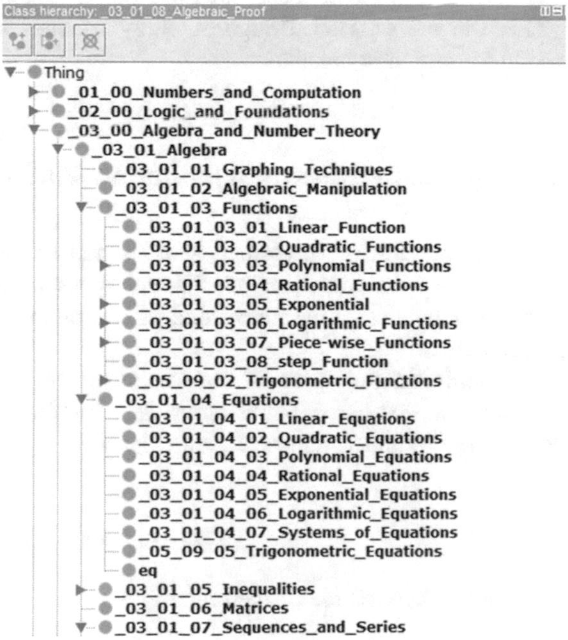

**Fig. 1** Ontology Math-Bridge in the editor Protege [9]

```
Class hierarchy: _03_01_08_Algebraic_Proof

▼─ ● Thing
    ►─ ● _01_00_Numbers_and_Computation
    ►─ ● _02_00_Logic_and_Foundations
    ▼─ ● _03_00_Algebra_and_Number_Theory
        ▼─ ● _03_01_Algebra
            ├─ ● _03_01_01_Graphing_Techniques
            ├─ ● _03_01_02_Algebraic_Manipulation
            ▼─ ● _03_01_03_Functions
                ├─ ● _03_01_03_01_Linear_Function
                ├─ ● _03_01_03_02_Quadratic_Functions
                ►─ ● _03_01_03_03_Polynomial_Functions
                ├─ ● _03_01_03_04_Rational_Functions
                ├─ ● _03_01_03_05_Exponential
                ►─ ● _03_01_03_06_Logarithmic_Functions
                ►─ ● _03_01_03_07_Piece-wise_Functions
                ├─ ● _03_01_03_08_step_Function
                ►─ ● _05_09_02_Trigonometric_Functions
            ▼─ ● _03_01_04_Equations
                ├─ ● _03_01_04_01_Linear_Equations
                ├─ ● _03_01_04_02_Quadratic_Equations
                ├─ ● _03_01_04_03_Polynomial_Equations
                ├─ ● _03_01_04_04_Rational_Equations
                ├─ ● _03_01_04_05_Exponential_Equations
                ├─ ● _03_01_04_06_Logarithmic_Equations
                ├─ ● _03_01_04_07_Systems_of_Equations
                ├─ ● _05_09_05_Trigonometric_Equations
                └─ ● eq
            ►─ ● _03_01_05_Inequalities
            ├─ ● _03_01_06_Matrices
            ▼─ ● _03_01_07_Sequences_and_Series
```

**Fig. 2** Extract from the ontology of polynomial functions

**Fig. 3** Extract from the ontology of rational functions

This function has the canonical form: $Ax^2 + Bx + C$. So we say that $F(x)$ is a polynomial function of degree 2.

For the trigonometric functions, we can neglect the degree and keep only the canonical form and name. For example the canonical form of a cosine function is:

$$F(x) = Cos(x). \tag{2}$$

According to the previous examples, we can create other concepts such as Degree, Canonical_form, and Name (Fig. 2).

The rational functions also have a canonical form $(P(x)/(Q(x))$ with $P(x)$ and $Q(x)$ are polynomials. Since the numerators and denominators are polynomials, we can link them to the concept "polynomial" of ontology (Fig. 3).

Each theme of Math-Bridge ontology contains specific attributes that can be useful for semantic annotation.

## 4    Representation and Annotation Mathematical Formulas

The interpretation of mathematical texts and annotations is a complex process implementing the treatment of different types of information: data acquisition, segmentation data, the structural description of an expression, symbol recognition...

To minimize the work of the treatment, we just interpret the logic part of mathematical text based on abstract syntax tree (According to formalism). It's very close to the MathML representation (Fig. 4).

Let:

$$F(x) = (1 + x^2)/2. \tag{3}$$

The abstract syntax tree of the function is:

**Fig. 4** Abstract syntax
tree of the function (3)

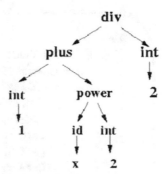

Each node is represented by a mathematical function (div, int, plus…) can be a starting point to bring closer the sub-trees with canonical forms (Fig. 5).

From the abstract syntax tree, we can generate two canonical forms,

- The first is a polynomial function of degree 2:

$$(x^2 + 1 \rightarrow ax^2 + bx + c) \quad \text{and} \quad (b = 0).$$

- The second represents a rational function:

$$((x^2 + 1)/2 \rightarrow (ax^2 + bx + c)/(ex + f)) \quad \text{and} \quad (e = b = 0).$$

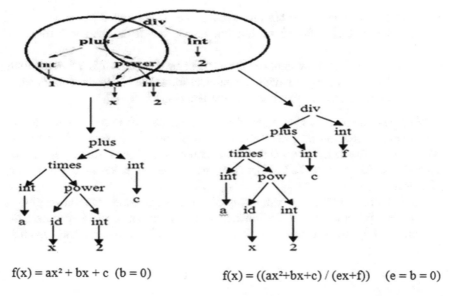

$f(x) = ax^2 + bx + c \ (b = 0)$        $f(x) = ((ax^2 + bx + c)/(ex + f)) \quad (e = b = 0)$

**Fig. 5** Generation of canonical forms from the function (3)

Debut
  While there is a formula / variable (fv)
      Construction of the abstract syntax tree of (fv)
          While there is a node
              Compare semantically and syntactically each subtree
              of the node with the trees canonical forms of ontology.
                  If a subtree is found, we annotate the exercise by the attribute
              "name" of the ontology
              Choose another node
          End while
          Choose another formula Statement
      End while
  End

**Fig. 6** Algorithm for semantic annotation

A semantic and syntactic comparison helps to bring closer the sub-trees with canonical forms, which allows annotating the exercises in several theme of ontology.

The following algorithm examines the canonical form of each formula found in a given exercise. Let (fv) a variable or formula found in the exercise.

As shown in Fig. 6, once the variable or formula (fv) is found, we build its abstract syntax tree:

- If there is a node in the abstract syntax tree, we compare semantically and syntactically each sub-tree of the node with the trees canonical forms of ontology.
- If a sub-tree is found, we annotate the exercise by the name of function (attribute "name" of the ontology) having a canonical forms similar to the one found.
- Or go to the next noeud and repeats the process.

We repeat the process with other formula or variable (fv) used in the exercise.

Semantic annotation requires the extraction of logical expression from mathematical text. The latter is crucial for the annotation, it was studied in several scientific works such as [10, 11]. They are based on labeling, segmentation, classification...

Furthermore, to reduce the complexity of logical expressions, we can develop or use existing patterns to simplify mathematical expressions, which facilitate the semantic and syntactic comparison between the tree MathML and the canonical forms.

# 5    Conclusion and Perspective

In this report, we presented a new method for annotating mathematical exercises based on ontology Math-Bridge. It represents a new support tool for students to target evaluation exercises. The approach uses the canonical form of variables and formulas in the annotation process and not the text as studied in other research. Since we work in mathematical content (logical expression, natural language), the method requires several essential steps to get to the stage of annotation. We hope in future works:

- Conceive patterns for the simplification of complex mathematical formulas.
- Develop a module that allows the semantic and syntactic comparison of mathematical formulas with canonical forms.
- Enrich the ontology by other extensions to facilitate extraction exercises.
- Realize a first prototype of our approach.

# References

1. Ma, Y., Audibert, L., Nazarenko, A.: Ontologies étendues pour l'annotation sémantique. In 20es Journées Francophones d'Ingénierie des Connaissances (2009, May)
2. Sojka, P.: Exploiting semantic annotations in math information retrieval. In: Proceedings of the Fifth Workshop on Exploiting Semantic Annotations In Information Retrieval. ACM (2012, November)
3. Kristianto, G.Y., Topić, G., Nghiem, M.Q., Aizawa, A.: Annotating scientific papers for mathematical formula search. In: Proceedings of the Fifth Workshop On Exploiting Semantic Annotations in Information Retrieval. ACM (2012, November)
4. Doush, I.A., Alkhateeb, F., Maghayreh, E.A., Alsmadi, I., Samarah, S.: Annotations, Collaborative Tagging, and Searching Mathematics in E-Learning. arXiv preprint arXiv:1211.1780 (2012)
5. RDFa Primer: Bridging the Human and Data Webs, http://www.w3.org/TR/xhtml-rdfa-primer/ (2010)
6. Prud'hommeaux, E., Seaborne, A.: SPARQL Query Language for RDF, http://www.w3.org/TR/rdf-sparql-query/. (Recommendation): W3C (2008)
7. Hibou, M., Labat, J.M., Spagnol, J.P.: Génération de feuilles d'exercices de géométrie à l'aide d'énoncés indexés automatiquament. In: Environnements Informatiques pour l'Apprentissage Humain 2003. ATIEF;INRP (2003, April)
8. Durand-guerrier, V., Mercat, C., Zsidó, J.: Math-Bridge. Gazette des Mathématiciens, Issue 131 (2012)
9. Noy, N.F., Sintek, M., Decker, S., Crubezy, M., Fergerson, R.W., Musen, M.A.: Creating semantic web contents with protege-2000. IEEE Intell. Syst. **2**, 60–71 (2001)
10. Kacem, A., Belaïd, A., Ahmed, M.B.: Embedded formulas extraction. In: Proceedings of 15th International Conference IEEE Pattern Recognition, 2000 (2000)
11. Kacem, A., Belaïd, A., Ahmed, M.B.: Extraction de formules à partir de documents mathématiques. In: Reconnaissance des Formes et Intelligence artificielle-RFIA'2000 (2000)

# Design of Educational Games: The Evolution from Computers to Mobile Devices

Ahmed Tlili, Fathi Essalmi and Mohamed Jemni

**Abstract** Due to the rapid growth of technologies, a new wave of mobile devices appeared and started replacing computers. Thanks to their features, mobile devices became important in different fields including education. This paper investigates the influence of mobile technology on designing educational games. Our goal is to identify the role of these new features that help all interested people (researchers, teachers, designers…) in designing and delivering mobile educational games. Furthermore, the paper compares various educational games dedicated to computers and mobile devices. The obtained results highlighted the importance of the new embedded mobile devices' technologies in making educational games more immersive and fun. Finally, a new set of recommendations that could enhance the development of new mobile educational games is presented.

**Keywords** Game design · Mobile technology · Computer educational games · Mobile educational games

## 1 Introduction

Learning methods and mediums have been adapted over time to deliver the learning content in an interactive way and to make learners more attached to the learning process. This evolution started from the classic learning through the e-learning systems, computer game based learning and finally mobile game based learning.

A. Tlili (✉) · F. Essalmi · M. Jemni
Research Laboratory of Technologies of Information and Communication & Electrical Engineering (LaTICE), Tunis Higher School of Engineering (ENSIT), University of TUNIS, 5, Avenue Taha Hussein, B.P. 56, 1008 Tunis, Tunisia
e-mail: ahmed.tlili23@yahoo.com

F. Essalmi
e-mail: fathi.essalmi@isg.rnu.tn

M. Jemni
e-mail: mohamed.jemni@fst.rnu.tn

© Springer International Publishing Switzerland 2016
A. El Oualkadi et al. (eds.), *Proceedings of the Mediterranean Conference on Information & Communication Technologies 2015*, Lecture Notes in Electrical Engineering 381, DOI 10.1007/978-3-319-30298-0_16

During the last few years, one of the problems that teachers have been struggling with is learners' boredom in class [1]. Prensky claimed in [2] that the motivation factors used before are not effective with the present generation of learners. Therefore, researchers have thought of using computer[1] game based learning as a method of delivering learning content to learners. According to [3], this method adds the immersion criterion which doesn't exist in classrooms. In addition, computer game based learning made learning more fun and engaging for learners [4].

Thanks to the rapid growth of technologies, different mobile[2] devices appeared and became indispensable in people's life. These devices compared to computers are smaller, wireless and cheaper [5, 6]. Therefore, they are used in various fields such as press [7], medicine [8] and education. Kambourakis et al. defined in [9] mobile learning or m-learning as the combination of two fields which are electronic learning and mobile computing. According to [10], the different features of mobile devices mentioned previously allowed the use of mobile games in the learning field.

This paper investigates the evolution of educational games from being used on computers to being used on mobile devices by studying and comparing the design of these two learning methods. Also, this paper helps all interested people (researchers, teachers, designers...) to better design their mobile educational games.

This paper is structured as follows: Sect. 2 defines the game design dimensions used in both computer and mobile educational games. Section 3 presents a comparative study between mobile and computer educational games based on their pedagogical design. Sections 4 and 5 present also a comparative study between mobile and computer educational games based on their software and hardware design. Section 6 discusses the obtained results. Besides, it lists a set of recommendation to help design mobile educational games. Section 7 concludes the paper and gives potential future work.

## 2 Game Design Dimensions

This section presents different game design dimensions used in both mobile and computer educational games. Table 1 lists these dimensions and their definitions that this paper took into consideration during the comparative study.

From Table 1, we can observe that designing an educational game is much bigger than a set of images. It needs also some other dimensions which are able to manage the learning process (between the learner and the device). Also, we can observe that these dimensions can be classified into 3 groups which are: pedagogical, software and hardware. *The pedagogical group* covers the basics of the game such as the methods and ways it uses to deliver the course content. Not only that, but it also includes the game/learner relation and how the game allows learners

---

[1]In this paper "computers" refers to personal computers and laptops.

[2]In this paper "mobile devices" refers to mobile phones, tablets and PDA.

**Table 1** Game design dimensions and their definitions

| Dimensions | Definition |
| --- | --- |
| Learning strategies | The set of techniques and actions taken by learners in order to make the learning process efficient |
| Game immersion technique | The used techniques to make the playing and learning process less boring and more interesting and enjoyable |
| Game controller | The input devices which link learner's actions to the game |
| Game communication | The design techniques used to facilitate learners' communication with the game elements |
| Game platform | The operating systems used on computers and mobile devices to handle different applications including games |
| Game display | The output devices used to display the game content |
| Internet connection | The used technology within games to connect to the internet |

to interact with the learning content. This is assured by 4 dimensions which are: learning strategies, game immersion techniques, game communication and game controller. *The software group* is the framework land used to develop a game. It includes the game platform dimension. *The hardware group* is the basic part of the game. It is needed to display the game and make it excitingly more advanced. The dimensions featuring in this group are: Game display and internet connection.

# 3  Pedagogical Group

## 3.1  Learning Strategies

Various strategies are used to deliver learning contents. These strategies can be classified as formal and informal. They can be achieved by computer and mobile educational games [10]. According to [11], these two learning strategies differ on where the learning process is taking place. Usually, a formal learning takes place in schools or universities while an informal learning goes beyond that and occurs within our daily life. Besides, mobile devices exploit the wasted time in learning a particular subject by allowing the "Just-In-Time" learning strategy. This gives learners the chance to instantly access information and learn when it's needed without too much effort. For example, when a learner is waiting in a queue, he/she can make use of that time and use his/her mobile device to play and learn [12]. In addition, mobile educational games made "situated" learning strategy possible. In this strategy, learners can play and learn in an environment context similar to the one in the game. For example, in [13], learners learn history by playing an educational game within the archaeological park.

## 3.2 Game Immersion Techniques

Computer educational games include different immersion techniques such as virtual environment and sound. While, mobile games kept these techniques and used other features to make the learning process more immersive and interesting. These features are the use of GPS (Global Positioning System) and digital cameras embedded within mobile devices. For example, in Frequency1550 [14], a mobile game for teaching the history of medieval city of Amsterdam, two teams are competing to win. Each team has a member equipped with GPS walking through the streets of Amsterdam. Thus, both teams can track the position of their team member on the game map and guide him/her to win. In [13], students learn history by identifying within the archaeological park different sites to accomplish missions. To do so, they take pictures of these sites using their mobile devices' camera. Thanks to the 3-D reconstruction functionality defined within the game, these students can see how probably the scene looked like two thousand years ago.

## 3.3 Game Communication

Different techniques were used while designing computer educational games. These techniques allowed learners to communicate with each other and with the game elements while learning (microphones, chat box and dialogue items...). For example, in [15] learners have to talk to other Non Player Characters (NPCs) and players to learn English and finish the game. However, when it comes to mobile games, designers have thought of making use of the technologies defined within mobile devices to make educational games more interesting. One of these technologies is SMS (Short Message System). This technology is widely used in mobile learning fields because it's simple, easy to use and very friendly [16]. For example, during the SMS Crossword puzzle game, learners had to send their answers (set of words) of a puzzle displayed on a projector screen by the instructor using the SMS technology [17].

## 3.4 Game Controller

Computer educational games are designed to use as input devices mouse and keyboard. For example, Khenissi et al. developed an educational version of Pacman for teaching programming [18]. It uses a keyboard and a mouse as inputs to control the Pacman through a maze and answer correctly different programming questions. Mobile technologies took game design to a higher level where the input devices became a touch screen with a stylus. For example, in the adaptive geometry game, learners use a stylus to mark the polygon from a list of given forms in order to win [19].

Besides, fingers can be used to replace stylus [20]. In a game designed to enhance children's knowledge regarding water cycle, learners control their game character to move or pick up items by touching the screen [21]. However, this technology can be tricky especially for learners who have fat fingers [22].

# 4 Software Group

Computers and mobile devices are equipped with different installed platforms (Linux, Windows, Android...). Each platform has its own unique features and characteristics. This highlights the problem of deploying educational games to run on all platforms which is almost impossible (e.g. games which are deployed on Android platform are not compatible with iOS platform and vice versa). For example, we can mention TuxMath [23] an arcade computer game initially created for Linux platform which aims to teach mathematics. On the other hand, Sandberg et al. [24] designed a mobile educational game for Android platforms. It aims to teach primary school pupils English. While, Aslan [25] developed an iOS game called "Candy Factory" for teaching middle school pupils fraction in mathematic.

# 5 Hardware Group

## 5.1 Game Display

Another point that must be taken into consideration while designing mobile educational games is the size of the screen. Computers' screen size ranges from 15 to19 in. while mobile devices' screens are smaller [6, 26]. Lavín-Mera et al. [27] insisted that the design experience for mobile devices is not the same for computers. This is because the content displayed easily on computers may not fit mobile devices' screen. In addition, some of the problems caused by small sized screens and found in [28] are: the graphical representation is small due to the screen size and that can make graphics unclear and incomprehensible. Also, the amount of information presented on a regular screen can be hard to read when they are presented on smaller screens. For example, in the game "Detective Alavi" for teaching vocabulary, researchers took into consideration the size of the avatar to be played with [6].

## 5.2 Internet Connection

Internet connection is important when it comes to games in general and educational ones in particular. This can be seen if the game can only be played online or in case of multiplayer mode. In computer educational games, learners use ADSL with its both forms wireless and wired to access game servers. However, in mobile games, in addition to the ADSL technology, designers made use of the 3G/4G mobile technology embedded within mobile devices. According to [29], 3G mobile devices allowed the use of mobile educational games in both formal and informal learning environments. For example, Tam et al. [30] created a quiz game where learners can evaluate their general knowledge by competing against each other. They had to log into the game server using a 3G network.

# 6 Discussion and Recommendations

The comparative study (presented in Sects. 3, 4 and 5) highlighted that computer and mobile educational games share some design elements. In addition, it showed that the new technology embedded within mobile devices can be used during the design process. Table 2 summarizes the results of this study.

We can deduce from Table 2 that in addition to the features already used in computer educational games, the evolution to using mobile devices offered a new set of embedded technologies (camera, GPS...) which led to a flexible design procedure. This served a more fun, immersive and interesting learning experience. Furthermore, based on the evolution of educational game design (from computers to mobile devices) highlighted in Table 2, a set of design recommendations is found:

- Designers should take into consideration the mobile devices features (small, wireless...) while designing mobile educational games. This can inspire them to deliver a more fun and interesting learning experience using different learning strategies (learning strategy dimension).

**Table 2** Computer educational games versus mobile educational games

| Groups | Game | | |
|---|---|---|---|
| | Dimensions | Computer educational game | Mobile educational game |
| Pedagogical | Learning strategies | Formal, informal | Formal, Informal, just in time, situated |
| | Immersion techniques | Virtual environments, sound | Virtual environments, real environments, sound, GPS, digital camera |
| | Game communication | Microphone, chat box | Microphone, chat box, SMS |
| | Game controller | Keyboard, mouse | Touch technology, stylus |
| Software | Game platform | Multiple | Multiple |
| Hardware | Game display | Big screen | Small screen |
| | Internet Connection | ADSL | ADSL, 3G/4G |

- The game buttons should be designed with large sizes. This facilitates controlling the game even by learners with fat fingers (game controller dimension).
- The type of platform that the designed educational game will be running on should be studied previously. This can increase the number of learners playing and learning using the game (game platform dimension).
- The tool to be used for designing mobile educational games should be studied previously. In fact, some tools allow generating multiple versions of the designed game to different platforms (game platform dimension).
- The amount of displayed information should be short. This makes it easier to read and more understandable (game display dimension).
- The clarity of educational games' images should be taken into consideration when resizing them to fit mobile devices' screen (game display dimension).


# 7 Conclusion

This paper investigated the design evolution of educational games from computers to mobile devices. This investigation was based on 3 groups (pedagogical, software and hardware). The obtained results demonstrated the important role of the new technology embedded within mobile devices in making mobile educational games more fun and interesting. Also, these results showed the important role of the mobile devices' features (small, wireless...) in delivering new learning strategies which are not found in computer educational games. Besides, this paper highlighted a set of design recommendations that all interested people (researchers, teachers, designers...) should take into consideration while designing a mobile educational game.

Future work will focus on evaluating mobile educational games which use the proposed design recommendations. Also, it will focus on applying the metrics presented in [31] to select the suitable personalization strategies for mobile educational games.


# References

1. Daschmann, E.C., Goetz, T., Stupnisky, R.H.: Testing the predictors of boredom at school: development and validation of the precursors to boredom scales. Br. J. Educ. Psychol. 421–440 (2011)
2. Prensky, M.: Computer games and learning: digital game-based learning. In: Handbook of Computer Game Studies 97–122 (2005)
3. Carbonaro, M., Cutumisu, M., Duff, H., Gillis, S., Onuczko, C., Schaeffer, J., Schumacher, A., Siegel, J., Szafron, D., Waughb, K.: Adapting a commercial role-playing game for educational computer game production. In: Game on North America, pp. 54–61 (2006)
4. Prensky, M.: Fun, play and games: what makes games engaging. In: Digital Game-Based Learning, pp. 1–5(2001)

5. Fotouhi-Ghazvini, F., Earnshaw, R.A., Robison, D., Moeini, A., Excell, P.S.: User interface design within a mobile educational game. In: International Conference on Internet Technologies and Applications (2011)
6. Fotouhi-Ghazvini, F., Earnshaw, R.A., Robison, D., Excell, P.S.: The MOBO City: A Mobile Game Package for Technical Language Learning, pp. 19–24 (2009)
7. Väätäjä, H., Koponen, T., Roto, V.: Developing Practical Tools for User Experience Evaluation—A Case From Mobile News Journalism, pp. 240–247 (2009)
8. West, D.M.: Improving Health Care through Mobile Medical Devices and Sensors. Brookings Institution Policy Report (2013)
9. Kambourakis, G., Kontoni, D.P.N., Sapounas, I.: Introducing attribute certificates to secure distributed E-learning or M-learning services. In: IASTED, pp. 436–440 (2004)
10. Koutromanos, G., Avraamidou, L.: The use of mobile games in formal and informal learning environments: a review of the literature. Educ. Media Int. **51**(1), 49–65 (2014)
11. Sefton-Green, J.: New spaces for learning: developing the ecology of out-of-school education. Research Institute for Sustainable Societies, McGill, South Australia (2006)
12. Lavín-Mera, P., Torrente, J., Moreno-Ger, P., Vallejo-Pinto, J., Fernández-Manjón, B.: Mobile game development for multiple devices in education. Int. J. Emerg. Technol. Learn, pp. 19–26 (2009)
13. Ardito, C., Buono, P., Costabile, M.F., Lanzilotti, R., Pederson, T., Piccinno, A.: Experiencing the past through the senses: an m-learning game at archaeological parks. MultiMedia IEEE 76–81 (2008)
14. Huizenga, J., Admiraal, W., Akkerman, S., Ten Dam, G.: Learning history by playing a mobile city game. In: The 1st European Conference on Game-Based Learning, pp. 127–134 (2007)
15. Tsai, M.J., Pai, H.T.: Exploring students' cognitive process in game-based learning environment by eye tracking. In: WorldComp12, pp. 211–214 (2012)
16. Abas, Z.W., Lim, T.S.K., Tai-Kwan, W.: Mobile learning initiative through SMS: a formative evaluation. ASEAN J. Open Distance Learn. 49–58 (2009)
17. Goh, T.T., Hooper, V.: To TxT or not to TxT: That's the puzzle. J. Inform. Technol. Educ. Res. 441–453 (2007)
18. Khenissi, M.A., Essalmi, F., Jemni, M.: A learning version of Pacman game. In: Information and Communication Technology and Accessibility (ICTA), pp. 1–3 (2013)
19. Ketamo, H.: mLearning for kindergarten's mathematics teaching. In: Proceedings IEEE Workshop Wireless and Mobile Technologies in Education, 2002, pp. 167–168 (2002)
20. Gu, Y., Tian, N.: Research on Transplanted Design of Mobile-Terminal-Based Educational Games. Open J. Soc. Sci. 17–21 (2014)
21. Furió, D., González, S., Juan, M., Seguí, I., Costa, M.:. The effects of the size and weight of a mobile device on an educational game. Comput. Educ. 24–41 (2013)
22. Siek, K.A., Rogers, Y., Connelly, K.H.: Fat finger worries: how older and younger users physically interact with PDAs. In: Human-Computer Interaction, pp. 267–280 (2005)
23. Tedesco, A., Furtado, B.: Mapeamento de jogos educacionais. In: Revista Espaço Pedagógico (2013)
24. Sandberg J., Maris M., Geus, de, K.: Mobile English Learning: An Evidence Based Study With Fifth Graders Computers & Education, pp. 1334–1347 (2011)
25. Aslan, S.: Game-based Improvement of Learning Fractions Using iOS Mobile Devices (Doctoral dissertation, Virginia Polytechnic Institute and State University) (2011)
26. Molnar, A., Frías-Martínez, V.: Educamovil: mobile educational games made easy. In: Educational Multimedia, Hypermedia and Telecommunications, pp. 3684–3689 (2011)
27. Lavín-Mera, P., Moreno-Ger, P., Fernández-Manjón, B.: Development of educational videogames in m-Learning contexts. In: Digital Games and Intelligent Toys Based Education, pp. 44–51 (2008)
28. Kim, L., Albers, M.J.: Web design issues when searching for information in a small screen display. In: The International Conference on Computer Documentation, pp. 193–200 (2001)

29. Demirbilek, M.: Investigating attitudes of adult educators towards educational mobile media and games in eight European countries. J. Inform. Technol. Educ. Res. 235–247 (2010)
30. Tam, V., Cheung, S.W., Fok, W., Lui, K.S., Wong, J., Yip, B.: Turning mobile phones into a mobile quiz platform to challenge players' knowledge: an experience report. In: IEEE International Conference Advanced Learning Technologies, pp. 943–945 (2008)
31. Essalmi, F., Ayed, L.J.B., Jemni, M., Graf, S.: Kinshuk: generalized metrics for the analysis of E-learning personalization strategies. Comput. Hum. Behav. **48**, 310–322 (2015)

# Toward a Measurement Based E-Government Portals' Benchmarking Framework

Abdoullah Fath-Allah, Laila Cheikhi, Rafa E. Al-Qutaish and Ali Idri

**Abstract** E-government benchmarking is the process of classifying e-government according to agreed best practices or standards. It can help agencies enhance their portals' quality by identifying the missing best practices, and providing guidelines to implement them. The aim of this paper is to introduce a benchmarking framework for e-government portals based on measurement of best practices. We have first identified and presented two examples of the benchmarking frameworks available in the literature. Based on the comparison conducted, the findings show that although the benchmarking frameworks are serving their intended purposes, they still suffer from some limitations. The paper also highlights how the new framework differs from the other frameworks and overcomes their limitations.

**Keywords** E-government · Portal · Best practice · Benchmarking framework · Benchmark

## 1 Introduction

E-government benchmarking has acquired an enormous attention over the last years [1]. Benchmarking is the process of comparing one's performance with another. It can be used to both: provide directions for improvements and determine good from

A. Fath-Allah (✉) · L. Cheikhi · A. Idri
Department of Computer Science, ENSIAS—University Mohammed V, Rabat, Morocco
e-mail: Hazgour.Abdoullah@gmail.com

L. Cheikhi
e-mail: Cheikhi@ensias.ma

A. Idri
e-mail: Idri@ensias.ma

R.E. Al-Qutaish
Department of Software Engineering & IT—École de Technologie Supérieure,
University of Québec, Montreal, Quebec, Canada
e-mail: Rafa.Al-Qutaish@etsmtl.ca

© Springer International Publishing Switzerland 2016
A. El Oualkadi et al. (eds.), *Proceedings of the Mediterranean Conference on Information & Communication Technologies 2015*, Lecture Notes in Electrical Engineering 381, DOI 10.1007/978-3-319-30298-0_17

bad practices [1]. For this purpose, e-government portals' benchmarking should not be done only to benchmark. However, it should allow agencies to benchmark their e-government portals on the one hand, and provide directions and recommendations to help agencies improve their portals' quality on the other hand.

Many benchmarking frameworks exist in the literature such as: Brown University [2, 3] that focuses on ranking e-government Websites for 198 countries, European Union (EU) [4–6] that focuses on ranking the European countries in e-government, United Nations (UN) [7] that focuses on ranking the UN member states countries in the capacity to use ICT, Waseda [8] that focuses on ranking 55 countries in e-government, and Accenture [9–12] that focuses on ranking 22 participating countries in e-government.

In this paper we are going to investigate two of the e-government benchmarking frameworks, including: Brown University [2, 3] and EU [4–6]. Afterwards, a new benchmarking framework for e-government portals based on measurement of best practices is proposed. The framework is based on a best practice model that we have built in a previous published research [13]. Eventually, a comparison between all the three frameworks is conducted to highlight how the new framework is different than the other frameworks, and how this framework will allow agencies to both: benchmark the e-government portals and provide guidelines and recommendations to improve the portals' quality.

This paper is structured as follow: Sect. 2 provides an overview of two e-government benchmarking frameworks, whereas, Sect. 3 presents the proposed framework, which we will refer to as 'Measurement Based e-government Portals' Benchmarking Framework' (MBeGPBF). Section 4 provides a comparison between the benchmarking frameworks presented in the literature and the MBeGPBF, followed by a discussion and analysis. Finally, Sect. 5 concludes the paper and gives directions for future work.

## 2  Literature Review

In this section we provide an overview of two benchmarking frameworks, including: Brown University and EU.

The purpose of the Brown University framework is to rank e-government Websites for 198 countries, it is considered straight forward and based on objective measures. The framework [2, 3] calculates the e-government rank by multiplying the number of features (a range of 27 features with 18 features as a max) by four and adding the number of executable services in the e-government portal (28 services as a max). This means that the framework is counting the services' number and including only the services that allow complete transactions.

The European commission has published several e-government surveys in 2009, 2010, and 2012 [4–6]. The framework focuses on ranking the European countries in e-government.

In the 2009 survey, e-government was assessed by measuring the availability and sophistication of 20 services (12 services for G2C and 8 services for G2B), user experience and e-procurement [4]. The EU framework is counting both the service number and sophistication; this means that the framework is both qualitative and quantitative. The availability of services is measured on the basis of two levels (i.e. available or not available). The service is considered available if its sophistication is 4 or above, while it is considered not available if its sophistication is between 0 and 3. The sophistication of services is calculated by translating the 5 maturity stages into percentages. Furthermore, user experience is related to users' interaction with the portal, and is considered a pre-requisite for repeated visits and inclusiveness of e-government services.

In the 2010 survey, the scope of the benchmark was increased by life event (LE) measurement, regional and local service analysis, and status across horizontal IT enablers [5]. In the 2012 survey, the benchmark was more about measuring how Europe is aligned with the 2011–15 action plan, rather than measuring the quality of e-government in general [6].

## 3 Proposed Benchmarking Framework

In this section we explain the components of the MBeGPBF, which is composed of the best practice model and survey questions. Then we provide the procedural view and calculation methodology of the MBeGPBF.

### 3.1 Best Practice Model

We have collected e-government portals' best practices after an extensive literature review [13]. These best practices were collected from research papers, industry, international standards, and case studies. After collecting those best practices, we have noticed that they are not logically grouped; hence we have proposed a structured way that allows researchers and practitioners to directly find e-government portals' best practices. This model is composed of five best practice categories. Each category contains sub categories and each subcategory contains best practices related to it. We have defined those best practices in a previous published research paper [13]. The categories of the best practice model are as the following:

- Back-end that can be defined as the best practices that run in background and usually the users do not see them. This includes the system, data processing and business logic. This includes the following best practice subcategories: Interoperability, Use of standards, Modularity, Security, Privacy, Single sign on, Delegation and Reusability.

- Front-end Web design that can be defined as the best practices that the user usually interacts with and sees and are related to the interface or design of the portal. This includes the following best practice subcategories: One stop shop, Ease of navigation, Personalization, Industrialization, and Structuration.
- Front-end Web content that can be defined as the best practices that the user usually interacts with and sees and are related to the information and content of the portal. This includes the following best practice subcategories: Relevancy, Accessibility, Search engines, Periodical change, Rich content, Interactive games, Mobile application, Statements, Translations, Understandability, Help and Responsiveness.
- External that can be defined as the best practices that are loosely coupled with the technical aspects of the portal and are mostly related to the marketing of the portal and to the inclusion of the citizen in the e-government process. This includes the following best practice subcategories: E-participation, Sociability, Advertising, Referencing, Incentives, Contests, Emailing, Data analytics, Contact information and Performance ratings.
- Service that can be defined as the best practices that are related to the services of the e-government portal. This includes: Service customer centricity, Service interoperability, Service help, Payments, Workflows, Service performance ratings, User forms and Service responsiveness.

## 3.2   Survey Questions and Best Practices

Each best practice subcategory of the best practice model contains best practices, and each best practice can be measured using one survey question that measures its presence/absence.

As an example, we took the "Relevancy" and the "Periodical change" of the Web content category. Which are defined as follows:

- Relevancy: Organizing the portal's Web content according to the citizens' needs. This means that information should not be organized from an organizational perspective; however the information should be relevant to the citizen or citizen centric. An example is grouping information by theme and target groups.
- Periodical change: The process of the changing and updating the e-government portal's content periodically. This can be achieved via: having expiry dates and team of editors responsible of changing the content of the portal periodically.

Many best practices should be achieved as a way to satisfy one subcategory by measuring the best practices' presence or absence in the portal. Table 1 shows best practices and survey questions for the 'Relevancy' and 'Periodical change' best practice subcategories that belong to the Web content category.

**Table 1** Best practices and survey questions for 'relevancy' and 'periodical change' subcategories

| BP subcategory | Survey questions | Best practices |
|---|---|---|
| Relevancy | • Is the information grouped by theme or life event (ex. mothering, marriage, birth…)? [4, 5] (yes/no binary scoring) | • Grouping information by theme [4, 5] |
| | • Is the information grouped by target groups (ex. students, workers, job seekers…)? [3–5] (yes/no binary scoring) | • Grouping information by target groups [3–5] |
| | • Is the information at the right level of detail? [14] (yes/no binary scoring) | • Have information at the right level of detail [14] |
| Periodical change | • Are there expiry dates or review dates of the portal's pages? [15] (yes/no binary scoring) | • Having expiry dates or review dates [15] |
| | • Is there a date of the last update available in the portal's pages? [16] (yes/no binary scoring) | • Having date of the last update [16] |
| | • Is there a team of editors responsible of the portal's timeliness? [17] (yes/no binary scoring) | • Having a team of editors responsible of the portal's timeliness [17] |

## 3.3 Procedural View

The survey questions (as described in Table 1) of the best practice model can be executed by the Webmaster of the e-government portal. The output of this survey can be used to give a final grade (compliance with best practices' grade) to each portal and generate a report for agencies to figure out the missing best practices to enhance the portals' quality.

The portal's grade can be calculated by equally weighting the grades of each best practice category (backend, Web design, Web content, external and service). The grade of a best practice category can be calculated by equally weighting the grades of its corresponding best practice subcategories. Except for the service category; as e-government portals may have many services, the grade of each service needs to be evaluated separately. Afterwards, the service category grade can be calculated by averaging the grades of all the portal's services (this way both service number and sophistication are included in the benchmark). Finally, the grade of each best practice subcategory can be calculated using the output of the survey. This might be programmed in a web based application to automate the data collection process and output.

## 4 Comparison and Analysis

In this section, we provide a comparison between the two benchmarking frameworks presented in the first section and the MBeGPBF presented in the second section of this paper. This comparison is then followed by a discussion and analysis to highlight some of the drawbacks of those frameworks and how the MBeGPBF will tackle them.

Table 2 provides a comparison between the three frameworks (Brown University, EU and MBeGPBF) in terms of many criteria including: purpose, scope of the benchmark, benchmarking criteria, if the framework is including the service number, if the framework is including the service sophistication, if the framework is not limiting the measured services, if the framework is not overweighting measures, if the framework is not discriminating services (for example not including services with low sophistication) and if the framework can identify gaps and areas of improvement for the e-government portals.

**Table 2** Comparison of the frameworks

| Criteria | Framework | | |
|---|---|---|---|
| | Brown University | EU | MBeGPBF |
| Purpose | Rank e-government websites for 198 countries | Rank the European countries in e-government | Can be used by any agency to benchmark and enhance the portals' quality |
| Scope | E-government portals | E-government | E-government portals |
| Benchmarking criteria | • Presence of 27 Features<br>• Number of services | • The 20-service method (availability and sophistication)<br>• User experience (accessibility, usability, user satisfaction monitoring, one stop shop approach, user focused portal design) | • Backend: (Interoperability, use of standards, modularity, security, privacy, single sign on, delegation, reusability)<br>• Web design: (One stop shop, ease of navigation, personalization, industrialization, structuration)<br>• External: (E-participation, sociability, advertising, referencing, incentives, contests, emailing, data analytics, contact information, performance ratings)<br>• Service: (Service customer centricity, service interoperability, service help, payments, workflows, service performance ratings, user forms, service responsiveness)<br>• Number and sophistication of services |

(continued)

**Table 2** (continued)

| Criteria | Framework | | |
|---|---|---|---|
| | Brown University | EU | MBeGPBF |
| Including service number | Yes | Yes | Yes |
| Including service sophistication | No | Yes | Yes |
| Not limiting the measured services | No (28 services as max) | No (20 services) | Yes |
| Not overweighting measures | No (features over weighted) | No (G2C services over weighted) | Yes |
| Not discriminating services | No (services not allowing transactions) | No (services with maturity between 1 and 3) | Yes |
| Identification of areas of improvement | No | No | Yes |

From this table we can notice that the MBeGPBF can be used by any agency for benchmarking and enhancing the portals' quality. Besides that, the scope of both Brown University and MBeGPBF frameworks is e-government portals, while the scope of the EU framework is e-government. Furthermore, the benchmarking criteria of the MBeGPBF are more structured, based on a best practice model and include all the aspects of e-government portals (back-end, Web design, Web content, external and service) compared to Brown University and EU frameworks (presence of 27 features for Brown University and user experience for EU). Besides that, all the frameworks are including the service number in their rankings, while only the EU and MBeGPBF frameworks are including the service sophistication in their rankings; for the EU framework, the service sophistication is the percentage corresponding to the service maturity, and for the MBeGPBF framework, the service sophistication is the subcategories of the service category (for instance, payments which offer the possibility to pay for the services online, or workflows which offer the possibility to track the status of the service after its execution are all aspects related to the sophistication of services). Furthermore, both frameworks; Brown University and EU are limiting the services measured (28 services for Brown University and 20 services for EU), however the MBeGPBF is including all the services in the benchmark. Moreover, the MBeGPBF is not overweighting measures, however, Brown University framework is overweighting the features over executable services (multiplication by 4) and the EU framework is overweighting the G2C over G2B services (12 services for G2C and 8 services for G2B). Furthermore, the MBeGPBF is including all the services in the assessment and is not discriminating services, however the Brown University framework is not counting services not allowing transactions and the EU framework is not counting services with a maturity between 1 and 3. Eventually, only the MBeGPBF is allowing agencies to generate reports to identify areas for improvement and missing best practices after the benchmark. This is because the framework is based on measurement of best practices (presence or absence) using survey questions; hence all the best practices identified as not existing in the survey should be highlighted.

To summarize, the MBeGPBF framework is based on a best practice model that measures the portals' compliance with e-government portals' best practices. This allows both: the calculation of a final grade for e-government portals in an easy and structured way, and the generation of a report that locates performance gaps and areas of improvement for agencies. Moreover, the model includes all the aspects of e-government portals without overweighting any aspect over the other (back-end, Web design, Web content, external and service). Furthermore, the framework is including both: the service number; since it calculates the grade of the service category by averaging the grades of all the services in the portal, and the service sophistication; since all the subcategories in the service category are related to the sophistication of services (such as service customer centricity which relates to ensuring that the services are designed for the citizen). Moreover, the framework is not limiting the services nor discriminating them. This is because the framework counts all the services present in the e-government portal. However, this represents a limitation of the new framework, since it might be time consuming to evaluate every service independently for portals with a large number of services.

# 5  Conclusion

In this paper, a new framework for benchmarking e-government portals has been proposed. This framework is based on measurement of e-government portals' best practices. For this purpose, we have analyzed two examples of e-government benchmarking frameworks, namely, Brown University [2, 3] and EU [4–6].

Afterwards, we have explained the components of the MBeGPBF, which are the best practice model and the survey questions. The best practice model is composed of five best practice categories including (back-end, Web design, Web content, external, and service), each category contains subcategories and each subcategory contains best practices. The survey questions are used to measure the presence/absence of those best practices. Then we have explained the procedural view of the MBeGPBF and how it can be used to calculate the portals' compliance with best practices grade. This grade is an equally weighted average of the grades of each best practice category (backend, Web design, Web content, external and service).

Eventually, we have conducted a comparison between the two benchmarking frameworks presented in the literature review and the MBeGPBF. This comparison was followed by an analysis to highlight the drawbacks of the two benchmarking frameworks raised in the literature review with respect to these criteria, and an explanation on how the MBeGPBF satisfy all of them. Based on this analysis and the features that the proposed framework provides, we believe that the MBeGPBF moves beyond the benefits of the two benchmarking frameworks discussed in the literature review. This is because the MBeGPBF is not only benchmarking to benchmark, but it helps agencies enhance their portals' quality by identifying the gaps and missing best practices, and providing guidelines to implement them.

As a future work, our next step is to define the best practice model (categories, subcategories and best practices) and the survey questions in a detailed way.

# References

1. Janssen, M.: Measuring and benchmarking the back-end of e-Government: a participative self-assessment approach. In: Electronic Government, pp. 156–167. Springer (2010)
2. West, D.M.: E-government and the transformation of service delivery and citizen attitudes. Public Adm. Rev. **64**(1), 15–27 (2004)
3. West, D.: Global e-government, 2007 (2007)
4. Capgemini: eGovernment Benchmark 2009 (2009)
5. E. Commission: Digitizing public services in Europe: putting ambition into action, 9th Benchmark Measurement. Rep. Prep. Capgemini IDC Rand Eur. Sogeti DTi Eur. Comm. Dir. Gen. Inf. Soc. Media (2010)
6. E. Commission: Public Services Online 'Digital by Default or by Detour?' Assessing User Centric eGovernment performance in Europe—eGovernment Benchmark 2012. Rep. Prep. Capgemini IDC Sogeti-Pract. Indigov RAND Eur. Dan. Technol. Inst. Dir. Gen. Commun. Netw. Content Technol. (2013)
7. United Nations: UN E-government survey 2012: E-government for the people (2012)
8. Waseda University Institute of e-Government: Waseda University International e-Government Ranking 2013. March 2013
9. Hunter, D., Jupp, V.: E-Government Leadership. Rhetoric vs Reality Closing the Gap. Accenture (2001)
10. Rohleder, S.J., Jupp, V.: eGovernment Leadership: High Performance, Maximum Value. Accenture (2004)
11. Cole, M., Jupp, V.: Leadership in Customer Service: New Expectations, New Experiences. Accenture (2005)
12. Roberts, D.: Leadership in Customer Service: Delivering on the Promise. Accenture (2007)
13. Fath-Allah, A., Cheikhi, L., AL-Qutaish, R., Idri, A.: E-government portals best practices: a comprehensive survey. Electron. Gov. Int. J. **11**(1/2), 101–132 (2014)
14. Sørum, H.: An empirical investigation of user involvement, website quality and perceived user satisfaction in eGovernment environments. In: Electronic Government and the Information Systems Perspective, pp. 122–134. Springer (2011)
15. The European Centre for Total Quality Management: E-Government A Best Practice Perspective. July 2002
16. Choudrie, J., Ghinea, G., Weerakkody, V.: Evaluating global e-government sites: a view using web diagnostics tools. Electron. J. E-Gov. **2**(2), 105–114 (2004)
17. United Nations: Compendium of Innovative E-government Practices, vol. IV. UN (2012)

# Part IV
# Networking, Cloud Computing and Security

# Neural Networks and PCA for Spectrum Sensing in the Context of Cognitive Radio

Abdessamad Elrharras, Rachid Saadane, Mohammed Wahbi
and Abdellatif Hamdoun

**Abstract** Cognitive radio has been proposed in order to benefit opportunistically from the unused portions of the spectrum, knowing that the first and the critical phase of this approach is the spectrum sensing, wherein the cognitive user must sense his external environment, to detect and profit dynamically from the free channels. One of the most used methods to detect the holes in the frequency spectrum is the energy detection; this technique does not need any prior knowledge about the primary signal. It is simpler and it requires less sensing time. However, there are a lot of problems that decrease the performance of the energy sensor; it is susceptible to the uncertainty in noise power. In this respect, we propose in this work, hybrid architecture which combines the simplicity of the energy detector, and the robustness of artificial neural networks ANN. The Principal Component Analysis is suggested as a pre-processing module in order to extract signal features.

**Keywords** Spectrum sensing · Cognitive radio · Energy detection · Primary and secondary user · AWGN channel · Probability of detection · Artificial neural networks ANN · Principal component analysis PCA

A. Elrharras (✉) · R. Saadane · M. Wahbi · A. Hamdoun
Engineering System Laboratory, SIRC/LaGeS-Hassania School of Public Works,
BP 8108 Oasis-Casablanca, Morocco
e-mail: a.elrharras@gmail.com

R. Saadane
e-mail: rachid.saadane@gmail.com

A. Elrharras
Information Processing Laboratory, Ben M'sik Faculty of Sciences Hassan II
Casablanca University, BP.7955 Sidi Othmane, 20702 Casablanca, Morocco

© Springer International Publishing Switzerland 2016
A. El Oualkadi et al. (eds.), *Proceedings of the Mediterranean Conference
on Information & Communication Technologies 2015*, Lecture Notes
in Electrical Engineering 381, DOI 10.1007/978-3-319-30298-0_18

# 1    Introduction

The recent evolution of wireless technologies causes a great demand in terms of the spectral resources. However, the current management of the frequency spectrum is a static allocation that cannot support this growth [1]. The Spectrum Utilization is not uniform: depending on time and space, we can find a frequency bands over-loaded while others remain unused. In this context, the approach of the cognitive radio was introduced by Mitola to end these problems, and increase the spectral efficiency by using the spectrum in an opportunistic manner.

The cognitive radio approach aims to better exploit the frequency spectrum. For this, smart radio equipment must be equipped with sensors allowing it to collect electromagnetic information on the one hand, learning resources and analyzing information collected on the other hand [2]. It can then deduce how it should reconfigure, in order to better adapting to its environment.

Current researches on cognitive radio are primarily concerned with the detection of free resources for assisting the opportunistic spectrum access. This dynamism brought to the spectral management allows a secondary user (SU) to take advantage from the Primary User (PU) absence, on condition that the PU should not suffer harmful outside interference, and does not make any changes to allow co-existence with SU.

The first step and the most crucial in SU equipment, is the fact to sense a wide spectral band to detect free bands. There are several methods for doing this detection in the literature [3–5], but the energy detection remains the most chosen, not only for its simple implementation, but also for what it does not need to know any information in advance about the PU.

In this method, the energy of the received signal is measured, and by making a comparison with a predetermined threshold, which presents the energy of the noise on the channel, if the captured signal energy exceeds the threshold, we declare that the PU is present, otherwise it is absent.

The uncertainties of noise, shadowing and channel fading are problems that limit the performance of the energy detection technique. In our context, we propose hybrid architecture to detect free bands, which combines the simplicity of the energy detector, and the robustness of the artificial neural networks.

This paper is organized as follows, Sect. 2 deals with the mathematical formulation of classical energy detection method, and explains the MatLab implementation of this detector. While Sect. 3 gives more detailed description of the artificial neuronal network. Then, Sect. 4 describe the mathematical background of the Principal Component Analysis and it crucial role to minimize the space dimension of signal used. And, Sect. 5 presents the results of ANN and PCA implementation and compares the performance of detection between the simple energy detector and the other architectures suggested in this work. Finally, Sect. 6 concludes the paper.

# 2  Energy Detection

## 2.1  Classical Energy Detection Model

Spectrum sensing consists in detecting the PU presence that emits an unknown signal x(t). This detection can be formulated as a binary hypothesis test as follows:

The hypothesis $H_0$: presents the existence of noise alone. And, the hypothesis $H_1$: shows the presence of the signal.

$$\begin{cases} y(t) = n(t) & : \quad H_0 \\ y(t) = \varepsilon * x(t) + n(t) & : \quad H_1 \end{cases} \quad \text{with} \quad 0 < \varepsilon \le 1 \tag{1}$$

Respectively, n(t): Is the noise present in the canal. y(t): The received signal. x(t): The signal transmitted currently being unknown and deterministic.

In the energy detection technique [6], we receive a signal in which the energy is measured; the detection of the primary signal is done by comparing measured energy with a threshold $\lambda$ that presents the noise energy.

$$E = \frac{1}{T} \sum_{n=1}^{N} |Y(n)|^2 \quad \begin{cases} E < \lambda & : \quad H_0 \\ E \ge \lambda & : \quad H_1 \end{cases} \tag{2}$$

## 2.2  Improved Energy Detection Model

In low SNR there are a lot of problems that decrease the performance of energy detection method [7], like the uncertainty of the noise, shadowing, and channel fading, and the big challenge in this technique is to estimate the threshold of detection, in the classic method we see that a static threshold is used, but as we know that the threshold depends on the environmental noise, we propose in this work a dynamic threshold to increase the probability of detection [8] (Fig. 1).

The implementation has been done in MATLAB environment. The proposed simulation model is divided into two parts. In the first part, the primary user generates FM signals at 100 MHz. To simulate a real communication, we add to the

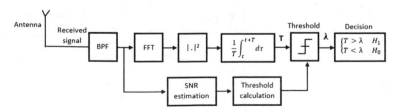

**Fig. 1** Energy detection model

generated signals a Gaussian noise. In the second part, we find the secondary user that contains the energy detection algorithm to detect the presence or the absence of primary signal.

# 3    Artificial Neural Networks

## 3.1    Formal Neuron

Studying artificial neural network was inspired mainly from the biological learning system [9], the biological model is composed of complex layers of interconnected neurons. In effect, the human brain is composed of approximately $10^{11}$ neurons, each one have an average of $10^3$ connections. It is believed that the considerable calculation power of the brain is a result of the parallel and distributed processing performed by these neurons.

The artificial neuron has generally multiple inputs and a single output. Actions of excitatory synapses are illustrated by the coefficients called synaptic weights; these weights are coupled with all inputs. The numerical values of these coefficients are adjusted in the learning phase (Fig. 2).

The formal neuron that is given in the figure above has n inputs denoted as $\{X_1, X_2...X_n\}$. Each line that connects these inputs to the summation junction is assigned a weight, denoted as $\{W_1, W_2...W_n\}$. The neuron activation function $F(a)$ in McCulloch-Pitts model is a threshold function:

However, linear and sigmoid functions are also used in different situations. The output y of the neuron is given by the formula (3):

$$f = \sum_{i=1}^{n} X_i * W_i + b \tag{3}$$

One of the most important parts of a neuron is its activation function. In this work, we have chosen a sigmoid function, because it's nonlinearity of making it possible to approximate any function.

**Fig. 2** Mathematical model of the formal neuron

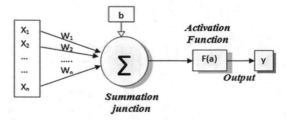

## 3.2 Multi-layer Perceptron

A Multi-Layer Perceptron (MLP) consists of a number of artificial neurons inter-connected, this network is organized in the form of layers, such as the layer 'i' receive as input the outputs of the previous layer 'i − 1' and feed its outputs to the next layer 'i + 1'. This model is called a direct neural network or a feed forward neural network. The first and last layers are called, respectively, the input and output layers. Layers that are neither input nor output are known as hidden layers.

MLP are especially trained using the back-propagation algorithm [10], which aims at minimizing the global error measured at the output layer, by the relation below:

$$e(t) = y_d(t) - y_m(t) \tag{4}$$

where $y_d(t)$ denotes the desired output, and $y_m(t)$ the measured output of the neuron.

The BP algorithm uses an iterative supervised learning procedure, where the MLP is trained with a set of predefined inputs and outputs. And, the global error $E_g(t)$ is calculated by Eq. (5), this error can be minimized by the gradient descent technique.

$$E_g(t) = \frac{1}{2} \sum_{i=1}^{n} \left( y_{d,i}(t) - y_{m,i}(t) \right)^2 \tag{5}$$

The most efficient algorithm to optimize the global error $E_g(t)$ is that of gradient descent [11].

## 4 Dimensionality Reduction with PCA

PCA for dimension reduction technique use all the information contained in indi-viduals keep (signals in our case), to compress them and produce a vector of smaller dimension [11]. This techniques project a signal vector from the space represen-tation to another space with smaller dimension (Fig. 3).

The purpose of the PCA method is condensing the original data into new group so that they have no correlation between them and are ordered in terms of the percentage of variance contributed by each component. Thus, the first principal

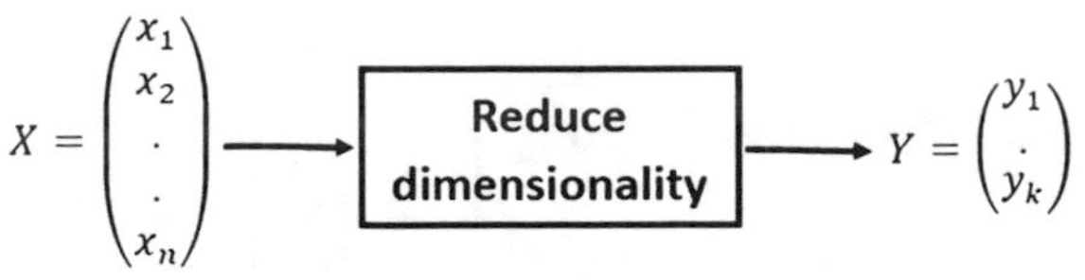

**Fig. 3** Dimensionality reduction by PCA

$$X = \begin{pmatrix} x_1 \\ x_2 \\ \cdot \\ \cdot \\ x_n \end{pmatrix} \longrightarrow \boxed{\begin{array}{c} \textbf{Reduce} \\ \textbf{dimensionality} \end{array}} \longrightarrow Y = \begin{pmatrix} y_1 \\ \cdot \\ y_k \end{pmatrix}$$

component contains information about the maximum variance, the second principal component contains information about the following variance, and the process is repeated until the nth and the last component principal. Information loss decreases by a step to the next.

## 5  Results and Discussion

### 5.1  Database Used

The database of signals used in our simulation was generated using MatLab. It contains 2000 FM signals for different values in low SNR $[-25; -12$ dB] and 400 that contain only Gaussian noise. The first table shows how we have used these data in both the learning and the test phases (Table 1).

### 5.2  Proposed Architecture

In this part, we have proposed hybrid architecture to detect the presence of the primary user, which combines the simplicity of the energy detector, and the robustness of the artificial neural networks [9]. In this architecture, we have replaced the comparison of the signal energy calculated with the threshold, by an ANN block that will make a decision (Fig. 4).

### 5.3  PCA Results

The principal components obtained on the covariance matrix give that the first 14 components with percentage of eigenvalue superior than 3 % explain together 97 % of total the signal data-base variance.

The first principal component describes the greater percentage (19 %), while the second component illustrate about 15 % of total variance. The following graph gives the percentage of the first 22 eigenvalues (Fig. 5).

**Table 1** Data-base used in learning and testing

| Signal | Learning phase | Testing phase |
|---|---|---|
| Primary signal | 1000 | 1000 |
| Noise | 200 | 200 |

**Fig. 4** Proposed architecture

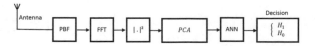

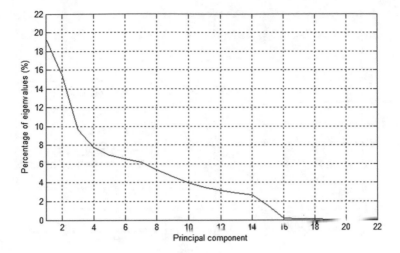

**Fig. 5** Percentage of the first 22 eigenvalue

## 5.4 ANN Training Results

In order to determine the most convenient value of the number of neurons in the hidden layer, an intuitive method has been used. This method consists in testing experimentally several architectures, with different hidden layer size.

The second table summarizes the results obtained for these different architectures (Table 2).

Analyzing the results shown in the table above, it is clear that the best results are obtained for the tow architectures with 15 and 50 hidden neurons, for both the

| Signal | Probability of detection in testing phase | |
|---|---|---|
| | Absence of PU | Presence of PU |
| 4 | 0.78 | 0.59 |
| 6 | 0.81 | 0.61 |
| 10 | 0.93 | 0.76 |
| 15 | 0.97 | 0.93 |
| 20 | 0.78 | 0.82 |
| 25 | 0.91 | 0.75 |
| 30 | 0.90 | 0.83 |
| 35 | 0.94 | 0.73 |
| 40 | 0.89 | 0.85 |
| 50 | 0.97 | 0.92 |
| 60 | 0.71 | 0.85 |

**Table 2** Probability of detection in testing phase

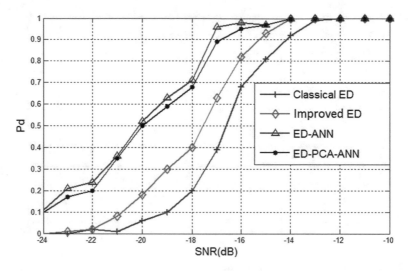

**Fig. 6** $P_d = f(SNR)$ for different method used for the detection

probabilities of detection are respectively: $P_d = 0.97$ in the case of absence of PU, and $P_d = 0.93$ in the case of presence of PU for the first and $P_d = 0.92$ for the second architecture.

Comparing the performance of the proposed scheme with the classical and the improved energy detection method has given the graph below:

According to the Fig. 6, we can make three remarks. The first is that the detection with dynamic threshold has a little performance improvement from the classical energy detection method. The second prove that the decision with the ANN has a great performance improvement from both classical and improved energy detection. The third and the last remark, the dimensionality reduction with PCA does not have a great influence on the performance of detection.

## 6   Conclusion

In this paper, we propose hybrid architecture based on the simplicity of energy detection, and the robustness of the artificial neural networks. This method was implemented in MatLab; the obtained results show that the probability of detection will be higher for the hybrid architecture and even for the reduced data with PCA. As a future work we suggest to implement this architecture in FPGA to profit from the parallel computing in this kind of chips, this parallelism will decrease the sensing duration.

# References

1. FCC: Spectrum policy task force report. In: Proceedings of the Federal Communications Commission (FCC"02), Washington, DC, USA, Nov 2002
2. Mitola, J.: Cognitive radio for flexible mobile multimedia communications. In: IEEE International Workshop on Mobile Multimedia Communications, pp. 3–10. Nov 1999
3. Zeng, Y., Liang, Y., Hoang, A., Zhang, R.: A review on spectrum sensing for cognitive radio: challenges and solutions. EURASIP J. Adv. Signal Process. **2010**, 2 (2010)
4. Yücek, T., Arslan, H.: A survey of spectrum sensing algorithms for cognitive radio applications. IEEE Commun. Surv. Tutorials **11**(1), First Quarter (2009)
5. Axell, E., Leus, G., Larsson, E., Poor, H.: Spectrum sensing for cognitive radio: "State-of-the-art and recent advances". IEEE Sig. Process. Mag. **29**(3), 101–116 (2012)
6. Urkowitz, H.: Energy detection of unknown deterministic signals. Proc. IEEE **55**, 523–531 (1967)
7. Herath, S.P., Rajatheva, N., Tellambura, C.: Energy detection of unknown signals in fading and diversity reception. IEEE Trans. Commun. **59**(9), 2443–2453 (2011). Elrharras, R.S.: Spectrum sensing with an improved energy detection. In: 2014 International Conference on Multimedia Computing and Systems (ICMCS), pp. 895, 900. 14–16 April 2014
8. Fausett, L.: Fundamentals of Neural Networks: architectures, algorithms, and application. Prentice Hall, New Jersey (1994)
9. Fiete, R.: Learning and coding in biological neural networks. Ph.D thesis, Harvard University, Cambridge, Massachusetts (2003)
10. Shlens, J.: A tutorial on principal component analysis. arXiv preprint arXiv:1404.1100 (2014)
11. Pattanayak, S., Ojha, M., Venkateswaran, P., Nandi, R.: Spectrum hole detection in TV band using ANN model for opportunistic radio communication. In: 2014 Annual IEEE India Conference (INDICON), (pp. 1–6). Dec (2014)

# Scalable Framework for Live Data Sharing Through 802.11

Eduardo Soares, Pedro Brandão and Rui Prior

**Abstract** We propose a multi-platform framework for easy development of applications that share live or recorded data of any type in a classroom. It is especially aimed at training in the medical area, where it can make the learning process much more interactive and enriching, but is equally well suited for use in any type of workshop, tutorial, or other learning environment. The framework is browser-based, for better portability. In order to scale well to a large audience, the framework uses multicast for communication. It provides configurable reliability that is adaptable to data flows with different requirements, real time (RT) or not. It also provides security, privacy and access control features that are necessary in medical training environments. Finally, it allows session discovery and management, and multi-sender support.

**Keywords** Multicast · Network · Reliable multicast

## 1 Introduction

Technology has been reshaping our concept of teaching; the educational content became richer with the usage of elements like ultrasounds, electrocardiograms, auscultations or even 3D models [2]. These materials need to be shared between teachers and students and in the process cannot be transformed in any way that

E. Soares (✉)
Faculdade de Engenharia da Universidade do Porto,
Instituto de Telecomunicações, Porto, Portugal
e-mail: easoares@fe.up.pt

P. Brandão · R. Prior
Faculdade de Ciêcncias da Universidade do Porto,
Instituto de Telecomunicações, Porto, Portugal
e-mail: pbrandao@dcc.fc.up.pt

R. Prior
e-mail: rprior@dcc.fc.up.pt

© Springer International Publishing Switzerland 2016
A. El Oualkadi et al. (eds.), *Proceedings of the Mediterranean Conference on Information & Communication Technologies 2015*, Lecture Notes in Electrical Engineering 381, DOI 10.1007/978-3-319-30298-0_19

makes major changes irreversible. Compression with too much loss, to reduce transmission size, is not an option, making it difficult to send to many users.

Current options for solving this problem use third party services, such as cloud storage or a content management system. In education, it is common to use Moodle [5]. These options are not the best solution; they mandate a connection to a server, possibly outside the local network, for all users. This places unnecessary load on the network nodes, as in a classroom scenario origin (lecturer) and destination (audience), are under the same access point (AP), and thus on the same physical link.

Some solutions are also device dependent. Applications for a single platform create barriers, and make it hard to use in the real world, where multiple platforms abound. To avoid this, solutions should be device independent and open. A good option is using the web browser as a multi-platform enabler. All the modern operating systems have a variety of web browsers to choose from, and most of them support the latest standards of HTML, CSS and JavaScript.

## 1.1  Scenario

Our working scenario focuses on a classroom or conference, where all the participants are in a room over the same Wi-Fi 802.11 link of a single AP. We assume that all the participants should be able to access the transmitted content. In addition, no one without physical access to the room should be able to access the shared content, even if the wireless link is accessible. The accessed content can be a live stream or previously recorded. This last point creates a special case, as it needs to look like a real time (RT) stream.

## 1.2  Objectives

As per the scenario and its requirements, the work in this paper aims to accomplish the following objectives: (i) **Reliable delivery**: all packets sent should be received by all users that are the intended recipients; (ii) **Scalability**: the framework should accommodate from a small number of users up to a few dozens; (iii) **Security**: it is crucial that only authorized users receive the shared content and that receivers can validate the senders; (iv) **Work without Internet**: there should be no need for a reliable Internet connection. An example could be in a conference, to access the Internet some login and credentials could be needed but it can be easy to access the local Wi-Fi 802.11 network; (v) **RT content**: even if the platform does not capture content, it is important to adapt when the content needs to appear as RT to the receivers; (vi) **Multi-platform**: not everyone has the same device or uses the same platform. The solution needs to work within the maximum number of platforms in order to cover the maximum number of users.

## 2  Proposed Solution

At the time of this writing, no web browser Application Programming Interface (API) allows to send data in a multicast connection. Furthermore, only a small sub-set of them is prepared to communicate directly between each other with little usage of an intermediate web server. As such, the creation of a pure, in-browser solution is currently impossible. To overcome these problems, we developed the communications part of the solution in pure Java. This part offers a connection to the web browser so anyone can use HTML tools to create the user interface.

Figure 1 shows a simplified structure of the architecture developed. The components are a library that implements a service to create sessions for multicast communication in a secure and reliable way; a web server to expose the data received and the library API for the browser where the application runs

The service creates a representation of a session for sharing data; a group of channels forms each session. A user owns each channel, and each user can have multiple channels. A central coordinator, called session manager, manages and controls the communication. The session manager is also a fundamental part of the search protocol and authentication described respectively in Sects. 2.1 and 2.3.

## 2.1  Search and Discovery

For the search protocol, the session manager periodically sends a packet with the sessions' information to a predefined multicast channel. Each user that wants to know what sessions are available listens to that channel. To avoid too much noise its session manager sends this every 20 s.

Users that are listening and waiting to receive can send a packet requesting this information, so as to speed up the process. When a session manager receives this request, it waits a short, random time, from 0 to 10 s before sending it. This strategy tries to avoid collisions of packets from multiple session managers, each with its own session.

All the waiting time values used in this protocol are based on the protocols studied that are presented in Sect. 3.

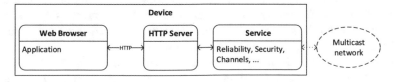

**Fig. 1** Architecture of the solution

## 2.2   Reliable Multicast

To provide reliability on the multicast connection, we chose a NACK strategy. With this approach, a receiver issues a request when it determines that it lost a packet. These requests are sent unicast to the original sender.

In order to avoid request collisions from multiple receivers reporting a missing packet, each one waits a random time between 0 and 10 s.

To optimize request sending, we can ask for several lost packets in the same NACK message. This avoids extra packets for each lost packet in a receiver, and cooperates with the random waiting for sending the NACK.

When the sender receives a NACK it resends the packets that were lost. A possible optimization is using network coding [4] to minimize the number of packets sent. This provides the opportunity to send a network-coded packet that contains information for multiple receivers.

Given the nature of a live stream environment, waiting for a lost packet can break its liveness. To improve the user experience, we have an option to forget about a missing packet after a chosen time has elapsed.

The application that creates the sending channel defines the value for forgetting the packets, because it can create a buffer at a higher level appropriate to the content and adjust the waiting time for missing packets.

## 2.3   Security

Security is one of the key points given the nature of the data. Only users that are supposed to get the data should be able to see it. It should not be possible for an outsider, which is not allowed in the room where the session is happening, to obtain any important data, even if he can capture all the transmitted packets. For this, we need authentication of the users and secure data transmission.

**Authentication**. We need to block users that should not have access to the data. For this, and given the environment described in Sect. 1.1, we use a protocol based on public key (PK) cryptography. It allows users to prove they are in the room and have access to a token given by the session creator. It work as follows:

1. The creator of the session generates a public/private key pair;
2. The PK is advertised in the session announcement as described in Sect. 2.1;
3. The PK fingerprint and an authentication token to authenticate the users to the session creator is transmitted by the user who created the session through analogue means (e.g. writing on the classroom board);
4. Users validate the PK and perform authentication to obtain the session information and cipher keys. Defining the client as the user that wants to connect to the session, and the server as the session creator, we have:

(a) Client opens a connection to the Authentication Server (AS);
(b) AS replies with the PK that must match the one advertised;
(c) Client generates a symmetric key (SK) and sends it to the server ciphered with the PK received. From this point onwards, all the messages are encrypted with this SK;
(d) Server challenges the client with a nonce;
(e) Client answers the challenge with a hash of the nonce concatenated with the authentication token;
(f) If the client correctly answers the challenge, the server sends information about the session (SK to cipher the stream data, known channels of data and session management channel). Otherwise an error is replied.

**Secure Data Transmission, Key and User Management**. From the authentication protocol, users receive a key that ciphers all the data shared in all the channels. This causes a security problem, as an authenticated user could introduce fake packets inducing corruption of the shared data. We assume that this is not a problem, and authenticated users are not inside attackers.

If a user leaves the session, the SK does not change. We use this approach for two reasons. The first is that in a connectionless environment it may be hard to detect if a user leaves. The second reason would be that even if the user notified he was leaving, the session would not be more secure from knowing that and changing the key. As per defined protocol for authentication, the user could redo it and get the new key. We follow a security policy similar to the one used in classrooms. Whoever is attending can access the session. Of course, if the channel to transmit the authentication token is more private we can restrict more the allowed users.

## 2.4 Session Management

To keep a coherent vision of the session between users, and inform them of creation deletion of data channels, we developed a session management protocol.

To create a channel, a user issues a request to the session manager. Then, the library that is running the protocol asks who is using it if this user should be authorized. The answer is transmitted to the user and if affirmative, the session manager sends this information to the management channel.

The management channel is a channel that all users must be listening in order to receive information of created and deleted channels. To avoid fake packets in this channel, all packets are signed with the private/public key of the session manager (also used in the authentication protocol).

## 2.5   Web Server and HTTP API

As described in the start of this section, we need a web server to load an application built in HTML, JavaScript and CSS and to provide an HTTP API of the created library.

Given the multitude of devices and the simple requirements, we built a small, custom solution. The most complex requirement was to fully support the HTTP API. This API exposes the session feature and reliability from the created library. It has three main components: **Session**: Information about surrounding and connected sessions. Ability to search for sessions and/or create new ones; **Channels**: Information about existing channels in the currently connected session. It also allows to create new channels, remove existing ones, and send to or receive from a channel; **Notifications**: Built over Server-Sent Events (SSE) [10], with the option to reply to a notification that needs an answer.

In addition to the described API, we developed a solution to surpass the web browsers' limitations. Nowadays, web browsers have rich options in order to capture video or audio and use files. However, they cannot go much further to talk with other types of hardware, like Bluetooth or USB devices. This causes limitations in the described scenario, where many acquisition devices interface with the operating system over Bluetooth and with custom protocols on top.

The data generated by these types of hardware has to be pipelined into the multicast channel. A possible solution could be the usage of PhoneGap [1]. This framework allows developing an application in HTML, CSS and JavaScript and then porting it to different OSes. It provides a JavaScript API to access lower capabilities of the hardware that are normally blocked in the browser environment. However, in our scenario, this would mean that the information would flow from the Java Virtual Machine (JVM) (adaptation library from PhoneGap), to the web viewer (that PhoneGap uses to host the HTML, CSS and JavaScript code) and back to the JVM (our library for the multicast sessions). This would create a round trip time that could become noticeable, an extra overhead in the application in order to include the PhoneGap wrapper and creating a dependency, thus making it impossible for a native application to reuse our library.

To avoid this, we introduce the plugins component. Plugins are modules that have a single function and can connect to other plugins in order to form a graph. They can be a source of data, which means that they connect to an external device, collect its data and send it to another plugin. In order to send the data, plugins can be of the output type, where they send from one or more data sources (multiplexing) to an output. This destination can be a channel of the multicast library or other location, like a local file or a remote server. To operate over the collected data, we use a processing plugin. This makes it possible to create filters or encoders for the collected data before sending it to an output.

The graph with input, processing and output plugins operates within the JVM, thus avoids transporting the data through different layers.

Viewers follow the same idea of reusable components across applications, similar to plugins, to allow data visualization. This component runs in the web browser and is self-contained without external requirements. Each viewer displays a particular type of data. Given the passive nature of this component, we can add new viewers uploading them via an HTTP API. The HTTP API also exposes the loading of a viewer, listing existing viewers and their information (data type supported, parameters accepted, name and description) and removing viewers.

# 3 Related Work

In this section, we discuss some solutions to part of the problems that we enumerated previously.

## 3.1 Reliable Multicast

Through the years, there have been some proposals to achieve reliability over multicast. They can be divided in two categories The first tries to copy TCP, having the receivers acknowledge the sender every packet received. This causes much overhead in the communication, and does not scale well. Protocols that do this usually have optimizations, like using intermediate nodes to receive acknowledgements and do repairs in case of failure. These nodes can be other receivers or specialized network equipment. The other option is to assume that every node receives the packets, and when some do not, they inform about the failure and try to have it resent. This mechanism is called NACK.

Both solutions add some extra information to the packets sent to identify them. The minimum data that needs to be added is a sequence number.

The Tree-Based Reliable Multicast (TRAM) [3] protocol was built to send data from a source to multiple receivers at more than a hop of distance. This protocol builds a tree of receivers where the sender is the root, and works like the TCP adaptation mentioned previously. The parent of a node in the tree is the one that receives the acknowledgement and replies when packets are missing. The protocol defines extra packets, to inform when a session is still running but the source is not sending data.

The Lightweight Reliable Multicast Protocol (LRMP) [8] was designed for environments that can handle package delay, as long as there are guarantees that all of them arrive. It recovers missing packets by sending NACK messages to the multicast group. An element of the group that has the lost packet sends it to the sub-group of the requesting node. The protocol uses NACK suppression to avoid NACK messages collision. It also employs a strategy of trying to recover the packet in a close node. To achieve this, it sends the NACK with a small Time To Live (TTL). After a time out, it resends the NACK after increasing its value. This mandates that each node must keep a queue of received packets to answer requests. The protocol also has some optimizations to avoid congestion, adapting the transmission rate. Moreover, it gives the possibility to use Forward Error Correction (FEC) packets [9].

The Pragmatic General Multicast (PGM) [6] protocol uses some of the previous techniques to achieve reliable multicast. It uses NACK sent via unicast to the sender, and has optional usage of FEC packets, among others. It also has the option of using specialized network equipment to improve the packet repair on reception failure. In this protocol, routers can suppress NACKs that other receivers already asked for and are waiting for replies.

## 3.2 Multi-platform Development

We chose to start by testing in Microsoft Windows and Android because they are the most used platforms on computers and mobile devices.

While there currently exist options that allow for the development of a single solution for the targeted platforms, most of these are based on developing a web application. This implies an application coded in JavaScript, along with the usage of some platform-specific wrappers added at compile-time. Examples of this can be found in [7]. Our approach, however, overcomes the limitations of the web browser APIs for communication.

## 4 Conclusion

The work presented aims to achieve a solution to enable easy development of multi-platform applications that can share data using multicast communications without the need of a remote service.

Multicast allows achieving the best scalability possible in an environment where the network equipment cannot scale at the same rate as that of the number of devices. While we are still limited by the AP that connects the users, the fact that we do not go outside the local network alleviates congestion problems that would otherwise affect other network nodes. Multicast also avoids duplication of data per receiver, removing overload of packages at the AP.

The objectives set forth in the introduction were achieved, although with some constraints.

**The reliable delivery cannot be assured in RT**. Waiting for a lost packet can take a variable time, and if the request for it is also lost, the process starts again. Given the nature of a wireless environment with several users, this is prone to happen. So we trade-off to support soft RT.

**Scalability by default**: we built the solution to always use multicast to send data. With a small number of users, it could be a better option simply to use TCP; the overhead of the communication would not be a problem. TCP would give reliability but it could affect the RT aspect of the stream.

**Multi-platform is only tested on Android and MS Windows**: but all the software should run without problems in any other platform that supports the Oracle JVM and a web browser (including but not limited to Linux and Mac OS). The web browser requirements are for the SSE and XMLHttpRequest API to be present, which is the case in all the modern web browsers (e.g. Mozilla Firefox or Google Chrome). Stricter requirements only depend on the applications created using this framework.

An important point that was achieved was the capability to function without an Internet connection. This required extra steps for establishing a connection in a secure way.

In the future, we should support easier authentication via previous exchange of PKs between the users involved in the session. This removes the need to share key fingerprints and authentication tokens, making it possible to authenticate specific recipients. Also notably missing and currently under development are tests against real-live scenarios and comparison with TCP based solutions.

**Acknowlegments** This work was funded by Project I-CITY—ICT for Future Health, NORTE-07-0124-FEDER-000068, funded by the Fundo Europeu de Desenvolvimento Regional (FEDER) through the Programa Operacional do Norte (ON2) and by national funds through FCT/MEC (PIDDAC).

# References

1. Adobe: PhoneGap|about: http://phonegap.com/about/. Accessed 29 Aug 2014
2. Body, V.: Visible body|3D human anatomy: http://www.visiblebody.com/index.html (2014). Accessed 03 Sept 2014
3. Chiu, D.M., Hurst, S., Kadansky, M., Wesley, J.: TRAM: a tree-based reliable multicast protocol (1998)
4. Costa, R.A., Ferreira, D., Barros, J.: FEBER: feedback based erasure recovery for real-time multicast over 802.11 networks. CoRR abs/1109.1265 (2011)
5. Dougiamas, M., Taylor, P.: Moodle: Using learning communities to create an open source course management system. In: World Conference on Educational Multimedia, Hypermedia and Telecommunications. vol. 2003, pp. 171–178 (2003)
6. Gemmell, J., Montgomery, T., Speakman, T., Crowcroft, J.: The PGM reliable multicast protocol. IEEE Netw. **17**(1), 16–22 (2003)
7. LeRoux, B.: PhoneGap|PhoneGap, Cordova, and what's in a name? http://phonegap.com/2012/03/19/phonegap-cordova-and-what%E2%80%99s-in-a-name/ (March 2012). Accessed 29 Aug 2014
8. Liao, T.: Light-weight reliable multicast protocol specification. Internet-Draf: draft-liao-lrmp-00.txt 13 (1998)

9. Luby, M., Vicisano, L., Gemmell, J., Rizzo, L., Handley, M., Crowcroft, J.: The use of forward error correction (FEC) in reliable multicast. Technical report RFC 3453, December (2002)
10. W3C: Server-sent events. http://ww1w.w3.org/TR/eventsource/ (December 2012). Accessed 29 Aug 2014

# Congestion Avoidance in AODV Routing Protocol Using Buffer Queue Occupancy and Hop Count

**Bouchra El Maroufi, Moulay Driss Rahmani and Mohammed Rziza**

**Abstract** Owing to network congestion, data could be loosed when its transmitted from source to destination. The reactive routing algorithm Ad hoc On-Demand Distance Vector Protocol (AODV) uses number of sequences and hops to select an optimal path. However, it doesnt consider nodes loads to choose the suited route. In addition, its efficiency declines sharply when dealing with high load and fast mobility. To face this problem, this paper presents an improved protocol Buffer Queue Occupancy-AODV (BQO-AODV) which selects optimal routes by considering the minimum value of BQO metric instead of minimum hop count. The results of tasting have shown better performance of our proposed protocol (BQO-AODV) comparing to AODV routing protocol in terms of average end to end delay, packet delivery ratio and normalized routing load.

**Keywords** Manet · AODV · Congestion avoidance · Buffer · Queue length

## 1 Introduction

It is well known that congestion is one of the most critical problem in routing in Ad Hoc networks [1]. It was defined as an overcrowding or blockage result of overloading which is similar to traffic jam caused by many cars on a narrow road. Congestion occurs mainly, at any intermediate node, when the packets number exceeds buffer capacity. Thus, node becomes congested and starts losing packets. Routing protocols [2], during transmission, are needed in order to define routes and

B.E. Maroufi (✉) · M.D. Rahmani · M. Rziza
LRIT, Associated Unit to CNRST (URAC 29), Faculty of Sciences,
Mohammed V University, Rabat, Morocco
e-mail: el.maroufi.bouchra.smi@gmail.com

M.D. Rahmani
e-mail: mrahmani@fsr.ac.ma

M. Rziza
e-mail: rziza@fsr.ac.ma

© Springer International Publishing Switzerland 2016
A. El Oualkadi et al. (eds.), *Proceedings of the Mediterranean Conference on Information & Communication Technologies 2015*, Lecture Notes in Electrical Engineering 381, DOI 10.1007/978-3-319-30298-0_20

deliver packets from the source to the final destination. In this work, we will focus on avoiding congestion for AODV distance vector routing algorithm [3].

AODV protocol is a reactive routing algorithms characterized by the use of serial number to identify whether the routing is new or old then avoiding routing loop. Moreover, each intermediate node could save the routing request and response results while remaining update to network structure changes. Since Mobile ad hoc network nodes are highly dynamic in nature, congestion is the main factor for more packet loss and longer delay. AODV uses hop count as route selection metric which is not an efficient method under this situation in ad hoc networks. In fact, it doesn't take mandatory precautions to handle the nodes which become congested under heavy network traffic.

We can find in the literature, many routing algorithms in mobile ad hoc networks for routing and congestion free networks. The authors in [4] have proposed a new mechanism for selecting routes called ETR-AODV which optimizes node energy and traffic by adding a congestion field and energy field in RREQ or RREP packet. Then the destination node select route which contains a maximum value of energy and minimum value of congestion and unicast a reply with this path selected. Another approach was presented in [5] which is an adaptive load-balancing approach, which presents an effective scheme to balance the load in ad hoc network. It is implemented in the process of route request. When route request (RREQ) messages are flooded to acquire routes, only the qualified nodes, which have a potential to serve as intermediate forwarding nodes, will respond to these messages, so that the established path will not be very congested, and the traffic will be distributed evenly in the network. In this scheme, a threshold value, which is used to judge if the intermediate node is overloaded, is variable and changing along with the nodes interface queue occupancy around the backward path. The authors in [6] have proposed Dynamic Load Aware Routing Protocol (DLAR), which considers the load of intermediate nodes as the main route selection metrics. The load of a route is defined as the number of packets buffered in the queue of the node. The authors in [7] have presented a survey of various congestion aware and congestion adaptive routing protocols. Some of such routing algorithms discussed are congestion aware distance vector (CADV), congestion aware routing protocol for mobile adhoc networks (CARM), hop-by-hop congestion aware routing protocol for heterogeneous mobile ad hoc networks, congestion adaptive routing protocol (CRP), etc. The paper suggests that the problem of congestion is associated with the network and it has to be solved by having compromised solution rather than elimination. AOMDV routing protocol in [8] used Queue Length and Hop Count value together to select a route from source to destination that avoids congestion and load balancing. A threshold value is defined after a threshold alternate path is chosen. Intermediate nodes avoids broadcast of RREQ if the routes are already congested.

In this paper, we propose a modified AODV protocol designed to avoid congestion in network taking in consideration the route which has the smallest maximum node Buffer Queue Occupancy (BQO) metric.

The remainder of the paper is organized as follows: Sect. 2 presents the proposed routing protocol. Sect. 3 shows the simulations and analysis of the obtained results. Finally, we conclude the paper in Sect. 4.

## 2 Proposed Approach

Most on-demand routing protocols include route request, route reply and route maintain procedure. Changes in our approach BQO-AODV is carried out in route request and route reply procedure, where a reserved forward BQO field and backward BQO field of node are inserted in the RREQ and RREP respectively. Whenever a node receives a RREQ or RREP, it compares its $BQO_j$ with the forward BQO and backward BQO. If it is greater it sets the forward BQO or backward BQO to $BQO_j$.

### 2.1 Buffer Queue Occupancy (BQO)

To calculate the buffer queue occupancy we define several terms:
$Len_j$:    Number of data packets in interface queue of node j.
$Limit_j$:    Total interface queue size of node j.

The Buffer queue occupancy of node j is calculated by the following equation:

$$BQO_j = \frac{Len_j}{Limit_j} * 100. \tag{1}$$

Buffer queue occupancy in Eq. (1), is the percentage of the network interface queue that is occupied.

### 2.2 Enhanced Route Discovery Mechanism

Whenever a source node requires communicating with another node for which it does not have a route, it initiates the route discovery phase by broadcasting a Route Request (RREQ) packet to all its neighbors, where the forward and backward BQO values are initialized to zero.

When a neighboring node received the route request message, it calculates the parameter $BQO_j$ defined in (1) and stored it in the forward BQO field of the ROUTE REQUEST packet. As the RREQ is broadcasted in the whole network, upon receiving the RREQ, an intermediate node first checks whether it has received this RREQ before. If so, it drops the RREQ. The intermediate node compares its $BQO_j$ with forward BQO field if it is greater it sets the Forward BQO to $BQO_j$.

When the destination receives a RREQ packet, if there is no route in its routing table, it sends a RREP using the reverse route. But, if there is a route to the source, the destination compares the forward BQO value in the RREQ received with the forward BQO value of the route currently in use. If it is smaller, the destination replaces its current route information with the RREQ packet route and sends a RREP to the source.

Each intermediate node that receives a RREP packet checks to see if the RREP is the first reply or not. If it is, the intermediate node compares its $BQO_j$ to the backward BQO value contained in the RREP header, updates its route table and forwards the RREP to the next hop towards the source. If the received RREP is not the first reply, the intermediate node checks the freshness of the route in the RREP. If it is fresher than the current one, it updates its route table and forwards the RREP to the next hop towards the source. But, if the route in the received RREP is as fresh as that of the current route, the intermediate node compares the backward BQO with the backward BQO value of the current route, and uses the best one (the least congested one) for sending data.

When a source node receives a RREP packet, it checks its route table to determine if it is the first reply to a route request or not. If it is, the source starts using the route received otherwise it checks the route freshness. If the new route received is fresher than the one currently in use (has a destination sequence number larger than that of the route is being used), the source updates its route table and starts using the new route for sending data. But, if the two routes have the same freshness, the source compares the backward BQO value in the RREP with that of the current route, and starts using the best one (the least congested one) for sending data. If the two routes have the same backward BQO the path with minimum number of hops will be selected.

# 3   Performance Evaluations

In this section, we evaluate BQO-AODV and AODV performances under the same environment. We will start by displaying configuration of the used environment parameters. Next, various traffic load conditions and nodes mobilitys are explored in order to study the effectiveness of our approaches.

## 3.1   Simulation Environment

To evaluate the performance of BQO-AODV, Each simulation is run for 200 s and repeat 10 times. These performances are evaluated by increasing the maximum number of connections and pause time. We compare it with AODV by using NS2.34 [9]. In the process of simulation, we assume every protocol shares the same model and node configuration. Their initial parameters are shown in Table 1.

**Table 1** Simulation parameters

| Parameter | Value |
|---|---|
| Dimensions | 1200 m * 1200 m |
| Number of nodes | 50 |
| Source type | CBR |
| Antenna type | Omni directional |
| Spread type | Two ray ground |
| Wireless channel capacity | 2 Mb/s |
| Communication radius | 250 m |
| Packet size | 512 bytes |
| Packets rate | 8 Packets/s |
| Buffer size | 50 |
| MAC layer | IEEE802.11 |
| Transport layer | UDP |
| Max. number of connections | 10, 20, 30, 40, 50 |
| Pause time | 10, 20, 30, 40, 50, 60, 70, 80, 90, 100 |

We evaluate the performance of BQO-AODV through three metrics: packet delivery ratio, average end to end delay and normalized routing load.

## 3.2   The Effect of Mobility

To analyze the effect of mobility, pause time was varied from 10 s (high mobility) to 100 s (low mobility). The number of nodes is taken as 50 and the maximum number of connection as 50. Graphs showed in Figs. 1a, 2a and 3a shows the effect of mobility for AODV and BQO-AODV protocols with respect to various performance metrics. The average end to end delay of BQO-AODV in Fig. 1a shows a better

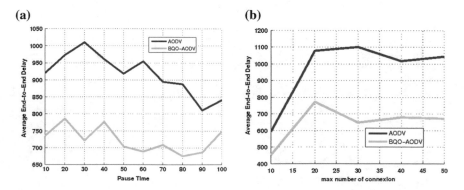

**Fig. 1** Average end-to-end delay: **a** Pause time with 26 sources. **b** Maximum number of connections

results than AODV for almost all mobility levels (pause times of 10, 20, 30, 40, 50, 60, 70, 80, 90, and 100 s). Figure 2a shows the packet delivery ratio of the tow protocols AODV and BQO-AODV. The pdr of BQO-AODV is greater than AODV At low and high mobility. The protocol BQO-AODV generates less normalized routing load than AODV protocol as shown in Fig. 3a. The reason is that the protocol BQO-AODV uses the route that contain less congested node accordingly it uses RREQ packets often.

## 3.3 The Effect of Traffic Load

In this simulation, the number of connection (source and destination) was varied from 10 to 50, CBR sending rate 8 packets per second, maximum node speed 10 m/s and pause time 0 s. Figures 1b, 2b and 3b had shown the average end-to-end

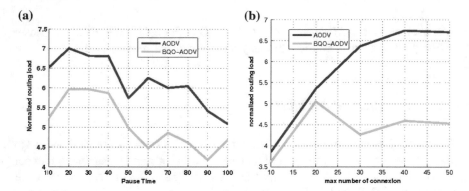

**Fig. 2** Normalized routing load: **a** Pause time with 26 sources and **b** maximum number of connections

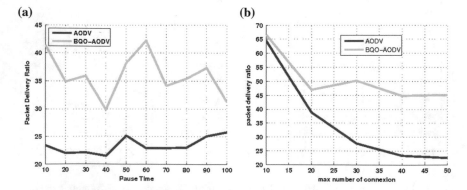

**Fig. 3** Packet delivery ratio: **a** Pause time with 26 sources. **b** Maximum number of connections

delay, the normalized routing load and the packet delivery ratio for BQO-AODV and AODV respectively. Figure 1b shows the average packet arrival time for each protocol. We can see that BQO-AODV has a lower delay than AODV, when the number of flows increased from 20 to 50. The end-to-end delay had increased from 1078 to 1044 s. The corresponding variation for BQO-AODV was from 773 to 672. The ratio of packets delivered by AODV and BQO-AODV is affected by the traffic load. When the numbers of flows were less than 20, BQO-AODV did not show much improvement over AODV. When the number of flows increases from 20 to 50 the packet delivery ratio of AODV had a sudden fall from 39 to 23 % when compared with the above packet delivery ratio, the figures of the BQO-AODV fall only a gradual fall from 47 to 45 % as shown in Fig. 2b. In Fig. 3b when the numbers of flows were less than 20, BQO-AODV did not show much improvement over AODV. When the number of flows increases from 20 to 50 the normalized routing load incurred by BQO-AODV is very less when compared to AODV routing protocols.

## 4 Conclusion

In this paper, we have proposed a new routing protocol take into account the buffer queue occupancy (BQO) of intermediate nodes and try to avoid congestion using routes that go through congested nodes, here the BQO is the number of packets in interface queue of intermediate nodes. Source and destination make routing decisions by selecting the least congested route, where the congestion of a route is determined as the maximum BQO value at the intermediate nodes. Detailed simulations were used to evaluate the performance of the proposed congestion mechanism and to compare it with AODV routing protocol. The simulation results show that the proposed protocol can result in substantial improvement in the packet delivery ratio, average end-to-end delay and normalized routing load.

## References

1. Gerla, M.: Ad hoc networks. In: Ad Hoc Networks, pp. 1–22. Springer (2005)
2. Bansal, M., Rajput, R., Gupta, G.: Mobile Ad Hoc Networking (manet): Routing protocol performance issues and evaluation considerations. The Internet Society (1999)
3. Perkins, C.E., Royer, E.M.: Ad-hoc on-demand distance vector routing. In: Second IEEE Workshop on Mobile Computing Systems and Applications, 1999. Proceedings. WMCSA'99, IEEE (1999), pp. 90–100
4. Kumar, K., Prakash, S., Singh, S.K.: An energy and traffic aware routing approach as an extension of aodv. Int. J. Comput. Appl. 27(7), 6–10 (2011)
5. Yuan, Y., Chen, H., Jia, M.: An adaptive load-balancing approach for ad hoc networks. In: 2005 International Conference on Wireless Communications, Networking and Mobile Computing, 2005. Proceedings, vol. 2, IEEE (2005), pp. 743–746

6. Lee, S.J., Gerla, M.: Dynamic load-aware routing in ad hoc networks. In: IEEE International Conference on Communications, 2001, IEEE (2001). ICC 2001, vol. 10, pp. 3206–3210
7. Shrivastava, L., Tomar, G., Bhadauria, S.S.: A survey on congestion adaptive routing protocols for mobile ad-hoc networks. Int. J. Comput. Theory Eng. 3(2), 189–196 (2011)
8. Puri, S., Devene, S.R.: Congestion avoidance and load balancing in aodv-multipath using queue length. In: 2009 2nd International Conference on Emerging Trends in Engineering and Technology (ICETET), IEEE (2009), pp. 1138–1142
9. Issariyakul, T., Hossain, E.: Introduction to Network Simulator NS2. Springer (2011)

# OLSR-RAIP5: Optimized Link State Routing with Redundant Array of Independent Paths 5

Adel Echchaachoui, Fatna Elmahdi and Mohammed Elkoutbi

**Abstract** The improvement of data routing performances and security in mobile networks mainly depends on the quality and type of message transmitted and also on the routing algorithms used. In Mobile Adhoc Networks (MANET), there are a lot of constraints that must be considered to improve quality and security of mobile communications. In this paper, we propose a new approach to enhance quality, reliability and security of mobile communications based on a multipath routing protocol inspired on the RAID5 technology. We propose a new algorithm for packet segmentation and routing data via multiple disjoint paths. This provides better security and ensures greater reliability of the information exchanged in a MANET. We have tested our proposition and done some simulations on the OLSR protocol which is widely used in context of MANETs. Through conducted simulations, we have shown that our proposed solution offers better reliability and security than the standard OLSR protocol.

**Keywords** Mobile adhoc routing · Ad Hoc · Multi-path · Security · Quality of service

## 1 Introduction

In MANETs, data routing is generally done in a decentralized manner. Mobile nodes communicate continuously with each other to recognize the network topology and to maintain routing paths. Routing protocols are responsible for choosing the best path from a source to a destination. The mainly used routing protocols in the context of MANETs are: OLSR, AODV, DSR and TORA [1]. These protocols route packets based on a single best path approach. Nevertheless, some other

A. Echchaachoui (✉) · F. Elmahdi · M. Elkoutbi
SIME Laboratory, E.N.S.I.A.S, Mohammed 5 University, Rabat, Morocco
e-mail: adel.echchaachoui@um5s.net.ma

© Springer International Publishing Switzerland 2016                           201
A. El Oualkadi et al. (eds.), *Proceedings of the Mediterranean Conference on Information & Communication Technologies 2015*, Lecture Notes in Electrical Engineering 381, DOI 10.1007/978-3-319-30298-0_21

protocols like GZRP and AOMDV propose a multi-path manner to ensure performance or backup links. In this paper, we propose an extension of the OLSR protocol to support multi-paths routing combined by the integration of the RAID5 technique that is well known of enhancing reliability and performances in hard disks. The integration of the RAID5 technique in the multipath OLSR version will provide more immunity towards several attacks in MANETs.

This paper is composed of five sections. In Sect. 2, we summarize some related works. In Sect. 3, we briefly present RAID techniques and how they enhance performance and reliability of data access in hard disks. Our contribution is discussed in Sect. 4 where Sect. 5 presents some simulation results and compares the behavior of our solution with the standard OLSR protocol.

## 2 Related Work

Multi-path routing algorithms give more performance and reliability than single path ones but they can be difficult to maintain and can create more overhead when processing routing information. Several multi-path algorithms exit in the literature, we will present below some of them.

In the paper [2], the authors have proposed a multipath routing scheme to secure the dispersed data transfer. This allows coping with the node capture attacks. The approach is based on the MP-OLSR and the secure transfer method and routing structure [3]. The results show that this method is effective (in comparison with the MP-OLSR) when the network is very dense. However, no attack was used to test the level of resistance of this method.

Joshi and Rege presented in [4] a new method of selecting MPRs and computing multiple paths based on the notion of energy to improve the quality of service and routing in Ad Hoc networks. This method allows improving the QoS in Ad hoc Networks. However, it has a negative impact on traffic performances and it has not been evaluated in a complex environment to properly measure its performances.

In [5] the authors propose a new method for calculating the multipath routing by introducing the criterion of radio interference. Dijkstra's algorithm was used to calculate disjoint routes and links between the source and the destination. To improve the process of selecting routing paths, the radio interference levels were considered.

The authors of [6] studied two algorithms: FBMP and WC, in both cases: single-path and multi-path. The FBMP is based on three criteria: the residual bandwidth, the propagation delay and the computation of link, to calculate multiple routes between source and destination. The multi-path routing FBMP significantly improves the quality of service in the network. However, it is much more complex and requires an execution time much more important than the single-path routing process of the WC.

Routing strategies that use multi-path selection MPRs flooding are sensitive to disturbance attacks. To solve this problem, Cervera et al. proposed in [7] a new solution for multiple paths based on the additional coverage principle in the selection of MPRs and the algorithm of nodes and disjoint paths.

The authors of chapter [8] used levels of interference to study and evaluate the performance of two multipath routing algorithms: The NIA-OLSR that considers disjoint nodes and the LIA-OLSR that uses disjoint links. The results showed that the NIA-OLSR solution provides better performance in terms of rate of packet reception and transmission delay than the LIA-OLSR solution that can prevents network congestion by reducing traffic load.

# 3 RAID Technology (Redundant Arrays of Inexpensive Disks)

RAID technology was invented in 1987 by researchers at the University of Berkelcy in order to improve performance and provide fault tolerance in disks data access [9]. Five basic architectures of RAID (from 1 to 5) were defined to ensure the security, availability and reliability of data access. The principle of this system is based on the distribution of data across multiple storage drives, providing high availability and good quality of read/write.

The five RAID levels differ depending on the type of data distribution. Our comparative study of different RAID architectures enabled us to draw the following table (Table 1):

The RAID 5 level is the most interesting solution that offers a high data transfer rate while maintaining very high fault tolerance [10].

# 4 Our Contribution: OLSR-RAIP5

In this section, we present our redundancy and security mechanism based on a new approach that we have called RAIP5 (for Redundant Array of Independent Paths).

**Table 1** Comparison table of the RAIDs

| RAID | Performance | Capacity | Reliability | Cost |
|------|-------------|----------|-------------|------|
| 1 | – | – | ++ | – |
| 2 | Obsolete | | | |
| 3 | Replaced by RAID 4 | | | |
| 4 | + | + | + | – |
| 5 | ++ | + | + | – |

## 4.1 RAIP5

RAIP5 is a new technique of redundancy independent links that computes disjoint paths between a source node and a destination node in an Ad hoc network. RAIP5 works as follows:

Using the routing table, the node will calculate different disjoint paths between a source and a destination. If the number of such paths is less or equal than two, the standard routing algorithm will be used. Otherwise, the redundancy technique is applied as follow: The data packet will be divided into n-1 parts where n is the number of disjoint paths. Then the algorithm will calculate the third part that consists of the parity of the n-1 previous parts. All parts will be sent simultaneously, each part on a distinct path. At the destination, the node will rebuild the original packet based only on n-1 parts. Even if we lose one part of the initial packet, the destination can recover the packet based on the RAID5 technique. In this paper and for simplicity we have considered only the three (n = 3) best disjoint paths between a source and a destination.

## 4.2 OLSR-RAIP5

We have implemented the RAIP5 in the OLSR protocol, which we called OLSR-RAIP5. It allows multi-path routing based on four steps as shown in Fig. 1:

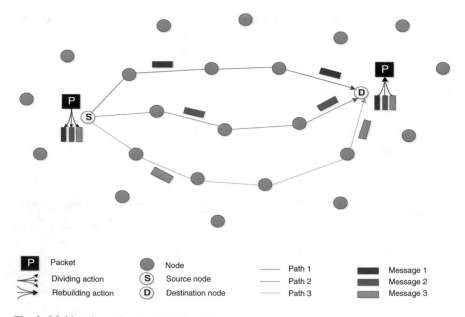

**Fig. 1** Multi-path routing by OLSR-RAIP5

- Calculate and find the three best disjoint paths.
- Divide the packet into two parts and calculate the parity part (redundancy).
- Transmit the three parts simultaneously via the three disjoint paths.
- Rebuild or recover the packet and then perform an integrity check.

**Step 1: Calculation of the disjoint multi-paths.**
Based on the node routing table and the neighborhood information, disjoint paths can be easily calculated. From the source node topology table, we calculate the set of the **last-nodes** that allow us to reach the destination node. For each element $e_i$ of the **last-node** set, we test if $e_i$ is a neighbor of the source node. If it is the case then our first path is found. Otherwise we apply recursively the procedure and the element $e_i$ becomes the new destination. When a path is found, all intermediate nodes will be deleted from the set of **last-nodes**. At the end of the algorithm, we get the three totally best disjoint paths between the source and the destination
**Step 2: Dividing the packet.**
If we consider a packet p, it will be divided into two parts $p_1$ and $p_2$. The node will calculate $p_3 = RAID5(p_1, p_2)$.

Once the three parts are created, we marked each one by the position in order to ease the reassembly operation at the destination node.
**Step 3: Sending the three parts on the three disjoint multi-paths.**
OLSR-RAIP5 protocol will use the concept and the technique of redundancy independent links (Sect. 4.1), and an identifier (ID) of the next node, to transmit packets (that constitute the original packet) through three different paths between source and destination. This information is stored in the fields "common header" and "IP header" of the CBR packet.
**Step 4: Rebuilding the packet.**
When receiving the three parts, the destination node checks the identification of each part before rebuilding the original packet and checking its integrity. If the integrity check is negative, this can be a sign of error transmission or potential attacks.

If at least one part of the packet has been lost or that the check is negative, the original packet can even if be regenerated and an alarm will be sent to the sender.

# 5   Simulation Results

To evaluate the effectiveness and the performances of the OLSR-RAIP5, we conducted via NS2 some simulations to compare between our solution and the standard OLSR protocol. We have chosen a MANET of 50 nodes dispersed over an area of 600 m × 600 m. Nodes communicate in a wireless environment using Omni-directional antennas based on the 802.11b standard. The CBR traffic was generated with different rates, from 5 to 30 Kbps. The time of each simulation was set at 50 s.

We studied the behavior of the two protocols; OLSR and OLSR-RAIP5 against several types of attacks. We also measured some performance metrics like: throughput, delivery ratio and end-to-end delay. We have deployed passive and active attacks to study and evaluate the robustness of our solution.

## 5.1   Passive Attacks

The objective of a passive attack is to listen and capture traffic without any disruption or modification of the content. The malicious node that analyzes traffic, can only obtain one part of the original packet generated by the OLSR-RAIP5 and it cannot have the complete data packet, because of the disjunction of routing links used by our solution. However, all data and information relating to the OLSR routing tables are divulged in the attack.

## 5.2   Active Attack

### 5.2.1   Integrity Attack

During this attack, a part of the transmitted packet can be altered. The standard OLSR accepted and treated all infected messages, while the OLSR-RAIP5, due to its integrity control, it will be able to detect the different changes in the wrong parts of the packet and reported the reception problem by sending an alarm to the source node.

### 5.2.2   Drop Packet Attack

The black hole is one of the fiercest attacks in MANETs. A malicious node can uses this attack to destroy all the messages it receives and can even generate erroneous messages in the network. Using this attack, the standard OLSR was unable to recover any lost of packets. In addition, it receives all erroneous messages sent by the malicious node. However, the OLSR-RAIP5 was able to rebuild all original packets even when a part of the packet was dropped, and it rejected the packets that have problem of integrity.

## 5.3   Performance Evaluation

In this section, we present and compare the simulation results done with OLSR and the OLSR-RAIP5. The performance metrics measured are: the delivery ratio (PDR), the end-to-end delay and the throughput.

### 5.3.1 Packet Delivery Ratio

Figure 2 shows the packet delivery ratio which is equal to **#packets_received/ #packets-sent** when using OLSR and OLSR-RAIP5 in the context of a black hole attack. In normal mode (without attack), the PDR is almost the same for OLSR and OLSR-RAIP5 about 96 %. In black hole attack mode, the PDR values of the OLSR vary between 20 and 33 %, while those of the OLSR-RAIP5 are better, between 60 and 65.5 %, when the number of nodes varies from 5 to 30. This can be explained by the fact that our solution allow is more robust against black attacks. Even if some packets are dropped by the black hole attack, the OLSR-RAIP5 can in some case regenerated the initial packet.

### 5.3.2 End-to-End Delay

Figure 3 shows the end-to-end delay for the two protocols. In normal mode (without attack), the delay of the OLSR increases from 1 to 22 ms when the network density increases from 5 to 30 nodes, whereas the values of the OLSR-RAIP5 vary between 2.2 and 36 ms. In the attack mode, delays of the OLSR grow at an average rate of 120 %, while those of the OLSR-RAIP5 increase slightly with 7 %.

The results show that in normal mode, the OLSR provides better delays of data transmission that our solution. This is certainly due to different steps of OLSR-RAIP5 and queuing operations. However, when the black hole attacks are generated, our solution shows a better behavior than the standard version.

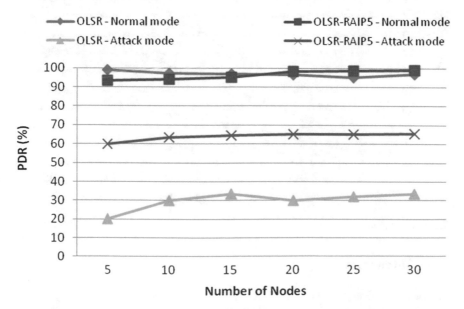

**Fig. 2** PDR

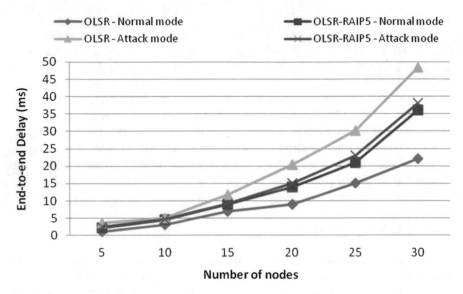

**Fig. 3** End-to-end delay

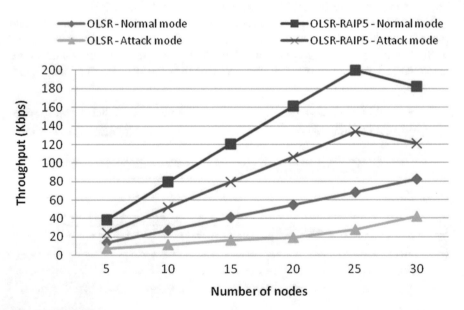

**Fig. 4** Throughput

### 5.3.3 Throughput

As shown in Fig. 4, the OLSR-RAIP5 shows a better behavior regard the throughput. In the normal mode (without attack), the throughputs of the OLSR vary between 14 and 82 Kbps, while those of the OLSR-RAIP5 vary between 38 and 182 Kbps. In the attack mode, the values of the OLSR dropped with an average of 128 %, while the throughputs of our protocol decreased by 51 %. Multiple path routing over OLSR-RAIP5 gives more bandwidth, three times bandwidth than the OLSR standard protocol.

# 6 Conclusion

This work aims to improve performance and security of routing protocols in MANETs by adopting a multi-path algorithm combined with the RAID5 technique that allows performance and reliability. We have developed and implemented an algorithm in the context of the OLSR standard protocol leading to a new multi-path routing called OLSR-RAIP5. This new approach is based on two main steps: 1— Find disjoint paths; 2—Segment the packet into multiple messages before transmitting them simultaneously; 3—rebuild the data packet at reception.

To evaluate performance and robustness of our solution, we have conducted some simulations on NS2 operating in normal and attack scenarios. The obtained results show that our solution provides globally good performances, security and reliability in insecure MANETs. We plan to focus on managing the queue to further reduce transmission delays of the OLSR-RAIP5 and also study the impact of some important mobility models on several performance metrics.

# References

1. Abusalah, L., Khokhar, A., Guizani, M.: A survey of secure mobile Ad Hoc routing protocols. Commun. Surv. Tutorials IEEE **10**, 78–93 (2008)
2. Uemori, T., Kohno, E., Kakuda, Y.: A node-disjoint multipath scheme for secure dispersed data transfer in ad hoc networks. In: First International Symposium on Computing and Networking (CANDAR), 2013, pp. 441–447 (2013)
3. Hiroaki, Y., Eitaro, K., Yoshiaki, K.: Tree structured group id-based routing method for mobile ad hoc networks. In: The thirteenth Conference on Networks (ICN 2014), pp. 33–37 (2014)
4. Joshi, R.D., Rege, P.P.: Implementation and analytical modelling of modified optimised link state routing protocol for network lifetime improvement. Commun. IET **6**, 1270–1277 (2012)
5. Phu Hung, L., Pujolle, G., Thi-Mai-Trang, N.: An interference-aware multi-path routing protocol for mobile ad hoc network. In: IEEE International Conference on Networking, Sensing and Control (ICNSC), 2011, pp. 503–507 (2011)
6. Adami, D., Callegari, C., Giordano, S., Pagano, M.: Single-path and multi-path label switched path allocation algorithms with quality-of-service constraints: performance analysis and implementation in NS2. Commun. IET **6**, 398–407 (2012)
7. Cervera, G., Barbeau, M., Garcia-Alfaro, J., Kranakis, E.: A multipath routing strategy to prevent flooding disruption attacks in link state routing protocols for MANETs. J. Netw. Comput. Appl. **36**, 744–755 (2013)

8. Phu Hung, L.: A Performance evaluation of multi-path routing protocols for mobile ad hoc networks. In: IEEE 15th International Conference on Computational Science and Engineering (CSE), 2012, pp. 484–491 (2012)
9. Patterson, D.A., Gibson, G., Katz, R.H.: A case for redundant arrays of inexpensive disks (RAID). SIGMOD Rec. **17**, 109–116 (1988)
10. Thomasian, A., Menon, J.: RAID5 performance with distributed sparing. In: IEEE Transactions on Parallel and Distributed Systems, vol. 8, pp. 640–657 (1997)

# A New Cluster-Based Paradigm for SIP Routing in MANET

Aïcha Aït Hacha, Slimane Bah and Zohra Bakkoury

**Abstract** We propose a new paradigm based on a time varying grid to handle SIP routing over MANET. This paradigm takes advantage of the geometric properties of the square grid in order to establish calls using the closest route to the callee in terms of physical distance. Normally, shorter paths and optimized latencies are reached by broadcast based approaches where bindings of the whole network are stored in every single node and where SIP REGISTER messages are flooded. Our approach utilizes a grid based clustering method to arrange MANET nodes in an organized structure during the initialization phase in order to mark the centroid of each cluster. Then, the structure is updated periodically using the initial coordinates of the centroids. We resort also in this work to the use of replication in an astute manner in order to sidestep to the need of flooding based solutions.

**Keywords** SIP routing · SIP registration · MANET · Replication · Square grid · P2P SIP · Overlay network · Localization technique · Landmarking

## 1 Introduction

MANET (Mobile Ad Hoc Network) has attracted widespread attention in the scientific community since the mid-'90s in furtherance of the ubiquitous computing concept, driven by the ease of its deployment and administration as compared to the conventional networks. It can be defined as a set of mobile nodes sharing temporarily the same wireless medium and having to collaborate together driven by the

A.A. Hacha (✉) · S. Bah · Z. Bakkoury
AMIPS Research Group, Ecole Mohammedia d'Ingénieurs (EMI),
University Mohammed V, Rabat, Morocco
e-mail: a.aithacha@gmail.com; aicha.aitha@gmail.com

S. Bah
e-mail: slimane.bah@emi.ac.ma

Z. Bakkoury
e-mail: Bakkoury@emi.ac.ma

© Springer International Publishing Switzerland 2016       211
A. El Oualkadi et al. (eds.), *Proceedings of the Mediterranean Conference on Information & Communication Technologies 2015*, Lecture Notes in Electrical Engineering 381, DOI 10.1007/978-3-319-30298-0_22

absence of a fixed infrastructure. Consequently, nodes have to perform the whole network functionalities such as routing, services, QoS and security tasks in a collaborative way. These idiosyncrasies make this kind of network very suitable for emergency battlefield communications and civilian applications such as emergency services and disaster recovery. However, when it comes to MANETs, we face a bunch of constraints imposed by the very nature of it. Hence, it's inconceivable to support existing protocols over it in a traditional way without altering the components of the network or the protocol itself. SIP (Session Initiation Protocol) does not make an exception to this rule. It is incontestably the preeminent signaling protocol for multimedia session establishment and control. It is used to provide services such as basic voice, presence, video conference, messaging and so on… As a matter of fact, handling SIP multimedia services over Mobile Ad hoc Networks can be incontestably exploited to afford new business opportunities. This brings us to conceive a new paradigm to handle SIP over MANET. The authors of [1] conducted an evaluation of SIP signaling over both IPv4 and IPv6 running in MANET, using different mobility models. They found poor performances and low efficiency of VoIP application. According to this study, several factors are impacting these performances such as the routing protocol, TCP versions, servers' capacities, the SIP server mobility, the SIP retransmission timers and so on… Many other studies demonstrated through simulations that traditional SIP framework is useless in MANET. Thus, several approaches and frameworks have been proposed by researchers endeavoring to handle SIP registration and user discovery in MANET in the face of the lack of a fixed infrastructure in such environment. The proposed approaches can be dichotomized in two categories: SIP without overlay network and P2P SIP on overlay network. The first category can also be divided into broadcast based and cluster based approaches. Broadcast based approach considers every node as a proxy/registrar server and relies on the flooding technique to register SIP users and to set up SIP calls, which leads inevitably to the introduction of a high overhead. In the cluster based approach, the SIP network is organized into clusters in which only cluster-heads act as proxy/registrar servers. SIP over MANET on overlay network can rely either on multicast routing or DHT table usage.

The present work proposes a new cluster-based paradigm to handle SIP over MANET independently from the underlying routing protocol and sidestepping the need to flood the network or to stock registration bindings of the whole network in every single device. Compared to existing solutions, our approach has also the advantage to take into consideration the physical proximity in SIP routing. This paper is organized as follows: in the first section, we endeavor to examine with a critical eye this variety of existing solutions. Then, we present the proposed paradigm for SIP routing in MANET while pointing out its advantages compared to the existing solutions. Finally, we discuss our future directions.

## 2  A Critical Overview of Related Work

The paper [2] intends to use a limited set of SIP functionalities in a distributed way. Every single device contains a user agent module and a server module handling proxy and registrar servers' functionalities. The paper divided SIP operations over MANET into two key operations: user discovery and SIP session initiation and management. It distinguishes between two methods to discover SIP users' binding which are: fully distributed SIP (dSIP) and SIP with Service Location framework (sSIP). dSIP registration is performed by broadcasting SIP REGISTER message, then the binding of SIP URI and IP address of that specific node is stored in every node receiving that REGISTER message, with a limited validity time. Each node is considered as an independent Registrar which prevents the single point of failure problem. Registration here relies on broadcast which obviously induces overhead and scalability problems. Moreover, the storage capacity constraint is not taken into account in this network; sometimes, it is infeasible to store binding of the whole network in every single device. Latter works overcome the aforementioned short-coming by DHT usage. To set up a call in dSIP, the calling user agent issues a SIP INVITE message and forwards it to the local server module which looks up for specified SIP user's binding. Then, it relays the INVITE to the corresponding IP address found from the binding. The second method mentioned in this paper named sSIP, makes use of the Service Location Protocol (SLP) which bypasses the need of registration, being replaced by SLP. The SLP request is sent containing the SIP URI as attribute filter and then all nodes in MANET receives this request but only the one that matches the SIP URI, returns its own IP address. Details provided in this work about sSIP are scare: we don't know when exactly the node should issue its SLP request. If it is issued when establishing the call, then the authors should ponder about having an increased call establishment time.

In [3], the authors proposed another middleware to set up and manage SIP sessions in MANET through the use of SLP. For the sake of efficiency, SIPHoc piggybacks SLP messages to routing messages; regardless of this latter's type being reactive or proactive. This piggybacking implies that service lookup and route discovery occur simultaneously and it reduces the number of packet transmissions. If the underlying routing protocol is reactive, service discovery occurs during the call set up. If it is proactive, the service discovery is done during registration. Each SIPHoc device running in an independent MANET (i.e. with no connection to Internet) is composed of two separate components: A SIPHoc Proxy and a MANET SIP layer. The first component acts as an outbound SIP proxy and a Registrar for local SIP applications and the second components is the one responsible of piggybacking service information onto routing messages, it also advertises the location over the network. For SIPHoc Registration, each user agent sends its binding to its local SIPHoc proxy contained in the same physical device. Then, this local SIPHoc proxy stores this binding as an entry of the local Location Service table. After, the SIPHoc proxy contacts the MANET SLP layer of the same device which registers an entry containing the address of the proxy responsible of that specific user agent. This entry is next advertised in the network by the MANET SLP. For SIPHoc call set up, the caller sends the INVITE to its SIPHoc proxy which looks for the callee SIP URI in its local Location Service table. If it does not find it, it looks for the callee proxy in the MANET SLP entry. Then, the INVITE message is forwarded to that proxy which checks if it has the required URI in its local Location Service table. According to this approach, each node is aware of most of other nodes' bindings in the network which leads to the problem of scalability and storage constraint.

It is worth to mention that both of overlay network and MANET shared fundamental properties such as decentralization, dynamic topology and self-organization. Accordingly, we believe that the overlay approach is the most suitable manner to handle SIP over MANET. That's why we thoroughly examine Thirapon Ph.D. thesis [3] which entirely focuses on the use of the overlay approach to handle SIP over MANET. The thesis proposed SIPMON, a P2P distributed framework to offer SIP user discovery on MANET. This framework utilizes a cross layer design based on a DHT which is continuously updated by OLSR routing protocol. By DHT usage, SIPMON provides a constant time for lookup which subsequently fosters a fast call set up time according to the simulations. SIPMON divided each P2P node into three modules: OLSR, SMON and P2P SIP. OLSR is the underlying routing protocol, which informs continuously SMON of any topology change. SMON (Structured Mesh Overlay Network) is an algorithm to form and maintain the overlay network. And, P2P SIP layer handles SIP requests and responses over the P2P SIP overlay network and uses DHT table via SMON API.

The framework ensures two key operations in SIP which are registration and call set up. For registration, the local address of the SMON node is configured as the node's outbound proxy, so the local P2P SIP intercepts the SIP REGISTER message. Then, it hashes the SIP_URI to obtain the Object_ID which is subsequently used by its DHT table to find the corresponding node_ID. Afterwhile, the P2P SIP forwards the SIP REGISTER message to this node, considering it as the registrar. The call set up process just like registration, relies on the same principle. P2P SIP module of the caller part extracts the target SIP_URI from the SIP INVITE message. Then, it hashes it to obtain the Object_ID. Next, it uses its DHT to look up for the node_ID which corresponds to the calling registrar. Finally, the caller forwards the INVITE message to this registrar to locate the address of the calling part. Compared to previous works, this framework reduces the overhead caused by broadcast messages. It also provides backward compatibility with OLSR standard because it uses a separate range of OLSR messages instead of altering existing messages like HELLO or TC.

The author highlights other advantages of his work such as the decrease of post dialing delay and the increase of call set up success ratio in comparison with previous works such as dSIP and CQSA. Nevertheless, some major points are missing in this work and must be reviewed in all their delicacy:

1. the author has merely described P2P lookup protocols such as Chord, Pastry, Tapestry an CAN without specifying which one is the most useful to create and maintain DHT in MANET. Below are listed some of the reasons:

   - Overlay forwarding routes differ from the underlying ad hoc routes. Thus, the chosen registrar through this mechanism is not necessarily the most appropriate in terms of proximity;
   - In case of concurrent joins and leaves, these protocols required a large convergence time to reach the consistency state.

2. Frequent Registration caused by mobility can overwhelm the network;
3. The DHT table is unstable during routing convergence. Thus, calls which are performed during routing convergence will be unsuccessful;
4. SIPMON assumes that each node maintains the list of all the other peers. Thus, the notion of overlay network (i.e. requiring the knowledge of only a subset of nodes) is not respected anymore in MANET,
5. The paper merely describes the join and leave operation. Although, concurrent joins and leaves must be taken into consideration.

## 3   The Proposed SIP Routing Paradigm

### 3.1   The Concept

In a very natural way, our approach views the MANET network as a cubic grid and utilizes a grid based clustering method to arrange nodes and to afford an optimized route to the destination, taking advantage of the geometric properties of the grid. Each cube represents a cluster, in which is elected a cluster-head lying in the center of the cube. All nodes belonging to a specific cluster register themselves in the cluster-head of their home-cluster. Their registrations (bindings between IP & SIP URI) are replicated in every cluster-head located at their home-cluster-head's horizontal and vertical. Accordingly, any caller willing to set up a call is able to locate quickly and directly the callee IP address by querying only the cluster-heads located at its diagonal. The structure of the MANET network is shown in Fig. 1.

In the following, we describe the SIP URI lookup process and demonstrate geometrically that our approach provides the closest route in terms of Euclidean distance. Note that this distance can be used only in presence of LoS (Line of Sight).

For that, let N be a node located in a cluster X. As stated above, N is registered in the cluster-head of the cluster X, denoted as O, as well in every cluster-head located at the horizontal and the vertical of the cluster-head O. N wants to establish a SIP session with another node M but it knows solely M's SIP URI and not its location. According to our approach, SIP URI location is performed as follows:

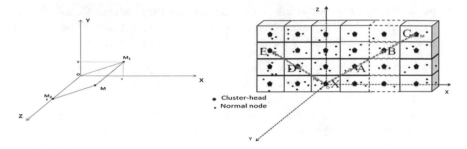

**Fig. 1** Nodes distribution over the cubic grid structure

1. Initially, the node N consults the cluster-head O to find the binding between SIP URI and IP address of the callee M. According to aforementioned replication method, if N doesn't find M's binding in the cluster-head O, it means that M doesn't belong to any cluster located at the horizontal and the vertical of the cluster X;
2. Then, O sends INVITE messages to the cluster-heads located at its diagonals (initially to A and D). If the said binding is not found in these diagonals, it means that it doesn't exist either in the clusters located at the horizontal and the vertical of these diagonals. This process is repeated until that a SIP INVITE message reaches M.

## 3.2 Assumptions

Some conditions must be fulfilled to build and maintains our paradigm based on the aforementioned techniques:

- All nodes are equipped with smart antennas;
- LoS (Line of Sight) is required between nodes that localize each other in order to avoid major error. It also permits the use of Euclidean distance;
- The physical network has the shape of a convex polygon so that the square-grid structure can be applied. This assumption is not very restrictive since the network can be seen as a set of convex regions;
- The network surface is assumed to be fixed during the lifetime of the network;
- Network is dense in such a way that $D \pm \varepsilon$ is the maximum distance between two adjacent nodes where D is the aforementioned maximum square size and $\varepsilon$ is a margin of tolerance.

## 3.3 Preliminaries

### 3.3.1 Localization Techniques in MANET

In the literature, we find a handful of coarse and fine localization techniques that are applicable in MANET. Fine techniques use TOA (Time of arrival) and/or DOA (Direction of Arrival) estimation; these techniques may be also combined to GPS signal when it's available. We did not consider GPS based method given that they are not accurate enough for small and moderate scale networks. We choose the Semi distributed Multimode TOA-DOA Fusion Localization Technique [4] as it distributed the processing power across multiple nodes. In this technique, base-nodes (i.e. nodes equipped with smart antennas) transmit periodically a signal to target nodes that respond by issuing a signal back which is used by based-node to calculate TOA & DOA which serve to calculate the range and the angle in polar coordinates. This technique requires a careful selection of a reference base-node whose local coordinates will serve as the main coordinates in which other nodes are localized. Accordingly, all nonreference base-nodes need to localize themselves with respect to the reference base-node. The localization scheme follows the steps below:

- The reference base node selection: it can be made based on some suboptimal algorithm which measures distances from base nodes to target nodes and find the appropriate base node that minimize the distance square summation;
- Nonreference base nodes localization: each nonreference base node has to find its position with respect to the reference base node. This position is broadcasted, so that base nodes know each other position;
- Target nodes localization: Given that each base node is in charge of a set of target nodes and that a target node could be localized by multiple base nodes, these localizations are fused in the responsible base node.

### 3.3.2 The Clustering Method

Previous research has already made use of the square-based infrastructure in MANET as a virtual infrastructure in order to improve energy efficiency [5], to conserve bandwidth and to provide location services by minimizing the need of flooding and avoiding transmission of location updates in all directions. The maximum square size 'D' can be determined by the forwarding method that is used. Given that R is the physical transmission range, $D = R/\sqrt{8}$ if packets can be transmitted to diagonal neighbors [5].

### 3.3.3 Replication

All lookup protocols resort to some replication strategies in order to ensure availability and fault tolerance. Replication can also reduce lookup latency. Our work also makes use of replication technique in an astute way to avoid the need of flooding and to reach the SIP destination through an optimal route.

## 3.4 Initialization Phase

The aim of this initialization phase is to identify network boundaries, square clusters and the Cartesian coordinates of each cluster centroid. In other words, this phase is a kind of landmarking, which is achieved through the following steps:

(a) Perform the localization technique:

1. Initially, we choose a random set of nodes as base nodes;
2. By the use of the Semidistributed Multimode TOA-DOA Fusion Localization Technique:

   - The reference base node is likely to be at the centroid of the network;
   - Each nonreference base-node knows all other nonreference base nodes' positions and target nodes' positions of which it is in charge. The message carrying the position information piggybacks also the node's SIP URI and IP address;
   - Each Nonreference bases node transmits its targets' nodes information to the reference base nodes.

(b) Square-grid Construction: Based on received information in the first step, the reference base node has to construct or approximate a regular square-grid: Suppose that there are "n" nodes in the network ($M_1$, $M_2$, ..., $M_n$) and each node $M_i$ has the Cartesian coordinates ($x_i$, $y_i$) which are all known by the reference base node according to the previous step. The square-grid is constructed following the steps below:

1. The reference base node calculates each $\alpha_i = y_i / x_i$;
2. Then, it applies a sorting algorithm such as Quicksort to the list ($\alpha_1$, $\alpha_2$,..., $\alpha_n$);
3. To cluster nodes belonging to the same straight line and passing through the reference base node, we calculate $d = | \alpha_i - \alpha_i + 1|$ based on the output of the step (2–(b)), if $d < \varepsilon$ then $M_i$ and $M_i + 1$ belong to the same straight line. If not, it is another straight line.
4. We choose the straight line that contains the maximum nodes:

$$Max(count(arg(|\alpha_i - \alpha_i + 1| < \varepsilon))).$$

5. This straight line is considered as the square grid generator since we can construct from it a square grid whose centroid is the reference base node and where the consecutive nonreference base nodes are spaced by $D \pm \varepsilon$.

(c) Broadcast of current structure to all nodes: The determined square structure is then broadcasted to all nodes, specifying the reference base node and nonreference base-nodes' coordinates.

## 3.5 Square-Grid Update

As stated above, the initialization phase allows nodes to identify the square grid boundaries and *centroids' coordinates* which will remain *fix* during the network lifetime. According to the aforementioned localization technique, the reference base node is periodically updated and is likely to be the centroid of the network. The node having the closest coordinates to a centroid coordinates is elected as the cluster-head of the cluster.

## 4 Conclusion and Future Directions

To handle SIP over MANET, this chapter has introduced a new paradigm with locality awareness. On the basis of some techniques such as the use of smart antennas, square-grid clustering, replication and landmarking, it constructs, maintains a physical square-grid and exploits geometric properties of this grid to afford an optimized route to the destination. Through our future work, we will alleviate the constraints imposed by our initial assumptions and try to maintain dynamically the square grid using a time-varying mobility model to predict the movement of the whole grid.

## References

1. Alshamrani, M., Cruickshank, H., Sun, Z., Fami, V., Elmasri, B., Danish, E.: Signaling performance for SIP over IPv6 Mobile Ad-Hoc network (MANET). In: Proceedings of the 2013 IEEE International Symposium on Multimedia; 12/2013, pp. 231–236
2. Leggio, S., Manner, J., Hulkkonen,A., Raatikainen, K.: Session initiation protocol deployement in Ad-Hoc networks: a decentralized approach. In: Proceedings of the International Workshop on Wireless Ad-Hoc Networks (IWWAN 2005), London, UK (May 23–26, 2005)
3. Thirapon, W.: P2P sip over mobile ad hoc networks. Ph.D. thesis, Telecom & Management SudParis. France, (October 2010)

4. Wang, Z., Zekavat, S.: Comparison of semidistributed multinode TOA-DOA fusion localization and GPS-Aided TOA (DOA) fusion localization for MANETs. EURASIP J. Adv. Signal Process. **439523** (October 2008)
5. Wang, Z., Zhang, J.: Grid based two transmission range strategy for MANETs. In: Proceedings of the 14th International Conference on Computer Communications and Networks (ICCCN2005), San Diego, CA, pp. 235–240, (October 2005)
6. Aithacha, A., Bah, S., Bakkoury, Z.: A new authentication scheme for SIP registration in a MANET environment. IJCSA **12**, 134–147 (2015)

# Particle Swarm Optimization Compared to Ant Colony Optimization for Routing in Wireless Sensor Networks

Asmae EL Ghazi and Belaïd Ahiod

**Abstract** Wireless Sensor Networks (WSNs) are an emerging technology that used to monitor various environments. Despite of WSN advantages, it suffers from intrinsic limitations related to communication failures, computational weaknesses and limited energy. Hence, many challenges are considered as NP-hard optimization problems, and resolved by metaheuristics. This paper, proposes a routing approach based on Particle Swarm Optimization (PSO). Compared to the ACO approach, PSO reduces the energy consumption and extends the life of WSN. Through performing many experimentations the PSO efficiency is validated.

**Keywords** Wireless sensor network · Metaheuristic · Routing · Ant colony optimization · Particle swarm optimization

## 1 Introduction

Wireless Sensor Network (WSN) is a new technology receiving increased interest. It's composed of sensors working in uncontrolled areas [1]. There are different kinds of sensor node according to physical parameters (temperature, humidity, pressure, …) [2]. Thus, WSN is used for many applications such as disaster relief, environmental control, precision agriculture, medicine and health care [3]. Nonetheless there are some intrinsic limitations for sensors like low processing capacity and low power [4]. Hence, new issues appear in operations research and optimization field [5, 6]. Rather than all problems many researches have tended to focus on routing problems. Routing in WSN is very challenging, as it has more different characteristics than that in traditional communication networks [7]. It's qualified as an NP-hard optimization problem [5]. That means we need robust and efficient techniques to solve this kind of problems, such as metaheuristics [8].

A.E. Ghazi (✉) · B. Ahiod
Faculty of Science, Department of Physics LRIT Associated Unit to CNRST (URAC 29), Mohammed V University, Rabat, Morocco
e-mail: as.elghazi@gmail.com

© Springer International Publishing Switzerland 2016
A. El Oualkadi et al. (eds.), *Proceedings of the Mediterranean Conference on Information & Communication Technologies 2015*, Lecture Notes in Electrical Engineering 381, DOI 10.1007/978-3-319-30298-0_23

Many metaheuristics, such as Genetic Algorithms (GA) [9], Artificial Bee Colony (ABC) [10], Particle Swarm Optimization (PSO) [11] and Ant Colony Optimization (ACO) [12] are used to solve routing problems [13]. The ACO metaheuristic has been successfully applied to solve routing problem in WSN [12, 14, 15]. Recently, also PSO becomes more used due to its performance.

We propose a new approach based on PSO for routing problem in WSN. The proposed approach gives better results compared to our advanced ACO [15].

The remaining of this paper is organized as follows: Sect. 2 gives the WSN routing problem and ACO. Section 3 introduces Particle Swarm Optimization (PSO). Section 4 presents our PSO-based algorithm for the routing problem. Section 5 shows the performance evaluation of our results. Finally, Sect. 6 concludes our work.

# 2 Ant Colony Optimization and Routing in Wireless Sensor Networks

Inspired from Ant behavior, M. Dorigo and G. Di Caro have developed in 1999 ant colony optimization algorithms [16]. Ants are insects that have a very high capacity to explore and exploit their environment despite their displacement way which is very limited (walking) compared to other species (flying). This moving inconvenience is offset by skills in manipulating and using environment. They use their environment as a medium of storage, processing and sharing information between all the ants in the colony. ACO basic steps are summarized in the Algorithm 1 [17].

---

**Algorithm 1** ACO

---

*Objective function* $f(x_{ij})$, $(i, j) \in \{1, 2, ..., n\}$
*Initialize the pheromone evaporation rate* $\rho$
**while** (criterion) **do**
   **for** *Loop over* $n$ *nodes* **do**
      *Generate the new solutions* (*using probabilistic rule* [15] )
      *Evaluate new solutions*
      *Mark the best routes with the pheromone* $\delta\tau_{ij}$
      *Update Pheromone* : $\tau_{ij} \leftarrow (1 - \rho)\tau_{ij} + \delta\tau_{ij}$
   **end for**
   *Daemon actions*
**end while**
*Output the best results and pheromone distribution.*

---

In our paper "Improved Ant Colony Optimization Routing Protocol for Wireless Sensor Networks" [15], we propose a routing protocol ranked among flat networks. The purpose is to find the optimal path, with minimal energy consumption and reliable links.

After detecting an event, source node splits data to $N$ parts. Each part is transmitted by an ant. Ant used the probabilistic decisions' rule, to travel between nodes from source until sink.

This approach gives good results, comparing to routing protocol EEABR and original ACO approach [14]. However, the novel optimization technique based on PSO, promises better performances.

## 3 Particle Swarm Optimization

Particle swarm optimization (PSO) [18] is originally attributed to Kennedy, Eberhart [19]. It is a computational method that optimizes a problem by iteratively trying to improve a candidate solution with regard to a given measure of quality. PSO optimizes a problem by having a population of candidate solutions, here dubbed particles, and moving these particles around in the search-space according to simple mathematical formulae over the particle's position and velocity (Eq. 1).

$$v_{i+1} = \omega v_i + \eta_1 rand()(Pbesti - x_i) + \eta_2 rand()(Gbest - x_i) \tag{1}$$

$$x_{i+1} = x_i + v_{i+1} \tag{2}$$

where $v_i$ is the velocity of particle $i$. $x_i$ is current position of particle $i$. $Pbesti$ is the best position of particle $i$ and $Gbest$ is the best position of the group.

Each particle's movement is influenced by its local best known position and global best in search-space. This is expected to move the swarm toward the best solutions. The Algorithm 2 describes PSO procedure.

---

**Algorithm 2** PSO

---

A population of particles with random values positions and velocities from $D$ dimensions in the search space
**while** Termination condition not reached **do**
   **for** Each particle $i$ **do**
      Adapt velocity of the particle using Equation 1
      Update the position of the particle using Equation 2
      Evaluate the fitness $f(x_i)$
      **if** $f(x_i) < f(Pbesti)$
         $Pbesti \leftarrow x_i$
      **end if**
      **if** $f(x_i) < f(Gbesti)$
         $Gbesti \leftarrow x_i$
      **end if**
   **end for**
**end while**

---

# 4   Routing in WSN and PSO

In [15] we present an improved routing protocol based on ACO where we maximized WSN lifetime and minimized energy consumption of sensors. In this paper we use PSO strategy with the same settings as ACO and we find better results. To search an optimal path from source to destination node, PSO routing approach initialization is needed, as described in [20].

1. *Initialization*:
   Initial paths, are formed by employing two agents; forward and backward [20]. The forward agent moves to determine the next node on the path, creates and maintains the routing table. At the same time, it collects many informations about nodes in the path. Backward agent is created by destination node and moves back from destination to source node.
2. *Optimal path*:
   The initialization of particle swarm (paths) is achieved, it is expressed as $\{x_1, \ldots, x_i, \ldots, x_m\}$. Every path $x_i$ is a potential best routing path. Note that the best $x_i$ as *Pbesti* when $x_i$ modified. The best path in all paths is expressed as *Gbest*. Assume that $x_i$ with $k$ hops consists of a set of nodes $\{n_s, \ldots, n_k \ldots, n_{sink}\}$. In order to find the best path we implement the PSO algorithm. The formula (Eq. 1) is useless in this case. Therefore, as described in [20] some permutations between *Pbesti* and $x_i$, *Gbesti* and $x_i$ are performed. For a credible comparison, the evaluation of particle is insured by the same metrics used for ACO approach [15].

# 5   Results

Performing many simulations of both approaches PSO and ACO, using Matlab and same experimentation conditions, we used a model of sensors based on "First Order Radio Model" of Heinzelman et al. [21]. To send and receive a message, power requirements are formulated as follows (Eqs. 3 and 4):

To send $k$ bits to a remote receiver by $d$ meters, transmitter consumes:

$$E_{Tx}(k, d) = (E_{elec} \times k) + (\epsilon_{amp} \times k \times d^2) \tag{3}$$

To receive $k$ bits, receiver consumes:

$$E_{Rx}(k) = E_{elec} \times k \tag{4}$$

where $E_{elec} = 50\,\text{nJ/bit}$ and $\epsilon_{amp} = 100\,\text{nJ/bit/m}^2$ are respectively energy of electronic transmission and amplification.

**Fig. 1** Simulation results for different WSNs. **a** 10 nodes, density is $225 \times 10^{-6}$ nodes/m², **b** 30 nodes, density is $187 : 5 \times 10^{-6}$ nodes/m², **c** 50 nodes, density is $222 \times 10^{-6}$ nodes/m², **d** 70 nodes, density is $222 \times 10^{-6}$ nodes/m²

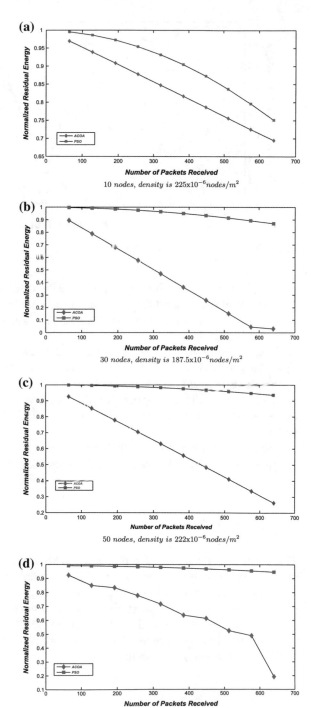

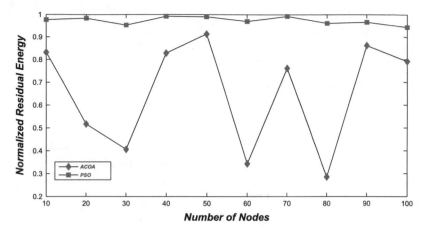

**Fig. 2** Residual energy for different WSNs

We prove that the PSO approach is better than the ACO one. By random deployment of different number of nodes in spaces, we simulate data transmission from a source node to known base station. The best approach appears in results presented in Fig. 1.

Figure 1 shows that PSO is higher than ACO by considering the residual energy. Thus, the power consumption is minimized, and the WSN lifetime is maximized.

In order to confirm the efficiency of our proposal, we simulated the transmission of 256 packets in different coverage areas where we deployed randomly a number of nodes (from 10 to 100 nodes). The shown results in Fig. 2 represent the residual energy.

The found results prove the efficiency of PSO for routing problems, in term of energy consumption and wireless sensors network lifetime.

## 6  Conclusion

This paper presents a routing protocol for WSN achieved by using particle swarm optimization, to optimize the node power consumption and increase network lifetime, while data transmission is attained efficiently. The evaluation of the protocol performance, is implemented in the same conditions. The conclusion is the performance of PSO is better than ACO approach [15], in terms of energy consumption and network lifetime. The future work could be investigate improvement of PSO and propose other comparisons.

# References

1. Potdar, V., Sharif, A., Chang, E.: Wireless sensor networks: A survey. In: International Conference on Advanced Information Networking and Applications Workshops. WAINA'09, IEEE, pp. 636–641 (2009)
2. Akyildiz, I.F., Su, W., Sankarasubramaniam, Y., Cayirci, E.: Wireless sensor networks: a survey. Comput. Netw. **38**(4), 393–422 (2002)
3. Xu, N.: A survey of sensor network applications. IEEE Commun. Mag. **40**, 102–114 (2002)
4. Masri, W.: QoS requirements mapping in TDMA-based wireless sensor networks. Ph.D. thesis, Toulouse University III-Paul Sabatier (2009)
5. Gogu, A., Nace, D., Dilo, A., Mertnia, N.: Optimization problems in wireless sensor networks. In: International Conference on Complex, Intelligent and Software Intensive Systems (CISIS), IEEE, pp. 302–309 (2011)
6. Ali, M.K.M., Kamoun, F.: Neural networks for shortest path computation and routing in computer networks. IEEE Trans. Neural Netw. **4**, 941–954 (1993)
7. Al-Karaki, J.N., Kamal, A.E.: Routing techniques in wireless sensor networks: a survey. Wireless Commun. **11**(6), 6–28 (2004)
8. Blum, C., Roli, A.: Metaheuristics in combinatorial optimization: overview and conceptual comparison. ACM Comput. Surv. (CSUR) **35**, 268–308 (2003)
9. Hussain, S., Matin, A.W., Islam, O.: Genetic algorithm for energy efficient clusters in wireless sensor networks. In: ITNG, pp. 147–154 (2007)
10. Saleh, S., Ahmed, M., Ali, B.M., Rasid, M.F.A., Ismail, A.: A survey on energy awareness mechanisms in routing protocols for wireless sensor networks using optimization methods. In: Transactions on Emerging Telecommunications Technologies (2013)
11. Kulkarni, R.V., Venayagamoorthy, G.K.: Particle swarm optimization in wireless-sensor networks: a brief survey. Syst. Man. Cybernet. Part C: Appl. Rev. IEEE Trans. **41**(2), 262–267 (2011)
12. Fathima, K., Sindhanaiselvan, K.: Ant colony optimization based routing in wireless sensor networks. Int. J. Adv. Netw. Appl. **4**(4) (2013)
13. Iyengar, S.S., Wu, H.C., Balakrishnan, N., Chang, S.Y.: Biologically inspired cooperative routing for wireless mobile sensor networks. Syst. J. IEEE **1**(1), 29–37 (2007)
14. Okdem, S., Karaboga, D.: Routing in wireless sensor networks using an ant colony optimization (aco) router chip. Sensors **9**, 909–921 (2009)
15. El Ghazi, A., Ahiod, B., Ouaarab, A.: Improved ant colony optimization routing protocol for wireless sensor networks. In: Networked Systems, pp. 246–256. Springer (2014)
16. Dorigo, M., Di Caro, G.: Ant colony optimization: a new metaheuristic. In: Proceedings of the 1999 Congress on Evolutionary Computation CEC 99, IEEE (1999), pp. 1–8
17. Yang, X.S.: Engineering Optimization: An Introduction with Metaheuristic Applications. Wiley (2010)
18. Tillett, J., Rao, R., Sahin, F.: Cluster-head identification in ad hoc sensor networks using particle swarm optimization. In: IEEE International Conference on Personal Wireless Communications, 2002, IEEE (2002), pp. 201–205
19. Eberhart, R.C., Kennedy, J.: A new optimizer using particle swarm theory. In: Proceedings of the Sixth International Symposium on Micro Machine and Human Science, vol. 1, New York, NY, pp. 39–43 (1995)
20. Zhang, X.H., Xu, W.B.: Qos based routing in wireless sensor network with particle swarm optimization. In: Agent Computing and Multi-Agent Systems, pp. 602–607. Springer (2006)
21. Heinzelman, W.R., Chandrakasan, A., Balakrishnan, H.: Energy-efficient communication protocol for wireless microsensor networks. In: Proceedings of the 33rd Annual Hawaii International Conference on System Sciences, IEEE, pp. 1–10 (2000)

# An Energy Efficient Cooperative MIMO Routing Protocol for Cluster Based WSNs

Alami Chaibrassou and Ahmed Mouhsen

**Abstract** Energy efficiency and quality of service are foremost concerns in Wireless Sensor Networks (WSNs). Among the methods used to achieve these requirements there is Virtual multiple input multiple output (MIMO) technique, where sensor nodes cooperate with each other to form an antenna array. These multiple antennas can be used to improve the performance of the system (lifetime, data rate, bit error rate…) through spatial diversity or spatial multiplexing [1]. In this paper, we propose a distributed energy efficient cooperative MIMO routing protocol for cluster based WSNs (EECMIMO) which aims at reducing energy consumption in multi-hop WSNs. In EECMIMO, sensor nodes are organized into clusters and each cluster head utilizes a weighted link function to select some optimal cooperative nodes to forward or receive traffic from other neighboring clusters by utilizing a cooperative MIMO technique. Simulation results indicate that virtual MIMO based routing scheme achieves a significant reduction in energy consumption, compared to SISO one for larger distances.

**Keywords** Wireless sensor networks (WSNs) · Cooperative multiple input multiple output (MIMO) · Cooperative communication · Clustering algorithms

## 1 Introduction

The current progress in the field of wireless technology allows to develop small size sensors called nodes, communicating with each other via a radio link and are characterized by their limited resources (energy supply, limited processing, memory

A. Chaibrassou (✉) · A. Mouhsen
Faculty of Science and Technology, Research Laboratory in Engineering,
Industrial Management and Innovation, University Hassan I, Settat, Morocco
e-mail: alami70@yahoo.fr

A. Mouhsen
e-mail: mouhsen.ahmed@gmail.com

© Springer International Publishing Switzerland 2016
A. El Oualkadi et al. (eds.), *Proceedings of the Mediterranean Conference on Information & Communication Technologies 2015*, Lecture Notes in Electrical Engineering 381, DOI 10.1007/978-3-319-30298-0_24

storage...). Their flexibility of use makes them more utilized to form a wireless sensor network (WSN) without returning to a fixed infrastructure. Furthermore, these nodes typically deployed in inaccessible areas to control a definite phenomenon, ensure the transfer of data collected using a multi-hop routing to a BS, which is far from the monitored field.

Based on the above observations, for a WSN to accomplish its function without failure of the connection between the nodes, because of a depleting battery of one or more nodes, we need a data routing protocol which gives higher priority to the energy factor compared to other limitations, in order to provide stability of the network. At this point, many studies have been made; the most known ones are based on the MIMO technologies. However, the node cannot carry multiple antennas at the same time due to its limited physical size. Therefore, a new transmission technique called "Cooperative MIMO" has been proposed [2]. This technique is based on the cooperation principle where the existence of different nodes in the network is exploited to transmit the information from the source to a specific destination by virtually using the MIMO system [3]. The Cooperative MIMO allows to obtain the space-time diversity gain [4], the reduction of energy consumption, and the enhancement of the system capacity [5].

In this paper, we would like to investigate cooperative virtual MIMO for cluster based WSNs, with the objective of maximizing the network lifetime. We first introduce a novel approach to grouping sensors into clusters and electing cooperative MIMO links among clusters on such that intra-cluster messages are transmitted over short-range SISO links, while inter-cluster messages are transmitted over long range cooperative MIMO links [6]. Each cluster head selects one or multiple cluster members using a weighted link function to form a MIMO array together with itself. To transmit a message to a neighboring cluster, the cluster head first broadcasts the message to other members in the MIMO array. The MIMO array then negotiates the transmission scheme with the MIMO array in the neighboring cluster, encodes and sends the message over the cooperative MIMO link between them. Theoretical analyses show that, we can achieve high energy efficiency by adapting data rate and transmission made (SISO, SIMO, MISO, MIMO) [7]. Simulation results have proved that the proposed scheme can prolong the sensor network lifetime greatly, especially when the sink is far from the sensor area.

The remainder of the paper is organized as follows: In Sect. 2, the Energy Efficiency of MIMO Systems is described. Section 3 describes the proposed protocol. Section 4 describes the simulation and results. Finally, Sect. 5 concludes the paper and provides directions for future work.

## 2 Energy Efficiency of MIMO Systems

### 2.1 System Model

In our protocol we use the system model proposed in [1], the total average energy consumption of MIMO transmission in WSNs includes two parts: the power consumption of all power amplifiers $P_{PA}$ and the power consumption of other circuit blocks $P_C$. The transmitted power is given by

$$P_{out} = \overline{E_b}R_b \cdot \frac{(4\pi)^2 d_{i,j}^{k_{i,j}}}{G_t G_r \lambda^2} M_l N_f.$$ (1)

where $\overline{E_b}$ is the required energy per bit at the receiver for a given BER requirement, $R_b$ is the bit rate, $d_{i,j}$ is distance between nodes i and j, $k_{i,j}$ is the path loss factor from node i to j, $G_t$ is the transmitter antenna gain, $G_r$ is the receiver antenna gain, $\lambda$ is the carrier wavelength, $M_l$ is the link margin and $N_f$ is the receiver noise figure given by $N_f = N_r/N_0$ where $N_0$ is the single-sided thermal noise power spectral density and Nr is the power spectral density of the total effective noise at the receiver input.

The power consumption of the power amplifiers is dependent on the transmitted power $P_{out}$ and can be approximated as

$$P_{PA} = (1 + \alpha)P_{out}.$$ (2)

where, $\alpha = \left(\frac{\xi}{\eta} - 1\right)$ with $\eta$ the drain efficiency of the RF power amplifier and $\xi$ the peak-to-average ratio. The power consumption of the circuit components is given by

$$P_c = M_t\left(P_{DAC} + P_{Mix} + P_{filt} + P_{syn}\right) + M_r\left(P_{LNA} + P_{mix} + P_{IFA} + P_{filr} + P_{ADC} + P_{syn}\right).$$ (3)

$$Pc = M_t P_{ct} + M_r P_{cr}$$ (4)

where $P_{DAC}$, $P_{mix}$, $P_{LNA}$, $P_{IFA}$, $P_{filt}$, $P_{filr}$, $P_{ADC}$, $P_{syn}$, $M_t$ and $M_r$ are the power consumption values for the DAC, the mixer, the Low Noise Amplifier (LNA), the Intermediate Frequency Amplifier (IFA), the active filters at the transmitter side, the active filters at the receiver side, the ADC, the frequency synthesizer, transmitters number and receiver s number respectively. The total energy consumption per bit according to [1] is given by

$$E_{bt} = \frac{(P_{PA} + P_C)}{R_b}.$$ (5)

## 2.2  Variable-Rate Systems

Using MQAM modulation scheme, the constellation size b can be defined as
$b = \log_2 M$ Further, we can define constellation size in terms of number of bits L,
Bandwidth B, and duration radio transceiver is on $T_{on}$, and data rate $R_b$
(bits/second) [1].

$$b = \frac{L}{BT_{on}} = \frac{R_b}{B} \tag{6}$$

The total energy consumption for Variable-rate Systems per bit according to [1]
is given by Eq. (7), where $\overline{P_b}$ is the average bit error rate.

$$E_{bt} = \frac{2}{3}(1+\alpha)\left(\frac{\overline{P_b}}{4}\right)^{\frac{-1}{M_t}} \frac{2^b - 1}{b^{\frac{1}{M_t}+1}} M_t N_0 \frac{(4\pi)^2 d_{i,j}^{k_{i,j}}}{G_r G_t \lambda^2} M_l N_f + \frac{P_C}{Bb}. \tag{7}$$

Based on the Eq. (7) the optimal constellation sizes for different transmission
distances are listed in Table 1.

## 3  The Proposed Protocol

The Clustering algorithm consists of five steps:

**Step 1: 1-hop neighbor discovery**
Each node broadcasts a message including its residual energy (RE) to its 1-hop
neighbors once receiving a RE message from a neighbor, a node adds an entry to its
1-hop neighbor list including the neighbor's residual energy and the estimated distance.

**Step 2: 1-hop weight discovery**
The weight of each node is calculated and broadcast to 1-hop neighbor. Also as part
of optimizing the energy resources of a WSN, it will be better to affect the CH role
to a node with high residual energy and small average intra-cluster distance. In this
regard, the weight for cluster head selection at each node i can be defined by

$$\text{weight}(i) = \frac{E_i}{\frac{\sum_{j=1}^{N(i)} d_{i,j}}{N(i)}}. \tag{8}$$

**Table 1** Optimized
constellation size

| Distance (m) | $b_{SISO}$ | $b_{MISO}$ | $b_{MIMO}$ |
|---|---|---|---|
| 10 | 5 | 8 | 10 |
| 20 | 4 | 6 | 8 |
| 70 | 2 | 4 | 5 |
| 100 | 2 | 3 | 5 |

where, $N(i)$ is the 1-hop neighbors number of node i, $d_{i,j}$ denotes the distance between nodes i and j, and $E_i$ is the residual energy of node i.

**Step 3: cluster formation**

In this step, sensor nodes with the high weight in their 1-hop neighborhoods elect themselves as cluster heads. The cluster head election procedure is executed on each node as every node is aware of the weights of its 1-hop neighbors. Isolated nodes declare them self as cluster heads.

**Step 4: cluster neighbor discovery**

In this step, all cluster members send a Cluster Forward (CF) message to their cluster heads, in which the updated 1-hop neighbor list is included. After receiving all the CF messages from its cluster members, a cluster head knows all the neighboring clusters.

**Step 5: cooperative MIMO link selection**

In this step, each cluster head negotiates with the cluster heads of neighboring clusters to select the optimal cooperative MIMO links, as more than one such links may exist between two neighboring clusters. In general, on one hand, the cooperative MIMO link with high energy efficiency should be selected to save transmission energy; on the other hand, a link with low residual energy should not be selected even if it has high energy efficiency to avoid exhausting the link. We define Ef (l) as the energy efficiency of a cooperative MIMO link l, which is determined by Eq. (10). We use Ei (l) to represent the residual energy of link l, which is set to the least residual energy of all nodes involved. To balance the effect of both factors, an empirical influence factor $\beta$ ranging from 0 to 1 is introduced, which can be adjusted according to the type of application. Thus the weight of a cooperative MIMO link is defined as

$$\text{weight}(l) = \beta \, \text{Ef}(l) + (1 - \beta)\text{Ei}(l). \tag{9}$$

$$\text{Ef}(l) = \frac{1}{E_{bt}}. \tag{10}$$

# 4　Simulation Results

## 4.1　Simulation Environment

To illustrate the value added by our proposed EECMIMO algorithm on network behavior, we evaluated the EECMIMO performances in terms of energy consumption per bit, stability, lifetime and amount of data sent to the BS in three different scenarios (SISO, MISO and MIMO). The stability period and the lifetime are defined respectively according to the following metrics: FND (first node dies) and HND (half node dies). Simulation parameters used for these evaluations are

**Table 2** Simulation parameters

| Parameter | Value |
|---|---|
| $\sigma^2$ | N0/2 = −174dBm/Hz |
| k | 2–5 |
| Round number | 5000 |
| Nodes number (N) | 100 |
| Network area | 100 m × 100 m |
| Packet length ($N_B$) | 4000 bit |
| $G_t G_r$ | 5 dBi |
| fc | 2.5 GHz |
| B | 10 kHz |
| $N_f$ | 10 dB |
| $M_l$ | 40 dB |
| β | 0.5 |
| η | 0.35 |
| $\overline{P_b}$ | 10–3 |
| $P_{ct}$ | 0.0844263 W |
| $P_{cr}$ | 0.112497827 W |
| $N_0$ | −171 dBm/Hz |
| λ | 0.12 m |
| ξ | $3.\dfrac{\sqrt{M}-1}{\sqrt{M}+1}, M = 2^b$ |

listed in Table 2. Where, the base station is located at the center of the network and in order to illustrate the effect of distance on energy consumption the base station moves in the horizontal direction.

## 4.2    *Performance Evaluation Discussion*

Firstly, we compare the energy consumption of EECMIMO using different multi-hop transmission MIMO, MISO and SISO with variable data rate according to Table 1. Figure 1 shows the graphs of energy consumption per bit with respect to the distance from the base station. Initially, the base station is placed at the center of the network. Then, the base station is moved away from the center in the horizontal direction. As shown in Fig. 1, the energy consumption of SISO has more advantage in energy saving when the transmission distance is less than 10 m, but the MIMO has more advantage in energy-saving when the distance is more than 10 m, this is because, for small distances circuit block power consumption dominates and for large distances amplifiers power dominate Further, SISO is still better than MISO until the traversed distance equals 23 m, for distance exceeding this value, the

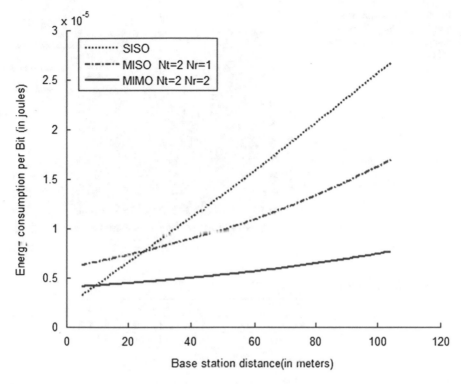

**Fig. 1** Energy consumption per bit for MIMO, MISO, SISO networks according to distance

standard deviation of SISO is also higher as compared to the other techniques. MIMO performs better than MISO in all the cases. Then the EECMIMO with MIMO technique and adapted bit rate is more energy efficient routing for large communication distance.

Secondly, we consider a fixed base station initially placed at position (50, 50) and we evaluate our proposed algorithm in terms of stability, lifetime and amount of data sent to the BS. Figure 2 shows the simulation results. From Fig. 2, we can see that $2 \times 2$ MIMO technique exceeds the other techniques in terms of stability, lifetime and amount of data sent to the BS. Furthermore, the $2 \times 2$ MIMO technique has a stability period (FND) considerably larger compared to other algorithms which allows the network to operate without fault for a very long time. Table 3 summarizes the simulation results of this scenario. From the simulation results, $2 \times 2$ MIMO is considered as an energy efficient routing technique. In fact, the stability period is increased approximately by 47, 14 and 12 % while the network lifetime is increased nearly by 11, 12 and 10 % compared with those obtained by SISO, MISO and $3 \times 3$ MIMO techniques respectively.

**Fig. 2** Distribution of alive nodes according to the number of rounds for each MIMO technique

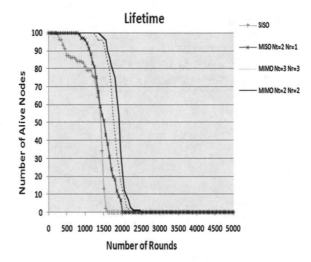

**Table 3** Simulation Results

| Protocol | FND | HND | Paquets number |
|----------|-----|-----|----------------|
| SISO | 282 | 1550 | $6.5 \times 10^7$ |
| MISO $2 \times 1$ | 931 | 1410 | $3.8 \times 10^8$ |
| MIMO $3 \times 3$ | 1070 | 1591 | $4 \times 10^8$ |
| MIMO $2 \times 2$ | 1336 | 1698 | $4.2 \times 10^8$ |

## 5   Conclusion

In this paper, an Energy Efficient Cooperative MIMO routing protocol for cluster based WSNs, called EECMIMO in which, sensor nodes are organized into clusters such that intra-cluster messages are transmitted over short-range SISO links, while inter cluster messages are transmitted over long-range energy-efficient cooperative MIMO links. To reduce energy consumption and prolong the network lifetime, an adaptive cooperative nodes selection strategy is also designed. After that, we investigate the use of multiple transmitters and multiple receivers in virtual MIMO, considering the case of variable data rate. Further, we investigate the impact of distance on the choice of MIMO, MISO and SISO, We demonstrate that in large range applications, by optimizing the constellation size MIMO systems may outperform MISO and SISO systems. Also The MIMO $2 \times 2$ technique is more suitable for any application WSN since it exceeds the other techniques tested in terms of stability, lifetime and the number of packets sent to the BS. EECMIMO is designed for stationary WSNs, in future works, our algorithm can be extended to handle the mobile wireless sensor networks under the platform NS2.

# References

1. Cui, S., Goldsmith, A.J., Bahai, A.: Energy efficiency of MIMO and cooperative MIMO techniques in sensor networks. IEEE J. Sel. Areas Comm. **22**, 1089–1098 (2004)
2. Nguyen T., Berder O., Sentieys O.: Cooperative MIMO schemes optimal selection for wireless sensor networks. In: IEEE 65th Vehicular Technology Conference, pp. 85–89 (2007)
3. Sendonaris, A., Erkip, E., Aazhang, B.: User cooperation diversity-part I: system description. IEEE Trans. Commun. **51**, 1927–1938 (2003)
4. Winters, J.: The diversity gain of transmit diversity in wireless systems with rayleigh fading. IEEE Trans. Veh. Technol. **47**, 119–123 (1998)
5. Belmega E.V., Lasaulce S., Debbah M.: A Survey on Energy-Efficient Communications. International Symposium on Personal, Indoor and Mobile Radio Communications Workshops, Turkey, pp. 289–294 (2010)
6. Dawei, G., Miao, Z., Yuanyuan, Y.: A multi-channel cooperative MIMO MAC protocol for clustered wireless sensor networks. J. Parallel Distrib. Comput. **74**, 3098–3114 (2014)
7. Sajid, H., Anwarul, A., Jong, H.P.: Energy efficient virtual MIMO communication for wireless sensor networks. J. Telecommunication Systems **42**, 139–149 (2009)

# Solving the Vehicle Routing Problem on GPU

Abdelhamid Benaini, Achraf Berrajaa and El Mostafa Daoudi

**Abstract** In this work we present a new parallel implementations on GPU of an heuristic based on Clarke and Wright algorithm for solving the Vehicle Routing Problem (VRP) with single and multi depots. To our knowledge, this is the first GPU implementation of such class of heuristics that solve the VRP. Indeed, our solution computes in parallel an initial solution (tours) in one step and then iteratively it improves the costs of all pairs of neighbor tours in one step. Obtained experimental results under CUDA show that the proposed implementations exploit efficiently the parallelism of the GPU.

**Keywords** VRP · Multi depots · Parallel · GPU · CUDA

## 1 Introduction

The Vehicle Routing Problem (VRP) is an important NP-hard combinatorial optimization problem for which there are many heuristic methods. The basic VRP that we study in this paper consists in: Several vehicles having a fixed capacity C have to deliver order known quantities $q_i$ of goods to n cities (numbered from 1 to n) from a single or from a multiple depots. Knowing the distance $d_{ij}$ between cities i and j (city 0 is the depot), the problem is to find tours for the vehicles. Each tour starts and ends at the same depot, so that each city is visited exactly once and the total quantity carried on any tour does not exceed the vehicle capacity C. Several

A. Benaini · A. Berrajaa
Laboratory LMAH, Normandie Université, FST, 25 Rue Philippe Lebon,
76063 Le Havre, France
e-mail: abdelhamid.benaini@univ-lehavre.fr

A. Berrajaa (✉) · E.M. Daoudi
FSO, Laboratory LaRI, Université Mohammed Premier, Oujda, Morocco
e-mail: berrajaa.achraf@gmail.com

E.M. Daoudi
e-mail: m.daoudi@fso.ump.ma

© Springer International Publishing Switzerland 2016
A. El Oualkadi et al. (eds.), *Proceedings of the Mediterranean Conference on Information & Communication Technologies 2015*, Lecture Notes in Electrical Engineering 381, DOI 10.1007/978-3-319-30298-0_25

variants of this traditional VRP have well studied due to their important applications in areas of logistics and transportation. Among them VRP with time windows [1], with fleet size and mix vehicles, with split delivery (for more details, see for example reference [2, 3]) and recently the searches are focused on electric and green VRP [4]. The literature on the methods that solve VRP is vast and large started with exact methodologies like linear programming, dynamic programming or branch and bound algorithms [3]. Because of the NP-hardness of the problem, many heuristics and metaheuristics are proposed, some examples are the Tabu method, Genetic algorithms, Ants colony algorithm [5]. In this paper, we are interested by VRP with single depot and its extension to multiple depots. Let D be the $(n + 1)$ x $(n + 1)$ distance matrix where $d_{ij} = d_{ji}$ is the traveling cost (distance) between cities i and j. The man goal is to find a minimal cost feasible solution which consists in finding a set of tours that minimize the total traveling distance while satisfying all cities demand and capacity constraint. One tour is a set of ordered cities served by the same vehicle during the same tour. We propose two-stage method that solves this problem and we describe its implementation on GPU (Graphics processing Unit). To our knowledge, few GPU implementations that solve the VRP are proposed in the literature: Genetic method on GPU [5], local search for TSP [6], Constraint-based local search [7].

The remainder of the paper is organized as follows. In Sect. 2 we present the technic for partitioning in sectors. Section 3 presents the CW algorithm. Section 4 is devoted to the parallel algorithm and its implementation on GPU. Section 5 consists in presenting experimental results. In Sect. 6 we extend our solution to the multi depots case. Conclusion and perspectives are given in Sect. 7.

## 2 Partitioning in Sectors

Each city is defined by its polar coordinates $(\theta_i, \rho_i)$ with the depot at $(0,0)$. We renumber the cities by increasing angle $\theta_i \leq \theta_{i+1}$, for each i. In the case where $\theta_i = \theta_{i+1}$, then cities are numbered by increasing radius $\rho_i$. A more general study of the partition in sectors can be found in [2].

Figure 1 shows an example of such numbering. In this example, cities 4 and 5 have the same angle $\theta$ and the city 4 is the nearest to the depot. After this renumbering, we partition the set of cities into sectors where the demand of each sector is less than the capacity C. Let $S_1, S_2, \ldots S_m$ these sectors while $S_1 = \{0, 1, \ldots, r_1\}, \ldots, S_i = \{0, r_{i-1} +1, \ldots, r_i\}$. We determine sector $S_i$ as the set of cities $\{0, r_{i-1} + 1, \ldots, r_i\}$ where $r_i$ the unique integer satisfying: $\sum_{k=r_{i-1}+1}^{r_i} q_k \leq C < \sum_{k=r_{i-1}+1}^{r_i+1} q_k$ for $1 \leq i < m$ with $r_0 = 0$ and $\sum_{k=r_{m-1}+1}^{r_m} q_k \leq C$ for $i = m$. Here, $s_i = r_i - r_{i-1}$ is the number of cities in sector $S_i$. Each sector is then treated as an independent VRP and solved by TS (Traveling Salesman) process which produces a single tour. For this study, we use the Clarke and Wright heuristic CW [2] which will be detailed bellow. It is clear that this initial partition in sectors has an influence

**Fig. 1** Example of
numbering and partitioning in
sectors

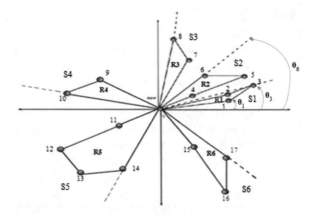

on the obtained final solution for the global VRP. Figure 1 shows an example of partitioning cities into 6 sectors: $S_1 = \{1, 2, 3\}$, $S_2 = \{4, 5, 6\}$, $S_3 = \{7, 8\}$, $S_4 = \{9, 10\}$, $S_5 = \{11, 12, 13, 14\}$ and $S_6 = \{15, 16, 17\}$. For this example, we have assumed that:

$$q_1+q_2+q_3 \leq C < q_1+q_2+q_3+q_4, \qquad q_4+q_5+q_6 \leq C < q_4+q_5+q_6+q_7, \qquad q_7+q_8 \leq C < q_7+q_8+q_9,$$
$$q_9+q_{10} \leq C < q_9+q_{10}+q_{11}, \quad q_{11}+q_{12}+q_{13}+q_{14} \leq C < q_{11}+q_{12}+q_{13}+q_{14}+q_{15}, \quad q_{15}+q_{16}+q_{17} \leq C.$$

## 3 Steps of CW Algorithm

CW is based on the computation of saving for combining two cities into the same tour. Initially, each city is considered to be on separate tour. The saving $s_{ij}$ for combining cities i and j into a single tour is then computed as: $s_{ij} = d_{i0} + d_{0j} - d_{ij}$ for $1 \leq i, j \leq n$. The coefficients $s_{ij}$ are ordered in non increasing fashion. Starting from the top of the saving list, $s_{ij}$ determining if there exist two tours one containing (0,j) and the other (i,0) that can feasibly emerged. So, combining these two tours by replacing (0,j) and (i,0) by (i,j). Many technics that improve this algorithm are proposed where the definition of saving is modified [1].

## 4 The Proposed Parallel Algorithm

Each sector is treated as an independent VRP solved by a single tour obtained by CW process. We construct these tours in parallel, the tour $R_i$ for the sector $S_i$, $i = 1 \ldots m$, using CW algorithm to compute each tour $R_i$. m is the number of obtained sectors and this is an upper bound of the number of vehicles used to solve this problem.

Let *CW(S)* the function that computes the tour R for the sector S. This function will be called in parallel, for $S = S_1, ..., S_m$ to produce m tours $R_1, ..., R_m$. The tours $R_i$ and $R_{i+1}$ are neighboring, ($R_1$ and $R_m$ are also neighboring), so moves of cities from $R_i$ to $R_{i+1}$ or from $R_{i+1}$ to $R_i$ could reduce the cost of these two tours. Similarly, swaps of cities between $R_i$ and $R_{i+1}$ could reduce the cost of these two tours. This process of moves/swaps of cities between $R_i$ and $R_{i+1}$, $i = 1 ... m-1$, is iterated in order to reduce the global cost of the VRP. These operations are done between pairs $(R_1,R_2)$, $(R_3,R_4)$, ..., $(R_{m-2},R_{m-1})$ in parallel. Assume that m is odd then since $(R_2,R_3)$, $(R_4,R_5)$, ..., $(R_{m-1},R_m)$ are also pairs of neighbor tours, we apply the same process to pairs $(R_2,R_3)$, $(R_4,R_5)$, ..., $(R_{m-1},R_m)$. Finally, we apply alternatively these two steps in order to reduce the global cost of VRP by reducing the sum of costs of each pair of neighbor tours.

## 4.1 Improvement of VRP Algorithm

The *k-opt* concept [3, 8] can be applied to sets of k tours by removing cities from one tour and inserting them into another for a saving in travel distance. In our work, we do this between pairs of adjacent tours $R_i$ and $R_{i+1}$ and we define two basic functions *move()* that moves one city and *swap()* that swap two cities. We call *Opt (R,R')* the function that consists in move or swap cities from R and R' tours. The *Improve(VRP) (or Improve($R_1, ... R_m$))* function that improve the initial solution $(R_1, ..., R_m)$, is defined as follows (like the steps of the Odd-Even sort algorithm):

```
Improve(VRP)=Improve(R₁,...Rₘ)
  if m is odd :
    repeat
       1. do in parallel Opt(R₁,R₂),Opt(R₃,R₄)… Opt(Rₘ₋₂,Rₘ₋₁)
       2. do in parallel Opt(R₂,R₃),Opt(R₄,R₅)… Opt(Rₘ₋₁,Rₘ)
    until the global cost is improved.
  if m is even:
    repeat
       1. do in parallel Opt(R₁,R₂),Opt(R₃,R₄)… Opt(Rₘ₋₁,Rₘ)
       2. do in parallel Opt(R₂,R₃),Opt(R₄,R₅)…Opt(Rₘ₋₂,Rₘ₋₁),
Opt(Rₘ,R₁)
    until the global cost is improved.
```

In the following we precise these two operations (*move()* and *swap()* functions):

A move of city $c \in R$ to R' is done if the operation is gainful that is if the total cost of the new tours is less than that the old one i.e.:

$\text{cost}(R - \{c\}) + \text{cost}(R' + \{c\}) < \text{cost}(R) + \text{cost}(R')$ and the capacity constraint is satisfied where cost(tour) is the tour length.

**Fig. 2** Move c from R to R'

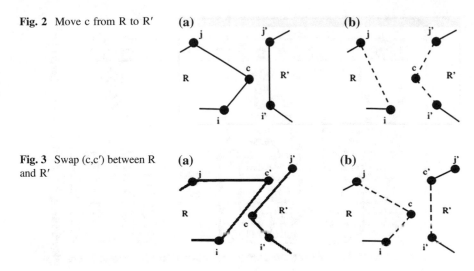

**Fig. 3** Swap (c,c') between R and R'

Figure 2 shows how the move of c from R to R' is realized: (a) before the move and (b) after the move. This amounts to remove c from R (say R–{c}) and to insert c in R' (say R' + {c}). A swap c ∈ R and c' ∈ R' is done if the operation is gainful that is if the total cost of the new tours is less than that the old one i.e.: cost (R − {c} + {c'}) + cost (R' − {c'} + {c}) < cost (R) + cost (R') and the capacity constraint is satisfied.

Figure 3 shows how the swap of (c,c') ∈ R x R' is realized: (c) before the swap and (d) after the swap. Finally the *Solve_One_Depot(VRP)* function that solves the VRP for one depot consists in the following two steps:

```
1. for all i, 1≤i≤m, do in parallel R_i=CW(S_i).
2. do in parallel Improve (R_1,… R_m).
```

## 4.2 The GPU Architecture

Graphics Processing Units GPU are actually available in most of personal computers. They are used to accelerate the execution times of variety of problems such as image processing, scientific computing, combinatorial optimization, the smallest unit in GPU that can be executed is called thread. Threads (all executing the same code and can be synchronized) are grouped into of equally sized and blocks and blocks are grouped in grid that represent one kernel (function) (blocks are independent and cannot be synchronized). Figure 4 explains the GPU architecture. The memory hierarchy of the GPU consists of three levels: (1) the global memory that is accessible by all threads. (2) the shared memory accessible by all threads of a block

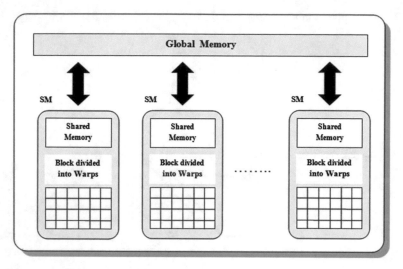

**Fig. 4** Simplified GPU architecture [7]

and the local memory (register) accessible by a thread. Shared memory has a low latency (2 cycles) and is of limited size (48 Kb for the GeForce). Global memory has a high latency (400 cycles) and is of large size (1−2 Gb for the GeForce). An entire block is assigned to a single SM (Stream Multiprocessor). Several blocks can run on the same SM. Each block is divided into Warps (32 threads by Warp) that are executed in parallel. The programmer must control the block sizes and the number on Warps.

### 4.3 Implementation of CW() on GPU

Let $r_i$ the number of cities in the sector $S_i$, $1 \leq i \leq m$. The needed data for computing $R_i$ is the $r_i \times r_i$ diagonal bloc of the distance matrix D composed of the columns $r_i + 1$, .... $r_{i+1}$. The tours $R_i = CW(S_i)$, for $1 \leq i \leq m$ are computed in parallel. Each $R_i$ is computed on a bloc of $s_i \times s_i$ threads where $s_i$ is number of cities in sector $S_i$. Since, in CUDA we can define only uniform size's blocs of threads, we use blocs of $s \times s$ threads where $s = \max(s_i)$, implying that $n \leq rm$. Finally we execute *the kernel* $CW(S_i)$ on m blocs noted $B_1$, ..$B_m$, each of $s \times s$ threads. Using CUDA syntax, the blocks, grid and *the kernel CW()* are defined as follows:

```
dim3 dimBlock(s,s); dim3 dimGrid(m,1);
CW<<<dimGrid,dimBlock>>>(S_i);
```

$S_i$ is represented by $s_i \times s_i$ bloc diagonal of the distance matrix D.

## 4.4  Improvement of the Initial Solution on GPU

The basic operation of this function is *move(c)* from R to R' or *swap(c,c')* between R and R'. Since all these basic operations are performed on bloc of r x r threads then they are done in one step. More precisely, each $Opt(R_i,R_{i+1})$, $1 \leq i < m$, is executed on bloc $B_i$ with shared memory $M_i$ and requires all distances $d_{cc'}$ between cities c and c' of the tours $R_i$ and $R_{i+1}$, which are stored in blocs $D_i$ and $D_{i+1}$ of the matrix D as shown in Fig. 5. So $D_i$ and $D_{i+1}$ must be stored in the shared memory $M_i$ of the bloc $B_i$. Now, if a city c is moved from $R_i$ to $R_{i+1}$, then we must update the shared memories $M_{i-1}$, $M_i$ and $M_{i+1}$ as shown on Fig. 6. Similarly, a swap of c and c' between $R_i$ and $R_{i+1}$ needs an update of $M_{i-1}$, $M_i$ and $M_{i+1}$ as shown in Fig. 6.

Using the CUDA syntax, the blocks, grid and the kernel Opt() are defined as follows:

```
dim3 dimBlock(s,s); dim3 dimGrid(m,1);
Opt<<<dimGrid,dimBlock>>>(D_i,D_{i+1});
```

Finally the *Solve_One_Depot(VRP)* kernel is executed on GPU in two sequential steps (Fig. 7):

*1. for all i, i≤i≤m, do in parallel $R_i$=CW($S_i$)* // on m blocs each of s x s threads
*2. do in parallel Improve($R_1$, $R_2$, ...),* // on m blocs each of s x s threads

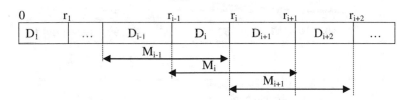

**Fig. 5**  Data repartition on the shared memory

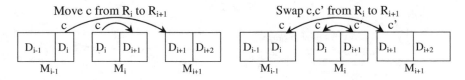

**Fig. 6**  Updating the shared memory in the move/swap

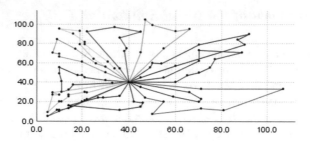

**Fig. 7** Example of obtained GPU solution with n = 110 cities

## 5 Experimental Results

This section is devoted to the comparison of the execution times of the CPU vs GPU of our GPU implementation presented above. We have used Nvidia GeForce, 1 Gb, with one SM of 48 CUDA cores (1,17 Ghz) and i3 with 4 core (2,4 Ghz) as CPU. So the optimal acceleration, or this hardware, cannot exceed 6. Our algorithms are implemented under CUDA V.6.0. environment. For tests we have fixed C while data ($d_{ij}$, $q_i$) are randomly generated for number of cities varying from 5 to

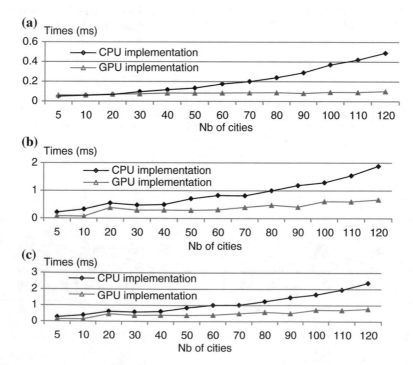

**Fig. 8** **a** GPU vs CPU of CW() implementation. **b** GPU vs CPU of Improve() implementation. **c** GPU vs CPU of Solve_One_Depot() implementation

120 (limited by the capacity of the shared memory of the card). To compare the CPU and the GPU implementations, we used the same data ($d_{ij}$, $q_i$ and C) and we solve the corresponding VRP on CPU and on GPU. Data transfers between CPU and GPU are done only at the beginning of *Solving_One_Dept()* program (before the *CW()* and *Opt()* kernels). Therefore, our executions times do not include the time of data transfers between CPU and GPU. In Fig. 8. (a), we compare the CPU vs GPU execution times for the CW kernel. The obtained times show that the acceleration factor increases with n and it is close to 5 for n = 120. The next figure (b) gives the execution times of the CPU and GPU for Improve() kernel. Here the acceleration factor is close to 3 for n = 120. The last figure (c) compares the execution times of CPU vs GPU for the VRP problem. The obtained times show that the acceleration factor is close to 3. Clearly, the increase of data size (number of cities) might increase the acceleration factor.

# 6    VRP with Multi Depots

The Multi depot VRP is defined as: Given d depots and n cities that must be served from theses depots, the problem is to both (a) assign each city to one depot and (b) find the set of tours that solve the problem. In our case, after assigning cities to depots, we solve d independents VRP (say $VRP_1$, ..., $VRP_d$) each associated to one depot ($VRP_k$ is associated to $depot_k$).

- The first step (a) is solved by assigning cities to their nearest depots. For each city i, compute distance(i,$depot_k$) and assign i to nearest $depot_j$ where distance(i, $depot_j$) = $\min_k$(distance(i,$depot_k$)). So, we obtain d VRP independent problems. Note that there are other approaches to assign cities to depots (see Refs. [9, 10] for more details).
- The next step (b) solves these d VRP by calling in parallel *Solve_One_Depot (VRP_i)* function for $1 \leq i \leq d$.

# 7    Conclusion

In this paper, we have presented a GPU implementation of an heuristic based on Clarke and Wright algorithm that solves the VPR single depot and multi depots. These implementations exploit efficiently the parallelism of the GPU. Indeed, our solution computes in parallel an initial solution (tours) in one step and then itera-tively it improves the costs of all pairs of neighbor tours in one step. Our CUDA implementations seem to be efficient according to the obtained results. To our knowledge, this is the first implementation of such class of heuristics that solve the traditional VRP. For the future work, we will extend our approach (heuristic and GPU implementation) to other variant of VRP: with time windows, with non homogeneous vehicle. Also, we generalize the *Opt()* to k tours rather than 2 tours that we have used, and we exploit the texture memory rather than the global and shared memories of the GPU.

# References

1. Solom, M.: Algorithm for the vehicle routing and Scheduling problems with time Window Constraints. Oper. Res. **35**(2) (1987)
2. Faulin, J., Juan, A.: On the use of monte carlo simulation, cache and splitting techniques to improve the Clarke and wright saving heuristics. J. Oper. Res. Soc. 62, 1085–1097 (2011)
3. Laport, G.: The vehicle routing problem: an overview of exact and approximate algorithms. Eur. J. Oper. Res. **59**, 345–358 (1992)
4. Sassi, O., Oulmara, A.: Joint scheduling and optimal charging of electric vehicles problem. ICCSA 2014, Part II, LNCS 8580, pp. 76–91 (2014)
5. Talbi, E.G., Hasle, G.: Metaheuristics on GPUs. J. Par. Dist. Comput. **73**(1), 1–3 (2013)
6. Fosin, J., Davidovic, D., Caric, T.: A GPU implementation of local serach operators for symmetric travelling salesman problem. Traffic Trans. **25**(3), 225–234 (2013)
7. Arbelaez, A., Codogne, P.: A GPU implementation of parallel constraint-based local search. In: Proceedings of the PDP'2014, 22nd Euromicro International Conference, pp. 648–655
8. Park, N., Okano, H., Imai, H.: A path-exchange-type local search algorithm for vehicle routing and its efficient search strategy. J. Oper. R. Soc. Jpn. **43**(1),197–208 (2000)
9. Cardon, S., Dommers, S., Eksin, C., Sitters, R., Stougie, A., Stougie, L.: L.A PTAS for the multiple depot vehicle routing problem. TUE—SPOR, 2008–03, TR (2008)
10. Tansini, L., Urquhart, M., Viera, O.: Comparing assignment algorithms for the multi-depot VRP. Technical Report, University of Montevideo, Uruquay (2001)

# Performance Analysis of Content Dissemination Protocols for Delay Tolerant Networks

Sara Koulali and Mostafa Azizi

**Abstract** Delay-Tolerant Networks (DTNs) have been proposed to interconnect devices in regions where permanent connections are not available. Due to long duration disconnections, devices in a DTN adopt the store-carry-forward paradigm to transport the generated packets to their respective destinations. Thus, intermediate devices take custody of the data being transferred and forward it as contact opportunities arise. Several content dissemination protocols have been designed to address the DTNs specificities. This paper surveys the proposed content dissemination protocols for delay-tolerant networks, and provides a comparative study of the Epidemic and Prophet content dissemination protocols on the basis of several performance criteria.

**Keywords** Delay tolerant networks · Content dissemination protocols · Prophet routing · Epidemic routing · The one

## 1 Introduction

The last decade witnessed increasingly growing research interests in the field of Delay Tolerant Networks [1]. This class of networks is mainly formed by intermittently connected mobile devices that exploit their mobility to transport produced data packets to their respective destinations. Thus, the well-functioning of the network fully relies on available contact opportunities governed by the devices mobility pattern.

Several applications with high latency degree tolerance could be satisfied by this communication model. Therefore, DakNet project [2] has been successfully deployed in remote parts of both India and Cambodia to provide villagers with

S. Koulali (✉) · M. Azizi (✉)
MATSI Laboratory, ESTO, Mohammed Ier University, Oujda, Morocco
e-mail: s.koulali@ump.ma

M. Azizi
e-mail: azizi.mos@ump.ma

© Springer International Publishing Switzerland 2016                    249
A. El Oualkadi et al. (eds.), *Proceedings of the Mediterranean Conference on Information & Communication Technologies 2015*, Lecture Notes in Electrical Engineering 381, DOI 10.1007/978-3-319-30298-0_26

asynchronous access to internet services. The project exploits public buses infrastructure to physically carry information between kiosks in villages and the remote internet hub. The buses are acting as mobile access points in a ferry like communication scheme. The ZebraNet [3] project for environmental monitoring objective is to track the wildlife of zebras. To collect gathered data, zebras are equipped with wireless collars that communicate via pairwise connections and a manned vehicle serves as a base station when occasionally it comes within the communication range of one of the collars. In [4] the authors propose an optimal ferry route design mechanism to enhance the connectivity of sparse ad hoc networks. The OPWP method is proposed to determine a set of ferry's way-points along with their corresponding waiting times. Those way-points are then ordered to form the ferry's tour. The optimal route design in presence of multiple ferries is considered in [5].

Delay Tolerant Networks have their own specificities and limitations that should be accounted for while designing novel protocols. Namely:

- Disconnection: As the devices move, the connection becomes intermittent. Thus, it is fair to assume that no end-to-end communication path will exist. Besides, disconnection could result also from devices becoming faulty or having their batteries depleted.
- High latency: Data Transmission in Delay Tolerant Networks is governed by the devices mobility model. Thus, generated packets will express high delay transmission that varies from several hours to several days.
- Resources scarcity: The devices in a DTN are generally battery powered and hence energy resources should be optimally consumed to extend the network lifetime.

Since efficient content dissemination is the corner stone of any reliable DTN deployment [6], several DTN content dissemination protocols have been proposed. Due to the devices mobility and wireless link volatility, frequent connection disruptions happens and End-to-End content dissemination solutions are doomed to behave poorly in the context of DTN. Thus, the focus shifted to protocols based on the store-carry-forward paradigm. Among the considerations that should be addresses by DTN content dissemination protocols is whether information about future contacts is available or not. Such information could be either deterministic for the case of predefined mobility or modeled by a given probability distribution function. For instance, the time between subsequent contacts of any pair of nodes for several mobility models (Random Walk, Random Direction, Random Waypoint) follows an exponential distribution with parameter $\lambda > 0$ [7].

In this paper we provide a comparative study of the epidemic and prophet content dissemination protocols for delay tolerant networks in terms of: Packet Delivery Ratio, Energy Efficiency and Energy Standard Deviation. We begin by reviewing the proposed content dissemination protocols for DTN in Sect. 2. Then we detail the realized experiments and discuss the obtained results in Sect. 3. We finally conclude the paper and announce future works in Sect. 4.

## 2   DTN Content Dissemination Protocols

Content dissemination protocols for Delay Tolerant Networks fall within two major categories, depending on whether the protocol creates replicas of messages or not, namely: forwarding-based and replication-based [8]. The protocols of the first family do not replicate messages and consequently consume less energy resources. Only a single copy of a message exists in storage within the network at any given time [9]. The protocols of the second family allow greater message delivery rates as multiple copies of the same packet may coexist in the network. Thus, the probability that at least one replica will reach the destination is increased. However, as several copies of the original message travel the network, communication overhead and energy consumption increase. Since network resources may quickly become constrained, deciding which message to transmit first and which message to drop happens to be a critical issue. Various replication-based protocols have been proposed. Therefore, epidemic routing [10] message transmission is compared to the spread of an infectious disease; when a device carrying a message encounters a device that does not have a copy, the carrier passes on a message copy. The infected nodes act similarly till the destination gets infected and receives the message. The Probabilistic Routing Protocol (Prophet) [11] relies on the history of encounters and transitivity. Prophet makes use of knowledge obtained from previous encounters with other relays to decide which node to choose in order to guarantee the message arrival to its destination. Prophet also uses transitivity mechanism when two devices rarely meet but there exists a device that frequently meets both of them. In two-hop relay routing protocol [12]: if there is no route between the source device and the destination, the source device transmits its packets to all neighboring nodes. A relay node is only allowed to send a packet to its destination node. MaxProp [13] establishes several mechanisms that aim to increase the delivery rate and lower delivered packets latency, along with mechanisms that allow to define the order for transmitting and deleting packets. A high priority is assigned to new packets by MaxProp while it attempts to prevent reception of the same packet twice.

## 3   Numerical Investigation

In this section we conduct a comparative study for the Epidemic and prophet routing protocols in terms of several averaged performance metrics. Table 1 summarizes the realized simulations parameters:

The retained performance metrics are:

- Packet Delivery Probability: the ratio of correctly received packets to number of emitted one.
- Residual energy: the remaining energy after simulation completion.

**Table 1** Simulation parameters

| Parameter | Value |
|---|---|
| Simulation duration | 10000 s |
| Simulation area | $1000 \times 1000$ m$^2$ |
| Number of nodes | {50, 100, 150, 200, 250, 300} |
| Mobility model | Random waypoint |
| Node movement speed | Min = 0.5 m/s Max = 1.5 m/s |

- Energy efficiency: the number correctly received packets to the total dissipated energy ratio.
- Standard energy Deviation: the average difference between individual devices residual energy and the average residual energy of the network.

We will be using the Opportunistic Network Environment simulator [14] (ONE) to run our experiments. Written in Java, ONE supports a variety of mobility models such as: Random Walk, Random Waypoint and bus movement. ONE can also import mobility data from real world traces or other mobility generators.

The PDR of prophet and epidemic content dissemination protocols is depicted in Fig. 1 for various topology sizes. We notice that prophet content dissemination achieves better delivery ratio than epidemic content dissemination with a maximal PDR of 25, 42 % for the networks with 100 devices.

In Fig. 2 the residual energy of the devices is reported. We notice that under prophet content dissemination protocol the devices consume less energy (higher residual energy). Reducing the communication load increases the level of residual (remaining) energy as compared to epidemic content dissemination protocol.

Figure 3 describes the energy efficiency for the considered topologies. The obtained results indicate that prophet manages to route more packets to their destinations with lesser energy consumption than the epidemic content dissemination protocol. This is justified by the selective approach of prophet content dissemination where the history of encounters is used as a criteria to forward packets which results in increased success probability.

**Fig. 1** Delivery probability for various topology sizes

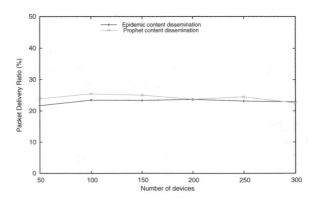

**Fig. 2** Average residual energy for various topology sizes

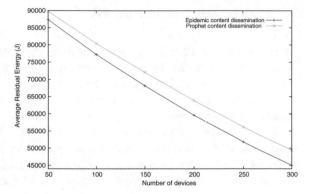

**Fig. 3** Energy efficiency for various topology sizes

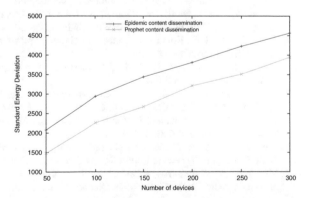

**Fig. 4** Standard energy deviation for various topology sizes

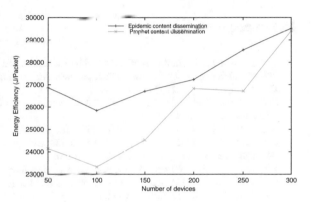

Figure 4 depicts the standard energy deviation of the two protocols for different topologies. The prophet content dissemination behaves better than epidemic content dissemination and manages to equilibrate the load over all the network such that the residual energy of each device is closer to the average residual energy of the network.

# 4 Conclusion and Future Works

In this paper we have made a comparative study of two well-known content dissemination protocols for delay tolerant networks: Epidemic and Prophet. This study compares the performances of the two protocols on the basis of three metrics: packet delivery ratio, energy efficiency and standard energy deviation. The experiments carried out show that prophet content dissemination outperforms epidemic content dissemination and manages to achieve higher PDR while consuming less energy and evenly dividing energy consumption over all the devices. As perspectives, we plan to extend our analysis to cover other available content dissemination protocols and consider the impact of selfishness on the system performance..

## References

1. Fall, K.: A delay-tolerant network architecture for challenged internets. In: Proceedings of the 2003 Conference on Applications, Technologies, Architectures, and Protocols for Computer Communications, pp. 27–34 (2003)
2. Pentland, A., Fletcher, R., Hasson, A.: Daknet: rethinking connectivity in developing nations. Computer 37(1), 78–83 (2004)
3. Juang, P., Oki, H., Wang, Y., Martonosi, M., Peh, L., Rubenstein, D.: Energy-efficient computing for wildlife tracking: design tradeoffs and early experiences with zebranet. In: ACM Sigplan Notices, vol. 37, pp. 96–107 (2002)
4. Bin Tariq, M.M., Ammar, M., Zegura, E.: Message ferry route design for sparse ad hoc networks with mobile nodes. In: Proceedings of the 7th ACM International Symposium on Mobile ad Hoc Networking and Computing, pp. 37–48 (2006)
5. Zhao, W., Ammar, M., Zegura, E.: Controlling the mobility of multiple data transport ferries in a delay-tolerant network. In: Proceedings of the 24th Annual Joint Conference of the IEEE Computer and Communications Societies (INFOCOM 2005), vol. 2, pp. 1407–1418 (2005)
6. Jones, E., Ward, P.: Routing strategies for delay-tolerant networks. ACM Comput. Commun. Rev. (CCR) [Submitted] (2006)
7. Groenevelt, R., Nain, P., Koole, G.: The message delay in mobile ad hoc networks. Perform. Eval. 62(1), 210–228 (2005)
8. Balasubramanian, A., Levine, B., Venkataramani, A.: Dtn routing as a resource allocation problem. ACM SIGCOMM Comput. Commun. Rev. 37, 373–384 (2007)
9. Jones, E., Li, L., Schmidtke, J., Ward, P.: Practical routing in delay-tolerant networks. IEEE Trans. Mob. Comput. 6(8), 943–959 (2007)
10. Vahdat, A., Becker, D., et al.: Epidemic routing for partially connected ad hoc networks. Technical report, Technical Report CS-200006. Duke University (2000)
11. Lindgren, A., Doria, A., Schelén, O.: Probabilistic routing in intermittently connected networks. ACM SIGMOBILE Mob. Compu. Commun. Rev. 7(3), 19–20 (2003)
12. Grossglauser, M., Tse, D.: Mobility increases the capacity of ad hoc wireless networks. IEEE/Acm Trans. Networking 10(4), 477–486 (2002)
13. Burgess, J., Gallagher, B., Jensen, D., Levine, B.: Maxprop: routing for vehicle-based disruption-tolerant networks. In: Proceedings of the IEEE Infocom, vol. 6, pp. 1–11. Barcelona, Spain (2006)
14. Keränen, A., Ott, J., Kärkkäinen, T.: The one simulator for DTN protocol evaluation. In: SIMUTools '09: Proceedings of the 2nd International Conference on Simulation Tools and Techniques (2009)

# Multi-hop Clustering Algorithm Based on Spectral Classification for Wireless Sensor Network

Ali Jorio, Sanaa El Fkihi, Brahim Elbhiri and Driss Aboutajdine

**Abstract** A Wireless Sensor Network (WSN) is composed of a large number of autonomous and compact devices called sensor nodes. This network can be an effective tool for gathering data in a variety of environments. However, these sensor nodes have some constraints due to their limited energy, storage capacity and computing power. Clustering is a kind of a technique which is used to reduce energy consumption and to extend network lifetime. Hence, multi-hop communication is often required when the communication range of the sensor nodes is limited or the number of sensor nodes is very large in a network. In this paper, we propose a multi-hop spectral clustering algorithm to organize the sensor nodes in a WSN into clusters. Simulation results show that the proposed algorithm performs better in reducing the energy consumption of sensors and effectively improves the WSN lifetime.

**Keywords** Wireless sensor network · Clustering · Spectral classification · Energy consumption · Multi-hop communication · Gateway nodes

A. Jorio (✉) · D. Aboutajdine (✉)
LRIT, Research Unit Associated to the CNRST (URAC 29),
FSR, Mohammed V University, Rabat, Morocco
e-mail: jorio.ali@gmail.com

D. Aboutajdine
e-mail: aboutaj@fsr.ac.ma

S.E. Fkihi (✉)
ENSIAS, Mohammed V University, Rabat, Morocco
e-mail: elfkihi@ensias.ma

B. Elbhiri (✉)
EMSI Rabat, Rabat, Morocco
e-mail: elbhirij@emsi.ac.ma

# 1 Introduction

A Wireless Sensor Network (WSN) consists of a large number of tiny sensor nodes, which are deployed over a hostile, an inhabitable, and a harsh environment, possibly for a limited period, with a common, objective, and collaborate to provide distributed sensing, storage, and communication services [1]. However, one of the most important constraints of the sensor nodes is the low power consumption requirement [2]. The sensor nodes are equipped with limited, generally irreplaceable power sources. WSNs become increasingly useful in many critical applications, such as nuclear, biological, home automation, battlefield surveillance, and environmental monitoring.

Traditional routing protocols for WSN may not be optimal in terms of energy consumption [3]. Thus, clustering can be used as an energy-efficient communication protocol. The objectives of clustering are to minimize the total transmission power and to balance the load among the nodes for prolonging the network lifetime. Clustering consists to organize the network into several clusters of sensors. Each cluster is managed by a special node or leader, called cluster head (CH), which is responsible for coordinating the data transmission activities of all sensors in its cluster. All sensors in a cluster communicate with a CH that acts as a local coordinator for performing data aggregation. CHs can transmit gathered data back to the Base Station (BS) directly or through multi-hop communication among CHs [4]. However, when we consider a general sensor network that may be deployed over a large region, the energy spent in the power amplifier related to distance may dominate to such an extent that using a multi-hop mode may be more energy efficient than single-hop mode [5]. Also, multi-hop communication is more realistic because nodes may not be able to communicate directly with the BS due to the limited transmission range.

In this paper we propose a Multi-Hop Clustering Algorithm based on Spectral Classification (MHCA-SC). MHCA-SC is based on spectral clustering and graph theory to cluster the WSN to a fixed and an optimal number of clusters. Each sensor node transmits sensing data to the BS using multi-hop communication. The CHs are selected periodically based on residual energy. MHCA-SC aims to prolonging network lifetime by distributing energy consumption and minimizing control overhead.

The remaining of the paper is organized as follows. Section 2 includes a survey of the related research. Section 3 describes the network model. Section 4 exhibits the detail of the proposed scheme. Simulation results and its discussion are presented in Sect. 5. Finally, Sect. 6 concludes the paper.

# 2 Related Work

In a WSN, as the energy supply for sensor nodes is usually extremely limited, it is essential to improve energy efficiency. Clustering is the most energy-efficient organization for wide application in the past few years, and numerous clustering algorithms have been proposed for energy saving [3, 4].

Single-hop and multi-hop communication are two basic communication models which are used in WSNs. In the case of single-hop communication the furthest member nodes or CHs tend to deplete their battery energy faster than other nodes in a network. In other words, in single-hop where data packets are directly transmitted to the BS without any relay, the nodes located farther away have higher energy burden due to long-range communication, and these nodes may die out first. Some examples of these single-hop clustering approaches are LEACH (Low Energy adaptive clustering hierarchy) and LEACH-C (LEACH Centralized) [6], DECSA (Distance-Energy Cluster Structure Algorithm) [7], and EECS (Energy-efficient clustering scheme) [8].

Since a large number of sensor nodes are densely deployed, multi-hop communications are prone to occur in WSNs. One approach is to cluster a WSN into clusters such that all members of the clusters are directly connected to the CHs. Sensor nodes in the same cluster can communicate directly with their CH without any intermediate sensor nodes. CHs can transmit gathered data back to the BS through multi-hop communication among CHs. There are many multi-hop clustering algorithms such as EEUC (Energy-Efficient Unequal Clustering) [9], EEM-LEACH (Energy Efficient Multi-hop LEACH) [10], MR-LEACH (Multi-hop Routing with LEACH) [11]. Previous research (e.g., [5]) has shown that multi-hop communication is usually more energy efficient than single-hop communication because of the characteristics of wireless channel.

In the other hand, The spectral methods for clustering have recently started to get a lot of attention in many research areas. These methods make use of the spectrum of the adjacency matrix of the data to cluster a considered set of elements. They are considered as powerful techniques in data analysis [12–14]. Many clustering algorithms [15, 16] based on spectral classification are proposed for WSN.

# 3 Network Model

## 3.1 Assumptions

We make some assumptions about the sensor nodes and underlying the network model, which are as follows:

- All the sensor nodes are uniformly dispersed within a square field.
- All the sensor nodes and the BS are stationary after deployment.
- Sensor nodes transmit their data periodically to the BS.
- Each member node communicates with its respective CH by using single-hop communication.
- CHs transmit gathered data back to the BS through multi-hop communication among CHs.
- A WSN consists of homogeneous nodes in terms of node energy.

## 3.2 Energy Model

The radio model utilized in our approach is similar to that of LEACH [6]. The energy consumed by the radio in transmitting $L - bits$ data over a distance $d$ is given by:

$$E_{Tx}(L, d) = \begin{cases} L \cdot E_{elec} + L \cdot \epsilon_{fs} \cdot d^2 & \text{if} \quad d < d_o \\ L \cdot E_{elec} + L \cdot \epsilon_{mp} \cdot d^4 & \text{if} \quad d \geq d_o \end{cases} \tag{1}$$

When receiving this data, the energy consumed by the radio is given by:

$$E_{Rx}(L, d) = L \cdot E_{elec} \tag{2}$$

where $E_{elec}$ is the energy dissipated per bit to run the transmitter or the receiver circuit. The parameters $\epsilon_{fs}$ and $\epsilon_{mp}$ depend on the transmitter amplifier model we use. They represents respectively the free space (with $d^2$ power loss) and the multipath fading (with $d^4$ power loss) channel models and depend on the distance $d$ between the transmitter and the receiver.

## 4 The MHCA-SC Mechanism

This section describes the detail of the MHCA-SC to meet the demands of a wide range of applications. Each step of our approach will be explained in the subsections.

## 4.1 Pre-processing Step

In this step we determine the graph corresponding to the considered WSN. First, the BS broadcasts a "hello" message to all nodes. Next, each node computes the approximate distance to the BS based on the received signal strength and determines its position by using a number of different technologies, e.g. exploiting radio or sound waves [17]. This localization helps nodes to define their zone location. After that, each sensor node transmits the derived position to the BS in a short message. Finally, the BS constructs the graph $G$ and derives the similarity matrix $A$ which represents the network. In our approach we consider the similarity function as a Gaussian [12]. Hence, similarity matrix is constructed as follows:

$$A = [a_{ij}] = \begin{cases} exp\left(\frac{-d^2(i,j)}{2\sigma^2}\right) & \text{if } i \neq j \\ 0 & \text{otherwise} \end{cases} \tag{3}$$

with $d(i, j)$ is the euclidian distance between nodes $i$ and $j$.

After that, the BS deduces the degree matrix and the Laplacian matrix. Thus, the out-degree of a node $i$ corresponds to the total number of incoming edges $(\sum_{j=1}^{N} a_{ij})$. The degree matrix $\mathbf{D} \in \Re^{N \times N}$ of G is a diagonal matrix with diagonal entries equal to the out-degree of each node. The $N \times N$ Laplacian matrix of the graph is, as introduced by [14], expressed as follow:

$$L = D^{-1/2} * A * D^{-1/2} \qquad (4)$$

## 4.2 Clustering Step

The objectives of the current step are to define the optimal number of clusters and to form them. Based on the Laplacian matrix $L$, we form a new matrix $U$ composed of the $k$ eigenvectors related to $k$ largest eigenvalues of $L$. In order to determine the $k$ clusters of the WSN, we apply the classification algorithm k-means to the matrix $U$. The sensor node $\mathbf{i}$ is assigned to cluster $C_j$ if and only if row $\mathbf{i}$ of the matrix $\mathbf{U}$ was assigned to cluster $C_j$ [14].

Nonetheless, the most important question raised by the proposed strategy concerns the optimal number of clusters ($k$) that must be considered. The objective function that allows to decide whether to reconsider the partitioning process or not of the WSN, is defined by the distance matrix $M_{dis}^{k}$ ($M_{dis}^{k} = [dis_{ij}^{k}]$); with $dis_{ij}^{k}$ is the distance between the node i and the node j of the cluster labeled $k$) of each cluster. The allowed threshold to this function is $d_0$. Hence, if at least one element of any $M_{dis}^{k}$ is greater than $d_0$, the considered number of clusters will be incremented ($k + 1$) and the k-mean algorithm will be reused. Otherwise, the optimal number of clusters is $k$. By this way, all member nodes operate in a free space model.

## 4.3 Cluster Head Election Step

Once the clusters are determined, the next step of the MHCA-SC consists to select the CHs. In our approach, the CHs are determined by taking into account the *id* of the nodes and their residual energy. For each cluster, we use the number $C_k = (r \bmod |S_k|)$ to elect a CH; $r$ is the round of the simulation and $|S_k|$ is the total number of nodes in the cluster $k$. The node with $id = C_k$ and residual energy $E_r$ greater than threshold $E_{rmin}$ will be the CH of the cluster $k$ in the round $r$. $E_{rmin}$ is the average energy of the cluster. Nevertheless if the residual energy $E_r$ is less than this threshold $E_{rmin}$ this node must broadcast a short message informing the node with $id = C_k + 1$ to its residual energy and so on. Thus, the energy consumption will be distributed with more equatability between all nodes.

## 4.4   Intra-cluster Single-hop Routing

Once the clusters are formed, a schedule TDMA (Time Division Multiple Access) will be created automatically to assign to each node a time when it can transmit its data to its own CH. This technique ensures not having collisions between data messages.

## 4.5   Inter-cluster Multi-hop Routing

When CHs deliver their data to the BS, each CH aggregates the data from its cluster members, and then sends the packet to the BS via multi-hop communication. For this, We introduce a threshold $d_0$ into our multi-hop forwarding model. If a node's distance to the BS is smaller than $d_0$, it transmits its data to the BS directly; otherwise it should find a relay node which can forward its data to the BS. The concrete scheme of choosing the best relay node is explained as follows.

First, CHs announces their zone location, distance with BS, and number of messages it relayed. Next, Source CH utilizes these details to find its relay node. The relay node is located in the adjacent zone of the current CH. Each CH choose the relay node as the next hop CH that requires the minimum communication value. Finally, when a CH sends data, it must sense the channel to see if anyone else is transmitting data, if so, the CH waits to transmit the data.

## 5   Simulation Results

In this section, we present the results of the evaluation experiments of the MHCA-SC algorithm. MATLAB software was used to simulate its performance. In this study, we have considered first order radio model simulation to LEACH and the simulation parameters for our model are mentioned in Table 1.

Figure 1 presents an example of WSNs with $N = 500$ nodes randomly distributed.

Figure 2 presents an example of the clustering step using MHCA-SC algorithm. We note that the network is subdivided into $k = 27$ clusters and there is no intersection between the different clusters. Figure 3 presents the clustering structure when inter-cluster communication was established.

In order to evaluate the performances of the new proposed protocol, we propose to compare it to the HCA-SC [16], the SCNOC [15], the DECSA [7], and the LEACH-C [6] algorithms.

Figure 4 gives the curves of the number of nodes alive over time. We show a significant improvement for the MHCA-SC approach in terms of numbers of periods relating to the First Node Died (FND).

**Table 1** Experimental
simulation parameters

| Parameter | Value |
|---|---|
| $E_{elec}$ | 50 nJ/bit |
| $\epsilon_{fs}$ | 10 pJ/bit/m$^2$ |
| $\epsilon_{mp}$ | 0.0013 pJ/bit/m$^4$ |
| $E_{DA}$ | 5 nJ/bit/message |
| Area of network | 300 m * 300 m |
| Zone 1 | $(x \in [0, 300], y \in [0, 100])$ |
| Zone 2 | $(x \in [0, 300], y \in [100, 200])$ |
| Zone 3 | $(x \in [0, 300], y \in [200, 300])$ |
| Sink coordination | (150 m, 350 m) |
| $d_0$ | 88 m |
| Message size | 4000 bytes |

**Fig. 1** Distribution of
wireless sensor network
$N = 500$

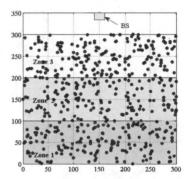

**Fig. 2** Network clustering
using MHCA-SC algorithm

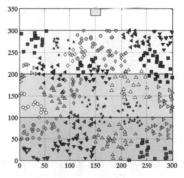

Figure 5 gives the total network energy remaining in every transmission round. This energy decreases rapidly in HCA-SC, SCNOC, DECSA, and LEACH-C protocols than in MHCA-SC protocol.

Table 2, presents the effects of the nodes density on the performances of the compared protocols. For different values of $N$ equal to 100, 300 and 500, our algorithm presents an improvement of performances compared to the other

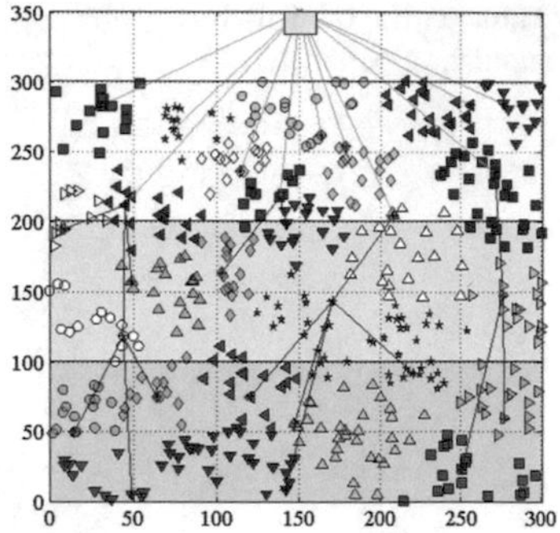

**Fig. 3** Clustering structure
when inter-cluster
communication was
established

**Fig. 4** Number of nodes
alive over time of the
compared protocols

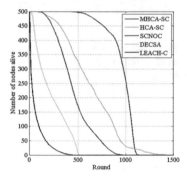

**Fig. 5** Evolution of the
remaining energy in the
network

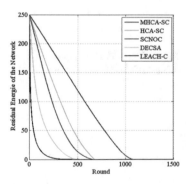

**Table 2** Impact of the node density N on the performances of the five compared algorithms

| Number of nodes | First node died | | | | |
|---|---|---|---|---|---|
| | MHCA-SC | HCA-SC | SCNOC | DECSA | LEACH-C |
| 100 | 223 | 86 | 50 | 8 | 19 |
| 300 | 352 | 157 | 115 | 9 | 17 |
| 500 | 465 | 182 | 120 | 19 | 8 |

algorithms. It follows that even if the node density increase the new proposed approach still gives best results compared to the other ones.

The best results of the MHCA-SC algorithm can be explained by: (i) the proposal starts by selecting the clusters before the election of the CHs. (ii) the approach considers the node's residual energies when electing CHs. And (iii) CHs transmit gathered data back to the BS through multi-hop communication among CHs.

# 6 Conclusion

Energy efficiency is one of the main challenges in the design of protocols for WSNs as battery power of sensors is limited. The main objective is to improve the network lifetime as much as possible. In this paper a Multi-Hop Clustering Algorithm based on Spectral Classification is proposed. In the algorithm, the network is clustered by spectral classification and the CHs choice another CH with shorter distance as its the next jump to transmit gathered data to the BS. From the simulation results, it is clear that the proposed algorithm gives a better lifetime and minimized energy consumption.

# References

1. Akyildiz, I.F., Su, W., Sankarasubramaniam, Y., Cayirci, E.: A survey on sensor networks. IEEE Commun. Mag. **40**(8), 102–114 (2002)
2. Anastasi, G., Conti, M., Francesco, M.D., Passarella, A.: Energy conservation in wireless sensor networks: a survey. Ad Hoc Netw. **7**(3), 537–568 (2009)
3. Al-karaki, J.N., Kamal, A.E.: Routing techniques in wireless sensor networks: a survey. IEEE Wirel. Commun. **11**, 6–28 (2004)
4. Boyinbode, O., Le, H., Mbogho, A., Takizawa, M., Poliah, R.: A survey on clustering algorithms for wireless sensor networks. In: 13th International Conference on Network-Based Information Systems, pp. 358–364 (Sept 2010)
5. Bhattacharyya, D., Kim, T.H., Pal, S.: A comparative study of wireless sensor networks and their routing protocols. Sensors **10**(12), 10506–10523 (2010)
6. Heinzelman, W.B., Chandrakasan, A.P., Balakrishnan, H.: An application-specific protocol architecture for wireless microsensor networks. IEEE Trans. Wirel. Commun. **1**, 660–670 (2002)
7. Yong, Z., Pei, Q.: A energy-efficient clustering routing algorithm based on distance and residual energy for wireless sensor networks. Procedia Eng. **29**, 1882–1888 (2012)
8. Ye, M., Li, C., Chen, G., Wu, J.: Eecs: an energy efficient clustering scheme in wireless sensor networks. In: 24th IEEE International Conference on Performance, Computing, and Communications, pp. 535–540 (2005)
9. Li, C., Ye, M., Chen, G., Wu, J.: An energy-efficient unequal clustering mechanism for wireless sensor networks. In: IEEE International Conference on Mobile Adhoc and Sensor Systems, pp. 604–612 (Nov 2005)
10. Antoo, A., Mohammed, A.R.: Eem-leach: energy efficient multi-hop leach routing protocol for clustered wsns. In: International Conference on Control, Instrumentation, Communication and Computational Technologies, pp. 812–818 (July 2014)

11. Farooq, M., Dogar, A., Shah, G.: Mr-leach: Multi-hop routing with low energy adaptive clustering hierarchy. In: Fourth International Conference on Sensor Technologies and Applications, pp. 262–268 (July 2010)
12. Luxburg, U.: A tutorial on spectral clustering. Stat. Comput. **17**(4), 395–416 (2007)
13. Nascimento, M.C., de Carvalho, A.C.: Spectral methods for graph clustering a survey. Eur. J. Oper. Res. **211**(2), 221–231 (2011)
14. Ng, A.Y., Jordan, M.I., Weiss, Y.: On spectral clustering: analysis and an algorithm. Proc. Adv. Neural Inf. Process. Syst. **14**, 849–856 (2001)
15. Elbhiri, B., El Fkihi, S., Saadane, R., Aboutajdine, D.: Clustering in wireless sensor networks based on near optimal bi-partitions. In: 6th EURO-NF Conference on Next Generation Internet, pp. 1–6. (June 2010)
16. Jorio, A., El Fkihi, S., Elbhiri, B., Aboutajdine, D.: A hierarchical clustering algorithm based on spectral classification for wireless sensor networks. In: International Conference on Multimedia Computing and Systems, pp. 861–866. April (2014)
17. Mao, G., Fidan, B., Anderson, B.: Wireless sensor network localization techniques. Comput. Netw. **51**(10), 2529–2553 (2007)

# Message Priority CSMA/CA Algorithm for Critical-Time Wireless Sensor Networks

Imane Dbibih, Imad Iala, Ouadoudi Zytoune and Driss Aboutajdine

**Abstract** IEEE 802.15.4 Wireless Sensor Network (WSN) is among of the most deployed networks for several applications. On the one hand IEEE 802.15.4 features provide a good solution for low-cost and low-power wireless communications. But, on the other hand, this standard does not process the quality of service in real time applications, where we should reduce the latency in parallel of the energy consumption. IEEE 802.15.4 standard is based on the CSMA/CA (Carrier Sense Multiple Access/Collision Avoidance) algorithm, which does not take into account any indications of nodes priority. However, in some applications, such event-monitoring networks, we need to ensure low latency, especially for real time messages. In this paper we present a new priority based scheme for managing node priority to get access to the channel. In our contribution, Message Priority CSMA/CA (MP-CSMA/CA) algorithm, the probability to gain channel access is based on the message priority. The main of our algorithm is to differentiate services; it gives more chance to nodes that have real-time messages, to access the medium firstly. The results show an efficient quality of service provided by our proposed algorithm.

**Keywords** WSN · MAC layer · Slotted CSMA/CA · Message priority · Medium access process · Real-time applications

I. Dbibih (✉) · I. Iala · D. Aboutajdine
LRIT Associated Unit to the CNRST-URAC N29, Faculty of Sciences,
University Mohammed V, 4 Avenue Ibn Battouta B.P. 1014 RP, Rabat, Morocco
e-mail: imane.it@gmail.com

I. Iala
e-mail: Imad.iala01@gmail.com

D. Aboutajdine
e-mail: aboutaj@fsr.ac.ma

O. Zytoune
LARIT, Faculty of Sciences, University Ibn tofail, Kenitra, Morocco
e-mail: zytoune@gmail.com

© Springer International Publishing Switzerland 2016
A. El Oualkadi et al. (eds.), *Proceedings of the Mediterranean Conference on Information & Communication Technologies 2015*, Lecture Notes in Electrical Engineering 381, DOI 10.1007/978-3-319-30298-0_28

# 1   Introduction

Today the wireless sensor networks are more and more deployed for emergency response applications (monitoring the movement of vehicles in a hostile zone [1], disaster monitoring [2], etc.). In this type of applications it is necessary to differentiate services in order to reduce latency. IEEE 802.15.4 standard is heavily used in WSN applications. The works [3–9] have tried to analyze the performance of both periods CAP (Contention access period) and CFP (Contention free period); they all conclude that the management of the quality of service is among of the limitations whose the standard suffers. In this paper we propose to manage the quality of service during the CAP period. We propose a new method based on the slotted CSMA/CA algorithm, which gives to nodes the priority to access the medium on the basis of the type of sent messages. The aim of our algorithm is to reduce the latency of real-time traffic.

The rest of this paper is organized as follows. Section 2 presents a summary of works that have enhanced the standard in terms of QoS. Section 3 details our contribution. Then, the performance analysis and the experiment results are shown in Sect. 4, while Sect. 5 summarizes the paper conclusion with referred prospects.

# 2   Related Work

The IEEE 802.15.4 features have been studied by a lot of researchers in order to evaluate and enhance wireless sensor network performances in terms of quality of services. The works [10–13] have tried to process the concept of priority for real-time applications. Boughanmi et al. [10] use a new mechanism based on CSMA/CA algorithm called Blackburst message, as a means to differentiate the medium access priority. When a node wants to access the medium, it waits a time interval called LIFS (long inter-frame space) before it sends its Blackburst message in order to see if there is another Blackburst message longer than its own. Blackburst length depends on data priority. The problem is that the transmission of Blackburst message will increase energy consumption. As well, the latency will be negatively impacted, due to the wait time of LIFS period.

In [11], the authors propose a new period at the beginning of each superframe called PAP (Priority Access Period) That is assigned by the coordinator. Its objective is to provide an opportunity for node to send its real-time data or asking for guarantee time slots from the coordinator. The disadvantage of this method is that the PAP period is not allocated according to the priority of nodes traffic; therefore it does not improve the QoS of the highest priority messages.

Severino et al. [12] suggest to assign a priority level to each traffic type, and for each one they adjust the value of *CW (Contention Window)* parameter. Also, in [13] the authors create an analytic model of medium access contention, during the CAP period of the IEEE 802.15.4 standard. This study was carried out to evaluate the *CW* impact on the medium contention of nodes having prioritized messages.

# 3 The Proposed Approach

## 3.1 Message Priority CSMA/CA Algorithm

MP-CSMA/CA algorithm proposes to improve the performance of WSN deployed for real-time applications. So, it intends to improve the quality of service in terms of prioritized messages latency. In this paper we aim to enhance the quality of services during the CAP period of the standard IEEE 802.15.4 [14], which does not consider the message priority to access the channel.

Our algorithm consists to assign the priority to access the medium to nodes that have the highest priority message. We consider two priority levels, high priority assigned to critical-time message and low priority assigned to non-real-time messages. So, we define the parameter $MsgP$ that can take 0 for the highest priority and 3 for the lowest priority. To differentiate services we make two intervals of which nodes can backoff for a $Wtime$ (wait time) to access the medium. We vary the minimum and the maximum of this interval based on the message priority. So, according to the value of $MsgP$ we calculate the $Wtime$ following the Eq. (1).

$$Wtime = (random([0, 2^{BE} - 1]) + MsgP) * BP \qquad (1)$$

knowing that BP is a unit of time called backoff period, where one backoff period is equal to $aUnitBackoffPeriod = 80$ bits $(0.32$ ms$)$ [14].

In our algorithm, nodes have two ranges to take their random values. For example if two nodes want to access the medium $N_1$ with a high message priority $(MsgP = 0)$ and $N_2$ with a low message priority $(MsgP = 3)$. The node that has the highest priority, $N_1$, will automatically take a small value of $Wtime$. So, it will necessarily get access to the medium and transmit its message in the first attempt, which will reduce the latency of the prioritized messages. Figure 1 shows the process of the MP-CSMA/CA algorithm proposed in this paper. Knowing that [14]:

- *BE* Backoff Exponent defines how many BPs a device shall wait before attempting to access the channel.
- *NB* defines the number of times the CSMA/CA algorithm was required to backoff when performing the current transmission.
- *CW* Contention Window is the number of times the node performs CCA (Clear Channel Assessment) procedure and during which the medium must be sensed idle before the transmission can start.

MP-CSMA/CA algorithm consists of six steps:

Step (1): initialize the parameters $NB = 0$, $CW = 2$ and $BE = macMinBE$, where *macMinBE* is the minimum value of *BE*.

Step (2): assign the appropriate value to the parameter $MsgP$ according to the message priority.

Step (3): calculate the $Wtime$ using (Eq. (1)).

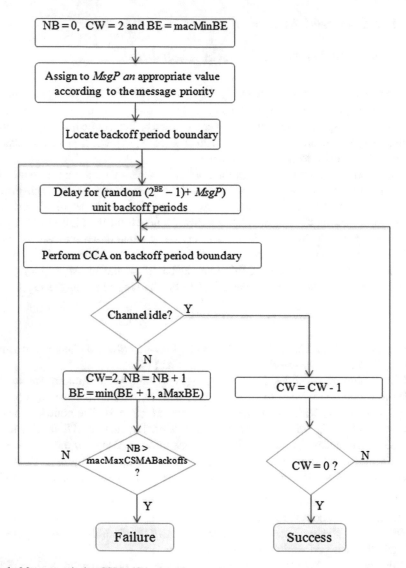

**Fig. 1** Message priority CSMA/CA algorithm

Step (4):  after the expiration of the wait time, the node performs CCA (determining whether the channel is free or busy), to assess whether the
channel is idle. If the channel is assessed to be busy the node goes to
Step (5), otherwise, it goes to Step (6).

Step (5):  if the channel is busy, the values of *NB* and *BE* are increased by one.
Then the node tests if the value of *NB* exceeds *macMaxCSMABackoffs*
(the maximum number of backoffs the CSMA/CA algorithm will attempt

before declaring a channel access failure), if this is the case it declares a transmission failure. Otherwise, it goes to Step (3).

Step (6): if the channel is not busy, the value of CW is decreased by one, and then the node tests whether CW reaches 0, if this is the case it access the channel, otherwise it goes to Step (4).

## 4 Simulation Results

### 4.1 Simulation Parameters

Our solution was implemented and tested using the model, denoted as WPAN, in the version 2.35 of NS-2 (Network simulator) [15]. It provides implementation of the standard IEEE 802.15.4 without sleep function (make devices in sleep state during the inactive period). So, we implement all the functions that put the transceiver on sleep state periodically. In these simulations we compare our algorithm with the algorithm CSMA/CA in the standard IEEE 802.15.4, and since we do not consider CFP period in our algorithm, so we will extend the CAP period along the active time also for the standard.

In this section, we present a comparison between MP-CSMA/CA algorithm results and those of the algorithm CSMA/CA. We set up our simulations in a beacon-enabled mode. The number of nodes randomly deployed in the network is fixed to 100 nodes, and we increase linearly the source nodes (nodes generating the traffic in the network) from 10 to 100 nodes, in order to vary the traffic loads in the network. We use one PAN (personal area network) coordinator, and all other nodes are devices. At the application layer we use the CBR traffic (Constant Bit Rate) produced by source nodes, which generate packets with 100 Bytes every 1 s. All results are the average of 10 simulations. The simulations parameters are cited in Table 1.

### 4.2 Prioritized Message Latency

In this experiment, we calculate the average latency of all prioritized messages transmitted between the source and the destination.

Figure 2 shows that in most traffic levels, our algorithm ensure a very low latency compared to that of slotted CSMA/CA algorithm. Theoretically this is logical, because our algorithm provides the highest priority to node that wants to access the medium for sending prioritized messages. This simulation shows the effectiveness of our algorithm, in terms of latency, against the slotted CSMA/CA algorithm, either in high traffic load (100 source nodes) or in low traffic load (10 source nodes).

**Table 1** Simulation parameters in NS-2

| Parameter | Value |
|---|---|
| BO | 4 |
| SO | 3 |
| macMinBE | 3 |
| aMaxBE | 5 |
| macMaxCSMABackoffs | 4 |
| Duty-cycle | 50 % |
| Transmission range | 150 m |
| Simulation time | 800 s |
| Number of nodes | 100 |
| Type of topology | Mesh |
| Routing protocol | AODV |
| Application layer | CBR |
| Data packet size | 100 Bytes |
| Space range | 300 × 300 m |

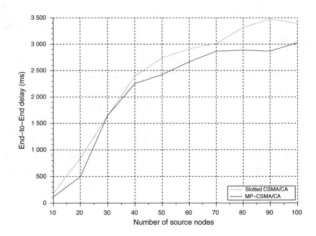

**Fig. 2** The latency of prioritized messages

We chose to simulate in the area $300 \times 300$ m, in order to create a multi-hop transmission. The objective was to show the effect of our algorithm on the behavior of the intermediate nodes. The message priority provides to nodes more chance to obtain the medium access, which will reduce latency along the path between the source and the destination.

## 4.3 Average Network Latency

In this experiment, we calculate the latency of all the sent data by the source nodes whatever their priorities.

According to the shown results in Fig. 3, we can confirm that our algorithm ensures an average latency reduced compared to that provided by the algorithm slotted CSMA/CA. in our algorithm the message priority will ensure to the intermediate nodes a high chance to gain the channel contention. This will positively influence on the average latency of all messages sent in the network.

### 4.4 Packet Delivery Ratio

In this experiment we evaluate the PDR of all massages transmitted in the network, for different traffic levels.

From the results shown in Fig. 4, we can confirm that our algorithm ensures an improved PDR compared to that of slotted CSMA/CA algorithm. We know that in

**Fig. 3** The average latency

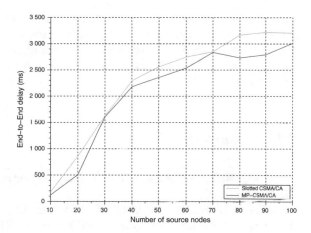

**Fig. 4** Packet delivery ratio

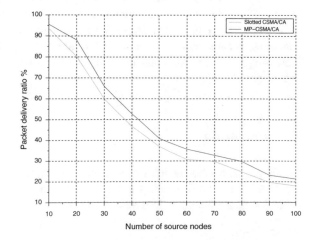

**Fig. 5** The throughput

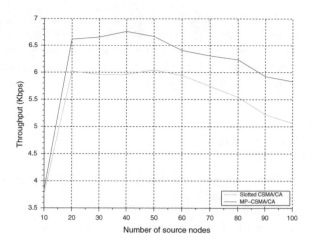

our algorithm there are two intervals to take the backoff time, which will efficiently manage the collision problem. More there is a high global priority more the value of *Wtime*, taken by the node, is reduced. This gives more chance to nodes, with a high message priority, to obtain the medium access and to transmit successfully in the first attempt.

## 4.5 Throughput

In this experiment we calculate the throughput achieved by all nodes deployed in the network.

In Fig. 5, we see that the results of MP-CSMA/CA algorithm are more efficient compared to those obtained by the algorithm slotted CSMA/CA. These results are proved by the latency results shown in Fig. 3. Theoretically, once we reduce the latency of transmission, automatically that allows to send more packets in a short time. Also, we remark that above 70 source nodes, the level of the throughput decreases, because when we increase the number of nodes in the network we risk increasing the traffic load and the congestion, which more increases the collision occurrence.

## 5 Conclusion

In this paper, a new medium access method, MP-CSMA/CA algorithm has been presented. This algorithm aims to improve the quality of service of WSN deployed for real-time applications, in terms of prioritized message latency. Our method take into account the message priority in the medium access process. The effectiveness of our algorithm has been proved by the obtained results in low and high traffic load. Simulations confirmed that through our algorithm, we are able to achieve improved packet delivery ratio and low prioritized message latency. We believe that the concept underlying the proposed solution can be adopted in different topologies, so, in our future work we intend evaluating this algorithm in different application environments.

# References

1. He, T., Krishnamurthy, S., Stankovic, J.A., Abdelzaher, T., Luo, L., Stoleru, R., Yan, T., Gu, L., Hui, J., Krogh, B.: Energy-efficient surveillance system using wireless sensor networks. In: Proceedings of the 2nd International Conference on Mobile Systems, Applications, and Services, pp. 270–283 (2004)
2. Tia, G., Pesto, C., Selavo, L., Yin, C., JeongGil, K., Lim, J.H., Terzis, A., Watt, A., Jeng, J., Chen, B., Lorincz, K., Welsh, M.: Wireless medical sensor networks in emergency response: implementation and pilot results. In: 2008 IEEE Conference Technologies for Homeland Security, pp. 187–192, May 2008
3. Anastasi, G., Conti, M., Di Francesco, M.: The MAC unreliability problem in IEEE 802.15.4 wireless sensor networks. In: Proceedings of the 12th ACM International Conference on Modeling, Analysis and Simulation of Wireless and Mobile Systems, pp. 196–203 (2009)
4. Jianping, Z., Tao, Z., Lv, C.: Performance evaluation for a beacon enabled IEEE 802.15.4 scheme with heterogeneous unsaturated conditions. Int. J. Electron. Commun. 66(2), 93–106 (2012)
5. Park, P., Di Marco, P., Soldati, P., Fischione, C., Johansson, K.H.: A generalized Markov chain model for effective analysis of slotted IEEE 802.15.4. In: MASS '09. IEEE 6th International Conference Mobile Adhoc and Sensor Systems 2009, pp. 130–139, Oct 2009
6. Koubaa, A., Alves, M.: Tovar, E.: A comprehensive simulation study of slotted CSMA/CA for IEEE 802.15. 4 wireless sensor networks. In: IEEE WFCS, vol. 6, pp. 63–70 (2006)
7. Liu, X., Leckie, C., Saleem, S.K.: Performance evaluation of a converge-cast protocol for IEEE 802.15.4 tree-based networks. In: 2010 Sixth International Conference Intelligent Sensors, Sensor Networks and Information Processing (ISSNIP), pp. 73–78, Dec 2010
8. Lee, J.S.: An experiment on performance study of IEEE 802.15. 4 wireless networks. In: ETFA 2005 10th IEEE Conference Emerging Technologies and Factory Automation 2005, vol. 2, 8 p. (2005)
9. Cuomo, F., Cipollone, E., Abbagnale, A.: Performance analysis of IEEE 802.15.4 wireless sensor networks: an insight into the topology formation process. Comput. Netw. 53(18), 3057–3075 (2009)
10. Boughanmi, N., Ye-Qiong, S., Rondeau, E.: Priority and adaptive QoS mechanism for wireless networked control systems using IEEE 802.15.4. In: IECON 2010—36th Annual Conference on IEEE Industrial Electronics Society, pp. 2134–2141, Nov 2010
11. Ding, G., Farley, R.: A MAC protocol for wireless personal area networks. In: 2013 International Conference Computing, Networking and Communications (ICNC), pp. 900–904, Jan 2013
12. Severino, R., Batsa, M., Alves, M., Koubaa, A.: A traffic differentiation add-on to the IEEE 802.15.4 protocol: implementation and experimental validation over a real-time operating system. In: 2010 13th Euromicro Conference Digital System Design: Architectures, Methods and Tools (DSD), pp. 501–508, Sept 2010

13. Ndih, E.D.N., Khaled, N., De Micheli, G.: An analytical model for the contention access period of the slotted IEEE 802.15.4 with service differentiation. In: IEEE International Conference Communications, 2009. ICC '09, pp. 1–6, June 2009
14. IEEE 802.15.4, Part 15.4: Wireless LAN medium access control (MAC) and physical layer (PHY) specifications for low-rate wireless personal area networks (LR-WPANs). Standard IEEE, Dec 2003
15. The Network Simulator—ns-2: http://www.isi.edu/nsnam/ns/ (2013)

# Part V
# ICT Based Services and Applications

# Evaluation of Communications Technologies for Smart Grid as Part of Smart Cities

Abdelfatteh Haidine, Abdelhak Aqqal and Hassan Ouahmane

**Abstract** The smart grid represents the core of the future smart cities because it has to guarantee a reliable, safe, economic and environment friendly energy. However, for the use of the smart grid applications, such as smart metering or distribution automation, a two-way communications infrastructure is necessary in parallel to the electricity grid. In this paper, we analyze and evaluate the different communication technologies that could build the communications platform for the exchange of the smart grid information. We first discuss the requirement of the smart grid applications, and then we compare the most adequate communication technologies that are gaining field in the practice; namely high-speed narrowband Power Line Communication (N-PLC), CDMA450 and WiMAX. However, it is clear that the utility has to build platform with mix technologies. Therefore, we discuss the most practical combinations/scenarios to build an optimal mixture of technologies.

**Keywords** Smart grid · Power line communications (PLC) · CDMA450 · Smart metering · Wimax · CAPEX/OPEX · Total cost of ownership

## 1 Introduction

The building of future smart grid brings several challenges for the utilities. One of these challenges is the deployment of a two-way communication infrastructure, or the purchase of the communication services. To make the right decision, utilities have to find an optimal tradeoff between different aspects, such as an acceptable quality of service, optimal communications cost, etc. Therefore, utilities started

A. Haidine (✉) · A. Aqqal · H. Ouahmane
National School of Applied Sciences, UCD University, 24000 El Jadida, Morocco
e-mail: a.h.haidine@ieee.org; abdel.haidine@gmail.com

A. Aqqal
e-mail: aqqal.a@ucd.ac.ma

H. Ouahmane
e-mail: ouahmane.h@ucd.ac.ma

© Springer International Publishing Switzerland 2016
A. El Oualkadi et al. (eds.), *Proceedings of the Mediterranean Conference on Information & Communication Technologies 2015*, Lecture Notes in Electrical Engineering 381, DOI 10.1007/978-3-319-30298-0_29

277

different pilot project, in order to be able to evaluate the technical performances in the field and to validate their business case for different communication technologies. Generally, most of the projects focused on GPRS, RF mesh, power Line Communications (PLC) or a mix of technologies, like PLC/GPRS or RF/GPRS. In the recent years, some utilities showed an increasing interest in the CDMA450 technology; (CDMA: Code Division Multiple Access). The CDMA450 is a form of CDMA2000 that operates across the 410–470 MHz cellular band. This band is often referred to as the 450 MHz band, and this gives rise to the name CDMA450; [1]. Furthermore, the use of low frequency allows a larger coverage and good wave propagation and penetration. The good wave penetration characteristic allows avoiding the reception problems that GRPS meets when the smart meters are located in the cellar of the buildings. The large coverage increases the economic feasibility, because fewer base stations/antennas are needed to cover the service area. This decrease CAPEX as well as OPEX, in term of maintenance and site rent. Therefore, the goal of this paper is the assessment of the CDMA450 and its comparison with other technologies; according to different implementation scenarios. On the other hand, WiMAX represents an alternative solution to CDMA450; even if it has shorter coverage per cell as it uses higher frequencies than 450 MHz. WiMAX has two major advantages compared to CDMA450, which are: (a) WiMAX is widely worldwide and this affects the equipment diversity/supply in the market and the prices; and (b) WiAMX Forum is still developing new WiMAX release to match better to the requirements of the new applications/service emerging in the market; and this is the case with WiGRID.

The rest of the paper is organized as following: in the second section, we present the requirements of smart grid applications which must be fulfilled by the proposed communication solutions. The section is closed by a qualitative comparison of the technologies used in the usual telecommunications market and we discuss their suitability to the smart grid. Third discuss presents a new form of solution adopted by the electricity/gas utilities to build a reliable communications infrastructure if form of utility-telecoms. Last part is devoted to the main characteristics of the CDMA450, WiMAX and their evaluation/positioning as candidate solutions to smart grid. Then, an assessment tool to evaluate the grid communications networks is shown.

## 2 Role of Communications in Smart Grid Landscape

### 2.1 Positioning of Communications in Smart Grid

Today, power grids are in a transformation period, to build the future smart grids, in order to sustain and improve the security, reliability and efficiency of the grid. In this grid of the future, Information and Communications Technologies (ICT) build the foundation to support different smart grid related applications, each with

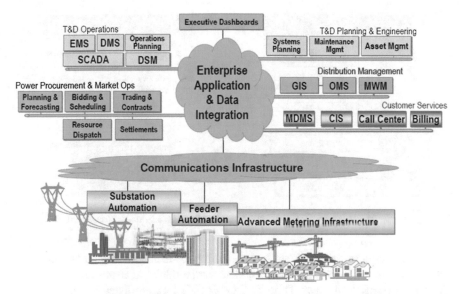

**Fig. 1** Communications infrastructure in smart grid (EPRI)

different quality of service requirements. This allows interconnecting and enabling the two-way flow of real-time information within the power utility, as well as between the utility and its suppliers, partners and consumers. The ICT covers three categories of technologies: sensing, control devices, applications, and the communications networks that bridge them all together; as illustrated in Fig. 1. The main challenge in building the two-communication network is to deploy a network with minimal investments, and which is preferment (capacity and low latency), easy to install/operate/maintain, reliable, etc.

## 2.2 Requirements from Smart Grid Applications

The communications solutions to be deployed in smart grid have to fill a set of requirements, in order to make such infrastructure economically viable and to guarantee reliable functioning of the whole system. Generally, the following requirements build the core design specifications; [2]: High reliability and long lifetime; Interoperability, Security and Cost effectiveness.

Qualitative representation of the key requirements from smart grid is presented as function of critical level in Fig. 2. However, before designing communications network, it is important to analyze the applications that will use this infrastructure. The applications should give a quantification of the required quality level of service (QoS) in terms of required capacity, type of traffic, maximal delay, etc. For example, in case of MV/LV distribution automation there are twelve key

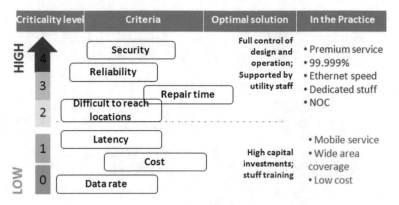

**Fig. 2** Criteria set for SG applications with criticality level and solutions (based on [6])

application services going from to fault detection to smart metering and others). The estimated required QoS; represented by data rate, traffic type, maximum latency, Bit Error Rate (BER) and Mean-Time To Repair (MTTR); are listed in [3, 4].

## 2.3 Candidate Communications Technologies

The communications equipment manufacturers have designed a large set of solutions to support the ordinary telecommunications services. However, their adaptability to the smart grid applications is not fully matched by one solution; as listed in Table 1 with their main advantages and drawbacks.

From the above table, it can be deduced that there is no single answer to the technology question: which communication technology shall be deployed for smart metering. However, the technical requirements of the smart metering application are moderate, in term of delay, jitter and bandwidth, the total cost of ownership (TCO) for the rollout deployment of the technology often becomes the decisive criteria in the decision making. The TCO is mostly depending on the deployment environments, which influence the resulting required investments. Therefore, the evaluation of the technology suitability has to be done according to the deployment environment, [5].

## 2.4 Selection of Adequate Communications Technologies

For the selection of adequate technology for smart grid applications, different factors must be taken into consideration. The main factors are illustrated by Fig. 3. Applications generate the traffic which should be transported through the networks.

**Table 1** Candidate communications for smart grid

| Technology | Main advantages | Main drawbacks | Deployment in field |
|---|---|---|---|
| DSL | Ready infrastructure<br>High bit rates (>2 Mbps/user) | Dependence on DSL service provider<br>Need a local network to connect devices to DSL | Access network<br>Backhauling |
| WLAN | Widely used technology, making it low-price solution | Limited coverage<br>Security issues | In-home |
| Satellite | 100 % coverage<br>Easy to deploy | High costs<br>Adequate as backup path | Backbone |
| Fiber optic | Very high capacity | Very high costs | Backbone |
| 2G/3G | Wide coverage<br>Mature technology | Low availability, overload<br>Poor indoor coverage | Radio access |
| LTE | Higher throughput<br>Sorter delay (for control plan as well as data plane) | Same as GSM<br>Costs still high<br>Coverage still limited | Radio access<br>Metropolitan |
| WiMAX | High bit rate<br>Large cell coverage | Must be owned by the utility (high investments) | Access<br>Backhauling for PLC data concentrators |
| CDMA450 | Spectrum: better coverage and penetration of radio wave (better indoor coverage) | Very limited number of brands in market | In-home coverage<br>Access coverage<br>Metropolitan coverage |
| PLC (LV, MV, HV) | • No new wiring is needed<br>• 100 % penetration | License for 450 MHz may be already allocated<br>Sensitive to noise<br>Performance fluctuations | Indoor, access and backhauling (HV) |

The applications define the traffic volume/profile and the required QoS. Regulatory regulates the use of the infrastructure and the standards allowed to be used in their territory. For example, when designing wireless networks, only a very limited spectrum of slots are available. Furthermore, the use of available frequencies will be allowed only after purchasing a license from regulatory authorities. The communication network to be built should fulfill the requirement for the applications with minimum costs; considering CAPEX (for hardware/software purchasing and deployment) as well as OPEX (from operation and maintenance).

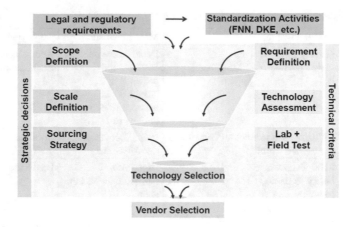

**Fig. 3** Factors influencing technology selection for smart grid [9]

## 3 U-Telco: Utility Telecoms Entity as Optimal Solution

In the practice, three options are possible for utilities to achieve their smart grid plans: (a) Drawing up a commercial contract with a Telco (b) Forming a partnership with a specialist Telco; and (c) Creating a utility Telco (*U-Telco*); [6]. Regarding the high sensibility of electricity/water/gas infrastructure; referred to as National Critical Infrastructure (NCI), some utilities have opted the third option even if this requires some investments to build their infrastructure; such as the example in the Netherlands [7]. In the Netherlands, the U-Telco is built around two major communications technologies; namely the Narrowband PLC and CDMA450; [8].

This option is evaluated to be right and correct, but generates high investments and new challenges for the utilities, which has not even the know-how to deal with such infrastructure called "*Third Grid*". Such Greenfield network deployment has the main inconvenient of the high costs; however, it offers a higher flexibility and freedom degrees to customize the network according to the needs of the utilities. Furthermore, a Greenfield design allows the possibility to hybridize the technologies, in order to reach the optimum of infrastructure utilization and also to implement the different level of security as much as needed. Example options for decreasing investments are: (a) share the infrastructure with third parties, like utilities for gas, transport companies, logistic companies; (b) Smart Cities (also referred as *Digitalized Cities*) is a new keyword that gather politicians, researchers and network operators to build societies an cities that are more adapted to the new era of digital (called Era of Big Data and Data Analysis); and (3) The utility can just buy the communications equipments and implement them (elements of CAPEX), while the OMC (Operations, Supervision and Maintenance Center) and other functionalities can be outsourced to the incumbent network operators of the country. This option has been implemented in Netherlands, where KPN (MNO) integrated supervision of the CDMA450 in his operation center.

# 4 Designing the Optimal Mix of Technology- When and Where?

## 4.1 Communications Platform

As stated above, there is no unique ideal solution to cover the smart grid applications. An optimal configuration will be a mix of technologies. When considering the solutions used in the practice, a platform of communication scenarios is possible; as it is illustrated in Fig. 4 and discussed further below.

## 4.2 CDMA450-WiMAX and Assessment of Optimal Solution

For the assessment of CDMA450, a tool is set composed of four blocks: (1) input block, which makes available all the necessary information to for the network costs calculation; (2) Output modules, which summarize the calculation results and makes them ready for plotted and comparison of different communications technologies over the dashboard; (3) Dashboard/control panel, which allows the selection of the scenarios to be calculated and analyzed, to select the communication technologies and their parameters and to visualize the different result diagrams; and (4) the calculation core, which builds the main part of CDMA450 business case. In the calculation core, the input values are used to calculate the costs of the different components of the network per year (e.g.: network access, core network, license, network management, project management, etc.). Interaction between different block is illustrated in Fig. 5.

**Fig. 4** Platform of possible communications scenarios for smart grid

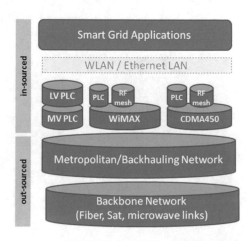

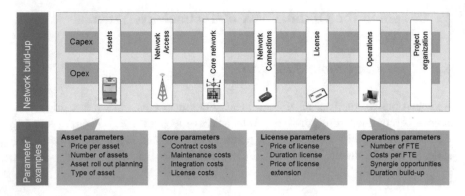

**Fig. 5** Components building the assessment tool and cost components; [10]

For the cost estimation of the network deployment in case of CDMA, the CAPEX and OPEX will be approximated as follows: Number of needed base stations; Connection of base station to the access network. The transmission of the read data from meters through the national transport network will be possible by buying the service from backbone network operator. The costs contain will a fix component (CAPEX) and a monthly component (OPEX).

The investment model for WiMAX installations must consider all aspects of design, deployment, and integration from the core through the systems architecture, service edge, access network and device. While the initial spend on a WiMAX deployment will have a large focus on capital components associated with procuring the necessary equipment throughout the network and systems architecture, as the WiMAX service is introduced and subscriber adoption and usage rates grow, the ongoing operating expenses will consume a growing share of the total cost of ownership. Thus, it is important that operator consider the end-to-end impact of deploying and operating a WiMAX system; Fig. 6.

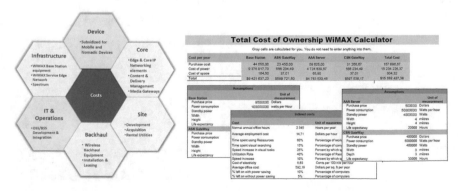

**Fig. 6** Cost components for WiMAX (*left*) its calculator for smart grid

# 5   Conclusions and Outlooks

In this paper, we discuss the role of communications infrastructure in the realization of future smart grid, which builds the core of smart cities. Different communications solutions used for ordinary telecoms services could be adopted for smart grid. However, some grid applications have different requirements. Three major solutions are considered in the practice; N-PLC, CDMA450 and WiMAX. The main advantages, drawbacks and costs components are discussed in the paper as well. WiAMX is a technology that is still issuing new release to adapt the standards to new emerging services; in the contrary to N-PLC and CMDA450. The new release called WiGRID is more adapted to the industrial applications; such smart grid, railways, etc.

As future steps, evaluation scenarios will be developed, where applications form smart grids are taken into consideration. In a second step, new applications from other puzzle parts of the smart cities will be included in the evaluations tool.

# References

1. Nedevschi, S., et al.: Potential of CDMA450 for rural network connectivity. IEEE Commun. Mag. **45–1**, 128–135 (2007)
2. Pauzet, O.: Cellular Communications and the Future of Smart Metering. Sierra Wireless, Inc., Sept 2010
3. CIGRE Working Group Del. D2.21, WG, D2-21: Broadband PLC applications, June 2008
4. DLC + VIT4IP Project: Scenarios and Requirements Specification. European Project from FP7, Deliverable D 1.1., Sept 2010
5. Le Peltier, V.: Smart Metering and Smart Grid: What can a Telco contribute? Telecoms for Smart Grids Conference 2012, London, UK, Sept 2012
6. Thomas, A.: Smart Grid Communications: the Key Decisions. Mott MacDonald WP, 2013
7. van der Meulen, A.: Private national radio network as a communication backbone for AMI. Metering Europe 2010, Vienna, Austria, 22–24 Sept 2010
8. Haidine, A., Müller, J.: Building utility communications through deployment of power line communications. CIRED'2013, Sweden, June 2013
9. Grandel, M.: Communications for the smart meter rollout in Germany. at the Conference Telecoms for Smart Grids, London, UK, Sept 2012
10. Haidine, A., Müller, J., Hurkx, J., Alons, K.: Evaluation of CDMA450 as communications technology for smart grid applications. In: IEEE EPS ISGT-2013, Brazil (2013)

# Context Modeling and Metamodeling: A State of the Art

Zineb Aarab, Rajaa Saidi and Moulay Driss Rahmani

**Abstract** Due to advances in sensor technology, sensors are becoming more powerful, cheaper and smaller. This evolution generates big amounts of data, which may have no value unless we have the ability of analyzing, interpreting and understanding it. Context-awareness computing (collection, modeling, metamodeling, reasoning and distribution of context) plays a significant role in this challenge. This paper tries to contribute by analyzing, comparing and consolidating works in this sense especially in context modeling and metamodeling.

**Keywords** Context-awareness · Context-modeling · Context-metamodeling · Context reasoning · Context-awareness computing

## 1 Introduction

A wide and unique accepted definition of the concept of the context is not available in the literature. However, many researchers have tried to find their own definition for what the concept of context actually includes. Thus, sophisticated and general context models have been proposed. The most popular context modeling techniques are surveyed in [1]. Furthermore, various systems depending on this concept were born, in particular mobile devices such as notebooks, PDAs, and smartphones, which have led to the introduction of pervasive or ubiquitous systems [2] that are

Z. Aarab (✉) · R. Saidi · M.D. Rahmani
LRIT Associated Unit to CNRST (URAC 29), Faculty of Sciences,
Mohammed V University, BP 1014, Rabat, Morocco
e-mail: aarab.zineb@gmail.com

R. Saidi
e-mail: r.saidi@insea.ac.ma

M.D. Rahmani
e-mail: mrahmani@fsr.ac.ma

R. Saidi
INSEA, BP 6217, Rabat, Morocco

© Springer International Publishing Switzerland 2016                287
A. El Oualkadi et al. (eds.), *Proceedings of the Mediterranean Conference on Information & Communication Technologies 2015*, Lecture Notes in Electrical Engineering 381, DOI 10.1007/978-3-319-30298-0_30

becoming increasingly popular. However, there are few works interested in context metamodeling that should provide generic metamodels to be applied on different application domains.

Our objectives in revisiting the literature are: firstly to learn how context-aware computing techniques (especially context modeling and metamodeling) have helped to develop solutions in the past. Secondly to highlight challenges and how can we apply those techniques to solve problems in future research directions. Thus, several requirements have to be taken into account when modeling context information, they are identified and explained in [3]: mobility and heterogeneity, timeliness (real-time and historic), deficiency or imperfection, dependencies and relationships, reasoning, usability of modeling formalisms and efficient context provisioning.

The goal of this paper is to present a state-of-the-art of context modeling and metamodeling in pervasive computing discussing current approaches. The paper is organized as follows: Sect. 2 presents a comparison between context modeling approaches. In Sect. 3 we will discuss context metamodeling approaches. And finally we conclude in Sect. 4.

## 2 Context Modeling

To introduce adaptation in context-aware ISs the consideration of the context is paramount. However, it must be built in an orderly and structured way for efficient and easy exploitation. This section presents research efforts summarizing most popular context modeling approaches. Our goal is not to study the implementation and support models context, but especially to analyze the structure and components of these models. Our goal will be to study the generic and proper construction of these metamodels to meet the needs of multiple domains without being overloaded and difficult to instantiate. Our discussion is based on the six most popular context modeling techniques: key-value, markup schemes, graphical, object based, logic based, and ontology based modeling. To avoid repeating what can be found in the literature about those approaches, they are synthesized and compared directly in Table 1, by giving their strength points and weaknesses and highlighting their applicability.

We have previously presented the main context modeling approaches. Despite each approach may provide an effective solution for a particular domain, and/or for a particular type of reasoning, none of them provides a solution from the data acquisition from sensors to the delivery to applications of high level context data. Also, none of them can satisfy all the requirements illustrated in the introduction. As a solution, some researchers have proposed hybrid modeling approaches. The process of how the multiple techniques can be combined together is presented in [3]. Two different hybrid approaches were proposed for context representation and reasoning [3]: (1) The *Hybrid fact-based/ontological model*, which combines the CML models with interoperability support and various types of reasoning provided

**Table 1** Synthesis through a comparison of context modeling approaches

| Approach | Works | Pros | Cons | Applicability |
|----------|-------|------|------|---------------|
| Key value | [1, 3–5] | The simplest data structure<br>• Easy to manage<br>• Flexible | Lack of capability for sophisticated structuring<br>• Not allowing the representation of types of context other than preferences<br>• No dependencies and relationships capturing<br>• No validation support<br>• No standard processing tool<br>• No evolutivity<br>• Difficult to recover information | They are usually used to represent preferences or points of interest that are part of the user's profile. Frequently used in the distributed service frameworks |
| Markup scheme (e.g. XML) | [1, 3, 5] | Use a variety of markup languages including XML<br>• Hierarchical data structures<br>• Schema validation<br>• Flexible<br>• Availability of processing tools | Not very good candidates for generic context information models<br>• Can become complex<br>• Application depended | Typically used for profiles representation. For more expressiveness the UAProf standard uses both standard RDFS and XML [1]. Can be very interesting in several fields (e.g. for storing sensor descriptions or to decouple data structures used by two components in a system...) |
| Graphical (e.g. UML) | [1, 5–8] | Strong graphic component<br>• Generic structure<br>• Association and relationships modeling<br>• Representation of heritage associations and encapsulation<br>• Implementations are available<br>• Constraints validation is possible [9] | Not standardized<br>• No tool support for CML<br>• Querying can be complex<br>• Interoperability could be difficult | Very useful as a tool for structuring a relational database for a large volume of permanent data. Also to store historical context |
| Object based | [1, 5] | Use encapsulation and reusability | Lack of standards and validation | As in [5] they can represent context in programming code |

(continued)

**Table 1**  (continued)

| Approach | Works | Pros | Cons | Applicability |
|---|---|---|---|---|
| | | • Relationships modeling • High level of formality for accessing to contextual information is provided by specified interfaces | • Difficult to recover information | level and support data transfer over network. Also enables context runtime manipulation with short term |
| Logic based | [1, 5] | Simple • Express policies, constraints and preferences • Allows logical reasoning and high-level context information to be extracted using low-level context | Lack of standardization • Reusability and applicability are reduced • Coupled with applications | Can be used in artificial intelligence and in multimedia system and where context is modeled as abstract mathematical entities with properties useful. Also modeling events and actions |
| Ontology based | [1, 3, 5, 10–14] | Powerful tool to represent concepts and interrelationships • Logical reasoning • Provide formal semantics of context data • Strong validation • Allows checking the consistency • Support for interoperability, heterogeneity, and representation of complex relationships | Representation can be complex • OWL-DL does not include very expressive constructors • Recover information can be complex | Modeling the context in ubiquitous environments and in the field of intelligent building. Also we can use ontology languages extended with rules, for example, the semantic e-wallet (e.g. the semantic eWallet) [14] |

by ontological models, and (2) *loosely coupled markup-based/ontological model*, which adopts a context modeling approach that is based on a loose interaction between a markup model and an ontological model.

# 3  Context Metamodeling

A metamodel defines the language and semantics to specify the particular model domains or applications. Thus, a context metamodel consists of organizing and presenting a more abstract view of context models in order to show their construction and manipulation. Metamodeling context is a major challenge for the identification of concepts involved in manipulating the context, their relationships, their formalization and presentation of their semantics [15]. The previous section presented context modeling approaches for context-aware applications while in this section we will present the five most popular context metamodeling approaches: metamodel centered ubiquitous web applications, UML Profile metamodels, metamodels supporting ontological models, high level abstraction and framework metamodel.

## 3.1 Metamodel Centered Ubiquitous Web Applications (UWAs)

The article [16] proposes a web application modeling language based on UML extension for the web called UWE (UML based web engineering). This approach was explicitly defined in terms of metamodel based on the Meta Object Facility (MOF). The metamodel is defined as a conservative extension of the UML meta-model. In addition, it provides an accurate description of the concepts used to model Web applications and their semantics. The set of metamodels for modeling web applications that require the manipulation of the context presented in [17] are based on WebUML and they are supposed to be generic. The metamodel of [17] consists of four views for a high-level context modeling, profile, rules and events modeling. Its context is a set of the following elements: the *User Agent*, it describes the capabilities of the devices and it is independent of the application, the *User*, the *Network* e.g. the maximum bandwidth, the *Location* of the user and the *Time*.

In brief, these metamodels form a coherent set of techniques that support context-aware web applications regardless of the domain. However, we believe that these metamodels views are inadequate and they may neglect some aspects such as the acquisition and manipulation of contextual information.

## 3.2 UML Profile Metamodeling

The authors of [18, 19] provide a UML profile to adapt the MOF to the context modeling for WebML. In [19], the UML profile is called Context Modeling Profile (CMP). This profile defines appropriate stereotypes for reasoning and context management. The proposed metamodel consists of three stereotypes that aim to: context typing, model the relationships between elements of the context and modeling of contextual situations.

Through stereotypes this approach remains abstract which guarantee meta-models genericity. However, the fact of representing the different concepts of the context of a single term "ContextItem" create an inconsistency since these concepts have different roles involved in the construction of context. Furthermore, despite the diversity of UML tools, these tools do not provide standard ways to access the defined stereotypes and force constraints.


## 3.3    Metamodels Supporting Ontological Models

Metamodels that support ontological models share properties such as:

**Simplicity**: translated by the small size of the metamodel, the limited number of concepts, the definition of metaclasses and high-level ontological relationships for classification and typing of context concepts.
**Expressiveness**: instantiate metaclasses to build networks of concepts.
**Abstraction**: the use of ontologies allows reasoning about concepts by defining properties and logic description rules, which does not indicate specific concepts to some areas or applications.

Fuchs et al.[20] proposes a metamodel that uses a constructor "Entity class" to represent the entities (people, places, objects and events) that share common properties. These properties are typed by another constructor "Datatypeclass". Furthermore, the article [21] proposes a metamodel divided into four parts respectively representing the logical structure of the profile data, the semantics of this structure, the data content and the semantics of this content. We note that the "ontological" metamodels are fairly generic but not specific enough to provide dynamic links between concepts and between concepts and components of the application. In addition, the dynamic structure is not supported.


## 3.4    Metamodel with a High Level of Abstraction

The paper [22] provides a set of metamodels for supporting needs of context oriented approaches while providing a high-level semantic description. The context metamodel is composed of four views represented separately but linked to one another, which represent an advantage of fragmenting aspects of the context depending on the level of abstraction and nature of concepts. The principle of this approach is the following: the context is considered as information characterizing an entity. This contextual information is provided by an acquisition module representing the context providers (sensors).

In this approach two interesting aspects were considered: the acquisition of context and the notification. However, the concentration on typing concepts was in favor of other aspects characterizing the context and context-awareness. In addition, the modeling language used is not a standard but an old object model [4]. Finally, the term sensor puts away other heterogeneous context acquisition modes (implicit) and constitutes a challenge to showcase by context models.

## 3.5  Framework Metamodeling

A model-driven approach for the design of context-aware applications is proposed in [23]. It is based on a generic metamodel used to produce the context model of an application with model transformations. The [23] metamodel describes the contextual elements, their properties and relationships between them. The metaclass "Context" is the contextual information. It is either atomic or composite. Atomic context has the following specializations: Identity, Time, Location, Activity, Preference and Secondary.

In this approach several interesting aspects for the construction of context models were highlighted such as the definition of relationships between entities, contextual situations and typing contextual information. However, we see the introduction of some elements considered modeling details with other more abstract elements such as time constraints information and enumeration type. In addition, the categorization of contextual information is strict, which is undesirable for the genericity of the metamodel. Finally, we can say that this metamodel is overloaded and difficult to instantiate.

## 3.6  Discussion

In this section, we compare context metamodeling approaches discussed in previous sections through several constraints deduced from previous works in this field, we represent each constraint by a code and check whether the metamodeling approaches support them or not (see Table 2). Those constraints are: 1: Time constraints, 2: Validity of context, 3: History awareness, 4: Consideration of future values,

**Table 2** Comparison of context metamodeling approaches

| Metamodeling approaches | Works | 1 | 2 | 3 | 4 | 5 | 6 | 7 |
|---|---|---|---|---|---|---|---|---|
| Centered ubiquitous web applications (UWAs) | [16, 17] | Yes | No | Yes | No | Yes | No | ECA rules |
| UML profile | [18, 19] | No | No | No | No | Yes | Yes | No |
| Supporting ontological models | [20, 21] | Yes | No | No | No | Yes | Yes | No |
| With high level of abstraction | [4, 22] | No | No | No | No | Yes | Yes (subscription) | Notification |
| Framework | [23] | Yes | Yes | No | No | Yes | Yes | No |

5: Acquisition of context, 6: Context access and 7: Use of context (the manner use of context is it specified by the metamodel or not).

It is clear that each approach has its strengths and weaknesses. Thus, the best way to produce qualified and effective results will be to merge multiple meta-modeling techniques, which will reduce each other's weaknesses. Therefore, no single metamodeling technique is ideal but there is a strong link between some context modeling and context metamodeling (e.g. some metamodeling techniques prefer some modeling techniques). We think that a hybrid context metamodeling will give better results.

# 4 Conclusion

Context-awareness is one of the drivers of the pervasive computing paradigm, since a well-designed model is a key in any context-aware system. In this paper, we have analyzed context-aware computing research efforts in context modeling and metamodeling to understand the challenges in this field. Based on our analysis we estimate that the solution, to ensure context awareness in the pervasive paradigm, will be to use a hybrid approach for both context modeling and metamodeling. Our essential goal now is to propose a context management framework that could raise up context-aware challenges.

# References

1. Strang, T., Linnhoff-Popien, C.: A context modeling survey. In: Workshop Proceedings (2004)
2. Weiser, M.: The computer for the 21st century. Sci. Am. **265**(3), 99–104 (1991)
3. Bettini, C., Brdiczka, O., Henricksen, K., Indulska, J., Nicklas, D., Ranganathan, A., Riboni, D.: A survey of context modelling and reasoning techniques. Pervasive Mobile Comput. **6**(2), 161–180 (2010)
4. Ben Cheikh, A., Front, A., Giraudin, J.P., Coulondre, S.: An engineering method for context-aware and reactive systems. In: IEEE 2012 Sixth International Conference on Research Challenges in Information Science (RCIS), pp. 1–12 (2012)
5. Perera, C., Zaslavsky, A., Christen, P., Georgakopoulos, D.: Context aware computing for the internet of things: A survey. IEEE Commun. Surv. Tutorials **16**(1), 414–454 (2014)
6. Henricksen, K., Indulska, J.: Developing context-aware pervasive computing applications: models and approach. Pervasive Mobile Comput. **2**(1), 37–64 (2006)
7. Cipriani, N., Wieland, M., Grossmann, M., Nicklas, D.: Tool support for the design and management of context models. Inf. Syst. **36**(1), 99–114 (2011)
8. Bauer, J., Kutsche, R., Ehrmanntraut, R.: Identification and modeling of contexts for different information scenarios in air traffic. Technische Universität Berlin, Diplomarbeit (2003)
9. Blackburn, P., Bos, J., Striegnitz, K.: Prolog, tout de suite. College Publications (2007)
10. Öztürk, P., Aamodt, A.: Towards a model of context for case-based diagnostic problem solving. In: Context-97; Proceedings of the Interdisciplinary Conference on Modeling and Using Context, pp. 198–208 (1997)

11. Chen, H., Perich, F., Finin, T., Joshi, A.: Soupa: standard ontology for ubiquitous and pervasive applications. In: Mobile and Ubiquitous Systems: Networking and Services. MOBIQUITOUS 2004. The First Annual International Conference, pp. 258–267 (2004)
12. Zhang, D., Gu, T., Wang, X.: Enabling context-aware smart home with semantic web technologies. Int. J. Human-friendly Welfare Robot. Syst. **6**(4), 12–20 (2005)
13. Gu, T., Wang, X.H., Pung, H.K., Zhang, D.Q.: An ontology-based context model in intelligent environments. In: Proceedings of Communication Networks and Distributed Systems Modeling and Simulation Conference, pp. 270–275 (2004)
14. Gandon, F.L., Sadeh, N.M.: A semantic e-wallet to reconcile privacy and context awareness. In: The Semantic Web-ISWC 2003, pp. 385–401. Springer, Berlin, Heidelberg (2003)
15. Vieira, V., Tedesco, P., Salgado, A.C.: Designing context-sensitive systems: an integrated approach. Expert Syst. Appl. **38**(2), 1119–1138 (2011)
16. Koch, N., Kraus, A.: Towards a common metamodel for the development of web applications. In: Web Engineering, pp. 497–506. Springer, Berlin, Heidelberg (2003)
17. Kappel, G., Pröll, B., Retschitzegger, W., Schwinger, W.: Modelling ubiquitous web applications-the wuml approach. In: Conceptual Modeling for New Information Systems Technologies, pp. 183–197. Springer, Berlin, Heidelberg (2002)
18. Moreno, N., Fraternali, P., Vallecillo, A.: WebML modelling in UML. IET Softw. **1**(3), 67–80 (2007)
19. Simons, C., Wirtz, G.: Modeling context in mobile distributed systems with the UML. J. Visual Lang. Comput. **18**(4), 420–439 (2007)
20. Fuchs, F., Hochstatter, I., Krause, M., Berger, M.: A metamodel approach to context information. In: PerCom Workshops, pp. 8–14 (2005)
21. Trætteberg, H., Krogstie, J.: Enhancing the usability of bpm-solutions by combining process and user-interface modelling. In: The Practice of Enterprise Modeling, pp. 86–97. Springer, Berlin, Heidelberg (2008)
22. Belotti, R., et al.: Modelling context for information environments. In: Ubiquitous Mobile Information and Collaboration Systems, pp. 43–56. Springer, Berlin, Heidelberg (2005)
23. Achilleos, A., Yang, K., Georgalas, N.: Context modeling and a context-aware framework for pervasive service creation: A model driven approach. Pervasive Mobile Comput. **6**(2), 281–296 (2010)

# Web Services for Sharing and Reuse Water Data in Morocco

Aniss Moumen, Hassan Jarar Oulidi, Fouad Nafis,
Badraddine Aghoutane, Lamiaa Khazaz, Moroşanu Gabriela Adina
and Bouabid El Mansouri

**Abstract** This study aims to present a contribution to sharing and reuse water data in Morocco through web services (SOA) and spatial data infrastructure (SDI). In same way, we clarify the need for national standardization of water data exchange in order to build a National Water Information System (NWIS) in Morocco. This approach uses a prototype based on geospatial web services and open source modules, in order to facilitate the assessment of "Filling level of large dams in Morocco" indicator.

**Keywords** Web service · SOA · NWIS · OGC · Dams · Morocco

## 1 Introduction

The sharing and reuse of water data is an important step to achieve a National Water Information System (NWIS). Indeed, Morocco has launched its NWIS project since the early 2000s [1], following the launch in 1996 of the Euro-Mediterranean Information System on know-how in the Water sector [2]. So far, NWIS project in

A. Moumen (✉) · B.E. Mansouri
Geosciences of Natural Resources Laboratory (GeoNaRes), Faculty of Sciences,
University Ibn Tofail, Maamora Campus, PO Box 133, 14000 Kenitra, Morocco
e-mail: amoumen@gmail.com

H.J. Oulidi
Hassania School of Public Works (EHTP), Casablanca, Morocco

F. Nafis
Mohammadia School of Engineers, University Mohammed V, Rabat, Morocco

B. Aghoutane
Faculty Poly-Disciplinary Errachidia, Moulay Ismail University, Errachidia, Morocco

L. Khazaz
Faculty of Sciences Ben M'Sik, University Hassan II, Casablanca, Morocco

M.G. Adina
Faculty of Geography, University of Bucharest, Bucharest, Romania

© Springer International Publishing Switzerland 2016                                                297
A. El Oualkadi et al. (eds.), *Proceedings of the Mediterranean Conference
on Information & Communication Technologies 2015*, Lecture Notes
in Electrical Engineering 381, DOI 10.1007/978-3-319-30298-0_31

Morocco has been delayed, due to several factors including the normalization and standardization of water data and metadata at the National level. One of the technical solutions for this situation is the development of Services Oriented Architectures using SOAP and Geographic web services [3, 4] to facilitate exchange and reuse of data [5], as demonstrate by the work of MOUMEN's team by the development of a geo-catalogue water data [15] based on geospatial Web services: OGC WMS, WFS and CSW. Or the gathering in real-time of quantitative and qualitative water indicators from piezometric sensors with geospatial OGC standard SOS [14].

Before the existence of Web services, GIS applications exchanged data using CORBA and DCOM language [6]. The major drawback of these two technologies is that they are not interoperable but competitive [7]. Besides these two technologies which are object-oriented on Internet networks, firewalls prevent the transmission of Object queries and there is no journal that records and identifies them. Hence the importance of using a web service, which is defined according to W3C as an application designed in order to use a protocol (e.g. SOAP) to facilitate the interaction between machines [8].

We want through this work to present the contribution of the Service Oriented Architecture (SOA) to sharing and reuse of water data in Morocco, making reusable data at the filling of large dams in Morocco. These data are provided in PDF format on the corporate website of the Ministry Delegate in charge of Water (MDE).

## 2 Material and Method

The results obtained in this work are based on a methodology organized in four steps:

Step 1: Automatic PDF files Loading, representing "the daily situation of the main dams", from the corporate website of the Moroccan administration in charge of Water (MDW) (http://www.water.gov.ma/).

Step 2: Convert PDF to HTML format. For each file loaded in the previous step, a script will automatically convert each PDF page in an HTML page. This PHP script uses the module "pdftohtml.exe", one of XPDF modules [9].

Step 3: Integration into the database of the generated HTML files. Indeed a third PHP script sculpts the content by searching the data on the fill level of the main dam, exploiting two factors: HTML tags and structure of the original PDF file, which generates a CSV.

Step 4: Development of Web services SOAP (Microsoft 2000) for searching the data from the database. These are client and server web services, developed by exploiting the capabilities of NuSOAP PHP [10]. To evoke SOAP client web service, the user makes a query "GetFeature Info" from an accessible Geoportal via a web interface. All the functionality of this

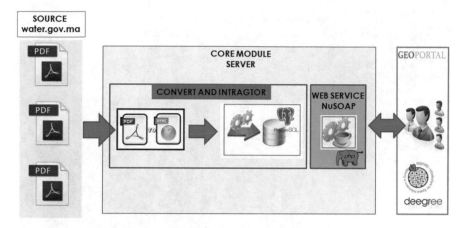

**Fig. 1** The loading process, integration, and data retrieval provided by web site of water department

geoportal is developed through the Open Source Deegree Igeoportal-std tool [11] and geographic web services Deegree-WMS [12] and Deegree-WFS. To summarize this process, a figure below indicated all the steps described above (Fig. 1).

# 3 Results

In order to make available the data of dams filling level in Morocco, the web service developed according to the approach described above is connected to a Geoportal [13, 14]. This is implemented as part of the work of this research on the study area "ZIZ-Rheris" which is justified by the following reasons:

1. Until 2014, this HBA (Hydraulic Basin Agency) still has not had its own website, like the case of other ABH of Morocco. Hence the idea to propose a free and open source solution, allowing the visualization of the data on a Web interfaces, accessible to the general public.
2. This area is at the heart of several projects in the framework of partnerships with international organizations, which mandate the use of recognized standards for the exchange of geospatial data.
3. This region counts more than 2 Million inhabitants (more than 7 % of Morocco's population) according to the last general census conducted in 2004 (HCP 2004), which requires to have the necessary means to extract and follow the fill level of the dam "HASSAN Addakhil", thus allowing the estimate of the requests and offers of water in the area.

**Fig. 2** Thematic map of the "ZIZ-Rheris" study zone, developed using the igeoportal-std Deegree
[13]

To consult the level of filling of "Hassan addakhil" dam, the user accesses the
Web Services SOAP Client, evoked with the query "GetFeature Info" of the OGC
WMS/WFS standard implemented by using the Open Source SDI Deegree. Below
there is the figure of the thematic map and the result of the query "GetFeature Info":

Once the user applies the query "GetFeature Info" on a point representing the
dam "HASSAN Addakhil" of the study zone (Fig. 2), he gets a Web window
containing a URL to launch the Web Services SOAP Client, that evokes the SOAP
Server Web Service in order to retrieve the data of the dam filling level stored in the
database and extract the PDF files loaded from the institutional website.

The user can easily consult the capacity of the dam, the current level and filling
rate and thus compare it with the one from the previous year. The comparison
column is automatically generated (either with a green arrow, orange or red)
depending on the level and degree of filling. This provides the user with quick
information on the status of the dam, if it is higher than last year (Green Arrow) or
below (Orange or Red).

On the same page, by clicking the "History" menu, the user can access the
curves representing the fill level for the current year and last year:

# 4  Discussion

Improve the availability and reuse of data loaded from an official source is the goal
of this work. In addition to the institutional portal of MDW, there are several portals
in Morocco that offer the opportunity to consult official documents, such as: the

portal of the National Documentation Center, of the High Commission for Planning (http://www.abhatoo.net.ma/), or the virtual center of documentation of the National Observatory of Human Development (http://www.albacharia.ma/), as well as the theses search engine (http://toubkal.imist.ma/). These three portals propose the research in the documentary fund associated with each portal using key words. In addition to these portals, The Moroccan administration launched in March 2011 the Moroccan Open Data platform (http://data.gov.ma), with the aim to allow the publication and consultation of administrative data.

In spite of the importance of these portals to easily find quality information, where all the data descriptive elements were considered, all of these portals do not offer the ability to view geospatial data in the form of thematic maps. Moreover, they do not allow the reuse of geographical data through web services, and finally because of the nature of these portals, they do not offer specialized water documentation, but rather a general data base.

Given this context, an initiative called the "SAWIS Initiative [15]" has been set up for sharing of data from research where the authors want to share content using the platform "GEOSAWIS" which is hosted, supported and managed by the Scientific Association for Water Information Systems (SAWIS) (http://sawis.org/) [16]. Similarly as it was announced by the authors of this initiative, the development of such a platform is conditioned mainly by the willingness of researchers to share data they have with the scientific community, which can be an obstacle for the development of such platforms [15]. Hence the interest to develop a prototype based on SOAP Web Services and geographic standards (OGC), to facilitate the reuse and sharing of water data from official sources.

Web services developed as part of this work, were coupled to form an open source Geoportal conforming to geographic standards, in order to complete the work done in the context of SAWIS Initiative. The use cases presented in this paper based on the data of the dam filling level is chosen to validate this prototype and is an example of possible cases that can develop and improve for better accessibility towards unprocessed data loaded from official sources.

Nevertheless, it should be noted that one of the main constraints for the development of such prototypes consists in the variety of formatting of PDF files loaded from the web sites. Indeed, the page layout of the PDF files depends on the authors; this format determines the conversion scripts and integration of data.

In spite of this situation, the development of these Web Services has shown the interest of standardization of data exchange, which is one of the major obstacles to the development of a National Water Information System (NWIS) [17], whose main function is to enable data sharing between the official actors of water sector in Morocco.

# 5  Conclusion

This chapter opens the way for a broad use of Service Oriented Architecture for the sharing and reuse of data generated from official sources. The work presented in this chapter, highlights the importance of standardization of water data exchange, leading to a National Water Information System (NWIS) between the different actors of the water sector in Morocco.

The use of Geo-spatial open source tools, according to international standards (SOAP and OGC), allowed to better assess the indicator "filling level of large dams in Morocco," which is a necessary factor for estimating the offer of water and an important indicator of the water resources management plans in Morocco.

# References

1. Ahmed SKIM, Système National d'Information au Maroc (2006)
2. SEMIDE: Euro-Med meeting of water directors. http://www.semide.net/documents/meetings/fol148169/fol079706 (1997)
3. Pornon, H., Yalamas, P.: Nouvelles architectures SIG et Webservices, In: Géoévénement. IETI, Paris, France (2008)
4. Henri, P., Pierrick, Y., Pelegris, E.: Services Web géographiques état de l art et perspectives. In: Géomatique Expert (2008)
5. Oulidi, H.J. et al.: HydrIS: an open source GIS decision support system for groundwater management (Morocco). Geo-spatial Inf. Sci. **12**(3), 212–216 (2009)
6. Tari, Z., Bukhres, O.: Fundamentals of Distributed Object Systems: The CORBA Perspective. Wiley (2004)
7. Peng, Z.-R., Tsou, M.-H.: Internet GIS: Distributed Geographic Information Services for the Internet and Wireless Networks. Wiley (2003)
8. Zhao, P., Di, L.: Geospatial Web Services: Advances in Information Interoperability. IGI Global (2010)
9. FOOLABS. Xpdf. 2014: http://www.foolabs.com/xpdf/download.html (2014)
10. NuSphere. PHP Web Services with NuSOAP 2010: http://www.nusphere.com/php_script/nusoap.htm (2014)
11. Judit, M., et al.: Deegree iGeoPortal—Standard Edition v2.5. 2006, lat/lon GmbH et Bonn University
12. Andreas, P., et al.: Deegree Web Map Service v2.5. 2006, lat/lon GmbH et Bonn University
13. Moumen, A., et al.: Géo-cataloguer les données sur les ressources en eau, un défi à relever. Revue International de Géomatique (2014)
14. Moumen, A., et al.: A sensor web for real-time groundwater data monitoring in Morocco. J. Geograph. Inf. Syst. (2014)
15. Moumen, A., et al., Initiative SAWIS: Une plate forme communautaire pour le partage des données sur l'eau au Maroc. Int. J. Innov. Sci. Res. (2014)
16. أ. ,مومن. ,et al., المغرب في المائية الموارد :المعلومات لتبادل جديدة هندسة نحو Arab. J. Earth Sci. (2014)
17. Geological Survey (U.S.): National water information system (NWIS). http://pubs.usgs.gov/fs/FS-027–98/ (1998)

# Multidimensional Project Metamodel Construction Methodology for a HIS Deployment

Farid Lahboube, Ounsa Roudies and Nissrine Souissi

**Abstract** Today, health facilities can be restructured only as part of a triple challenge: meeting the healthcare needs in a given area, meeting the requirements of safety and quality, and adapting to the constraints of scarcity of both human and financial resources. In this context, hospitals have undertaken full or partial reforms, which consist primarily in the introduction of Hospital Information Systems (HIS). However, digitalization projects of these structures encounter a wide range of issues due to the complexity and difficulty of such projects. This paper proposes a response to these difficulties according to a HIS project multidimensional metamodel. This one apprehends the dynamic nature of a HIS implementation considering its interactivity with other factors, phases and iterative nature.

**Keywords** Hospital information system · Complex system · Metamodel · Multidimensional · Project management

F. Lahboube (✉) · O. Roudies · N. Souissi
Mohammed V University in Rabat, EMI, Siweb Team, Avenue Ibn sina, B.P. 765, Agdal
Rabat, Morocco
e-mail: lahboube.f@gmail.com

O. Roudies
e-mail: roudies@emi.ac.ma

N. Souissi
e-mail: nissrine.souissi@enim.ac.ma

N. Souissi
Ecole Nationale Supérieure des Mines de Rabat, Rabat, Morocco

F. Lahboube
Hôpital d'Instruction Mohamed V, Rabat, Morocco

© Springer International Publishing Switzerland 2016                                       303
A. El Oualkadi et al. (eds.), *Proceedings of the Mediterranean Conference
on Information & Communication Technologies 2015*, Lecture Notes
in Electrical Engineering 381, DOI 10.1007/978-3-319-30298-0_32

# 1    Introduction

Hospital environment has utterly changed these recent years inducing high financial pressures and increasing demands from citizens. It seems mandatory then that healthcare institutions have to adapt their strategy, particularly the one related to the computerization. Indeed, the establishment of a Hospital Information System (HIS) [1, 2] is a real opportunity to achieve the performance objectives that must converge towards three main objectives, which are: meeting regional healthcare needs, ensuring the quality of hospital activity and optimizing the economic and organizational efficiency [3].

The hospital henceforth needs to build and sustain a performance management process, at all levels, in order to come up with a whole vision involving all healthcare actors. To achieve this goal, the healthcare processes IT strategy must take into consideration all the actors requirements, in each decision-making level of the hospital structure [4, 5]. The execution of such policy means:

- The implementation of a transversal supervision process that goes from requirements elicitation to the deployment of the solution.
- The adaptation of the supervision process to every level of decision (e.g. Executive Council, pole, service), by bringing together all healthcare skills.

However, such culture isn't embodied in the current HIS ecosystem [6]. Indeed, most public hospitals do not fall back on monitoring methods and tools (business or transversal), except within small project framework [7].

We believe that a malfunction of the supervision process during the deployment phase of a HIS will stray us from initial objectives. The informatization movements results confirm these statements and reveal a real deficiency in this supervision brick [8]. Our approach is to consider the HIS within the complex system paradigm and expand the engineering project concept according to this context. This paper intends to bring a formal solution to the problem of a HIS deployment supervision through a project multidimensional metamodel. The concepts introduced in this metamodel take into account the structural complexity and stakeholders diversity.

This work is organized in five sections: after introducing the objective of the current study and its context, the second section addresses the issue of deployment of HIS, according to the complex systems paradigm. The third one describes the methodology used to construct our meta-model project before presenting the proposed Project meta-model in the section four. Conclusions and perspectives are presented at the last section.

## 2 The Complexity of HIS Issue

As shown in [9] the complex system paradigm is a valuable approach that allows us to apprehend the HIS multidimensional aspect. The complexity of a HIS can be approached from several perspectives. Indeed, a HIS is considered a complex system if it reflects one of the following properties [10]: heterogeneity, processing flow, size, hierarchy and evolution. The HIS is also characterized by structural complexity due to the multiplicity of its processes and the strong correlation that results from their interaction [9]. Due to this complex and unpredictable quality, hospitals and clinics fall into the category of a complex and adaptive system (CAS) [11]. Furthermore, the non-linear interactions among heterogeneous agents, such as individuals, health care organizations, and governmental agencies [12] represent a risk for a HIS deployment. In other words, the complexity of health care processes is induced by an involvement of several professionals, disciplines, and departments. Multiple stakeholders and complex organizational arrangements are increasingly common features of information systems projects [13]. Thus, there is a need to design strategies that can cope with this complexity.

Given the fact that the HIS complexity has a considerable impact on the software project management; their implementation must be preceded by an all level evaluation that respects the diversity of points of view.

## 3 Project Metamodel Construction Methodology

The incorporation of a "Project Management" dimension into our supervision metamodel, aims to support the early stages of setting up a HIS. This Project metamodel will provide the necessary elements to be in line with the best practices in project management. Also, this metamodel will include standardized project management concepts to facilitate communication between the different stakeholders. Our Work is based on standards consistent with the PMI (Project Management Institute) Corpus Guide dedicated to Project Management Body of Knowledge (PMBOK® Guide—5th edition).

### 3.1 Construction Methodology

The proposed methodology is to precede the development of our project metamodel by a comparative study of selected models in the literature. The choice is motivated by the relevance of the models and points of view they represent. Indeed, the project metamodels found during our search did not cover all points of view at once; each metamodel being positioned on a particular point of view. Following this comparative analysis, we identified the concepts closely related to the process

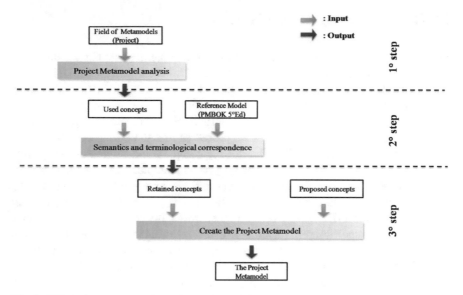

**Fig. 1** HIS project metamodel creation methodology

of a project. In order to complete this analysis, we compared all these concepts to those recommended by our reference model PMBOK 5.Ed [14]. Figure 1 summarizes the methodology adopted to design our metamodel.

The Methodology followed is to go through the literature concerning the project management metamodels and undertake a comparative analysis of the different used concepts. The ones chosen are compared, afterwards, to those recommended by the last version of the referential PMBOK based on semantics and terminological correspondence. The retained terms are completed with other concepts related to the HIS particularities.

## 3.2 Comparative Analysis of Project Metamodels

To conduct our comparative analysis, we chose the most significant metamodels in terms of expressiveness and conceptual coverage. The three metamodels used in this study are presented in chronological order of appearance. Each metamodel represents a point of view offering a schedule for the completion of a project.

The first metamodel [15] of this study proposes a methodology for a project management mainly based on deliverables, techniques and tools. In fact this metamodel is a technical approach for the completion of all project tasks. Note that this is a monolithic model that affects only a single and simple system.

The second metamodel [16] introduces a dynamic approach of a project characterized by several phases and iterations. It also introduces a categorization of the project into disciplines. This metamodel offers a better multi-user point of view for the establishment of an information system. However, it does not support all project resources apart from personal resources represented here by the metaclass "Person". Also, this metamodel considers only one project and doesn't allow its decomposition into sub-project/sub-processes.

**Table 1** Summary table of used concepts in the three metamodels

| Concepts used in metamodels | Metamodels | | |
|---|---|---|---|
| | 1 | 2 | 3 |
| Activity | | • | • |
| Activity relation project | | • | |
| Agreement | | | • |
| Assignment | | | • |
| Capability | | | • |
| Condition | | | • |
| Data | | | • |
| Deliverable | • | | |
| Deliverable package | • | | |
| Discipline project | | • | |
| Guidance | | | • |
| Information | | | • |
| Iteration project | | • | |
| Location | | | • |
| Materiel | | | • |
| Organization | | | • |
| Performer | | | • |
| Person | • | • | |
| Person role project | | • | |
| Person type | | | • |
| Phase project | • | • | |
| Project | • | • | • |
| Resources | | | • |
| Role project | • | • | • |
| Rule | | | • |
| Service | | | • |
| Standard | | | • |
| System | | | • |
| Task | • | | |
| Technique | • | | |
| Time record | • | | |
| Tool | • | | |
| Tool usage | • | | |
| Work product project | | • | |

The third and final metamodel [17] offers a wider vision of a project. It is activity oriented and provides a structured decomposition of the "resource" concept. Unlike previous models, this one allows a better structuring of resources and therefore a more rational allocation based on project activities. This model also allows to interface with the Zachman interrogative Framework [18] providing a more coherent construction of the system being modeled. The decomposition into sub-systems is supported here but is not significant since it appears as a reusable resource.

## 3.3  Discussion of Metamodels

Project metamodels presented above are monolithic, that is to say, they can be used without adding extensions or modifications. However, it is more than necessary to have a model that can adapt to the context of organizations and the problems inherent to the IT deployment process [19] (functional area, size of the project, experience of stakeholders…).

In order to build a metamodel able to withstand the constraints related to the complex environment of HIS, we conducted a terminological and semantic correspondence of the three metamodels to concepts based on the PMBOK referential through Table 1. This table shows that the different types of metamodels have common concepts. The gray lines in the table indicate that some concepts are included in different project metamodels. The retained terms after this analysis will be used to design our Project Metamodel presented in the next section.

## 4  Proposed Project Metamodel

Figure 2 shows the proposed project metamodel which involves several HIS specificities. It's an innovative model since it proposes a project management that is HIS characteristics centered. Furthermore, by using a complex paradigm system, this metamodel offers an efficient solution for managing IT projects within complex systems. First, we define a project by an *Objective* representing and describing the desired result of a system. A *Project* is based on a computerization *Process* project. Our metamodel incorporate several dimensions including the one related to the multi points of view (users). The metaclass *DisciplineProject* reflects the different views of the HIS users. It is associated to the activities accomplished in the same field.

The association of a *Role* to a *ProjectActivity* with cardinality 1 underlines the multi points of view dimension. Figure 3 describes this association.

The proposed metamodel provides more advanced semantics of a *Deliverable*. We have proposed a specialization of this *Deliverable* on *Capability*, *Result* and *Artifact*. *Capability* designates, as an agile software development, the ability to

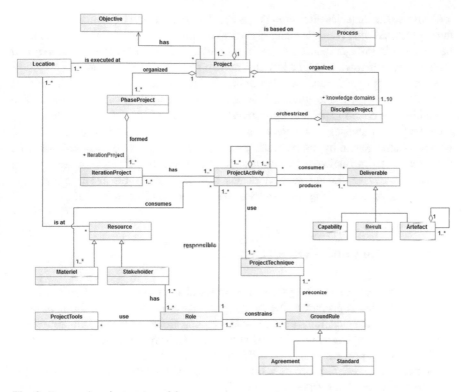

**Fig. 2** Proposed project metamodel

**Fig. 3** The multi points of view dimension

perform a service. The **Result, as** referred by PMBOK concept, is any output data resulting of the execution of a process and project management activities. **Result** includes outcomes (revised processes, restructured organization, tests, etc.) and documents (internal policies, plans, studies, procedures, specifications, reports, etc.). An **Artifact** is something that is produced or consumed by an activity. It can also be formed by a set of small artifacts. These can be a document, a model, a source code, requirements documents, test plans, test cases, etc.

In the context of HIS, regional [20, 21], national [22, 23], or even international dimension [24] requires taking space into consideration. Also, given the evolution of new technologies and groupware, the virtual dimension is henceforth to be observed.

Thus we propose the metaclass **Location** to designate the geographical or virtual whereabouts of the resources. In order to underline the dynamic aspect of a project, we will use the metaclass *IterationProject* to define a fraction of relatively short and clearly defined time (start date and end date) during which a flow of activities must be achieved. Project resources are represented by the metaclass *Resource*. It includes both hardware (*Material*) and human (*Stakeholder*) resources.

Complex projects such as deploying a HIS require the establishment of methodological rules and guides. We suggest the metaclass *GroundRule* with a specialization on *standard* and *agreement*. The latter means any official document including the premises and objectives of the project. The technical aspect of a project is represented by the metaclasses *ProjectTechnique* and *ProjectTools*. The first refers to formalized techniques to be used by a role to execute a *ProjectActivity*. While the second means the computer tools used to facilitate communication and sharing repositories between the project stakeholders.

## 5 Conclusion and Perspectives

The establishment of an IT system in a hospital environment is often slow and problematic. This is primarily due to complex interactions between the healthcare processes and the hospital financial and organizational environment and disciplinary diversity of the actors. Several recent studies have suggested addressing the implementation process of a new HIS as well as the costs and benefits of its implementation. Others propose a set of best practices that can make up for the shortcomings while implementing HIS projects.

This paper proposes a formal solution to complete the deployment process through a multidimensional project metamodel. The used concepts take into account the hospital environment specificities including those related to multidisciplinary actors.

Our metamodel is being experimented underway in a Moroccan hospital structure. Also, it will serve as a corner stone to construct a multi-aspects supervision metamodel. However, it doesn't take into consideration all the organizational levels of HIS such as regional and national levels. Future researches should suggest more extended approaches to this metamodel in order to incorporate the HIS multi-level aspect and thus meet the users requirements in an overall perspective.

## References

1. Rodrigues, J.J.P.C.: Health information systems: concepts, methodologies, tools, and applications, medical information science reference, pp. 35–64. ISBN 978-1-60566 988-5, Hershey (PA), (2010)
2. Haux, R., Ammenwerth, E., Winter, A., Brigel, B.: Strategic Information Management in Hospitals: An Introduction to Hospital Information Systems, Ed. pp. 25–43. New York [u.a.]: Springer (2004)

3. Winter, A., Haux, R., Ammenwerth, E., Brigl, B., Hellrung, N., Jahn, F.: Health Institutions and Information Processing. In Health Information Systems: Architectures and Strategies, pp. 3–23. ISBN 978-1-84996-441-8, Springer-Verlag London (2011)
4. Fichman, R.G., Kohli, R., Krishnan, R.: The role of information systems in healthcare: current research and future trends. Inf. Syst. Res. 22(3), 419–428. ISSN 1047-7047, September (2011)
5. Nemeth, C., Cook, R.: Healthcare IT as a force of resilience. In: Proceedings of the International Conference on Systems, Management and Cybernetics, pp. 1–12. Montreal, Canada, 15–19 Nov (2007)
6. Brender Harno, J.: Evaluation methods to monitor success and failure factors. In Health Information System's Development, Medical Information Science Reference, pp. 605–625. ISBN 978-1-60566-988-5, Hershey (2010)
7. Abirami, R., Dessa, D., Jigish, Z.: Challenges with adoption of electronic medical record systems. In: Medical Information Science Reference, pp. 986–993. ISBN 978-1-60566-988-5, Hershey (2010)
8. Direction générale de l'offre de soins, Atlas 2014 des SIH: Etat des lieux des systems d'information hospitaliers, Mai (2014)
9. Lahboube, F., Haidrar, S., Roudiès, O., Souissi, N., Adil, A.: Systems of systems paradigm in a hospital environment: benefits for requirements elicitation process. Int. Rev. Comput. Softw. 9(10), 1798–1806. ISSN 1828–6003 Oct (2014)
10. Bihanic, D., Polacsek, T.: Visualisation de Systèmes d'Information Complexes Une approche par un points de vue étendus, Studia Informatica Universalis 10, 235–262 (2012)
11. Patricia, A., Abbotta, B.: Joanne Fosterc, Heimar de Fatima Marind, Patricia C., Dykese, f.g., Complexity and the science of implementation in health IT-Knowledge gaps and future visions. Int. J. Med. Inf. 83(7), 12–22 (2014)
12. Moore, T.W., Finley, P.D., Linebarger, J.M., Beyeler, W.E., Davey, V.J., Glass, R.J.: Public health care as a complex adaptive system of systems. In: Eighth International Conference on Complex Systems, Quincy, MA, June 26–July 1 (2011)
13. Brear, M.: Organizational factors: their role in health informatics implementation. In: Medical Information Science Reference, pp. 1295–1303. ISBN 978-1-60566-988-5, Hershey (2010)
14. A guide to the project management body of knowledge (PMBOK® guide), Fifth edition. ISBN 978-1-935589-67-9 (2013)
15. Gane, C., Chapter 7: Process Management: Integrating, Project Management and Development New Directions in Project Management by Tinnirello, P.C. (ed.). ISBN: 084931190X; 26 Sept 2001
16. Martins, P.V., Rodrigues da Silva, A.: PIT-P2 M: ProjectIT Process and Project Metamodel, Software Engineering and Advanced Applications Conference SEAA (2005)
17. Department of Defense, Configuration Management Plan for The DoD Architecture Framework (DoDAF) and DoDAF Meta Model (DM2), Version 1.0, 3 Oct (2011)
18. Zachman, J.A., John Zachman's Concise Definition of The Zachman Framework (2008)
19. Breas, R., Guah, M.W.: Preparing healthcare organizations for new it systems adoption: a readiness framework. In: Medical Information Science Reference, pp. 1328–1341. ISBN 978-1-60566-988-5, Hershey (2010)
20. Harno, K.: Shared healthcare in a regional e-health network, health information systems: concepts, methodologies, tools, and applications. In: Rodrigues, J.J.P.C. (ed.) Medical Information Science Reference, pp. 554–568. ISBN 978-1-60566-988-5, Hershey (2010)
21. Pascot, D., Bouslama, F., Mellouli, S.: Architecturing large integrated complex information systems: an application to healthcare. Knowl. Inf. Syst. 27(2), 115–140. ISSN 0219-3116 (2011)
22. Heeks, R.: Information systems and developing countries. Inf. Soc. 18(1), 101–112. ISSN 1087-6537 (2002)
23. Brown, G.D.: Introduction: the role of information technology in transforming health systems, strategic management of information systems in healthcare. ISBN 1567932428, Chicago: Health Administration Press, (2005)
24. Rada, R.: Information Systems and Healthcare Enterprises, IGI Publishing, pp. 22–337. ISBN 978-1-59904-651-8, Hershey (PA) (2008)

# Generation System of Metadata Software Components: A Proposed Architecture

Yassine Aarab, Noura Aknin and Abdelhamid Benkaddour

**Abstract** We dedicate this paper to study the means necessary for the metadata filling. We assume that a software component has been developed in the state of the art and the owner wants to describe it in order to share it with the e-learning community. It should have for this, an artifact allowing it to inform its metadata component. The difficulty of this task requires the publisher is ergonomic, comfortable and easy to use so that all the cognitive effort is centered on the editorial activity. We present in this work the study of the operating principle and architecture of the system manager of the metadata information that we propose and baptize Interface Generator (IG).

**Keywords** E-learning · MVC · XML

## 1 Introduction

Several reasons led us to choose the development of a GUI instead of solutions, cheaper, presented in the state of the art.

We eliminated the first solution XML editors because few are actually used in the documentary field such as metadata. SGML has shown that the objective of a structured editor is not only allowing you to enter text and structure tags. This is to hide the complexity of this entry, in favor of a tool that allows the user to organize

Y. Aarab (✉) · N. Aknin (✉)
Information Technology and Modeling Systems Research Unit, Faculty of Science, Abdelmalek Essadi University, 93030 Tetuan, Morocco
e-mail: yacinorock@gmail.com

N. Aknin
e-mail: aknin@ieee.com

A. Benkaddour (✉)
Computer Science, Operational Research and Applied Statistics Laboratory, Faculty of Science, Abdelmalek Essadi University, 93030 Tetuan, Morocco
e-mail: ham.benkaddour@yahoo.fr

© Springer International Publishing Switzerland 2016
A. El Oualkadi et al. (eds.), *Proceedings of the Mediterranean Conference on Information & Communication Technologies 2015*, Lecture Notes in Electrical Engineering 381, DOI 10.1007/978-3-319-30298-0_33

313

their ideas and to focus on editorial content rather than procedures necessary for writing documents in line with a model. If tools like XMLSpy are certainly usable by computer, it is less clear to put them in the hands of editors not familiar with XML. So, except for temporary use, we eliminate this type of solution.

The second solution is to develop a static form has been removed for two reasons. First, it hasn't a great scientific interest. Second, because the model is not completely stabilized. We fear, indeed, having to repeatedly modify the program to meet each time a new version of the metadata schema.

## 2   Functional Principle of the IG

Before describing the functional principle of IG, we draw attention to the fact that it is not specifically reserved for the handling of metadata. The latter being a particular context of use of IG. So, in this paper, we are not talking about metadata but about elements in XML sense [1].

### 2.1   External View of the Functional Process

As shown in the Fig. 1, the IG component can be seen as a black box receiving input DTD, XSD schemas or applications increased as XML files profiles. Entries can be either files or URLs.

### 2.2   Internal View of the Functional Process

If we open the black box, see figure (Fig. 2), we see that the functional process of IG consists of five phases in cascade. Each phase performs an action in the process of form generation and supply, at the next stage, the intermediate data.

We study below, the components steps of each phase.

**Fig. 1** External view of IG

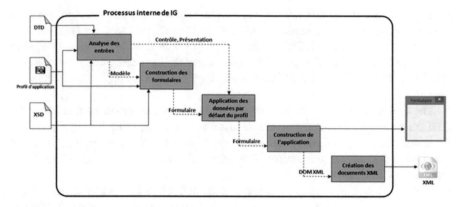

**Fig. 2** Internal view of IG

### 2.2.1 Input Analysis

The analysis process differs depending on the selected entry (Fig. 3). In the case where a DTD IG receives it first performs a test to check for syntax validation if the

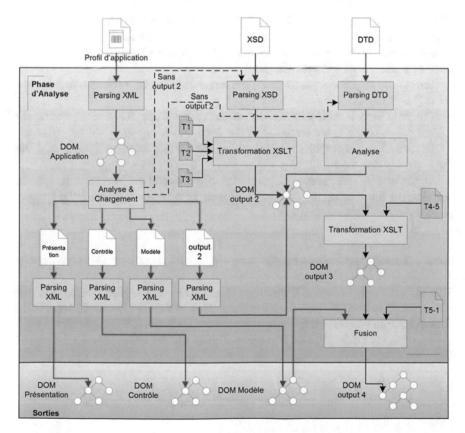

**Fig. 3** Phase "Input Analysis" of the functional process of IG

DTD is correct. If the test is positive, an analysis is performed to determine the structure of valid XML documents that can be instantiated from the DTD.

In case, IG receives an XSD schema, it proceeds in the same way, a validation test for an XSD parser. The aim is to check whether the scheme complies with the XSD standard W3C. As for DTD, the result of this transformation is a DOM of output-2.

The Output-2 is the junction point of the three types of input. At this level, the data are not yet ready to move to the second phase. The next step is to apply to each Output-2 obtained an XSLT transformation with T4-5 sheet to include two key descriptors for further processing. The result of this transformation is an XML stream we call Output-3.

The last step of the analysis phase is executed when the model is part of the MCP pattern. It is to merge some of its descriptors with the Output-3. This is performed through an XSLT transformation with T5-1 sheet. The result is an XML stream we call Output-4. He is remembered as a DOM to be passed to the next phase.

### 2.2.2 Building Forms

This phase is the central core of the generation process of the form. It is iterated for each received output-4 (Fig. 4). The first step begins by analyzing the Output-4 in search of candidate elements to be root XML instance document model of the document associated with the Output-4.

The second step is only activated if the template is an XSD schema. In this case, IG request to an API, named Castor [2].

The third step is to instantiate the model of the form. It's about creating the logical part of the graphic elements of the form we call widgets written in Java.

The fourth step is responsible for building the visual part of widgets. It analyzes the assembly and asked the group to use widgets graphics settings.

At the end of this phase, IG provides XML document we need to get the associate to the root of the document as well as a repository of all the widgets created for this document widget.

### 2.2.3 Default Application Profile

This phase is only activated if the input is an application profile. It takes place in two steps. The first, IG then retrieves the values of the descriptors defined in the default filter and assigns them to different form widgets (Fig. 5). The second step is the presentation part of the MCP pattern. This is the default defined in the default language is used.

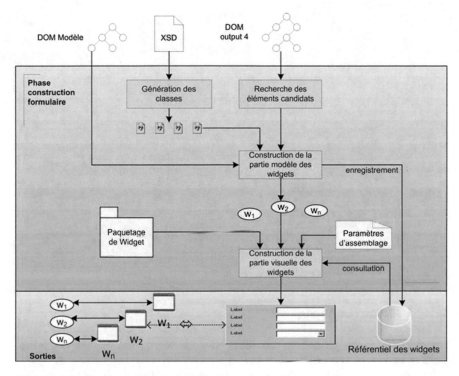

**Fig. 4** Phase "Building forms" of the functional process of IG

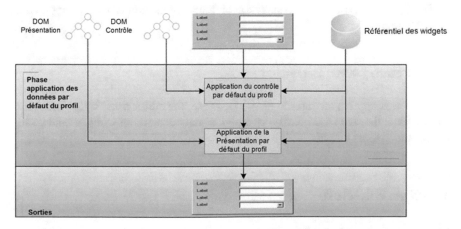

**Fig. 5** Phase "default Application profile" of the functional process of IG

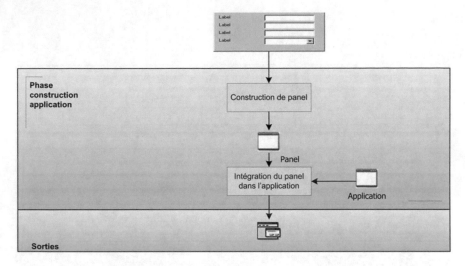

**Fig. 6** Phase "Construction of the application" of the functional process of IG

### 2.2.4  Building the Application

The objective of this phase is to provide users with a functional application. It takes place in two steps (Fig. 6).

The first is responsible for building a graphical panel with widget root obtained in the construction phase forms. This corresponds to a form template the application profile. Each panel is placed in a integrated tab associated at application panel.

The second step makes the connection elements of the editor with the application and its widgets.

### 2.2.5  Creating XML Documents

This phase is applied around the user. Therefore, the editor asks the application to serialize in a file, the DOM XML document obtained as filling widgets fields. Thus we get the system outputs.

## 3  IG Architecture

We just see the outline of the operating principle of IG. We now focus on the components of the system ensuring the functioning and implementation of technical choices. To realize the system, we chose the approach component, MVC architecture, Java and JavaBeans component model.

**Fig. 7** Components
constituting the IG

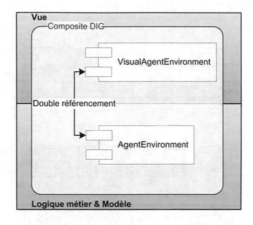

The IG system is a composite of order 3, because it is constructed by assembling components including an order 2. As we have said, all the components are built in accordance with MVC architecture.

Two types of components are distinguished in IG: Components that we call agents and standard components. The notion of agent is not related to the multi-agent systems [3]. An agent in IG is a component responsible for conducting a complex activity which he manipulates standard components.

As shown in the Fig. 7, IG consists of two components: *VisualAgentEnvironment* and *AgentEnvironment*. The first contains all the agents and components used in the view, the second contains the agents and components used in the model and business logic.

AgentEnvironment generator simplifies the reuse of the complexity of the encapsulating operation.

In the composite AgentEnvironment, there are mainly seven officers and three components. Each of them is responsible for providing specific services.

# 4   Widgets Used by IG

To build the form, IG needs a library of graphical components, called widgets, maintaining a certain API that we have defined. The model of this API is based on the principle of the MVC architecture which separates the graphical aspect of the functional aspect.

Building a data entry form valid XML documents with respect to an XSD schema requires, from a functional standpoint, six types of components classified into two categories:

1. Nodes components to represent a node in the XML document.
2. Sheets intended to represent a piece of the XML document components.

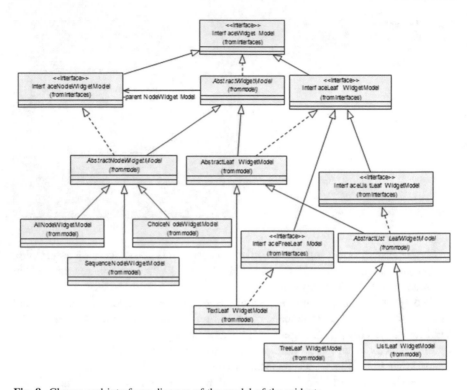

**Fig. 8** Classes and interfaces diagram of the model of the widgets

The node components are three in number. They must implement the interface InterfaceNodeWidgetModel. The special feature of this interface is that it provides methods for adding and removing components son, implements the interface InterfaceWidgetModel while performing the necessary checks on cardinalities son set in the XSD schema.

The attributes common to the six types of components are defined in the methods InterfaceWidgetModel interface. Abstract classes implement most of these methods so that the development of the six components that requires coding of very few specific methods (Fig. 8).

# 5 Results

The following screenshot (Fig. 9) shows the prototype of IG. We see the provision of the various components of the visual environment graphics. Part 1 displays the DOM tree of the current XML document. Part 2 contains the instructions posted by AgentConsign. Part 3 is dedicated to handling attributes of the asset, it is for

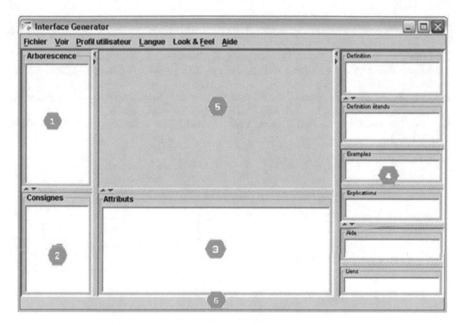

**Fig. 9** Screenshot of IG application

AgentAttribute. The frame 4 consists of six zones each displaying the contents of a descriptor support. It is managed by AgentHelp. Part 5 is used to display the result of form generation. It belongs to AgentForm. This is the main panel of the interface. Zone 6 is the status bar that displays its information AgentStatut. Finally, in the menu bar, users have controls to select entries from the application profile: the controller, presentation and language.

# 6  Conclusion

In this paper, we described the IG component designed to dynamically generate a metadata editing graphical form. We explained why we preferred to develop the IG compared to other solutions mentioned above. IG is able to analyze XSD schemas, DTDs or application profiles, build the corresponding forms and save in an XML document the data entered. We have seen how it is built and how it works. We showed an example of a result and we discussed the work and research opportunities to pursue. Actually, we are in the measuring to reuse IG in the development of a tool for describing and packaging components for their share.

# References

1. Graham, S., Daniels, G., Davis, D., Nakamura, Y., Simeonov, S., Brittenham, P., Zentner, C.: Building Web services with Java: making sense of XML, SOAP, WSDL, and UDDI. SAMS publishing (2004)
2. Wong, E., Burnell, L.J., Hannon, C.: An active architecture for managing events in pervasive computing environments. In: FLAIRS Conference (pp. 405–411) (2004)
3. Jacques, F. Les Systèmes Multi-agents, Vers une intelligence collective. InterEditions, Paris (1995)

# Part VI
# Mobile Agent Systems, Software Engineering, Data Mining and Big Data

# Toward Assistance for Planning Small IT Projects

Mohammed Ghaouth Belkasmi, Toumi Bouchentouf
and Abdelhamid Benazzi

**Abstract** In the case of small IT projects the project managers are tempted to not make a planning in order to save time. However, establishing a schedule is necessary to successfully achieve a project. To provide planning assistance we propose a platform based on solving the planning problem modeled as a constraint satisfaction problem (CSP). Which should help project managers to analyze the project feasibility and to generate useful schedule with just few informations as input. We also allow managers to supervise the project progressing and regenerate the schedule if an adjustment is made or a change has occurred at any phase of the project life cycle. Two versions of this platform were implemented and we are working on the third.

**Keywords** Project management · Planning · Feasibility · Schedule · Task assignment · CSP · Solver

## 1 Introduction

The IT project management is a process to organize from start to finish the project achievement. One key component of this approach is the planning which is based on a set of crucial operations: risk Management [1, 2], the cost estimation [2, 3] scheduling [2], assignment [2], and quality assurance [1].

M.G. Belkasmi (✉) · T. Bouchentouf
Team SIQL, Laboratory LSEII, ENSAO, Mohammed First University,
BP. 669, 60000 Oujda, Morocco
e-mail: ghaouth@gmail.com

T. Bouchentouf
e-mail: tbouchentouf@gmail.com

A. Benazzi
Laboratory MATSI, ESTO, Mohammed First University, BP.669,
60000 Oujda, Morocco
e-mail: benazzihamid@yahoo.fr

© Springer International Publishing Switzerland 2016
A. El Oualkadi et al. (eds.), *Proceedings of the Mediterranean Conference on Information & Communication Technologies 2015*, Lecture Notes in Electrical Engineering 381, DOI 10.1007/978-3-319-30298-0_34

However, in the case of small businesses, project managers are questioning the usefulness of establishing a rigorous schedule for small project (having size less than 1 year/man), especially as this task is time consuming [4]. Thus, these managers would often ignore planning and just try to assign tasks to resources. So to encourage managers to do planning we created a solution that provides planning assistance. This solution is a constraint programming [5] based platform [4]. Which allows, firstly, to simulate the feasibility of a project relative to its size, the human resources available to participate in its implementation and its duration. Then, if the simulation result is affirmative, a schedule is generated and submitted to be validating by the project manager before being distributed to the affected resources.

But, usually, as the project is progressing, some adjustments could be made. Like changing resources assignment (due to sick leave for example) or some deadlines could be changed. So, if any changes has occurred to the initial project situation, at any phase of the project life cycle, the previously generated schedule can be automatically updated according the the new situation. The new generated schedule has to be committed by the project manager before that the changes be visible to project's team members.

In the remaining sections of this article:

- We present the platform architecture in Sect. 2.
- We introduce the solver implementation in Sect. 3.
- In Sect. 4, we present some improvements that we made.
- Then we conclude with some trends of future works in Sect. 5.

## 2   The Platform Architecture

We designed the planning assistance platform, in its actual version, as shown in Fig. 1. It has three components: data, solver and interfaces.

As input, this platform is satisfied with a minimum of description informations. The project manager is asked to specify the size of the project, the resources (team

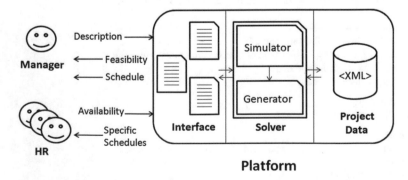

**Fig. 1** Platform architecture

members) and deadlines (generally imposed by the client) for this project. The assigned resources, which each one has his own profile, have to submit their availability and maintain it up to date. Based on these parameters passed as input, the platform analyses the feasibility of the project by ensuring the timeliness and assessing the adequacy of resources. If the project is deemed feasible, a schedule is generated.

## 2.1 Data Management Issue

Since we have a set of information and data to be processed, and especially to share with several employees (project managers, human resources) we chose XML as a data carrier [4]. Thus we modeled the description of a software project as an XML schema [6].

This model Fig. 2 will store the data provided by the project manager input and then contain the detailed generated schedule. And through transformation scenarios, this schedule can be visualized as Gantt chart which allows also showing the progress status.

## 2.2 The Solver

The solver is the kernel of the platform. It contains two components: a feasibility simulator and schedule generator.

**The simulator** The feasibility of a project is a question to answer before any further step in the project analysis. So, based on parameters passed as input, the simulator analyses the feasibility of the project by respecting the deadlines and assessing the adequacy of resources.

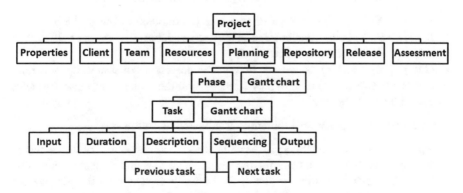

**Fig. 2** XML schema representing an IT project

**The generator** If the project is deemed feasible by the simulator, a schedule is generated in output as a matrix for assigning tasks to resources in a time grid covering the duration of the project.

More informations about the solver are given in Sect. 3.

## 2.3   Interfaces

There is two kinds of user interfaces:

**Project manager interfaces** helping him to create and describe a project (size, deadlines), assign resources, evaluate the feasibility, generate schedules and validate them. Also, he can visualize the achievement status of supervised projects. In this set of interfaces, some allow making adjustment of resources assignment or update of deadlines.

**Team members (resources) interfaces** allow each member to create his own profile (personal informations and skills), to submit his availability (vacancy, sick leave, training) and maintain it up to date. Members can also visualize a personalized schedule representing scheduled tasks that they have to do.

## 3   Implementing the Solver

### 3.1   Formalization of the Planning Problem

Insinuating, that planning takes account of all the constraints induced by the scheduling of tasks, the availability of resources and deadlines set by the client. All this led us to consider the planning problem as a constraint satisfaction problem [4] CSP (Constraint Satisfaction Problem) [7, 8]. A CSP is modeled as a set of constraints imposed on variables, each of these variables taking values in a domain. More formally, a CSP will be defined by a triplet $(X, D, C)$ such that:

- $X = \{X_1, X_2, \ldots, X_n\}$ is the set of variables (unknowns) of the problem.
- $D : X_i \mapsto D(X_i)$ is the function that maps each variable $X$ its domain $D(X)$, that is to say all the possible values of $X_i$.
- $C = \{C_1, C_2, \ldots, Ck\}$ is the set of constraints. Each constraint $C_j$ is a relationship between certain variables $X$, restricting the values that these variables can take simultaneously.

So the schedule generator problem is formalized as follows:

- $M_{RS}\{M_{rs}, 0 \leq r \leq R \text{ and } 0 \leq s \leq S\}$ is the matrix of allocation of tasks where the lines refer to resources and the columns represent the sequencing time (days or weeks depending on the desired affinity of schedule) the Mrs are the unknowns of the problem (Who will do what and when?).

**Fig. 3** Resolution principle
of CHOCO

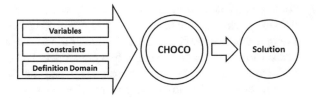

- $D : M_{rs} \mapsto D(M_{rs})$ is the function that maps each variable $M_{rs}$ its domain $D(M_{rs})$, that is to say the set of possible tasks that may be responsible for the resource $r$ during $s$.
- $C$ is the set of constraints. In our case we have identified two types of constraints constituents two subsets of $C$:

    - The subset of constraints ensuring compliance with the scheduling of tasks
    - The subset of constraints ensuring that all tasks are assigned, and thereafter the entire project is completed.

If the project is feasible, the problem thus modeled to experience two spells:

- A solution is calculated and a schedule is generated
- There is no solution and the cause is simply the availability of resources. Then the project manager is required to strengthen or redeploy resources allocated to the project, see negotiate their availability.

The schedule generation is performed by solving the CSP that we made to model the planning problem.

## 3.2 Implementation

The resolution of the CSP thus formulated, has been implemented in Java using the API CHOCO [9]. This solver requires presenting a description of the variables, their domains and the set of constraints, as illustrated in Fig. 3.

The implementation is about how to define, using CHOCO, the CSP model of the planning problem. Especially, expressing the set of constraints. Which requires a good representation of variables.

## 4 The Platform Improvement

### 4.1 Limits of the First Version

Within the first version of the platform, the simulator that we implemented had successfully judged the feasibility of projects among a range of testing projects including feasible ones and others not, because their durations was undersized compared to their sizes, or they have not enough members in the project team. To test our generator were taken as sample test project to be completed in three phases: Design, Development, and Test and Integration. We suppose that the project team has six members: two designers, two developers and two integrators, each one has his own availability. For the test we have varied the size of the project and we have seen if the generator arrives to give the desired schedule and also we calculate the required computation time, the number of nodes and backtracks made when searching for solutions. In all cases the generator has found a solution satisfying all the constraints and therefore it has managed to achieve planning. But we found that the generation becomes taking longer and longer time when the project size exceeds 45 days/man, which is not acceptable. But we kind of expected behavior since the size of the matrix [4], and therefore the number of variables and constraints that govern the values of these variables, increasing rapidly with the expansion of the project size. This led us to seek improvement in the modeling problem in CSP and the implemented schedule generator. Here comes the second version.

## 4.2 Improvements Made in the Second Version

To achieve the improved version of the platform, we made three enhancements:

1. **Encapsulation of the feasibility simulation**
   The feasibility simulation becomes part of the solver. So, we don't have to present the needed twice. The solver provides a service (a method) that let us simulate the feasibility of the project and the same instance will generate the schedule on demand and of course if the project is feasible.
2. **Reduction of the search space**
   The complete set of assignments is called the search space of the CSP [5]. So, in order to contain the search space expansion we made an efficient removal of values from the domains of the variable in order to prune portions that cannot possibly contain a solution of the problem. Therefore, we redefined the function D, which maps each variable Mrs its domain D(Mrs) in order to exclude incompatible values due not only to the resource availability but also to his profile: a designer should not have a development task even if he is available, or a developer should not have a development task out of the time slot fixed for the development phase.
3. **Reformulation of the constraints**
   The subset of constraints ensuring compliance with the scheduling of tasks is kept unchanged. The other subset of constraints ensuring that all tasks are assigned, have been simplified in one rule instead of as many rule as existing phases of the project.

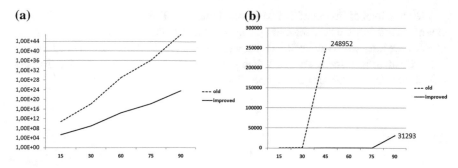

**Fig. 4**  **a** Search space, **b** running time

## 4.3   Results Within the Improved Version

**The search space** We implemented the improved $D$ function as a method named *getTasks()*, and to test the increasing of the size of the search space we made a comparison between the old platform and the new one with the improved solver. We obtained these results shown in the plot in Fig. 4a (the x axis represents the project size and the y axis represents the search space size with a base 10 logarithmic scale). We notice that the search space expansion is contained within the improved solver.

**The running time** Using the same range of project in the previous test [4], we compared the running time taken in the case of the old and the improved solver. We made this plot Fig. 4b to represent the results. Here we see that the old solution become quickly very long on response for project having a size greater than 45 days/man. But the improved solver is still responding in a good time until the size reaches 120 days/man, there the running time exceeds 5 min. For size over 245 days/man its about hours ($\simeq 9$ h). So we are still working on a third version that will ensure an acceptable running time for projects sized under 360 days/man.

## 5   Conclusion

In order to provide planning assistance to IT project managers, we proposed a platform [4] that evaluates the feasibility of a project and makes a planning within an assignment of tasks to available resources, having the suitable skills. To create such a platform we started by modeling an IT project using an XML schema in order to store information about the project and its planning. Then, we formalized the planning problem as a CSP. To solve that CSP, we implemented a solver using the API CHOCO. Thereafter, we developed many interfaces to allow, project managers and other team members, interacting with the platform and having a good and helpful visibility on the schedule, tasks assignment and the achievement progress.

Two versions of this platform were implemented. The second one brought some enhancements but we still seek more improvement to ensure an acceptable running time for projects sized under 360 days/man. So we are working on the third version that will include also a way of raising measures and informations for decision makers to evaluate the projects in progress.

# References

1. Mnch, J., Heidrich, J.: Software project control centers: concepts and approaches. J. Syst. Softw. **70**, 3–19 (2004) (Elsevier). doi:10.1016/S0164-1212(02)00138-3
2. Charette, R.N.: Software Engineering Environments. Concepts and Technology. McGraw-Hill International, New York (1987). ISBN-10:0070106452, ISBN-13:978-00701064
3. Boehm, B.: Software Engineering Economics. Prentice-Hall, Englewood Cliffs (1981). ISBN-10:0138221227, ISBN-13:978-0138221225
4. Belkasmi, M.G., Bouchentouf, T., Benazzi, A.: A constraint programming based platform for planning assistance and schedule generation: Case of small IT projects. In: 2011 International Conference Multimedia Computing and Systems (ICMCS'11), Ouarzazate, 7–9 April 2011. ISBN:978-1-61284-730-6. doi:10.1109/ICMCS.2011.5945575
5. Jussien, N.: Constraint programming for software engineering. In: Fifth Congress of Logic applied to Technology (LAPTEC'05), Himeji, Japan (2005)
6. Belkasmi, M.G., Bouchentouf, T., Azizi, M., Benazzi, A.: Modeling projects in e-learning course: a case of an information technology project. J. Comput. Sci. **6**(7), 823–829 (2010). ISSN:1549-3636 Â© 2010 Science Publications
7. Azouazi, M., Moussaid, M., Belaissaoui, K.: Modeling of a Problem of Time Jobs (FTE) with constraint satisfaction problems (CSPs): adapted to the new reform of higher education. In: Proceedings of International Symposium on Computer Applications and IA2006, pp. 184–191, Oujda (2006)
8. Belkasmi, M.G., Bouchentouf, T.: Implementing a generator schedules based CHOCO solver: the case of ENSA. JDTIC10, Fez, Morocco, 16 July 2010
9. Jussien, N., William, R., Lorca, X.: The CHOCO constraint programming solver. In: Workshop on Open Source Software for Integer Programming and Constrained (OSSICP'08), Paris, France (2008)

# Estimation and Optimization of Energy Consumption on Smartphones

**Khalil Ibrahim Hamzaoui, Wiame Benzekri, Gilles Grimaud, Mohammed Berrajaa and Mostafa Azizi**

**Abstract** The energy consumption into a smartphone is defined by the energy cost necessary for the components equipment to achieve their activities. These activities are induced by software executions related to the users' activity. In other words, the energy consumption could result from the execution of different interactions between hardware, software and users which by their behaviors trigger a workload on hardware components. The assessment and measurements of the cost of consumed energy, as well as problems in the methods used for evaluation and optimization of energy consumption, are the topic of this investigation. As result of this work, we conclude that some of the investigated techniques are more accurate than others for tracking the main sources or equipment responsible of consuming energy.

**Keywords** Smartphone · Mobile · Operating system · Kalimucho · AppScope · RAPL · BITWATTS · Energy · Consumption

K.I. Hamzaoui (✉) · G. Grimaud
IRCICA 2XS, Lille, France
e-mail: hamzaoui.khalil@gmail.com

G. Grimaud
e-mail: gilles.grimaud@lifl.fr

K.I. Hamzaoui · W. Benzekri · M. Berrajaa
Faculty of Sciences, LANOL, Oujda, Morocco
e-mail: wiame.benzekri@gmail.com

M. Berrajaa
e-mail: berrajaamo@yahoo.fr

M. Azizi
MATSI EST, Oujda, Morocco
e-mail: azizi.mos@gmail.com

© Springer International Publishing Switzerland 2016
A. El Oualkadi et al. (eds.), *Proceedings of the Mediterranean Conference on Information & Communication Technologies 2015*, Lecture Notes in Electrical Engineering 381, DOI 10.1007/978-3-319-30298-0_35

# 1   Introduction

Despite the success of smartphones on the market and the exponential growth of mobile applications, their usefulness has been and will be severely limited by the life or autonomy of their batteries. That is why the optimization of energy consumption in smartphone applications is of critical importance. However, most of mobile applications developed so far have been designed unconsciously of their consumption of energy [1].

The energy consumption results from a smartphone consumption of its components involved in running its applications. It is therefore important to measure and understand how the energy is consumed in these mobile devices.

In addition, a component of a smartphone can have one or more levels of power states:

- The active state: Where the application processor is operational.
- The inactive state: Where the application processor is inactive but the communication processor generates a low level of activity.
- The state of tail: Where the device is not in the inactive state but no application is activated.

The research presented in this chapter is part of the development of models. The second section will detail the main sources that are of energy consumption and their different impacts on battery life. The third section will focus on the evaluation, the importance as well as the measurement of energy cost. We conduct our survey on the following models: Kalimucho, AppScope, RAPL and BITWATTS. We will make an evaluation of different methods. The last section concludes this paper.

# 2   Hardware Components

Modern smartphones are equipped with a variety of integrated hardware components. These components have different impacts in mobile device power consumption.

Energy consumption models needs not only to consider the main hardware components of the device, but also its characteristics. Each hardware component can be in many states, each one consuming a different quantity of energy.

## 2.1   CPU

The power consumption of processor is strongly related to the frequency of the CPU. The energy consumption for a single CPU is calculated based on the formula in DevScope [2]:

$$P_{(component)} = P_{(Measured)} - P_{(Basic)} \tag{1}$$

$P_{(Basic)}$ represents the total power consumption of other hardwares including the CPU's like RAM. CPU Hardware has two states: idle and loaded [2].

## 2.2  Display

Multimedia applications, such as streaming, image, video capturing tools and games comprise now a considerable portion of the daily usage of smartphones. The power consumption of these applications depends heavily on the type of display technology being used, which also plays an important role in human-machine interaction. The power model for the display depends on the display technology used.

The Display type of smartphones are LCD, OLED and Active Matrix OLED (AMOLED). For example energy consumption for LCD is related to its brightness.

In Fig. 1, the brightness range of the screen is 0–255. The results show that the power consumption of LCD varies with the brightness level of the screen and shows also that the power consumption of the display is not linear with the brightness level. Pixel varies with the chromatic color composition that is being displayed and doing more intense colors that require more power to be displayed [2] (Fig. 2).

A smartphone display is composed of many individual pixels. The color of a pixel is constructed via the combination of basic RGB colors which are themselves the only colors directly emitted from the display unit via tiny sub-pixels. The display unit is defined by the arrangement of these sub-pixels.

Measurements power for AMOLED and OLED [3] show that when the screen is rocking level yellow (red + green), there is less energy consumed. Unlike LCD technology, the power consumption of an AMOLED screen is highly color

**Fig. 1** Power measurements for different colors on an AMOLED display [3]

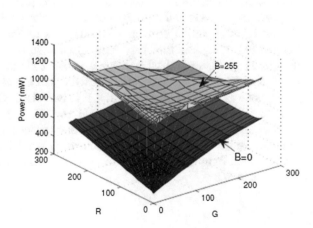

**Fig. 2** Change in power with color on OLED display [3]

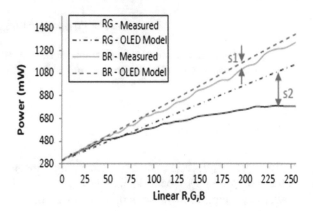

dependent. The LCD can be modeled using simply the brightness level as the variation with color is not significant.

## 2.3 Network Interface

The superscript **Wi-Fi**: The Wi-Fi is short for wireless connection standard and offers a bandwidth of 11 Mb/s for IEEE 802.11b standard. This component also has idle and loaded states. In loaded state, the energy consumption accordance with various frequencies of its data package delivery. The higher frequency of data package delivery, the higher consumption of energy is [4].

In idle state into both low and hight frequency statues there is the same energy consumption because they are under idle condition.

**3G**: Appeared in 2000, the third generation (3G) designates a generation mobile phone standards. It is represented mainly by the Universal Mobile Telecommunications System standard (UMTS) and CDMA2000, allowing rate of 2–42 Mb/s, which are much faster than with the previous generation.

The energy consumption of 3G depends on the state it is operating.

**4G**: 4G is the fourth generation of standards for mobile telephony. Succeeding 2G and 3G, it enables the "very high speed mobile" data transmission at higher theoretical speeds of 100 Mb/s or greater than 1 Gb/s.

4G smartphones can have two states: RRC_IDLE and RRC_CONNECTED. The smartphone transits from the RRC_IDLE to the "RRC_CONNECTED" when there is a reception/transmission of data (Table 1).

The energy consumed by the 3G network interface on receipt and transmission is minimal compared to the energy consumed by the 3G interface. However, during the transition from the IDLE state to the active state, 4G interface consumes more energy than 3G.

**Table 1** Power consumption of an HTC smartphone [5]

| Interface (3G/4G) | Emission (mW) | Reception (mW) | Passage to the active state (mW) |
|---|---|---|---|
| 4G | 438.39 | 51.97 | 1288.04 |
| 3G | 868.98 | 122.12 | 817.88 |

## 3 Experimentation

### 3.1 Kalimucho

Kalimucho is a platform of reconfiguration that implements a heuristic context deployment to find a configuration satisfying the conditions of context and of Quality of Service (QoS). This platform focuses on the management of the quality of services in mobile systems constrained by dynamically reconfiguring these applications are based on components that allow a modular flexibility of configuration [6].

The principle of Kalimucho is to capture the context elements necessary for making decision (Fig. 3).

In the example showed [7], this museum application offers visitors to view on their smartphone a description of works. The best QoS is to provide the video in color.

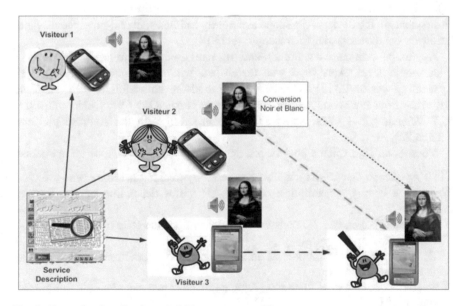

**Fig. 3** Example of application of visiting a museum [7]

A possible scenario of use of the application is when the visitor 3 moves. Two cases are presented:

Case 1: Low flow and the device are in the scope. To ensure the service, the solution is to reduce the number of data to be transmitted by a conversion of the video in black and white or by reducing the resolution of the image.
Case 2: The device is not in the scope of the server. The solution is to use the device of the visitor 2 or 1 to make a relay for device 3 under condition that the visitor Device 1 or 2 have the energy available.

For a good deployment of a suitable configuration, Kalimucho refers to four criteria to ensure a good quality of service: Energy, CPU, memory and network capacity.

## 3.2   AppScope

AppScope is an Application Energy Metering Framework for Android Smartphones using Kernel Activity Monitoring [8].

The objective of AppScope model is to measure the energy application system for applications that uses power equipment models and usage statistics for each hardware component.

The AppScope tool accurately estimates the energy consumption of Android applications. The system analyzes the traces of applications and the system AppScope uses an event-based approach [8] for collecting its information.

AppScope provides the energy consumption of Android applications automatically, being customized to the underlying system software and the hardware components in the device. AppScope accurately estimates, in real-time, the usage of hardware components at a microscopic level [8].

AppScope estimates accurately and in real time the use of hardware at a microscopic level. AppScope was developed with the Linux 2.6.35.7 kernel. SystemTap version 1.3 [9] also uses K probes and data collection for the evaluation. All evaluations are conducted on HTC Google Nexus One (N1; Qualcomm QSD 8250 Snapdragon 1 GHz, Super LCD 3.7 in.) [10] with the Android platform version 2.3.

Yoon et al. [8] explains how AppScope interact with the various components:

CPU: AppScope detects the switching process by tracking a wake event.
Wi-Fi: The energy consumption of the LAN varies depending on the flow of packet.
3G: The interface energy 3G consumption depends on the state of the CRR (CRR: The Radio Resource Control) [9].
LCD: The energy consumption of an LCD screen is proportional to the display brightness and display time.
GPS: AppScope monitors "Location Manager Calls" and calculate the duration of activation of GPS.

The experiments showed that AppScope generates less accurate when the presence of the tail energy of the influence on the accuracy of the results generated by AppScope results.

The precision of AppScope depends of the electrical aspect of the targeted unit. AppScope is operational in a limited number of processor architectures and does not take into account the multiple core architecture.

## 3.3 RAPL

In 2012, Hahnel et al., developed the RAPL (Running Average Power Limit). RAPL is an approach which allows a comparison of the two mobile applications and sectors shows that they show different energy consumption while offering similar services [11] introduced RAPL. Manual instrumentation is coarse and is at the expense of the power of the device, it remains inexact to calculate the energy consumed by tail applications. Snowdon et al. [12] tried to find approaches for energy efficiency and created templates that adapt the behavior of applications, for measure the energy required to decode an image and choosing the path for the result of a query.

The contribution of the paper is choice of path for the query result to calculate the energy consumed by the device drivers because an un-optimized driver will use more CPU resources, so there will be more energy consumption. Hardware devices such as hard disks or network interfaces consume energy, the proposed idea is to measure not only the energy consumption of the device, but also to measure the energy that the driver for this device consumes in terms of computing power.

RAPL is based on energy sensors available in the latest Intel processors to measure the energy consumption and to take account of energy consumption in the software components. The use of RAPL infrastructure serves to characterize the energy costs and decode video tranches and choice of path for the query result.

Example of coding and decoding image: The encoder delivers two versions: a high quality (choice1) and another version of lesser quality that can be decoded with low consumption (choice2). The decoder will decide based on budget battery for choice between 1 and 2 [13].

Choice of path for the query result: Generally the System Management database has several choices to achieve the result of a query. The goal is to enable the DBMS to select not only the combination of the operator that calculates the result as quickly as possible, but also one that consumes less energy. According to [13], the limitations of RAPL model are:

- RAPL provides sensors that measure power consumption at the CPU and memory, but the problem resides in the technics and the impossibility of calculating consumption of IOs devices.
- RAPL must have precise information on the video split for minimizing energy costs which render it difficult to put into practice.

## 3.4  BITWATTS

In 2014, Colmant et al. developed the BITWATTS, a fine grained supervision technique that provides a real time estimation of consumed energy power. The mean idea of BITWATTS is based on the use of processes and the deduction of consumed energy. Additionally, this conception can be used in distributed environments [14].

Current approaches, like JOULEMETER, can supervising virtually one application at a time (each Virtual Machine supervises one and only one application). In this case, Virtual machines (VM) are considered as black boxes.

BITWATTS is developed with Scala [15], an extension of Power API actor toolkit [16]. Comparing for its output to physical measurements performed with a power for justify its effectiveness meter. Results indicate that BITWATTS provides trustworthy power estimation within a few per cent of actual measures when configured with the appropriate power model for the underlying hardware [17].

The idea is based on the use of process and deduces the energy consumption in a virtual environment. for example we have not access to physical CPU but we can observe the emulated processor by the Virtual machines. The architecture of BITWATTS can be written in Fig. 4.

### 3.4.1  Consumption Estimation in Communication Channels

The exchange of data between BITWATTS instances requires 2 levels of communication:

- Exchange of data between host and VM to estimate the consumption of a process in the virtual machine
- In a distributed configuration, to estimate the power consumption, BITWATTS has to use another server Evaluation of the supply pattern:

To demonstrate that BITWATTS can handle applications with diverse load, we begin observation with a basic experiment on Intel core I 3.

In this experiment, the results are compared not only with respect to the performance of average power limit (RAPL), but also with external measurement tool

**Fig. 4** BITWATTS
middleware architecture [14]

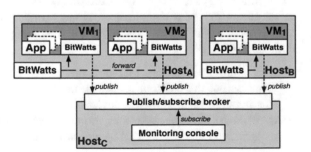

**Fig. 5** Decreasing load of stress on I3 in the host, compared to RAPL [14]

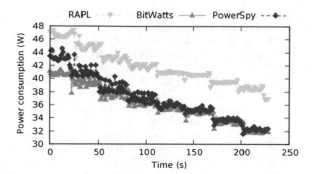

(Power spy) [17] which allow an information accurate consumption. In addition, for this experiment, the CPU frequency was put at 1.6 GHz (Fig. 5).

shows that energy is overestimated by RAPL, BITWATTS is near to PowerSpy which can be regarded as the ground truth. This indicates that BITWATTS performs an accurate estimate on the system under test.

# 4  Conclusion

In this paper, we studied the estimation and optimization problem of the energy consumed into a smartphone. First, we presented the assessment and measurement of the energy cost, then the problems encountered in the methods used for energy optimization. Precisely, four major models are considered: Kalimucho, AppScope, RAPL and BITWATTS.

Furthermore, we have discussed the energy costs of various components trying to correlate them with the aforementioned models. The obtained results allowed us to identify the main sources responsible for the energy consumption.

We noticed that the latest models do not take into account the energy of the tail which permanently influences the activities of the software, contrarily to BITWATTS that estimates the energetic consumption within a virtual environment.

In future works, we aim to suggest a hybrid model based on some of these models trying to take benefits from their advantages. We will perform measurements and tests to carry out the continuation of our work.

# References

1. Perrucci, G.P.: Energy saving strategies on mobile devices. In: Multimedia Information and Signal Processing—Aalborg University, Aalborg, janvier, pp. 1–150 (2009)
2. Jung, W., Kang, C., Yoon, C., Kim, D., Cha, H.: DevScope: a nonintrusive and online power analysis tool for smartphone hardware components. In: CODES + ISS'12 Proceedings, pp 353–362

3. Mittal, R., Kansal, A., Chandra, R.: Empowering Developers to Estimate App Energy Consumption Wattson Mobicom12
4. Ma, J., Yu, H., Gong, X., Zhang, X.: Research on Online Measurement Method of Smartphone Energy Consumption, pp. 443–447, 8–10 Nov 2014
5. Measurements from a HTC with a processor "Qualcomm MSM8655 1 GHz" 3G interfaces, Wi-Fi and 4G and a RAM 768 MBHTC, http://www.htc.com/us/smartphones/
6. Pathak, A., Hu, Y.C., Zhang, M.: Where is the energy spent inside my app? Fine grained energy accounting on smartphones with eprof. In: Proceedings of the 7th ACM European Conference on Computer Systems—EuroSys '12
7. Louberry, C., Roose, P., Dalmau, M.: Kalimucho: Plateforme d'Adaptation des ApplicationsMobiles, https://hal.archives-ouvertes.fr/hal-00593459/
8. Yoon, C., Kim, D., Jung, W., Kang, C., Cha, H.: AppScope: Application Energy Metering Framework for Android Smartphones using Kernel Activity Monitoring, Conference ATC12
9. SystemTap. http://sourceware.org/systemtap
10. Roy, A., Rumble, S.M., Stutsman, R., Levis, P., Mazieres, D., Zeldovich, N.: Energy management in mobile devices with the Cinder operating system. In: Proceedings of EuroSys, 2011 Information 2015, vol. 6, no. 15
11. Marcus, H., Bjorn, D., Marcus, V, Hermann, H.: Universit, Technische, Dresden, Dresden « Measuring EnergyConsumption for Short Code Paths Using RAPL », vol. 40, pp 13–17 2012
12. Snowdon, D.C., Petters, S.M., Heiser, G.: Accurate on-line prediction of processor and memory energy usage under voltage scaling. In: Proceedings of the 7th International Conference on Embedded Software, pp. 84–93, Salzburg, Austria, Oct 2007
13. Hamzaoui, K.I., Grimaud, G., Berrajaa, M., Azizi, M., Betari, A.: Survey on adaptation techniques of energy consumption within a smartphone. In: Science and information Conference 2014, London, UK, 27–29 Aug 2014
14. M.Colmant, M.Kurpicz, P.Felber, L.Huertas, R.Rouvoy, and A.Sobe. Process-level Power Estimation in VM-based Systems.European Conference on Computer Systems (EuroSys), Apr 2015, Bordeaux, France
15. The Scala langage Specification. http://www.scala-lang.org/
16. Power API (AGPL). http://powraepi.org
17. Kivity, A., Kamay, Y., Laor, D., Lublin, U., Liguori, A.: kvm: the linux virtual machine monitor. In: Proceedings of the Linux Symposium, vol. 1, pp. 225–230, June 2007

# Towards an Agent Based Model for Simulating Residential Mobility and Urban Expansion

**El-arbi El-alaouy, Khadija Rhoulami and Moulay Driss Rahmani**

**Abstract** Residential mobility and urban expansion are of great challenge to sustainable cities. Developing computer models for simulating such urban phenomena is a powerful tool to support informing urban decisions especially with the fact that half of the world's population now lives in cities. In this paper, we'll outline a mixed model of residential mobility and land use change. The model is based on bottom-up approaches i.e. Multi Agent System and Cellular Automata that allow to designing households and housing units. The model integrates Bayesian Belief Network and Markov Chain which permit respectively to design households' residential behaviors and to control forecasting land use change.

**Keywords** Residential mobility and land use change computer model · Multi agent systems · Cellular automata · Bayesian belief network · Markov chain

## 1 Introduction

Residential mobility is a research topic that attracts the attention of many disciplines such as sociology, demography, geography and psychology. The importance of the subject lies mainly in the challenge of understanding the residential choices of individuals in the city, to model and anticipate effects of mobility and in order to respond to the needs of society in terms of housing, transport and equipment infrastructure.

E. El-alaouy (✉) · K. Rhoulami · M.D. Rahmani
LRIT—Associated Unit to CNRST (URAC°29), FSR, Mohammed V University Of Rabat, 4
Av. Ibn Battouta, B.P. 1014, Rabat, RP, Morocco
e-mail: elalaouy.elarbi@gmail.com

K. Rhoulami
e-mail: rhoulamikhadija@gmail.com

M.D. Rahmani
e-mail: mrahmani@fsr.ac.ma

© Springer International Publishing Switzerland 2016
A. El Oualkadi et al. (eds.), *Proceedings of the Mediterranean Conference on Information & Communication Technologies 2015*, Lecture Notes in Electrical Engineering 381, DOI 10.1007/978-3-319-30298-0_36

Population growth could be seen as an engine for residential mobility. Population grows: number of households' individuals grows also. In fact, over time, individuals leave the household for one reason or another (employment, marriage etc.). Departing from their original households, they create new households. Thus, this creates a social demand for housing. In responding to this request, the population finds where to live. Therefore, green and cultivated areas become urbanized.

This urban expansion which is driven by residential mobility is situated mainly in the periphery of cities. Despite it responds to the housing access demand, it is the source of several environmental, social and economic effects i.e., Anarchic and rapid consumption of land, transport problems, access problems to basic services and equipment such as schools, universities, health centers, administrative centers, air pollution etc.

Such urban phenomena (i.e., residential mobility and urban expansion) get an increasingly interests and seems to be of great challenge for sustainable cities because of both their unexpected negative effects and new society needs that creates. Computer models are a powerful tool for responding to such problems. With prediction and simulation capabilities, these models could anticipate phenomena effects, support urban planners and decisions makers and test effects and effectiveness of urban politics based on scenarios.

In this chapter, we first present approaches we have chosen, secondly, we outline our developed residential mobility and land use change model that make simulations for over 10 years. Finally, results and discussion are then presented to highlighting usefulness of this technique for decisions making and urban planning.

# 2 Approaches

Residential mobility and urban expansion are socio-spatial processes that drive changes of both urban area and population's lifestyles at different scales. To model these phenomenons and predict future scenarios, researchers used and still use mostly the bottom-up approaches. Among these approaches, we could find Cellular Automata, Multi Agent Systems, Markov Chain and Bayesian Belief Network. But none of these approaches could solely produce accurate and better results. New approach which mix different approaches and benefit from advantages of each one could probably be a good alternative. CA is well known to represent urban space [1–3]. To improve its land use prediction, it needs to be merged with MC as reported by Yunyan et al [4]. [5, 6] are other works of CA-MC. MAS received much attention when modeling urban actors (e.g., households) [7–9]. BBN is a powerful classification method from which land use change models have benefited a lot [10–12]. MAS agents to act, they should be fed by human intelligence. This is where BBN intervene. BNN allow MAS agents to act intelligently in the urban space by extracting knowledge from socio economics census data.

# 3 Residential Mobility and Land Use Change Model

Land use change and development of built-up areas have been modeled differently in literature: some scholars used statistical concepts of land use transitions based on current and past remote sensing and GIS images [1, 2, 4]. Some others consider urban land use change as results of residential actions of urban actors e.g., population living in urban area [7, 8, 10]. Residential action which interests us is residential mobility. It expresses the movement question which leads households to decide whether to relocate or not. Many works have tried to answer to this question [13–15]. The model we're proposing combine the two above modeling tracks. The whole system is a set of households, housings, urban space and time. Considering system's specifications, we will now see the developed model that is similar to "Residential mobility" model of [16] and how each of CA, MC, MAS and BNN approaches has been used to design the whole system.

## 3.1 Cellular Automata

A cellular automaton consists of a discrete lattice of finite regular cells. Each cell has a state. The number of states is finite. Each cell has neighbors regarding the neighborhood's type (e.g. Van Neumann, Moore, etc). The lattice has two dimensions Thus, each cell could be identified as $C_{ij}$ where i is a column index and j is a line index. In our model, CA approach is used to represent the urban space and spatial dynamics. We'll expose here principal specifications that need to be considered when developing such CA approach:

**System's States**: each cell has a state that represents soil occupancy. States set is composed of static and dynamic states. The static elementary states are road or rail network, river or lakes, green space or forest, commercial building and equipment building. The dynamic elementary states are owned single house, owned apartment, rented single house, rented apartment, and building area.

**Neighborhood**: each cell has neighbors regarding Moore neighborhood. Using the formula $(2r + 1)^2 - 1$ of neighborhood's range $r$. we define the maximum number of neighbors. Given range = 1, each cell could have theoretically eight neighbors. But there are some exceptions (e.g. $C_{00}$, $C_{0j}$, $C_{i0}$ ...).

**Potential of attractiveness**: spatial interaction between cells is modeled by the potential of attractiveness $P_{C_{ij}}$ of a cell $C_{ij}$ that measures the ratio of number of cells that have the same state as $C_{ij}$ cell and the total number of neighbors. $\alpha_{ij}$ is an acceleration factor obtained by a sigmoid function where t is time and $\lambda$ is a predefined residential quality.

$$P_{C_{ij}} = \frac{\sum N_{C_{ij}} * \alpha_{ij}}{TNN} \text{ and } \alpha_{ij} = \frac{1}{1 + e^{-\lambda t}} \tag{1}$$

**Maps**: consist of the given maps of the studied area with which the model begins (e.g, 2006 and 2007). Maps are designed in our own format and consist of 2D lattice. The preparation of such maps requires different spatial data layers of the studied area (road and rail infrastructure, urban planning, building areas etc). The next section will outline how to predict future map t + 1 based on past maps (t − 1 and t).

## 3.2  Markov Chain

Markov Chain (MC) is a mathematical system that is widely used to forecast future land use dynamics. Markov Chain is formally a sequence of random variables $(X_n, n = 0, 1, \ldots)$ with values in a finite set called the MC state space. $(X_n, n > 0)$ is called a Markov Chain if:

$$\forall i_0, i_1, \ldots, i_{n-1}, i, j \in S, \quad \forall n \geq 0,$$
$$P(X_{n+1} = j | X_0 = i_0, X_1 = i_1, \ldots, X_{n-1} = i_{n-1}, X_n = i_n) = P(X_{n+1} = i_{n+1} | X_n = i_n) \tag{2}$$

Random variable $X_n$ represents the system's state at stage n. The MC process starts from one of the states and moves successively to another. The process move from i to j at the next step with a probability denoted by $P_{ij}$. State $X_{n+1}$ at n + 1 stage in the system could be computed by state $X_n$ at n stage and that using formula (3).

$$X_{n+1} = PX_n \tag{3}$$

$X_n$ is state vector at stage n, $P_{ij}$ transition probabilities compose the transition probability matrix which is computed by the formula (4).

$$P = \begin{bmatrix} P_{11} & \cdots & P_{1m} \\ \vdots & \ddots & \vdots \\ P_{m1} & \cdots & P_{mm} \end{bmatrix} \tag{4}$$

where, m is the number of soil occupancy type, $P_{ij}$ is the probability of changes from type i to type j. MC state space is nothing but the CA state system.

Based on the Markov stochastic process theory, we could calculate the transition probability from a state i to a state j in exactly n steps and that's by using the Chapman-Kolmogorov Eq. (5).

$$P_{ij}^n = \sum_{k \in S} P_{ik} P_{kj}^{(n-1)} \tag{5}$$

where, S is the MC state space, the nth transition probability matrix could be calculated by the initial transition probability matrix.

The initial probability distribution vector denoted by $\pi(0)$ designates states probabilities at stage 0. Probability distribution $\pi(n)$ at stage n could be calculated as follows:

$$\pi(n) = \pi(0)P^{(n)} \text{ or } \pi(n+1) = \pi(n) * P \tag{6}$$

Thus, transition rules that govern CA states are designed by MC. Based on this mixed approach. Cells state of CA lattice could move at any time of the simulation from one state to another. Formula (7) explains the cell state change principle.

$$P\left(C_{ij}^{t+1} = e\right) = \sum_{ij}^{\text{Neighborhood}} P(C_{ij}^{t} = e | C_{ij}^{t-1} = e') * P\left(C_{ij}^{t-1} = e'\right) e' \in E \backslash \{e\} \tag{7}$$

where, ij here are indices of row and column of CA lattice, e and e′ are MC-CA states system. This formula means the probability of cell $C_{ij}$ to move to state e at t + 1 stage depending on its Moore neighborhood and potential of attractiveness.

## 3.3 Multi Agent Systems

Multi Agent System is composed of number of agents that act and interact with each other in a shared environment. An agent is autonomous, sociable, reactive and proactive. Objective of MAS approach is to design City's population dynamics at an individual household scale. In our model, each household is represented by a a cognitive agent and is classified according to its family structure. Agents are feeding by household's properties (i.e., age, income, family structure and tenure status). Each agent has also a link with housing (CA Cell) where it lives. These characteristics give each agent a unique profile which could change at any time depending on demographic events and population growth principles. Actions translate agent residential choices. We've used BDI architecture which is sufficient and adequate to model agent's specifications. Agents decide autonomously based on their profiles. Two residential processes are of great importance which are relocation propensity and location choice. Both of these processes will be explained in the next section.

## 3.4 Bayesian Belief Network

Bayesian Belief Network (BBN) is a probabilistic graphical model that represents a set of random variables and their conditional dependencies using a directed acyclic graph. In our model, we've used BBN to implement the relocation process that's based on explanatory variables allows households to decide whether to relocate or not. Figure 1 illustrates the BBN model of relocation process.

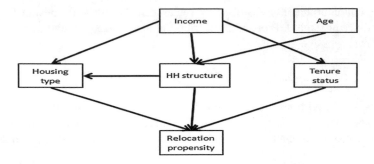

**Fig. 1** Representation of explanatory variables in the Bayesian belief network

A household computes its relocation propensity as given in the Eq. 8:

$$P(\chi_1 \ldots \chi_n) = \prod_{i=1}^{n} P(\chi_i | \Gamma_{\chi_i}) \text{ where } \Gamma_{\chi_i} \text{ are parents of } \chi_i \qquad (8)$$

The location choice process that permits to look for a new housing is influenced by residential quality $\lambda$ and actual and future housing characteristics. It is conditioned by the relocation decision of the household. The choice's process is as follows: Firstly, the household defines an investigation field based on Cellular Automata neighborhood. Secondly, it selects number of housing having a residential quality that satisfies its residential needs. Lastly, it chooses housing having the best score.

The household to be satisfied by housing, it should have an attractive potential $P_{C_{ij}}$ that is superior to its actual residential comfort $(P_k^t * \lambda)$. $P_k^t$ is the relocation propensity of household type k with features' vector K at time t.

$$P_k^t = P(relocation | K) \qquad (9)$$

The household computes the housing score using Formula (10) and choose the housing having the best score:

$$P_k^{t+1}(C_{ij}^{t+1} = e) = P_k(C_{ij}^{t+1} = e | C_{ij}^t = e') * P(C_{ij}^t = e') \qquad (10)$$

## 4 Results and Discussions

The simulation starts with a number of households, housings and maps that represents the studied area. As we are currently working on acquiring real data of Salé City, we will outline here some results of the residential mobility model we've developed by simulating a growth population scenario that uses a small synthetic

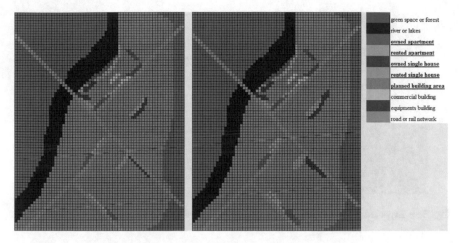

green space or forest
river or lakes
owned apartment
rented apartment
owned single house
rented single house
planned building area
commercial building
equipments building
road or rail network

**Fig. 2** City maps (2006 and 2007)

City. The number of housings and households is respectively 216 and 150 which means there are number of vacant housings at the simulation startup. Households have characteristics which form their unique profiles and which subsequently affect their residential choices regarding residential mobility and housing choice. Population grows yearly in terms of households number, using demographic events based on statistical data. These results are calibrated with a distribution of housing types by household categories similarly to [7].

Transition matrix and the probability distribution at initial stage are derived from maps at Fig. 2. Using Markov chain grounded principles we have described above, we compute transition matrix and probability distribution at next stage, and use it subsequently to forecast yearly land use change in terms of built up areas as explained in CA-MC section.

After development of built-up areas, households (including those generated by the population growth model) begin their residential choices as described in MAS section. New generated households look for vacant housings using the housing choice module that allows them to explore housings in their neighborhoods. Other households verify their residential satisfaction, if they're not satisfied with their actual residence, they use housing choice module to search for new residence. Thus, some households move to their new residence, while others stay because they couldn't find housings that correspond to their wishes and desires.

The simulation iterate yearly over ten years. But we could execute scenarios over less than 10 years. Figure 3 depicts the expected built-up areas and evolution of households and housings numbers over years of simulation. At the first year of the simulation, we note that number of households is less than housings. After the second year, in contrast to housings, households grow quickly. These results show a gap between population growth and houses construction. It means that annual built-up development rate is too slow regarding population growth rate. Such results, in addition to others that could be extracted by analyzing simulation

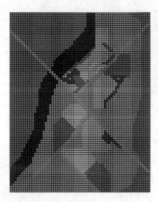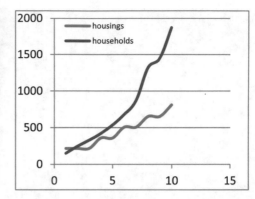

**Fig. 3  a** Expected built-up areas, **b** households and housings evolution

outputs, could support decisions makers and urban planners, to inform effective urban decisions (e.g., housing needs, urban regeneration, households' energy consumption etc).

## 5   Conclusion and Perspectives

The model presented here, can be used to give insights into the dynamics of residential mobility and the linked urban expansion. What is interesting about the modeling is that we used a bottom-up mixed approach. We've considered City, in which these two phenomena take place, as an emerging and auto-organized system where the global system dynamics are a set of individual behavior of smallest units i.e., households and houses.

We've simulated a population growth scenario. In this scenario, at first, there were more housings than households. As simulation progress, households' number grows more quickly than built-up areas. When at the end, more than half of households are without residence. This shows clearly that annual built-up development rate is too small comparing to households growth, and as to inform a decision, this rate should be revised by decisions makers and urban planners, in order to respond to the households needs. With the model framework we have developed, we could test and simulate a variety of scenarios of a given studied area, and this by varying model's inputs. For example, a shrinking City could be simulated by preparing a population decline scenario. In a futureless work, we would like to extend the model framework in way to extract more information such as spatial distribution of households' patterns, households' satisfaction and residential mobility rate. By acquiring spatial data of Salé City, further work will be proposed for the model validation which could be tested by comparing real and simulated urban area using validation techniques such as confusion matrix and kappa index.

# References

1. Dubos-Paillard, E., Guermond, Y., Langlois, P.: Analyse de l'évolution urbaine par automate cellulaire. Le modèle SpaCelle. L'espace géographique **32**(4), 357–378 (2003)
2. Mubea, K., Goetzke, R., Menz, G.: Simulating urban growth in Nakuru (Kenya) using Java-based modelling platform XULU. IEEE Model. Symp. (2013), 2013 European (pp. 103–108)
3. Mitsova, D., Shuster, W., Wang, X.: A cellular automata model of land cover change to integrate urban growth with open space conservation. Landscape Urban Plan. **99**(2), 141–153 (2011)
4. Du, Y., Wen, W., Cao, F., Ji, M.: A case-based reasoning approach for land use change prediction. Expert Syst. Appl. **37**(8), 5745–5750 (2010)
5. Moghadam, H.S., Helbich, M.: Spatiotemporal urbanization processes in the megacity of Mumbai, India: a Markov chains-cellular automata urban growth model. Appl. Geogr. **40**, 140–149 (2013)
6. Ma, C., Zhang, G.Y., Zhang, X.C., Zhao, Y.J., Li, H.Y.: Application of Markov model in wetland change dynamics in Tianjin Coastal Area. China. Procedia Environ. Sci. **13**, 252–262 (2012)
7. Haase, D., Lautenbach, S., Seppelt, R.: Modeling and simulating residential mobility in a shrinking city using an agent-based approach. Environ. Model. Softw. **25**(10), 1225–1240 (2010)
8. Gaube, V., Remesch, A.: Impact of urban planning on household's residential decisions: an agent-based simulation model for Vienna. Environ. Model Softw. **45**, 92–103 (2013)
9. Tan, R., Liu, Y., Zhou, K., Jiao, L., Tang, W.: A game-theory based agent-cellular model for use in urban growth simulation: a case study of the rapidly urbanizing Wuhan area of central China. Comput. Environ. Urban Syst. **49**, 15–29 (2015)
10. Kocabas, V., Dragicevic, S.: Bayesian networks and agent-based modeling approach for urban land-use and population density change: a BNAS model. J. Geogr. Syst. **15**(4), 403–426 (2012)
11. Bacon, P.J., Cain, J.D., Howard, D.C.: Belief network models of land manager decisions and land use change. J. Environ. Manage. **65**(1), 1–23 (2002)
12. Verstegen, J.A., Karssenberg, D., Van Der Hilst, F., Faaij, A.P.: Identifying a land use change cellular automaton by Bayesian data assimilation. Environ. Model Softw. **53**, 121–136 (2014)
13. Debrand, T., Taffin, C.: Les facteurs structurels et conjoncturels de la mobilité résidentielle depuis 20 ans. Economie et statistique. **381**, 125–146 (2005)
14. Dieleman, F.M.: Modelling residential mobility. a review of recent trends in research. J. Housing Built Environ. **16**, 249–265 (2001)
15. Coulombel, N.: Residential choice and household behavior: state of the art. Sustaincity working paper 2.2a, ENS Cachan (2010)
16. Agbossou, I.: Modélisation et simulation multi-agents de la dynamique urbaine: application à la mobilité résidentielle (Doctoral dissertation, Université de Franche-Comté) (2007)

# A New Distributed Computing Environment Based on Mobile Agents for SPMD Applications

Fatéma Zahra Benchara, Mohamed Youssfi, Omar Bouattane,
Hassan Ouajji and Mohammed Ouadi Bensalah

**Abstract** In this paper, we propose a new distributed environment for High Performance Computing (HPC) based on mobile agents. It allows us to perform parallel programs execution as distributed one over a flexible grid constituted by a cooperative mobile agent team works. The distributed program to be performed is encapsulated on team leader agent which deploys its team workers as Agent Virtual Processing Unit (AVPU). Each AVPU is asked to perform its assigned tasks and provides the computational results which make the data and team works tasks management difficult for the team leader agent and that influence the performance computing. In this work we focused on the implementation of the Mobile Provider Agent (MPA) in order to manage the distribution of data and instructions and to ensure a load balancing model. It grants also some interesting mechanisms to manage the others computing challenges thanks to the mobile agents several skills.

**Keywords** High performance computing · Parallel and distributed computing environment · Mobile agent · Image processing

F.Z. Benchara (✉) · M. Youssfi · O. Bouattane · H. Ouajji
ENSET Mohammedia, Hassan II University of Casablanca, Mohammedia, Morocco
e-mail: benchara.fatemazahra@gmail.com

M. Youssfi
e-mail: med@youssfi.net

O. Bouattane
e-mail: o.bouattane@gmail.com

H. Ouajji
e-mail: ouajji@enset-media.ac.ma

M.O. Bensalah
FSR, Mohammed V University of Rabat, Rabat, Morocco
e-mail: m.bensalah@fsr.ac.ma

© Springer International Publishing Switzerland 2016                                      353
A. El Oualkadi et al. (eds.), *Proceedings of the Mediterranean Conference
on Information & Communication Technologies 2015*, Lecture Notes
in Electrical Engineering 381, DOI 10.1007/978-3-319-30298-0_37

# 1 Introduction

Nowadays, everyone need to get information, results and achieve tasks in real time. So it is possible by the use of the computer science technologies which make the complex tasks easy in order to perform these. For example running an application of weather predictions which is based on a big number of data and complex simulations using just one or two processors can be a hard task for the machine and sometime impossible to achieve the results. So we need to introduce cooperation amongst processing power of different machines in order to overcome these problems. The parallel computing concept is widely used in order to overcome these challenges with its flexible and extensible architectures. Many fast parallel machines are created for different applications needs but they presented another challenges according to their high cost and to their limitation on the test and validation of new parallel algorithms. So the creation and the use of the Parallel Virtual Machine (PVM) [1] are considered as a suitable solution for these needs. This PVM machine is constituted over a grid computing using a set of heterogeneous machines connected with each other by the middleware. In [2], the authors proposed a virtual machine using mesh connected computer MCC mesh with multiple broadcast and which is recently improved by the use of GPUs and FPGAs in [3, 4]. And in [5] by designing and developing a parallel and distributed virtual machine in [6] by assigning a set of distributed VPEs (Virtual Processing Element) objects for each processing element in the grid. We can say that by introducing the concept of the grid and especially the middleware, the parallel computing are converged to the parallel and distributed one where the computing performance depends on the quality and the performance of the middleware. The question now is how to achieve the high performance computing. For the load balancing problem some algorithms have been designed for distributed systems [7, 8]. And in [9] where the authors used the mobile agents to move the loads in a distributed system from overloaded nodes to the lightly loaded ones and considering that all the grid nodes are homogeneous.

In this context, related to all these previous works we are focused on the use of the middleware which is based on the mobile agents. It is considered as a new grateful computer science technology which is used in [10] for proposing a new model for automatic construction of business processes based on multi agent systems. And in [11] for improving the management, the flexibility and the reusability of grid like parallel computing architecture; and the time efficiency of a medical reasoning system in [12]. So thanks to the several interesting mobile agents skills, we design and implement a distributed environment composed by the middleware which assigns and orchestrates a set of mobile agents as AVPUs (Agent Virtual Processing Unit) for each physical processor in heterogeneous grid computing. It implements some interesting mechanisms for maintaining the parallel computing challenges: load balancing, fault tolerance, and to reduce the communication cost. And ensure a high performance computing. This paper is organized as follows: We will describe the proposed model for parallel and distributed computing, its main components which are: the mobile agents team leaders and team workers and the Mobile Provider Agent in the Sect. 2. The Sect. 3 is focused on presenting several mechanisms used by the Mobile Provider Agent in order to perform a load balancing middleware and a high performance parallel and distributed computing. Some interesting results performed by implementing the c-means and the fuzzy c-means algorithm in this model will be presented in Sect. 4.

## 2 Distributed Computing Environment Architecture

### 2.1 Distributed Computing Environment Model

Distributed Computing Environment is a new scalable and robustness model for performing a high performance execution of the parallel programs in distributed system. It constitutes a parallel and distributed grid computing which is flexible with different topologies: 2D Mesh, 3D Mesh…and architectures: SIMD, SPMD, MIMD, MPMD… It is based on a cooperative mobile agent team works as (AVPUs) for performing parallel and distributed tasks. In Fig. 1 for example to perform a distributed segmentation of big data we need to implement the well known segmentation algorithm the c-means which is performed with SPMD architecture according to the distributed implementation presented in [13]. Each Team worker (AVPU) receives the distributed program and data from its Team leader (AVPU) and achieves tasks and sends the results to its Team leader (AVPU) for performing the image segmentation.

### 2.2 Model Main Components Overview

This distributed computing environment is based on the power of the middleware and the mobile agents. The HPC of the parallel programs is performed in this environment by cooperative main components which are created in these different environment states:

**Fig. 1** 2D Mesh Grid Computing for distributed image segmentation based on (AVPUs)

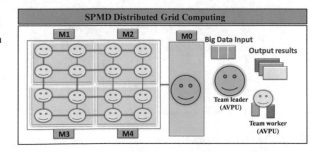

**Launching State**. The middleware creates (Node Agent Container) for each involved machine in the distributed computing and connects each of it in order to constitute the grid computing. We distinguish two particular nodes: the (Node Host Agent Container) for the first machine responsible for launching this environment and the (Node Provider Agent 'Container) where the Mobile Provider Agent (MPA) will be deployed.

**Deployment State**. When the parallel program is deployed, the middleware deploys the MPA agent and the (Team leader AVPU) for each node which encapsulates tasks and creates their (Team workers AVPUs). We can have one or two Team leader AVPU in the same node according to the number of the parallel programs deployed in this environment. Also each AVPU are autonomous and can decide to replicate itself in order to ensure a fault tolerance environment.

**Running State**. When the parallel program is running, the team leader AVPU sends the tasks and data to the MPA in order to manage and provide them to the Team workers AVPUs by ensuring the load balancing of tasks execution in the grid computing. At the end, the MPA agent sends the results to the Team leader AVPU in order to perform the final results and return it to the MPA agent in order to be broadcasted for different nodes in the grid.

## 3 From Parallel to Distributed Computing

### 3.1 A Fast Distributed Computing Middleware

As presented in Fig. 2, this environment is constituted over heterogeneous machines with different degree of performance. So we need to introduce an additional agent the MPA agent in our model for performing our main goal. The MPA agent is responsible for managing the pool of tasks and data and the results. It is an intelligent mediator between the Team leader AVPU and the Team workers AVPUs. This MPA agent has the knowledge of the number of team workers and its nodes

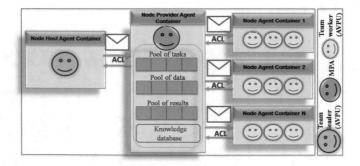

**Fig. 2** Distributed middleware mechanisms for computing management

performance. It manages a set of distributed pools by introducing the priority of the execution and the agent AID (Agent IDentifier) for each tasks and data in these pools. So the team workers AVPUs can easily follow their data and tasks when they move to the MPA agent container. The MPA agent has also the ability to decide according to the parallel program architecture when to send the tasks and data and when to keep the agents to move to the pools. So with the implementation of the MPA agent in our model we have a control about the load balancing problem, and we reduce at the same time the communication cost. Also MPA agent is autonomous, it can decide to move and to clone itself and resume its work in order to ensure a fault tolerance environment.

## 3.2 HPC Parallel and Distributed Computing Model

In this section, we make an overview about how this model can manage the main parallel computing challenges and grant an HPC model. In the load balancing communication diagram of Fig. 3, the AMS Agent performs the nodes performance monitoring and assembles and stores the results to the knowledge database. The MPA agent accesses the data and performs its tasks in order to ensure load balancing computing. In Fig. 4 we present the Fault Tolerance communication diagram where the MPA Agent has the ability to clone itself at a specific time. So when a problem happened the cloned MPAc agent starts. The MPAc resume the data and the MPA state and continues the tasks execution.

In Fig. 5, we show the Communication Cost part of this model. Each Agent in the grid can make its own decisions which reduce and manage the communication time. The asynchronous communication ability by exchanging ACL messages allows the agents to perform tasks and communicate at the same time.

In Fig. 6, we describe a scenario about the interaction between the different model main components in order to perform the execution of the parallel programs. This model is implemented using the JADE Framework [14].

**Fig. 3** Load balancing communication diagram

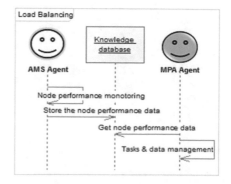

**Fig. 4** Fault tolerance
communication diagram

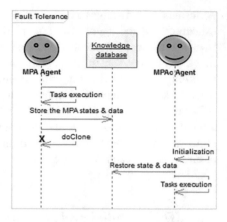

**Fig. 5** Communication cost
diagram

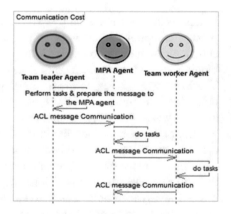

**Fig. 6** Cooperative and
distributed computing model

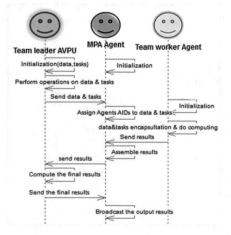

# 4 Applications

## 4.1 Distributed c-Means and FCM Implementation

The two algorithms: the c-means as defined in [15] and the Fuzzy c-means (FCM) which is proposed by Dunn [16] and extended by Bezdek [17] are implemented in this environment as distributed programs over SPMD architecture. Each one is encapsulated in a team leader agent. It cooperates with its team works in order to perform the big data image segmentation using the corresponding following steps presented under a sequence diagram in Fig. 6.

## 4.2 Experimental Results

In order to present the effectiveness features of the implementation of both programs in different team leader agents in our model we choose an MRI medical image with different class centers initialization which is segmented into 5 output images in Fig. 7a–g in order to detect the abnormal region in this MRI cerebral image.

In Tables 1 and 2, and in Fig. 8 we see clearly the dynamic convergence of c-means and fuzzy c-means algorithms running respectively at the same time in this model. And in Fig. 9, the variation of the Time of Segmentation (TSeg) according to the number of agents involved in the computing for these two algorithms.

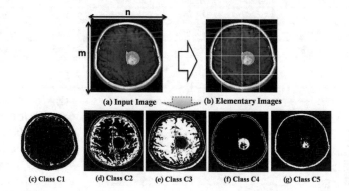

**Fig. 7** Output MRI images segmentation results

**Table 1** Different states of the distributed c-means (DCM) algorithm starting from class centers (c1, c2, c3, c4, c5) = (1, 49, 140, 240, 249)

| Iteration | Value of each class center | | | | | Error |
|---|---|---|---|---|---|---|
| | $C_1$ | $C_2$ | $C_3$ | $C_4$ | $C_5$ | $|J_n - J_{n-1}|$ |
| 1 | 1.000 | 49.000 | 140.000 | 240.000 | 249.000 | 8.19E+05 |
| 2 | 1.385 | 74.547 | 116.739 | 219.488 | 251.929 | 3.85E+05 |
| 3 | 3.082 | 77.002 | 111.825 | 201.473 | 249.149 | 3.98E+04 |
| 4 | 3.790 | 77.838 | 110.026 | 190.175 | 244.954 | 1.07E+04 |
| 5 | 3.790 | 76.132 | 107.894 | 183.859 | 242.084 | 6.94E+03 |
| 6 | 3.790 | 75.195 | 106.710 | 179.022 | 240.019 | 3.33E+03 |
| 7 | 3.790 | 73.316 | 105.261 | 176.592 | 239.212 | 2.31E+03 |
| 8 | 3.343 | 71.708 | 104.525 | 174.495 | 238.277 | 3.50E+03 |
| 9 | 3.082 | 71.343 | 104.403 | 172.825 | 237.385 | 2.12E+03 |
| 10 | 3.082 | 69.498 | 103.368 | 171.858 | 236.932 | 1.19E+03 |
| 11 | 2.876 | 68.258 | 102.866 | 171.382 | 236.932 | 1.70E+03 |
| 12 | 2.688 | 67.959 | 102.866 | 171.382 | 236.932 | 1.31E+03 |
| 13 | 2.688 | 67.959 | 102.866 | 171.382 | 236.932 | 0.00E+00 |

**Table 2** Different states of the distributed fuzzy c-means (DFCM) algorithm starting from class centers (c1, c2, c3, c4, c5) = (49.5, 50.5, 140.5, 240.2, 249.5)

| Iteration | Value of each class center | | | | | Error |
|---|---|---|---|---|---|---|
| | $C_1$ | $C_2$ | $C_3$ | $C_4$ | $C_5$ | $|J_n - J_{n-1}|$ |
| 1 | 49.500 | 50.500 | 140.500 | 240.200 | 249.500 | 5.67E+05 |
| 2 | 38.958 | 42.937 | 121.596 | 222.116 | 243.412 | 6.45E+04 |
| 3 | 20.897 | 34.294 | 108.416 | 206.890 | 244.013 | 8.97E+04 |
| 4 | 4.829 | 38.072 | 101.909 | 194.924 | 244.460 | 1.08E+05 |
| 5 | 1.083 | 48.186 | 99.567 | 186.355 | 243.646 | 5.24E+04 |
| 6 | 1.024 | 53.258 | 99.344 | 180.582 | 242.467 | 6.55E+03 |

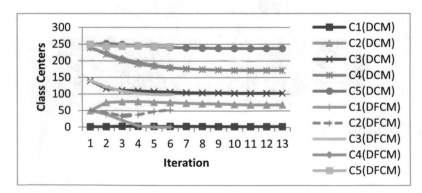

**Fig. 8** Dynamic convergence of (DCM) with initial class centers (c1, c2, c3, c4, c5) = (1, 49, 140, 240, 249) and (DFCM) with (c1, c2, c3, c4, c5) = (49.5, 50.5, 140.5, 240.2, 249.5)

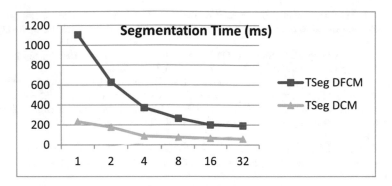

**Fig. 9** Time of Segmentation (TSeg) depending on the number of agents for DFCM and DCM applications

## 5 Conclusion

In this paper, we have presented a distributed environment for high performance parallel and distributed computing. It is based on a cooperative mobile agent grid computing flexible for different parallel architectures which is constituted over a distributed system. It is based on an innovated idea which is the use of Mobile Agents as AVPUs (Agent Virtual Processing Units). Each mobile agent associates its skills: autonomy, mobility and communication ability using ACL messages (Agent Communication Language) to provide the processing power needs to perform the parallel programs and to manage the parallel computing challenges. In this environment we distinguish the MPA agent (Mobile Provider Agent) which manages and orchestrates the computing between the different team works and grant a load balancing, and a fault tolerance environment. And it implements some interesting mechanisms which reduce the communication cost in the grid and perform the HPC.

## References

1. El-Rewini, H., Abd-El-Barr, M.: Advanced Computer Architecture and Parallel Processing. Wiley (2005)
2. Miller, R., et al.: Geometric algorithms for digitized pictures on a mesh connected computer. IEEE Trans. PAMI **7**(2), 216–228 (1985)
3. Wu, J., JaJa, J., Balaras, E.: An optimized FFT-based direct poisson solver on CUDA GPUs. IEEE Trans. Parallel Distrib. Syst. **25**(3), 550–559 (2014)
4. Rafique, A., Constantinides, G.A., Kapre, N.: Communication optimization of iterative sparse matrix-vector multiply on GPUs and FPGAs. IEEE Trans. Parallel Distrib. Syst. **26**(1), 24–34 (2014)
5. Youssfi, M., Bouattane, O., Bensalah, M.O.: A massively parallel re-configurable mesh computer emulator: design, modeling and realization. J. Softw. Eng. Appl. 11–26 (2010)

6. Youssfi, M., Bouattane, O., Benchara, F.Z., Bensalah, M.O.: A fast middleware for massively parallel and distributed computing. Int. J. Res. Comput. Commun. Technol. **3**(4), 429–435 (2014)
7. Hsiao, H.C., Chung, H.Y., Shen, H., Chao, Y.C.: Load rebalancing for distributed file systems in clouds. IEEE Trans. Parallel Distrib. Syst. **24**(5), 951–962 (2013)
8. Hsiao, H.C., Chang, C.W.: A symmetric load balancing algorithm with performance guarantees for distributed hash tables. IEEE Trans. Computers **62**(4), 662–675 (2013)
9. Liu, J., Jin, X., Wang, Y.: Agent-based load balancing on homogeneous mini-grids: macroscopic modeling and characterization. IEEE Trans. Parallel Distrib. Syst. 586–594 (2005)
10. García Coria J.A., Castellanos-Garzón J.A., Corchado J.M.: Intelligent business processes composition based on multi-agent systems. ELSEVIER. Expert Syst. Appl. **41**(4), 1189–1205 (2014)
11. Sánchez, D., Isern, D., Rodríguez-Rozas, A., Moreno, A.: Agent-based platform to support the execution of parallel tasks. ELSEVIER. Expert Syst. Appl. **38**(6), 6644–6656 (2011)
12. Rodríguez-González, A., Torres-Niño, J., Hernández-Chan, G., Jiménez-Domingo, E., Alvarez-Rodríguez, J.M.: Using agents to parallelize a medical reasoning system based on ontologies and description logics as an application case. ELSEVIER. Expert Syst. Appl. **39**, 13085–13092 (2012)
13. Benchara, F.Z., Youssfi, M., Bouattane, O., Ouajji, H., Bensalah, M.O.: Distributed C-means algorithm for big data image segmentation on a massively parallel and distributed virtual machine based on cooperative mobile agents. J. Softw. Eng. Appl. **8**, 103–113 (2015)
14. Bellifemine, F.L., Caire, G., Greenwood, D.: Developing Multi-Agent Systems with JADE. Wiley (2007)
15. Bouattane, O., Cherradi, B., Youssfi, M., Bensalah, M.O.: Parallel c-means algorithm for image segmentation on a reconfigurable mesh computer. ELSEVIER. Parallel Comput. **37**, 230–243 (2011)
16. Dunn, J.C.: A fuzzy relative of the ISODATA process and its use in detecting compact well-separated clusters. J. Cybern. **3**(3), 32–57 (1973)
17. Bezdek, J.C.: Pattern Recognition with Fuzzy Objective Function Algorithms. Plenum Press, New York (1981)

# Agent Story: An Agile Requirements Modeling Approach for Multi-agent Paradigm

Walid Dahhane, Jamal Berrich and Toumi Bouchentouf

**Abstract** Agents and multi-agent systems are currently one of the most interesting research fields in the computer science community; especially the natural way of capturing the structure and the behavior of complex systems has stimulated this huge interest Lind (Issues in agent-oriented software engineering, 2000). As this paradigm advance, systematic methods are needed to support the development of multi-agent systems. General approaches for identifying, modeling, and analyzing user requirements for specific multi-agent software systems based on goal models have been suggested DeLoach (Multiagent systems engineering a methodology and language for designing agent systems, 1999), Tveit (A survey of agent-oriented software engineering, 2001). Nonetheless, those approaches suffer from rigidity and complexity for a customer to validate and review his needs as gathered by business analyst. In this paper, we present an agile requirements engineering approach to model multi-agent system requirements in the form of agent stories capturing skills and capabilities from the agent perspective. The agent stories can then be estimated, prioritized and traced separately. To illustrate the approach benefits, a Weapon Mass Destructive (WMD) system example is presented. The example is also used to compare our suggested approach with a goal-based one.

**Keywords** Agent story · SMA · Requirement modeling approach · Agile · Multi-agent system · Goal model · User story

W. Dahhane (✉) · J. Berrich (✉) · T. Bouchentouf (✉)
Team SIQL, Laboratoire LSEII, ENSAO, Université Med I Hay Elquods, 60300 Oujda, Morocco
e-mail: dahhane.walid@gmail.com

J. Berrich
e-mail: jberrich@gmail.com

T. Bouchentouf
e-mail: tbouchentouf@gmail.com

© Springer International Publishing Switzerland 2016    363
A. El Oualkadi et al. (eds.), *Proceedings of the Mediterranean Conference on Information & Communication Technologies 2015*, Lecture Notes in Electrical Engineering 381, DOI 10.1007/978-3-319-30298-0_38

# 1 Introduction

Software technology and customer requirements have evolved rapidly over the last few decades. This evolution required software systems to be more intelligent to address problems with "custom" solutions adapted to the environment and context within which the system is executing. To meet these expectations, systems have been conceived with human skills and capabilities in mind to address real world problems. As a result, the abstraction of those competencies was defined in agent software [1, 2]. An agent has a collection of skills and capabilities making it able to address some specifics problems autonomously and some global problems when integrated in a multi-agent system [3].

A Multi-Agent System (MAS) is a specific system that is composed of multiple interacting intelligent agents [3, 4]. These systems can be used to solve problems that are difficult or impossible for an individual agent or monolithic system to solve. While this approach has yielded increasingly capable systems, system complexity has increased dramatically making it more difficult to understand, analyze, and build such systems. Even though many Agent Oriented Software Engineering methodologies have been proposed, few are mature or described in sufficient detail to be of real use [5, 6]. One of the issues still not mature and open in these methodologies is about expressing customer requirements. Requirements should be expressed in a comprehensive way for all stakeholders, estimated and prioritized according to the customer business value and time frame before even starting the real design and development of the desired system.

Most of MAS methodologies focuses only on the analysis and design and do not give support to the requirements phase [7]. This is the case of the Gaia methodology [8, 9] which was extended later by several researches to support requirements [10, 11] and the case of object-oriented based method MASSIVE [12] which covers analysis, design and code generation. The well know Tropos [13] method which is based on goal models is maybe the most developed and widely used approach in the community for dealing with requirements, analysis and design for MAS development.

Nonetheless, after investigations done trying to use goal model based approaches, we found that it tend to be rigid to change. We found also that goal models are not easy to understand by all stakeholders and specially system customer. Estimation and prioritization (and selection) of requirements to be done is not one of the strong points of this approaches either. In this paper, we present an agile approach to express requirements in the form of agent stories. The next section discusses this approach and our suggested formula to express system needs from the perspective of agents. To illustrate our purpose, a WMD application example is used next illustrating the difference between the Miller et al. goal model based approach [14, 15] and the agent story one.

# 2 Agent Story

## 2.1 Overview

As discussed in the previous section, general approaches for identifying, modeling, and analyzing user requirements for specific multi-agent software systems based on goal models have been suggested. Nonetheless, those approaches suffer from rigidity and complexity for a customer to validate and review his needs as gathered by business analyst. To consider an approach as agile, it should address the problem of change possibility. In real life, customer needs may be designed and implemented in a different way from his real need which was not clearly expressed at beginning. They can be also misunderstood by system designer and therefore should be changed and corrected easily. It should also give the customer a clear overview about his needs as understood by business analyst to be validated, prioritized and estimated as well. In this section we present agent stories which represent a more agile approach to identify, model and express requirements.

## 2.2 What Is an Agent Story

Agent story is a short, simple description of a capability or skill told from the perspective of the agent who has this capability/skill, and the goal for which this capability or skill is wanted. It is expressed using the following simple template (Fig. 1):

This definition is taken from the one of the well known user story template of Mike Cohn [16] and adapted to the agent oriented paradigm. The following table summarizes the main difference between them (Table 1).

**Why the perspective is the software agent instead of the system user**? The reason for this choice is autonomy. To design an autonomous agent, it requires from the designer to be a part of the system and think like the software agent will do. The system designer should do abstraction of technical requirements and think like if a real human-been agent will to resolve a specific problem. So, each agent should be autonomous in term of capacities and skills he has to reach some goals and objectives.

*As a <agent name>, I want be able to*

*<some capability or skill> so that <goal to be achieved (for*

*what the capability/skill is necessary)>.*

**Fig. 1** Agent story template

**Table 1** Difference between user story and agent story

|         | User story              | Agent story                                                          |
| ------- | ----------------------- | ------------------------------------------------------------------- |
| Who     | A real system user      | A software or real agent                                            |
| What    | A user goal             | A software agent capability or skill                                |
| Why     | A business value/A reason | The reason why this capability/skill is there (One of the agent goals) |

**Capability versus skill**: We make difference between agent capabilities and agent skills. With capabilities we mean mainly the sensors and effectors which let the agent interact with his environment. Skills are about using and exploiting these capabilities to form complex competencies. Another specificity of skills is they can be developed and improved within execution time and learning process from the environment, where capabilities are static and can't be changed without changing agent interface components.

Let's take the example of a security agent:

Capabilities: video records, access to cellular phone, open/close doors …
Skills: movement detection, face recognition, Police alarm (telephone call), sensible zones protection (using the open/close doors capability)…

We can summarize these agent stories in the following table (Table 2).

**Why this difference is made?** Capabilities are generally something pluggable to the agent, for the security agent example, video records can be achieved by plugging a camera recorder to the agent software. Skills are likely to be algorithms and techniques which can use agent capabilities to make it more intelligent. We can imagine for example, that movement detection will activate the face recognition skill to detect intruder which will trigger an alarm call to the police and a systematic close of all doors. We can say that capabilities represent the no intelligent part of the agent where the skills represent the intelligent one. And sometimes, to be more intelligent, agent should have a good set of no intelligent parts (capabilities) to be able to develop powerful intelligent abilities (skills).

**Table 2** The capability/skill/goal table

| Capability              | Skill              | Goal                                                    |
| ----------------------- | ------------------ | ------------------------------------------------------- |
| Video records           | Face recognition   | Recognize thieves and intruder                          |
|                         | Movement detection | Recognize suspect movement to reinforce surveillance    |
| Access to cellular phone | Police alarm       | Call police when unknown intruder are identified        |
| Open close door         | Zones protection   | Close automatically all doors of sensible rooms         |

# 3 Application Example

## 3.1 Overview

The application used to demonstrate our approach is a simulated Weapons of Mass Destruction (WMD) search system, where a team of robots attempts to find, detect, and remove WMDs [15]. In the system, several robots (agents) search an area looking for objects. When objects are found, the robots determine whether the objects are WMDs or inert. The team can detect three weapon types: biological, chemical, and nuclear. Each team has exactly one robot capable of detecting each type. If the object fails a test for one type, it must be tested for the other types. Only if an object fails all three tests can it be classified as inert.

First, we present the goal model applied on this example as suggested by DeLoach et al. [14], Miller et al [15] We discuss limitation of the model as applied on the example. Next we present our approach in form of agent stories expressing the example requirements.

## 3.2 Using Goal Model

Miller et al. has suggested the following diagram to model the WMD requirements (Fig. 2).

As we can notice in the goal suggested model, only one agent type is proposed. This design is likely to be based on the fact that each team has only one robot, so each robot is an agent. We think that we should not limit the design to only one software agent. It can be more sophisticated if different autonomous software agents are designed and each agent has a specific goal to achieve. It is like a small company where there is one agent employee playing different roles with different objectives. The employee is responsible for accounting, marketing, sales, hiring ...

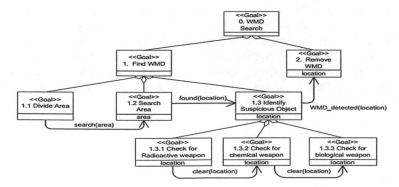

**Fig. 2** WMD search dynamic goal model

and once the company begin growing, the roles are dispatcher to other agent employees. The roles and objectives remains the same, it just deployment which changes. Same thing to WMD system, there are many different roles that should be designed separately even if they probably will be deployed to the same robot entity.

This leads us also to the question of prioritization. How can customer choose what functionality to be implemented first? How can we estimate the implementation and design cost before moving to other phases? What about communication capabilities between agents? What about human been agents which make part of the system?

All those questions remain without clear answer when using goal model diagrams. Unless if they are combined with others diagrams or approaches [17] which make the goal model less useful then what it should to be.

## 3.3    Using Agent Stories

Let's try to model the WMD requirements in the form of agent stories. First, we should gather capabilities and skills needed to address the problem:

Capabilities: communicate by wireless, object Manipulation and moving capability Skills: communicate current position, communicate zone to be searched, check for radioactive weapon, check for chemical weapon, check for biological weapon, move WMD to a quarantine zone, neutralize WMD objects, and search for suspicious objects

We can summarize those capabilities and skills with goal they should achieve in the following table (Table 3).

Based on this table we can gather agent stories as described in the Fig. 3. Notice that Chemical, Biological, Nuclear, Neutralizer Agents are specialization of a WMD Agent. The agent story allows us to model also the communication needs between agents, which is missing in the goal models.

As a result we have a backlog of stories representing the customer requirements as understood by business analyst. Each agent story can be traced, estimated, prioritized and discussed with the customer separately. Not all agent are necessarily software agents, we can imagine the Coordinator Agent as a human been, and all other agents software ones implemented in the one robot of each team. We can notice also that sometimes, the capability/skill's goal of one agent represents a skill for another one. That is mean that an agent goal can be transformed to a skill to achieve a more sophisticated and complex goal.

**Table 3** WMD system capabilities, skills and goals

| Capability | Skill | Goal |
|---|---|---|
| Wireless communication | Communicate position | Every agent is located |
| | Communicate zone to be searched | Every agent search a specific zone |
| | Communicate the check result | Destroy WMD objects once founded |
| Object manipulation | Check for radioactive weapon | Communicate the radioactive WMD position once found to be destroyed |
| | Check for chemical weapon | Communicate the chemical WMD position once found to be destroyed |
| | Check for biological weapon | Communicate the biological WMD position once found to be destroyed |
| | Move WMD to a quarantine zone | WMD object still harmful but in a quarantine zone |
| | Neutralize WMD object | Object can stay in place but harmless |
| Moving | Search for suspicious objects | Identify suspicious object to be checked if WMD or not |

*AS1#* As a Chemical Agent I want be able to **check for chemical weapon** so that i communicate the position of the founded chemical weapon to be processed

*AS2#* As a Biological Agent I want be able to **check for biological weapon** so that I communicate the position of the founded biological weapon to be processed

*AS3#* As a Nuclear Agent I want be able to **check for radioactive weapon** so that I communicate the position of the founded radioactive weapon to be processed

*AS4#* As a Neutralizer Agent I want be able to **move WMD objects to quarantine** so that scanned zones be clean

*AS5#* As a Neutralizer Agent I want be able to **neutralize WMD objects** so that objects be harmless

*AS6#* As a WMD Agent I want be able to **search for suspicious objects** so that WMD objects are found

*AS7#* As a WMD Agent I want be able to **Communicate the check result** so that WMD found objects be neutralized

*AS8#* As a WMD Agent I want be able to **move in an area** so that I can scan it looking for suspicious objects

*AS9#* As a WMD Agent I want be able to **communicate by wireless with other agents** so that information can be exchanged

*AS10#* As a WMD Agent I want be able to **communicate my current position** so I can be located

*AS11#* As a WMD Agent I want be able to **manipulate object** so that it can be checked if WMD or inert

*AS12#* As a Coordinator Agent I want be able to **communicate zone to be searched** so that zones are allocated to available agents

**Fig. 3** WMD system expressed using the agent story approach

# 4 Conclusions

We have presented a requirements modeling agile approach for the development of multi agent systems independently of any methodology. This approach suggests using a user story like template to capture functionalities to be developed from the perspective of the agent software. We presented the different formula parts used to express requirements and we presented an application example to illustrate our purpose. We believe that our approach fills the gap in the development of MAS between system designer and system customer. We think also that agent stories allow more control over the life cycle of the project by the facilities they offer on estimation, prioritization and tractability processes.

Nonetheless, this work only introduces the general principles of the approach which is at early stages and need more maturity and evaluation. We will work on this evaluation and on the integration of agent stories in exiting MAS methodologies using the JOCO v1.0 platform [18, 19]. We plan also to study the transition from RE, using agent stories and OOADA-RE [20, 21], to other design artifacts and final execution code.

# References

1. Franklin, S., Graesser, A.: Is it an Agent, or just a Program?: A Taxonomy for Autonomous Agents. University of Memphis, Institute for Intelligent Systems. Archived from the original on 1996 (1996)
2. Russell, S.J.: Rationality and intelligence. Artif. Intell. **94**(1), 57–77 (1997)
3. Ferber, J.: Multi-Agent Systems: An Introduction to Distributed Artificial Intelligence. Addison-Wesley (1999)
4. Ferber, J., Olivier, G., Fabien M.: From agents to organizations: an organizational view of multi-agent systems. In: Agent-Oriented Software Engineering IV, pp. 214–230. Springer, Berlin, Heidelberg (2004)
5. Sturm, A., Shehory, O.: A framework for evaluating agent-oriented methodologies. In: Agent-Oriented Information Systems, pp. 94–109. Springer, Berlin, Heidelberg (2004)
6. Bergenti, F., Gleizes, M.-P., Zambonelli, F. (eds.): Methodologies and software engineering for agent systems: the agent-oriented software engineering handbook, vol. 11. Springer (2004)
7. Dam, K.H., Winikoff, M.: Comparing agent-oriented methodologies. In: Agent-Oriented Information Systems, pp. 78–93. Springer, Berlin, Heidelberg (2004)
8. Wooldridge, M., Jennings, N.R., Kinny. D.: A methodology for agent-oriented analysis and design. In: Proceedings of the Third International Conference on Autonomous Agents (Agents-99). ACM, Seattle, WA, May 1999
9. Wooldridge, M., Jennings, N.R., Kinny, D.: The Gaia methodology for agent-oriented analysis and design. Auton. Agent. Multi-Agent Syst. **3**(3), 285–312 (2000)
10. Blanes, D., Insfran, E., Abrahão, S.: RE4Gaia: a requirements modeling approach for the development of multi-agent systems. In: Advances in Software Engineering, pp. 245–252. Springer, Berlin, Heidelberg (2009)
11. Juan, T., Pearce, A., Sterling, L.: ROADMAP: extending the gaia methodology for complex open systems. In: Proceedings of the First International Joint Conference on Autonomous Agents and Multiagent Systems: Part 1, pp. 3–10. ACM (2002)

12. Lind, J.: Iterative Software Engineering for Multiagent Systems: The MASSIVE Method. Springer (2001)
13. Bresciani, P., Perini, A., Giorgini, P., Giunchiglia, F., Mylopoulos, J.: Tropos: an agent-oriented software development methodology. Auton. Agents Multi-Agent Syst. **8**(3), 203–236 (2004)
14. DeLoach, S.A., Miller, M.: A goal model for adaptive complex systems. Int. J. Comput. Intell. Theory Pract. **5**(2), 83–92 (2010)
15. Miller, M.: A goal model for dynamic systems. Dissertation, Kansas State University (2007)
16. Cohn, M.: User Stories Applied: For Agile Software Development. Addison-Wesley Professional (2004)
17. Garcia-Ojeda, J.C., DeLoach, S.A., Oyenan, W.H., Valenzuela, J.: O-MaSE: A Customizable Approach to Developing Multiagent Development Processes. Springer, Berlin, Heidelberg (2008)
18. Berrich, J., Bouchetnouf, T., Benazzi, A.: oBDI2Jadex: An agent model based on O-MaSE methodology to design a BDI agents for Jadex. Int. J. Eng. Adv. Technol. **2**(6), 2249–8958
19. Berrich, J., Bouchetnouf, T., Benazzi, A.: Joco 0.1: Conteneur D'application Modulaire A Base Des Agents BDI De La Plateforme Jadex Suivant La Méthodologie O Mase. Int. J. Eng. Res. Appl. **3**(5), 558–564. Sep-Oct 2013
20. Zeaaraoui, A., Rougroun, Z., Belkasmi, M.G., Bouchentouf, T.: Object-oriented analysis and design approach for the requirements engineering. J. Electron. Syst. **2**(4), 147–153 (2012)
21. Dahhane, W., Zeaaraoui, A., Ettifouri, E., Bouchentouf, T.: An automated object-based approach to transforming requirements to class diagrams. In: The Second World Conference on Complex Systems (WCCS). ISBN: 978-1-4799-4647-1 (2014)
22. Lind, J.: Issues in agent-oriented software engineering. In: The First International Workshop on Agent Oriented Software Engineering (AOSE-2000) (2000)
23. DeLoach, S.A.: Multiagent systems engineering a methodology and language for designing agent systems. In: Proceedings of Agent Oriented Information Systems, pp. 45–57 (1999)
24. Tveit, A.: A survey of agent-oriented software engineering. In: NTNU Computer Science Graduate Student Conference, Norwegian University of Science and Technology (2001)

# MeTra-SPL for an Economic Analysis of Traceability in Software Product Lines

Zineb Mcharfi, Bouchra El Asri, Ikram Dehmouch, Asmaa Baya,
Fatima Zahra Hammani and Abdelaziz Kriouile

**Abstract** Traceability is mandated by several studies in Software Product Line engineering, as it helps assuring product quality in such complex environment. Yet, despite its importance, experience has shown that traceability is rarely established in Software Product Lines. Therefore, we compare in this paper the Return On Investment of three different traceability strategies adopted in a Product Line environment, through our model MeTra-SPL. We propose a quantitative analysis of cost-benefit of each traceability strategy to study the profitability of adopting a suitable traceability strategy.

**Keywords** Traceability strategy · Software Product Lines · MeTra-SPL · ROI

## 1 Introduction

The importance of traceability in software engineering has been highlighted during the NATO working conference in 1968 [1]. First introduced in safety-critical systems, such as USA Federal Aviation Authority [2], traceability is considered as

Z. Mcharfi (✉) · B. El Asri · I. Dehmouch · A. Baya · F.Z. Hammani · A. Kriouile
Mohammed V University In Rabat, Rabat, Morocco
e-mail: zineb.mcharfi@gmail.com

B. El Asri
e-mail: elasri@ensias.ma

I. Dehmouch
e-mail: ikram.dehmouch@gmail.com

A. Baya
e-mail: bayaasmaa@gmail.com

F.Z. Hammani
e-mail: fz.hammani@gmail.com

A. Kriouile
e-mail: kriouile@ensias.ma

© Springer International Publishing Switzerland 2016                                        373
A. El Oualkadi et al. (eds.), *Proceedings of the Mediterranean Conference on Information & Communication Technologies 2015*, Lecture Notes in Electrical Engineering 381, DOI 10.1007/978-3-319-30298-0_39

an important element for software quality assurance, and studies are conducted to best integrate this concept in complex systems, such as Software Product Lines (SPL). In fact, because of their reusability characteristics and the number of their artifacts, SPL require a good knowledge of the composing artifacts and relations between them, in order to better understand the system's architecture, evaluate the impact of changes on the system, and assure its maintenance [3].

However, traceability is still rarely adopted in practice, as it is laborious, usually manual, time and resource consuming, and error prone [4]. This motivated our work to study the impact of traceability strategy on SPL Return On Investment (ROI), by elaborating a model based on an adaptation of cost estimation model COPLIMO [5]. We named this model MeTra-SPL (Metrics for Estimating Traceability in SPL).

The remainder of this paper is as follow: Sect. 2 discusses traceability in SPL and cost estimation related works. Our proposed model METra-SPL is described in Sect. 3, and our case study and the discussion concerning the impact of traceability strategy on SPL cost estimation in Sect. 4. Conclusion and further lines of research are proposed in Sect. 5.

## 2  Product Lines Traceability and Cost Estimation Related Works

In this section we introduce traceability in SPL and software cost estimation techniques to present the motivations behind our present work.

### 2.1  Traceability in Software Product Lines

Traceability is "the ability to interrelate any uniquely identifiable software engineering artifact to any other, maintain required links over time, and use the resulting network to answer questions of both the software product and its development process" [6]. It deals with components and relations between them [1], producing and maintaining consistent documentation, verifying the completeness of requirements implementation [2], and being independent from individual knowledge [7]. Traceability improves system quality. It also helps choosing architectures and identifying errors, and facilitates communication between stakeholders [8]. It is very helpful for maintenance and evolution as it allows analyzing and controlling the impact of changes [3]. This is very useful in SPL context as traces help identifying elements impacted by changes in the Product Line (PL).

Traceability in SPL can be used either while developing, for short term purposes (e.g., to verify and validate requirements implementation), or in maintenance phase, for long term use (e.g., artifacts understanding, change management and components reuse) [1, 4, 9]. However, many difficulties can be faced when implementing traceability in SPL [10]: (i) larger documentation than for traditional software development; (ii) documents heterogeneity; (iii) need to link between different products and between them and the PL architecture. Also, unlike software engineering approaches for single systems, SPL introduces a complex dimension: variability. Variability represents an additional difficulty for traceability in SPL as one needs to understand its consequences during the development phases [10].

Some works handle traceability and variability issues in SPL while tracing relations between artifacts [8, 11, 12], others manage it throw a metamodel for SPL development [3]. Ghanam and Maurer [13, 14] use Acceptance Tests (AT) to generate test artifacts in an eXtreme Programing (XP) Agile SPL (ASPL) environment. The AT help implementing traceability through the entire development process. Díaz et al. [15, 16] address variability and traceability issues in a Scrum ASPL environment from an architectural view. In [17], a traceability reference model is introduced for an automated traceability between PL documents generated by feature-based object-oriented methodologies, and according to the type of traceability relations.

Despite the variety of those works, traceability is rarely used in practice as its implementation is surrounded by many difficulties and, in return, its benefits are not palpable in the short term. Therefore, measuring the ROI of tracing can help the project management making the decision of adopting a traceability approach, especially in complex systems like SPL.

## 2.2 Cost Estimation Techniques

Software engineering cost-benefits estimation helps managers to take decisions relative to [18]: (i) Budgeting, as cost models allow estimating the global cost of a project or software; (ii) Trade-off and risk analysis, as they help clarify decision making elements in a project, like scooping and human and technical resources; (iii) Project planning and control, by providing cost and schedule breakdowns by component, stage and activity; and (iv) Improving software investment analysis by estimating the benefits of the adopted strategy and choices made.

The importance accorded to cost-benefits estimation data justify the variety of proposed models in literature since late 1960's [18]. Those models can be categorized as follow [18, 19]: (i) algorithmic measurement like COCOMO 81 and SLIM models; (ii) expert judgment based measurement like Delphi; (iii) measurement by analogy like in neural networks; (iv) using dynamic techniques when the project effort estimation factors change in time, along the system development; (v) measurement based on regression techniques, used in conjunction with model-based techniques like for calibrating COCOMO II; or (vi) using composite techniques, which are a mixt of some of the previous presented techniques like the Bayesian analysis approach that combines analogy and expert judgment measurement.

Another possible categorization can be made depending on whether the measurement methodology is primitive, not based on another estimation model (e.g., SLIM and COCOMO 81), or derived from a primitive model, (e.g., COPLIMO which is a COCOMO II derivation adapted to SPL effort estimation) [20].

# 3 MeTra-SPL for Measuring Traceability Impact on SPL Return on Investment

Traceability cost-benefits estimation is a key element for decision making regarding the strategy to adopt. We propose therefore our MeTra-SPL model to study the impact of traceability on SPL cost-benefits.

## 3.1 COCOMO II and COPLIMO Effort Estimation Models

Since COPLIMO model is based on the wildly used software cost estimation model COCOMO II [29], and as it takes into consideration reusability in SPL effort estimation, we judged it as an ideal base for our study.

COCOMO II, for COnstructive COst MOdel, is an extension of the original model COCOMO 81 [19]. It was designed with new estimation factors, to better answer the evolving software development's needs [21]. COCOMO II is used in: (i) investment and risk management decisions involving software costs; (ii) project planning and budgeting; (iii) trade-off decisions relative to cost, schedule, performance or quality; (iv) implementation strategy [21].

Many derivations of this model are proposed in literature to discuss cost estimation of specific systems like SPL. The derived model object of our study is called COPLIMO. It addresses two cost-benefits factors [5, 22]: (1) the Relative Cost of Writing for Reuse (RCWR), which is the cost added to software development with reusing intention, and (2) the Relative Cost of Reuse (RCR) that represents the cost of reusing in the SPL. COPLIMO is based on COCOMO II factors and multipliers [5], which we adapt in our model as presented below.

## 3.2 MeTra-SPL: An Adaptation of COPLIMO Model

SPL approach is based on two major processes: Domain Engineering (DE) and Application Engineering (AE), in addition to maintenance process. In DE, we implement specially developed products (PFRAC). Those products are reused in AE (RFRAC), or adapted (AFRAC) and new ones integrated (PFRAC) in the maintenance phase.

Consequently, trace links are generated in DE process, and then used in AE process. Tracing requires an effort of generation in DE, an effort of use in AE phase, and has an impact on SPL maintenance and change management as it helps identify artifacts affected by SPL changes and links to be modified or created when updating the PL.

Regarding the different tracing objectives, we propose our model as below. $PMR_{DE}(1)$, $PMR_{AE}(1)$ and $PMR_M(1)$ are respectively DE, AE and maintenance effort estimation for implementing the first instance of the PL, and PM(1) the initial effort for developing without reusability.

$$PMR_{DE}(1) = PM(1) * (PFRAC). \tag{1}$$

$$PMR_{AE}(1) = PM(1) * [RCWR \ (RFRAC)]. \tag{2}$$

$$PMR_M(1) = PM(1) * [PFRAC + (RCWR * ATRAC)]. \tag{3}$$

$PMNR_{DE}(N)$, $PMNR_{AE}(N)$ and $PMNR_M(N)$ are respectively DE, AE and maintenance effort of developing N products under the SPL:

$$PMNR_{DE}(N) = N * A * (SIZE_P)^B * DOCU * \prod(EM). \tag{4}$$

$$PMNR_{AE}(N) = N * A * (SIZE_R)^B * DOCU * \prod(EM). \tag{5}$$

$$PMNR_M(N) = N * A * (SIZE_P + SIZE_A)^B * \prod(EM). \tag{6}$$

As documentation is a principal element for traceability, its impact has to be considered in measuring the SPL effort. Therefore, the Degree of Documentation multiplier (DOCU) impact is made in evidence in (4) and (5).

## 4 MeTra-SPL Case Study: The Impact of Traceability Strategy on SPL Return on Investment

Despite the benefits of traceability, there is still unwillingness on its implementation, especially in SPL where the environment is complex. The cost of traceability is easily noticeable at the beginning of the development, while the benefits can't be observed till advanced stages of development [1]. Therefore, we decided to analyze quantitatively, using MeTra-SPL, the impact of traceability on SPL's cost, and conclude on the traceability strategy to adopt in such a large scale system.

Starting form an environment of aircraft-spacecraft production [5], we considered three case studies with three different traceability strategies: Ad hoc, full and targeted. As described in literature [1, 23–27], tracing strategy adopted in a software, and consequently in a SPL, can be: (i) ad hoc tracing with no traces created during product development, but created when needed; (ii) full tracing where all traces identified and created from the beginning; and (iii) targeted tracing where only important and useful traces are created and maintained.

We applied our MeTra-SPL model and used rating tables provided by COCOMO II and COPLIMO models [5] to estimate the effort in the three cases. The same values were applied to unchanged factors: the initial SIZE, measured in KSLOC (thousands of Source Lines Of Code); the adjustment and calibration factors and multipliers A, B and $\prod$ (EM); the software portions factors PFRAC, RFRAC and AFRAC; and the Percent of Integration Required (IM) and Design (DM) and Code Modified (CM); while different values, depending on the case study, were used for factors impacted by traceability: DOCU, Assessment and Assimilation factor (AA) and Software Understanding increment (SU), as in traceability, documentation and software understanding are correlated: There is no traceability without maintaining efficient documentation and tracing helps better understanding the system and being independent from individual knowledge [7]. Other factors are consequently and logically impacted: RCWR and the Adaptation Adjustment Modifier (AAM) (Table 1).

The DOCU multiplier ratings are as follow: 0.91 for ad hoc traceability, which represents a low rating level as no traceability is implemented and consequently life cycle needs are not covered [5]. Full traceability and targeted traceability have respectively 1.23 and 1 for the DOCU rate as the documentation in the first one is quite excessive, while it is right sized for the second one.

As there is no traceability for the first case study, considerable efforts are needed to understand the relations between the artifacts and validate the choice of the most appropriate one. Therefore, AA increment takes the value of 6 for the ad hoc case study, and 2 for the two other traceability strategies as we only need basic search and documenting effort. For the SU increment, and regarding the documentation level for each case study and consequently the understanding of the software, we attribute the values of 40 for ad hoc traceability, 10 for full traceability and 20 for targeted traceability. We consider that the programmer is completely unfamiliar with the system in the three cases (Programmer's Unfamiliarity with the software UNFM = 1).

Our first observation is that, considering the cost of DE development, full traceability is expensive (144.65 KSLOC), while ad hoc traceability is the cheapest (107.02 KSLOC). Conversely, the cost of products' instantiation (AE) in the SPL is higher in the ad hoc traceability approach (4.82 KSLOC), comparing to full traceability (2.17 KSLOC) and targeted traceability (1.76 KSLOC) ones. The same tendency is observed for maintenance costs: 165.76 KSLOC for ad hoc traceability, while full and targeted traceability are cheaper with respectively 152.70 KSLOC and 147.94 KSLOC.

Our observations are in phase with literature: full traceability is expensive, while targeted one, regarding traceability objectives, costs less [1, 23–25]. This conclusion is consolidated in Figs. 1 and 2 that compare the ROI of targeted and full traceability in both AE and maintenance phases, and regarding the number of

**Table 1** COPLIMO parameters values regarding traceability strategy

| | DOCU | RCWR | PMR$_{DE}$(1) | PMR$_{AE}$(1) | PMR$_M$(1) | AA | SU | PMNR$_{DE}$(N) | PMNR$_{AE}$(N) | PMNR$_M$(N) |
|---|---|---|---|---|---|---|---|---|---|---|
| Ad hoc | 0.91 | 1.02 | 117.6 | 101.53 | 219.13 | 6 | 40 | 107.02 | 4.82 | 165.76 |
| Full | 1.23 | 1.56 | 117.6 | 137.23 | 254.83 | 2 | 10 | 144.65 | 2.17 | 147.94 |
| Targeted | 1.00 | 1.27 | 117.6 | 111.57 | 229.17 | 2 | 20 | 117.60 | 1.76 | 152.70 |

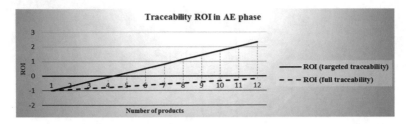

**Fig. 1** Traceability ROI in AE for full and targeted traceability

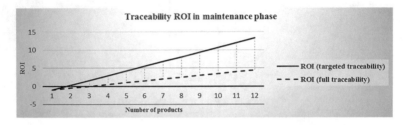

**Fig. 2** Traceability ROI in maintenance phase for full and targeted traceability

elements produced. A positive ROI is quickly reached for targeted traceability (starting from 5 products in AE and 2 products in maintenance) compared to a full one.

$$\mathrm{ROI}_{\mathrm{full/targeted}}(N) = \left[\mathrm{Cost}_{\mathrm{adhoc}}(N) - \mathrm{Cost}_{\mathrm{full/targeted}}(N)\right] / \left[\mathrm{Cost}_{\mathrm{adhoc}}(1) - \mathrm{Cost}_{\mathrm{full/targeted}}(1)\right].$$

(7)

## 5 Conclusion

Tracing is very helpful in SPL. However, its costs are noticeable earlier than its benefits [1]. Traceability is therefore rarely implemented while developing the PL.

In this paper, we provided quantitative values, based on MeTra-SPL model, an adaptation of SPL effort estimation model COPLIMO [5], to prove the benefits of adopting a traceability strategy on SPL cost and ROI. As a result of our work, we found that the implementation of an ad hoc traceability is cheaper in short term (DE process) than in long term (AE and maintenance phases), where full and targeted traceability shows positive ROI, especially with growing number of implemented products.

For our future work, we will focus on the elements that can guide the choice of traceability strategy and their impact on SPL cost estimation.

# References

1. Cleland-Huang, J., Gotel, O., Zisman, A.: Software and Systems Traceability (2012)
2. Cleland-Huang, J., Gotel, O., Hayes, J.H., Mäder, P., Zisman, A.: Software traceability: trends and future directions. In: Proceedings of the on Future of Software Engineering, pp. 55–69. ACM (2014)
3. Cavalcanti, Y.C., do Carmo Machado, I., da Mota, P.A., Neto, S., Lobato, L.L., de Almeida, E.S., de Lemos Meira, S.R.: Towards metamodel support for variability and traceability in software product lines. In: Proceedings of the 5th Workshop on Variability Modeling of Software-Intensive Systems, pp. 49–57. ACM, (2011)
4. Ramesh, B., Jarke, M.: Towards reference models for requirements traceability. software engineering, IEEE Trans. (2001)
5. Boehm, B., Madachy, R., Yang, Y.: A software product line life cycle cost estimation model. In: Proceedings of the 2004 International Symposium on Empirical Software Engineering, pp. 156–164. IEEE (2004)
6. Center of Excellence for Software Traceability. http://www.coest.org/index.php/what-is-traceability
7. Lindvall, M., Sandahl, K.: Practical implications of traceability. Softw. Pract. Exp. (1996)
8. Anquetil, N., Kulesza, U., Mitschke, R., Moreira, A., Royer, J.C., Rummler, A., Sousa, A.: A model-driven traceability framework for software product lines. Softw. Syst. Model. (2010)
9. Spanoudakis, G., Zisman, A.: Software traceability : a roadmap. handbook of software engineering and knowledge engineering (2005)
10. Jirapanthong, W., Zisman, A.: Supporting product line development through traceability. In: 12th Asia-Pacific Software Engineering Conference, pp. 506–514 (2005)
11. Anquetil, N., Grammel, B., Galvão, I., Noppen, J., Khan, S., Arboleda, H., Rashid, A., Garcia, A.: Traceability for model driven, software product line engineering. In: ECMDA Traccability Workshop Proceedings, pp. 77–86. Berlin (2008)
12. Berg, K., Bishop, J., Muthig, D.: Tracing software product line variability—from problem to solution space. In: Proceedings of the 2005 Annual Research Conference of the South African Institute of Computer Scientists and Information Technologists on IT Research in Developing Countries, pp. 182–191. South African Institute for Computer Scientists and Information Technologists (2005)
13. Ghanam, Y., Maurer, F.: An iterative model for agile product line engineering. In: SPLC, pp. 377–384 (2008)
14. Ghanam, Y., Maurer, F.: Extreme product line engineering : managing variability and traceability via executable specifications. In: Agile Conference, AGILE'09, pp. 41–48. IEEE (2009)
15. Díaz, J., Pérez, J., Garbajosa, J.: Agile product-line architecting in practice: a case study in smart grids. Inf. Softw. Technol. (2014)
16. Díaz, J., Pérez, J., Yagüe, A., Garbajosa, J.: Tailoring the scrum development process to address agile product line engineering. In: Proceedings of the 16th Conference on Software Engineering and Databases, JISBD (2011)
17. Jirapanthong, W., Zisman, A.: XTraQue: traceability for product line systems. Softw. Syst. Model. (2009)
18. Boehm, B., Abts, C., Chulani, S.: Software development cost estimation approaches—a survey. Ann. Softw. Eng. (2000)
19. Boehm, B.: Software Engineering Economics. Prentice-hall, Englewood Cliffs (NJ) (1981)
20. Heradio, R., Fernandez-Amoros, D., Torre-Cubillo, L., Garcia-Plaza, A.P.: Improving the accuracy of COPLIMO to estimate the payoff of a software product line. Expert Syst. Appl. (2012)
21. USC center for software engineering. http://csse.usc.edu/csse/research/COCOMOII/cocomo_main.html
22. Poulin, J.S.: Measuring software reuse: principles, practices, and economic models (1996)

23. Egyed, A., Biffl, S., Heindl, M., Grünbacher, P.: Determining the cost-quality trade-off for automated software traceability. In: Proceedings of the 20th IEEE/ACM International Conference on Automated Software Engineering, pp. 360–363. ACM (2005)
24. Egyed, A., Biffl, S., Heindl, M., Grünbacher, P.: A value-based approach for understanding cost-benefit trade-offs during automated software traceability. In: Proceedings of the 3rd International Workshop on Traceability in Emerging forms of Software Engineering, pp. 2–7. ACM (2005)
25. Egyed, A.: Tailoring software traceability to value-based needs. In: Value-Based Software Engineering, pp. 287–308. Springer, Berlin, Heidelberg (2006)
26. Egyed, A., Grünbacher, P., Heindl, M., Biffl, S.: Value-based requirements traceability : lessons learned. In: Design Requirements Engineering: A Ten-Year Perspective, pp. 240–257. Springer, Berlin, Heidelberg (2009)
27. Heindl, M., Biffl, S.: A case study on value-based requirements tracing. In: Proceedings of the 10th European Software Engineering Conference Held Jointly with 13th ACM SIGSOFT International Symposium on Foundations of Software Engineering. pp. 60–69. ACM (2005)

# Scrumban/XP: A New Approach to Cover the Third Level of CMMI Model

Zineb Bougroun, Adil Zeaaraoui and Toumi Bouchentouf

**Abstract** The company takes care more and more of their product quality, and this by introducing new methodologies that allow them to have a regular feedback of their customers satisfaction. This interest is developed by the adoption of agile methods and the respect of the CMMI model. In this paper we study the weaknesses of Scrum compared to CMMI, and we complete these defects by agile practices to cover the third level of CMMI.

**Keywords** CMMI · Scrum · XP · Kanban · Agile methods

## 1 Introduction

Quality takes more interest in the software development life. This is the reason why there are different aspects that seek to improve software quality, among these, there is the aspect of product quality using metrics, that we discussed in previous articles [1, 2], and the aspect of processes and methodologies aimed at organize work to improve the quality, in this section we have been interested to the second aspect.

Before talking about methods and methodologies we must necessarily pass on CMMI model, which is a process framework that is widely adopted by software and systems development companies. and in order to organize the development to

Z. Bougroun (✉) · A. Zeaaraoui · T. Bouchentouf
(Laboratoire des Systemes Electronique, Informatique et Images) LSE2I, (Système Informatique et Qualité Logiciel) SIQL Team Group, (National School of Applied Sciences—Oujda) ENSAO, Mohammed First University, 60000 Oujda, Morocco
e-mail: bougroun.zineb@gmail.com

A. Zeaaraoui
e-mail: adilzeaaraoui@yahoo.fr

T. Bouchentouf
e-mail: tbouchentouf@gmail.com

© Springer International Publishing Switzerland 2016
A. El Oualkadi et al. (eds.), *Proceedings of the Mediterranean Conference on Information & Communication Technologies 2015*, Lecture Notes in Electrical Engineering 381, DOI 10.1007/978-3-319-30298-0_40

produce quality software the company has adopted various kinds of method such us agile methods. CMMI is basically a process improvement framework which mandates a set of processes for software development management [3]. The agile methodologies can be thought of as an iterative software project management framework for development activities, among the most known agile methods is Scrum, XP and Kanban. CMMI has a wider scope and different aims to those of agile methodology [4].

In this paper we study the possibility to add some practices of XP and Kanban to Scrum method to cover the third level of CMMI model. And we present this work as follows, in the first part we will present the related works in the second we will explain CMMI model and the three methods, we continue our study with section that illustrates the CMMI practices not covered by Scrum and explain how to complete this method to cover the third level of this model, then we end our article by a discussion.

## 2   Related Works

Software Quality is a broad field that is divided into different axes, which include the quality of the product and the quality of process. In this work we will be interested in those concerning the development process. This axis has a different research carried out and among these researches many treated the CMMI model and agile methods. Several authors have discussed the compatibility of CMMI and agile methods. Julio [5] discusses the possibility for software companies of getting a CMMI certication of their processes by applying agile practices. Fritzsche and Keil [6] describe the limitations of CMMI in an agile environment and show that level four or five are not feasible under the current specications of CMMI and XP.

Kane and Ornburn [7] analyze which CMMI process areas are covered by XP and Scrum. Especially those areas related with process management are not considered by these two methods. Turner and Jain [8, 9] show how CMMI can help to successfully implement agile methods. And the last study we will mention is that of Diaz, Garbajosa, Calvo-Manzano who present the Mapping between CMMI Level 2 to Scrum Practices [10].

In continuation of this work we have established a mapping between the specific practices of the third level of CMMI and the practices of agile methods: Scrum XP and Kanban [11]. In which we have specified for each CMMI practice how it can be applied with agile practices, this work is related to the previous insofar it will be used to know the CMMI practices not covered by Scrum. In this work we study the weakness of Scrum method and we try to know the practices of other agile methods: Xp and Kanban, which can be added to Scrum in order to complete the cover of the third level of CMMI.

# 3    CMMI, Scrum XP and Kanban

## 3.1    CMMI

CMMI Family (2006) is a collection of American standards (SEI Software Engineering Institute) providing a framework for process improvement software development in particular those relating to the field of project management activity [12].

CMMI is a model that makes sense to the improvement of the project management in five levels of maturity. That contain twenty two areas, arranged into four categories [13, 14]: Engineering, Project Management, Process Management, and Support. Each level of the model covers different areas of activity that are used in the implementation or the production support of software products. Each area contains specific practices and generic practices. In this article we are interested to the specific practices of the third level of CMMI. The CMM was developed from 1987 until 1997. In 2002, CMMI Version 1.1 was released, Version 1.2 followed in August 2006, and CMMI Version 1.3 in November 2010. In this article we will study the last version.

## 3.2    Scrumban/XP

Scrum is an iterative, incremental framework for managing projects [15–18]. Scrum structures work in iterations called Sprints, which are between 2 and 4 weeks in length. At the beginning of each Sprint, functional team selects items from a prioritized list of requirements called the Product Backlog; this list is created and maintained by the Product Owner. Each sprint has a sprint planning meeting at the sprint beginning where the Product Owner and Team plan together about what to be done for the next sprint; the result is the sprint backlog. During the execution of each sprint, the team meets daily in 15-min meetings Daily Scrum Meeting to track the work progress [15]. At the end of the Sprint, the team holds a Sprint Review, where they demonstrate what they have built, to generate feedback which the Product Owner can incorporate into the Product Backlog for later Sprints. The team also holds a Sprint Retrospective, where they review their way of working, and agree on changes to improve their effectiveness.

Kanban is a lean approach to agile software development. Kanban uses cards as visual symbols to trigger and control flow through a production process [19]. The practices of Kanban can be added to Scrum to form the Scrumban method.

The core of Kanban means:

- Visualize the workflow
- Limit WIP (work in progress) assign explicit limits to how many items may be in progress at each workflow state.
- Measure the lead time (average time to complete one item).

The XP method was implemented by Kent Beck 1999. It is considered as a collection of software engineering practices and came to solve the problems of long development cycles. This method takes its name from its basic principle which is reducing the cost of changes in the extreme [20]. The method can be summarized on twelve practices:

- Planning game: Customers decide the scope and timing of releases based on estimates provided by programmers.
- Small releases.
- Metaphor.
- Simple design.
- Tests: Programmers write unit tests minute by minute.
- Refactoring: The design of the system is evolved through transformations of the existing design that keep all the tests running.
- Pair programming: All production code is written by two persons.
- Continuous integration.
- Collective ownership.
- On-site customer: A customer sits with the team full-time.
- 40-h weeks: No one can work a second consecutive week of overtime.
- Open workspace.

## 4   CMMI Level 3 and Scrum Practices

CMMI describes what to do to achieve a level of maturity, while agile methods explain how to improve the quality of a product to satisfy the customer. In previous work we have compared the CMMI practices and those of agile methods, and we have shown that Scrum practices cover 44 % of the specific practices of CMMI, the question we will try to answer in this part is: what is the percentage that can Scrum cover of the third level of CMMI by adding other agile methods practice?

Scrum is a project management method, so all the practices that CMMI has on the development and implementation of the project are not covered by this method, otherwise XP is also an agile method, but the aspect of this method is oriented implementation of the application, furthermore Kanban is oriented flow control. That is why the addition of a practice of these methods does not contradict Scrum.

Based on the work that makes the mapping between the third level of CMMI and the agile methods we discuss the process area not covered by Scrum and the possibility of the supplemented with agile practices.

## 4.1  Requirements Development (RD)

This process area contains the practices to produce and analyze customer, product, and product component requirements. This process area has the following practices that are not covered:

SP2.2:  Allocate Product-Component Requirements, this practices is not clearly defined in the Scrum method, and not in any other method, but the most logical way to implement this practice in the process of Scrum by discuss it in the meeting of review of product Backlog (requirements) with Product owner and team

SP2.3:  Identify Interface Requirements, this practices is not defined in the agile method. This practice should be also implemented in the meeting of review of product Backlog (requirements) with Product owner and team

SP3.1:  Establish Operational Concepts and Scenarios, not exist

## 4.2  Technical Solution (TS)

This process area contains all the skills required with respect to software architecture, design, and coding.

SP1.1:  Develop Alternative Solutions and Selection Criteria, this practice does not exist in XP method but if we merge the simple design in scrum this practice will be discuss in the meeting of the review of the product backlog with the customers

SP1.2:  Select Product Component Solutions, this practice has the same reasoning as the previous

SP2.1:  Design the Product or Product Component, this practice is in XP as simple design

SP2.2:  Establish a Technical Data Package, not exist

SP2.3:  Design Interfaces Using Criteria, this practice is in XP as simple design

SP2.4:  Perform Make, Buy, or Reuse Analysis, not exist

SP3.1:  Implement the Design, this practice is in XP as pair programming

SP3.2:  Develop Product Support Documentation, not exist

## 4.3  Product Integration (PI)

Processes in this part concern the ability to integrate multiple components to form a complete product. The practices not cover by Scrum are:

SP2.2:   Manage Interfaces; this practice is discussed in the daily cooperation
          between business people and developers
SP3.1:   Confirm Readiness of Product Components for Integration. Not exist

## 4.4   Verification (VER)

Practices of this part aim to ensure that the product being built is envisaged by the
client.
SP1.1:   Select Work Products for Verification, this practice can be discussed in the
          daily cooperation between business people and developers assured by XP
          method
SP1.2:   Establish the Verification Environment; the test practice requires us to go
          through this step
SP1.3:   Establish Verification Procedures and Criteria; the same us the previous
          practice
SP2.1:   Prepare for Peer Reviews; this practice is provided by the pair
          programming
SP2.2:   Conduct Peer Reviews; this practice is provided by the pair programming
SP2.3:   Analyze Peer Review Data; it can be discussed in the daily meeting of
          Scrum if it is any problem or in the meeting of the end of sprint
SP3.1:   Perform Verification; this practice is discussed in the daily cooperation
          between business people and developers
SP3.2:   Analyze Verification Results; this practice exists neither in Scrum nor in
          XP but the merge of the two methods will lead us to discuss the
          verification problems at the daily meeting and at the meeting of the end of
          sprint

## 4.5   Decision Analysis and Resolution (DAR)

This process records the decision and the reasons for the choice.
SP1.1:   Establish Guidelines for Decision Analysis; this practice is implemented in
          Kanban as Visualize the workflow, Limit WIP (work in progress),
SP1.2:   Establish Evaluation Criteria; the Kanban method use the Limit WIP (work
          in progress) to evaluate the workflow of the process
SP1.3:   Identify Alternative Solutions. This practice can be discussed in the
          meeting of Scrum based on the result of the use of Kanban practices
SP1.4:   Select Evaluation Methods. This practice can be also discussed in the
          meeting of Scrum based on the result of the use of Kanban practices

SP1.5:   Evaluate Alternatives; the Kanban method use the Limit WIP (work in progress) to evaluate the workflow of the process
SP1.6:   Select Solutions. This practice can be also discussed in the meeting of Scrum based on the result of the use of Kanban practices

## 5   Discussion

The Fig. 1 shows as the percentage of coverage of Scrum practices in CMMI model and also it shows the percentage of the merge of Scrum with some practices of XP and Kanban method. The scrum method covers 44 % of the third level of CMMI; XP method covers 45 % of the level and kanban 6 % [11]. Scrum covers more area process involving project management and organizational processes while XP is cover the two aspects, the management project and the implementation process. Kanban covers only 6 % because this method is based primarily on control and decision appearance. As we can see in this work and since XP brings together the two aspects, the management of project which already exists in Scrum, and the product development (the how). To complement the Scrum practices to cover the third level of CMMI, only practices interested in development can serve. These practices are; the simple design and the pair programming which can help us to implement the process area of Technical Solution, also the practices test and daily cooperation serve to implement the process area of the verification, this four practices will be discussed in the meeting of the Scrum, that helps us to do analysis and consequently implemented other specific practices. So the integration of XP practices cover 16 % of the third level of CMMI. The Decision Analysis and Resolution process area is covered by Kanban practices and also with the discussion

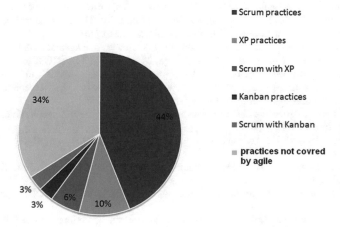

**Fig. 1**  Agile practices and CMMI level 3

of these practices in the meeting of Scrum, than it cover 6 % of the third level of CMMI specific practices. Only the process area Organizational Training is not implemented in agility practices and some specific practices of Integrated Project Management, Organizational Process Focus, Organizational Process Definition and Risk Management [11]. As a conclusion we can say that the addition of the four XP practices and those of Kanban can lead us to cover 66 % of the third level of CMMI.

# 6   Conclusion

The scrum method is a good implementation practices CMMI level three however this method only covers 44 % of the third level of CMMI. The question we wanted to answer in this work is the following: can we use other agile practices to complete the cover of CMMI level three? And what are these practices? Adding practices of XP or Kanban to Scrum is possible as much as these practices are in different areas that project management, which is the area of Scrum. Then the addition of the simple design, the pair programming, test and daily cooperation improves the percentage of the cover of CMMI level three by 16 %, and the merge of Kanban with the Scrum practices improve it by 6 %. Therefore Scrumban and the four practices of XP: simple design, the pair programming, test and daily cooperation will allow us to cover 66 % of CMMI level three.

# References

1. Bougroun, Z., Zeaaraoui, A., Belkasmi, M.G., Bouchentouf, T.: Joining ISO model with metrics using design quality properties. J. Inf. Syst. Manag. 2(4), 184–195 (2012)
2. Bougroun, Z., Zeaaraoui, A., Saber, M., Bouchentouf, T.: Enhancement of the taxonomy of metrics by ISO model using design quality properties. In: The International Syposium on Securiy and Safety of Complex Systems. Agadir, Morocco (2013)
3. Ramanujan, S., Kesh, S.: Comparaison of knowledge management and CMM/CMMI implementation. J. Am. Acad. Bus. Cambridge; Mar 2004; 4, ABI/INFORM Global p. 271
4. Pikkarainen, M., Mantyniemi, A.: An approach for using CMMI in agile software development assessments: experiences from three case studies. In: Accepted for SPICE 2006 Conference, that Will be in Luxemburg. 4–5 May 2006
5. Alegra, J.A.H., Bastarrica, M.C.: Implementing CMMI using a combination of agile methods. Clei Electron. J. 9(1), 7 (2006)
6. Fritzsche, M., Keil, P.: Agile methods and CMMI: compatibility or conflict? E-Inf. Softw. Eng. J. 1(1) (2007)
7. Kane, D., Ornburn, S.: Agile Development: Weed or Wildflower?. J. Defense Softw. Eng., CrossTalk (2002)
8. Turner, R.: Agile development: good process or bad attitude? In: PROFES, pp. 134–144 (2002)
9. Turner, R., Jain, A.: Agile meets CMMI: culture clash or common cause? In: XP/Agile Universe, pp. 153–165 (2002)

10. Diaz, J., Garbajosa, J., Calvo-Manzano, J.A.: Mapping CMMI level 2 to scrum practices: an experience report. O'Connor, R.V., et al. (eds.): EuroSPI 2009 CCIS 42, pp. 93–104 (2009)
11. Bougroun, Z., Zeaaraoui, A., Bouchentouf, T.: The projection of the specific practices of the third level of CMMI model in agile methods: Scrum, XP and Kanban. In: Colloquium in Information Science and Technology. Tetouan, Morocco (2014)
12. Capability Maturity Model Integration (CMMI): Version 1.1 CMMI for Software Engineering, Continuous Representation, August 2002, CMMI Product Team
13. CMMI; CMMI SM for Systems Engineering/Software Engineering/Integrated Product and Process Development/Supplier Sourcing, Version 1.1, Staged Representation (CMMI-SE/SW/IPPD/SS, V1.1, Staged (2002)
14. Chrissis, M.B., Konrad, M., Shrum, S.: CMMI Guidlines for Process Integration and Product Improvement (2003)
15. Beedle, M., Devos, M., Sharon, Y., Schwaber, K., Sutherland, J.: Scrum: a pattern language for hyperproductive software development. In: Harrison, N. (ed.) Pattern Languages of Program Design, vol. 4, pp. 637–651. Addison-Wesley, Boston (1999)
16. Sutherland, J., Viktorov, A., Blount, J., Puntikov, N.: Distributed scrum: agile project management with outsourced development teams. In: HICSS'40, Hawaii International Conference on Software Systems, Big Island, Hawaii (2007)
17. Schwaber, K., Beedle, M,; Agile Software Development with Scrum, Prentice Hall (2002)
18. Sutherland, J.. Agile development: lessons learned from the first scrum. Cutter Agile Project Manag. Adv. Service Executive Update 5, 1–4 (2004)
19. Ingason, H.T., Gestsson, E., Jonasson, H.I.: The project kanban wall: combining kanban and scrum for coordinating software projects. PM World J. II(VIII) (2013)
20. Beck, K.: Cynthia Andres; Extreme Programming Explained: Embrace Change (2nd Edition)

# MDA Approach: Refinement and Validation of CIM Level Using SBVR

Najiba Addamssiri, Abdelouhaed Kriouile, Sara Boussaa
and Taoufiq Gadi

**Abstract** Any change in the software based on business logic affects all MDA levels: CIM, PIM and PSM. This paper provides an approach to refine and validate CIM for enabling a premature customer's validation. We propose a set of rules for generating BPMN Model from use cases model which represent both of them our CIM. This transformation uses SBVR as an intermediate step. Firstly, we generate SBVR model from use case model and its textual description and secondly we transform it to BPMN model. SBVR represents the axe of the CIM refinement cycle, which is based on a natural language that is easily comprehensible and usable by business people and easily can be machine processed. Validated and refined CIM's achieves a high customer satisfaction, and allowing a better vision of the product as the customer expects.

**Keywords** MDA · CIM refinement · BPMN · SBVR · Model transformation

## 1 Introduction

Model Driven Architecture (MDA) is an approach uses models in process development. It emphasizes the development of software systems based on visual models as the primary software artifacts. The highest level in MDA; CIM defines the business process and aims to get the key business activities; this level is designed to enable the connection between the business analyst with a set of requirements and

N. Addamssiri (✉) · A. Kriouile (✉) · S. Boussaa (✉) · T. Gadi (✉)
LAVETE Laboratory, Université Hassan 1er, 2600 Settat, Morocco
e-mail: addam.naji@gmail.com

A. Kriouile
e-mail: kriouile1970@gmail.com

S. Boussaa
e-mail: sr.boussaa@gmail.com

T. Gadi
e-mail: g.taoufiq@yahoo.fr

© Springer International Publishing Switzerland 2016
A. El Oualkadi et al. (eds.), *Proceedings of the Mediterranean Conference on Information & Communication Technologies 2015*, Lecture Notes in Electrical Engineering 381, DOI 10.1007/978-3-319-30298-0_41

393

the IT architect with technical solutions [1]. In [2] business processes are termed CIM models and are typically constructed using the Business Process Modeling Notation (BPMN).

In our work [3] we have built CIM architecture around three views the static view, the dynamic view and the functional view. We have represented the business processes models with the BPMN and transforming it horizontally to detailed use cases model. In [4] we have presented the refinement of the CIM level using horizontal transformation from Use case model to BPMN model. Thus in order to complete this architecture and integrate the no IT person to validate their specifications we suggest to add the business rule language in this proposal.

Business rule languages are declarative languages that allow the specification of the constraints that apply to business objects during business activities [2], without prescribing how and by whom these constraints are imposed. Also in [5], business rules can be seen as requirements for business processes. Since business processes are supposed to implement business rules, we propose to refine the CIM level by transforming the use case model and its textual description to Semantic Business Vocabulary and Rules (SBVR) [6]. Then we propose a premature validation by involving the customer who can verify, change and validate the SBVR, which should conform to specifications. And finally we generate automatically the business process from the business rules obtained.

These transformations steps allow the domain experts to involve in the modeling requirements, which keeps the knowledge domain and adapt with any software product all over the levels modeling lifecycle of the development. Also, they confirm premature validation of the requirements and specifications by the customer; moreover they allow the sustainability of business specifications which is the goal of MDA and the capitalization of the knowledge. This approach make a system less responsive to changes in technology at low levels, and likewise presents a great benefit in term of time saving and the reducing the cost of maintenance.

The remainder of this paper is structured as follows: The second section discusses some related works. The third section presents the proposed architecture and the mapping rules. And the last section presents an evaluation of the proposal based on a case study.

## 2 Related Works

In literature, we could not find any published research that defines the validation and the refinement of the CIM level using a complete architecture. The authors of [7, 8] suggest approaches to transform SBVRSE (SBVR Structured English) to UML activity diagrams and BPMN diagrams, respectively. The both approaches focus on finding a correct process sequence. They use the "if condition then action" construct of business rules to determine the process dependencies as in [5]. At the other hand, there are only few papers presenting other kinds of transformations to obtain SBVR specifications from UML models or other knowledge representations, e.g. Cabot et al. [9] does that by transforming UML/OCL conceptual data models; in [10, 11], unstructured natural language is used as a source of knowledge for the extraction of business rules and business vocabulary. Currently, there is only the research of Thakore and Upadhyay [12], the research of Tomas et al. [13], which directly deal with SBVR to UCD (Use Cases Diagram) transformation. However, the extraction remains only the basic UCD elements by a semi-automatic transformation that requires human intervention.

Our related works [3, 4] focuses on the CIM modeling and the automating transformation from BPMN Model to Use Case Model with its Textual Description, and the refinement of the CIM from Use case model to BPMN model.

# 3 Overview of Our Proposed Approach

According to our previous work [3, 4], the CIM level is modeled by both the BPMN model and the use cases model with its textual description formalized by SBVR. In this paper, for refining and validating the CIM level, we propose three steps: firstly, to generate the business rules and the business vocabulary from existing CIM models, we use an automatic transformation from use case model and its textual description, which conforms to their metamodels, into SBVR model that conforms to SBVR metamodel. Secondly, in order to ensure a premature validation. The customer verifies, changes and validates the SBVR model. This early validation helps saving time and guarantees the generation of the product as the domain user expects. Thirdly we establish an automatic transformation from SBVR model that conforms to SBVR metamodel into BPMN model which conforms to BPMN metamodel. Thus greatly reduces the development time of the transformation and reduces errors that can occur during the manual specifications of the transformations.

## 3.1 Why SBVR?

In the modeling of a typical business application, a business analyst and the business owners define the requirements for the new business application. At this level of abstraction, business domain must be specified in a language that could specify business vocabulary and rules unambiguously, and use these specified expressions in model transformations in such a way to be easily comprehensible and usable by business people. To enable business rules to be machine processed, the plain English business rules are represented in some formal representation such as SBVR [14].

Some characteristics of SBVR are presented as follow: SBVR is an OMG's standard [6] that provides formal representation for business rules written in natural language, and the vision of SBVR is to express business knowledge in a controlled natural language which is unambiguous and understandable by human as well as to computer systems [15]. Also a SBVR rule is the key constituent of SBVR standard. Moreover SBVR rule can easily be machine processed to perform object rule modeling [7], perform rule consistency analysis or generating formal representations such as OCL constraints [16], databases, business rules repositories, business blueprints, business object models and software components. At last, the SBVR is fully integrated into the OMG's Model Driven Architecture via Meta Object Facility (MOF) or Eclipse Meta-modeling Framework (EMF).

Consequently, the best and the simplest architecture used for non-technical people are the semantic business rules and vocabulary.

In the next step we establish the mapping rules between SBVR and BPMN model presented in the CIM level.

## 3.2  Mapping Rules from UC to SBVR: UC2SBVR

This transformation is defined by seven transformation rules, which specify how one or more elements from the UC model (source model) and their textual description are transformed to one or more elements from the SBVR model.

While UML does not require or recommend any particular format to the textual description of the use cases, we have formalized it, in a previous work [4], by SBVR.

These rules are defined as follows:

- Rule 1: The actor represents the giver of the action of the verb. It will be transformed to the subject of the fact type.
- Rule 2: The use case represents some action. For example, the use case 'inserts card' will be transformed to the fact type.
- Rule 3: The association between the actor and the use case will be transformed to a relationship between a subject and a fact type. The use case that wasn't associated with an actor should be associated to its subject in the TD.
- Rule 4: the include relationship will be transformed to a sentence as follow: 'it is obligatory that' or 'it is necessary that' followed by the actor with its use case (included) then 'if' and the actor with its user case (including).
- Rule 5: Using 'in the case of' followed by the extension point and the actor with its use case.
- Rule 6: the successful scenario presented in the textual description will be transformed to a sentence as follow: Using 'in the case of' followed by the extension point and successful scenario sentence in TD.
- Rules 7: Using 'in the case of' followed by the extension point and the sentence of the error scenario in TD.
- Rule 8: The error scenario will be transformed to a sentence as follow: Using 'in the case of' followed by the extension point and the sentence of the error scenario in TD.

## 3.3   SBVR Validation

The customer verifies the business vocabulary and business rules to determine their compatibility with the specifications. If there is any modification the analyst should modify it and the customer verifies it another time. Once satisfied, the customer validates the model. The main advantage of this validation is to encourage customer involvement in developing his software, and to offer flexible communication and a consensus between customers and analysts, who have quite different backgrounds; also it permits accelerating the production and generating the product as the customer expects.

## 3.4   Mapping Rules from SBVR to BPMN: SBVR2BPMN

The mapping rules uses SBVR constructs as source elements, BPMN constructs as target elements. We give a detailed definition for each rule.

The representation of fact type can be of two types, active form and passive form. The fact type can be represented as term-verb-term and a verb can be of two types, transitive verbs and intransitive verbs [7].

- The statements involving the transitive verbs will represent some action. For example, the fact type "client inserts card" is a statement which includes a transitive verb "inserts" and shows an action "insertion of card into machine". These types of fact types can be transformed to activities.
- When a sentence has a transitive active verb, the giver of the action of the verb is the subject of the sentence. It will be transformed to swimlane.
- If SBVR rule has one transition coming and multiple parallel transitions going out of it, it will be transformed to a Parallel gateway. E.g. according to this rule "It is obligatory that machine retains card, and client contacts the bank if client enters the PIN after three tries"; it represents two activities 'machine retains card' and 'client contacts the bank' in case that he fails to enter successfully the PIN after three tries.
- When SBVR rule has conditions and actions which allow the transition from one activity to another activity. It will be transformed to XOR gateway. E.g. the rule 'it is obligatory that each client receives card if machine ejects card'; the activity 'client receives card' should be generated only if the activity 'ATM ejects card' is generated.
- If SBVR rule has multiple parallel transitions coming and only one transition outgoing it will be transformed to OR gateway, E.g. the rule 'it is obligatory that the client inserts the card and enters the PIN in order to the machine opens her account' the two activities 'client insert the card' and 'client enters the PIN' should be generated in order to generate the activity 'machine opens her account'.
- When two successive activities have the same subject with a transitive verb in the fact type it will generate a sequence flow between those activities, e.g. in the

rule 'the machine displays main-screen and requests PIN", the sequence flow will be established between 'displays main-screen' and 'requests PIN'.

- At last when two successive activities have different subjects with transitive verb in the fact type, it will generate a message flow between those activities. E.g. in the rule 'machine displays main-screen and user enters PIN', the sequence flow will be established between "machine requests password" and "user enters PIN".

QVT [17] transformation built will be able to extract swimlane, activities, gateways, sequence flow and message flow from SBVR. The transformation receives as input the SBVR in active form and generates as output the elements of BPMN diagram. The QVT code of the both transformations will be represented in a future work.

## 4 Case Study

To evaluate the approach we propose to use a case study that is known as an important business workflow. Therefore we have decided to choose the ATM (Automatic Teller Machine) system case study. We suggest that the CIM models are established and we want to refine it and to validate it by customer. Firstly we transform the use case diagram and its textual description to SBVR: Our use case diagram and TD formalized by SBVR are presented in Figs. 1 and 2.

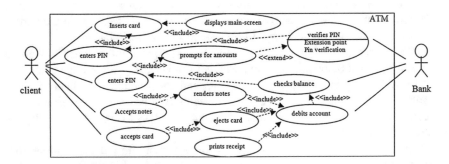

**Fig. 1** The use case diagram of our case study

**Fig. 2** TD-SBVR of our study case

| Machine *displays* main-screen |
| Client *inserts* Card |
| Client *enters* PIN |
| Bank *verifies* PIN |
| ErrScenario: PIN *is* incorrect |
| Machine *retains* card |
| Client *contacts* Bank |
| SuccScenario: PIN *is* correct |
| Machine *prompts* for amount |
| Client *enters* withdraw |
| Bank *checks* balance |
| Bank *debits* account |
| Machine *prints* receipt |
| Machine *renders* notes |
| Machine *ejects* card |
| Client *accepts* notes |
| Client *accepts* card |

```
Machine displays main-screen
Client inserts Card
Client enters PIN
Bank verifies PIN
PIN is incorrect
Machine retains card
Client contacts Bank
PIN is correct
Machine prompts for amount
Client enters withdraw
Bank checks balance
Bank debits account
Machine prints receipt
Machine renders notes
Machine ejects card
Client accepts notes
Client accepts card
```

**Fig. 3** Generated SBVR business vocabulary

```
It is obligatory that machine displays main-screen if client inserts card
It is obligatory that client enters PIN if client inserts card
It is obligatory that Bank verifies PIN if client enters PIN
In the case of PIN is incorrect, machine ejects card
In the case of PIN is correct, machine prompts for amount
It is obligatory that client enters withdrawal if machine prompts for amount
It is obligatory that bank checks balance if client enters withdrawal
It is obligatory that bank debits account if bank checks balance
It is obligatory that machine renders notes, ejects card and prints receipt if bank debits account
It is obligatory that client accepts notes if machine renders notes
It is obligatory that client accepts card if machine ejects card
```

**Fig. 4** Generated SBVR business rules

After applying the mapping rules of UC and SBVR proposed in the Sect. 3.4, we generate automatically the business vocabulary and business rules. Figure 3 is snapshot of vocabulary showing the terms and the fact types, and Fig. 4 presents the business rules for the ATM and the client interaction.

The second step is the customer validation: while in our approach there is an opportunity to constantly refine and reprioritize, new and changed rules may be planned by the customer. In this example some changes are planned by the customer, for example the following rule: "in the case of PIN is incorrect, machine ejects card" will be changed with: "in the case of PIN is incorrect after three tries, machine retains card". After the last customer's validation of the business vocabulary and rules, Fig. 5 presents a snapshot of the validate business rules.

The third step is applying the SBVR2BPMN mapping rules proposed in the Sect. 3.4 to generate automatically the new BPMN model.

Thus we obtained the swimlane client with the following activities: inserts card, enters PIN, enters withdraw, contacts bank, accepts notes, and accepts card.

```
It is obligatory that machine displays main-screen if client inserts card
It is obligatory that client enters PIN if client inserts card
It is obligatory that Bank verifies PIN if client enters PIN
In the case of PIN is incorrect, client enters PIN
In the case of PIN is incorrect after three tries, machine retains card
It is obligatory that Client contacts Bank if machine retains card
In the case of PIN is correct, machine prompts for amount
It is obligatory that client enters withdrawal if machine prompts for amount
It is obligatory that bank checks balance if client enters withdrawal
It is obligatory that bank debits account if bank checks balance
In the case of requested amount more than account balance, machine ejects card and prints receipt
In the case of requested amount less or equal to account balance, bank debits account
It is obligatory that machine renders notes, ejects card and prints receipt if bank debits account
It is obligatory that client accepts notes if machine renders notes
It is obligatory that client accepts card if machine ejects card
```

**Fig. 5** Validate business rules

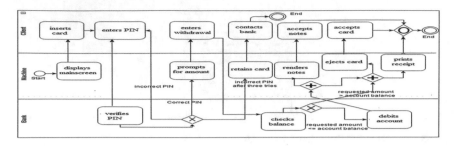

**Fig. 6** BPMN diagram from validate SBVR vocabulary and rules

And the swimlane machine with its activities: displays main-screen, prompts for amount, retains card, renders notes, ejects card, and prints receipt. And finally, the swimlane bank with the activities: verifies PIN, checks balance, and debits account. Figure 6 illustrates the generate BPMN model.

## 5 Conclusion

The paper proposes a transformation approach for validating and refining the CIM level represented by both BPMN and uses cases models. We have established an automatic transformation from the use case model to SBVR model in order to ensure a primitive validation by the customer, who verifies the SBVR model, and can change or add any new specifications according to his needs. After the validation of the SBVR we generate the new BPMN Model by transforming automatically the SBVR.

The proposal completes our previous works [3, 18] that subscribe under a global approach that aims to automate the whole development process. We can see that our approach lead us to a very important outcome which ensures high customer satisfaction by involving him regularly, that allows mitigate one of the most consistent problems in a software project, that's the client find at the end of the project differs from what he told us the beginning.

Some future works are strongly suggested to be done such as developing a tool that supports all the transformations performed in our earlier works.

## References

1. Brown, A.W., Iyengar, S., Johnston, S.: A Rational Approach to Modeldriven, Riverton, NJ, USA. IBM Systems Journal (2006)
2. Hay, D., Healy, K., Hall, J.: Defining Business Rules—What Are They Really?. The Business Rules Group (2000)

3. Kriouile, A, Gadi, T., Addamssiri, N., El Khadimi, A.: Obtaining behavioral model of PIM from the CIM. In: 2014 International Conference on Multimedia Computing and Systems (ICMCS), pp. 949–954. IEEE (2014)
4. Addmassiri, N., Kriouile, A., Balouki, Y., Gadi, T.: Generating the PIM behavioral model from the CIM using QVT. J. Comput. Sci. Inf. Technol. (2015)
5. Steen, B., Pires, L.F., Iacob, M.E.: Automatic generation of optimal business processes from business rules. In: 2010 14th IEEE International Enterprise Distributed Object Computing Conference Workshops (EDOCW). IEEE (2010)
6. OMG: Semantics of Business Vocabulary and Rules. OMG Standard, v. 1.0. (2008)
7. Raj, A., Prabhakar, T.V., Hendryx, S.: Transformation of SBVR business design to UML models. In: Proceedings of the 1st India Software Engineering Conference, ACM, pp. 29–38 (2008)
8. Eder, R., Filieri, A., Kurz, T., Heistracher, T.J., Pezzuto, M.: Model-transformation-based software generation utilizing natural language notations. In: 2nd IEEE International Conference on Digital Ecosystems and Technologies, DEST 2008, pp. 306–312 (2008)
9. Cabot, J., Pau, R., Raventós, R.: From UML/OCL to SBVR specifications: a challenging transformation. Inf. Syst. 35(14), 417–440 (2010)
10. Feuto, N.P., El Abed, W.: From natural language business requirements to executable models via SBVR. In: International Conference on Systems and Informatics (ICSAI), Yantai (2012)
11. Project, N.: NL2OCLviaSBVR—A Natural Language to OCL Transformation via SBVR [En ligne]. http://www.cs.bham.ac.uk/~bxb/NL2OCLviaSBVR/NL2OCLviaSBVR.html
12. Thakore, D., Upadhyay, A.R.: Development of use case model from software requirement using in-between SBVR format at analysis phase. Int. J. Adv. Comput. Theory Eng. (IJACTE) 2(12), 86–92 (2013)
13. Tomas, S., Paulius, D., Rimantas, B.: Approach for Semi-automatic Extraction of Business Vocabularies and Rules from Use Case Diagrams, vol. 174, pp. 182–196 (2014)
14. Bajwa, I.S., Lee, M.G., Bordbar, B.: SBVR Business rules generation from natural language specification. In: AAAI Spring Symposium: AI for Business Agility, pp. 2–8 (2011)
15. Skersys, T., Butleris, R., Kestutis, K.: Approach, extracting business vocabularies from business process models: SBVR and BPMN standards-based. In: 11th International Conference of Numerical Analysis and Applied Mathematics 2013: ICNAAM 2013, vol. 1558, no. 11, pp. 341–344 (2013)
16. Bajwa, I., Behzad, B, Mark, G.L.: OCL constraints generation from NL Text. In: IEEE International EDOC, Vitoria, Brazil (2010)
17. OMG: QVT, Meta Object Facility (MOF) 2.0 Query/View/Transformation Specificatio, January 2011 [En ligne]. http://www.omg.org/spec/QVT/1.1/
18. Kriouile, A., Addamssiri, N., Gadi, T., Balouki, Y.: Getting the static model of PIM from the CIM. In: 3rd Colloquium IEEE on Information Science and Technology (CiSt'14), Tetuan, pp. 168–173 (2014)

# ZeroCouplage Framework: A Framework for Multi-supports Applications (Web, Mobile and Desktop)

El Hassane Ettifouri, Abdelkader Rhouati, Walid Dahhane and Toumi Bouchentouf

**Abstract** The companies are currently confronted with the implementation problem of their applications on several supports (Web, mobile and desktop). The Responsive Web Design [1, 2] partially answers to this problem as it does not allow having a mobile-native version, nor a desktop one. So, we propose a new approach which relies on the use of meta-model MDA (Model Driven Architecture) [3, 4] for the CIM and PIM models. Yet instead of having a PSM for each support, we propose a new open source framework [5] based on the adaptation of the MVC model [6] entitled as ZeroCouplage framework in order to have only one PSM, in which we will conceive and develop the same application and deploy it on several supports.

**Keywords** Responsive Web Design · Zerocouplage Framework · MDA · MVC · M2VC

## 1 Introduction

With the massive use of smartphones and tablets (mobile support) simultaneously with the computers and PCs (web and desktop support), the companies must represent their solutions, services and offers on the various existing supports in order to reach the maximum possible users and customers. Thus, for each support, we have

E.H. Ettifouri (✉) · A. Rhouati · W. Dahhane · T. Bouchentouf
Team SIQL, Laboratory LSEII, ENSAO, Mohammed First University,
Hay Elquods, BP 669, Oujda, Morocco
e-mail: h.ettifouri@gmail.com

A. Rhouati
e-mail: abdelkader.rhouati@gmail.com

W. Dahhane
e-mail: dahhane.walid@gmail.com

T. Bouchentouf
e-mail: tbouchentouf@gmail.com

© Springer International Publishing Switzerland 2016                    403
A. El Oualkadi et al. (eds.), *Proceedings of the Mediterranean Conference on Information & Communication Technologies 2015*, Lecture Notes in Electrical Engineering 381, DOI 10.1007/978-3-319-30298-0_42

to develop a dedicated application, which necessitates a high cost in terms of development and maintenance. Indeed, the need change or anomaly correction must be operated on the various supports. Capitalization will be also difficult to reach if there is only a one dedicated team for each support. Moreover, the different technologies of each support must be mastered (HTML, CSS and Javascript for the Web and the web-mobile; Swing architecture or javaFX for the desktop support and mobiles architecture for the native-mobile applications).

How can conceive and develop only once application then to be able to deploy it on several supports?

The Responsive Web Design is a concept which gathers a set of principles of design and technologies, which allow a Web site to auto-adapt according to the screen size of the support used to consult it from a computer, a tablet or a smartphone.

The Responsive Web Design thus ensures an ergonomics of quality by optimizing the interface for each terminal [7]. However, by exploring the details, this technique also involves some disadvantages. It is in particular the time of download. Most of the time, the users are obliged to download a HTML/CSS code in an useless way [7, 8]. Moreover, the images resizing requires the CPU calculations, which can also slow down the support and the total loading rate of the page. There is definitively more difficulties in downloading the contents of a "Responsive" site than the contents of a specialized mobile site.

On the other hand, the Responsive Web Design requires a quite higher time of design than a classical Web site. It is also more difficult to innovate with the Responsive Web Design, which imposes its constraints related to its adaptability character [7]. In reply, the design is based on the HTML and CSS, and it remains valid only for the two Web and mobile-web supports.

Like alternative of the Responsive Web Design, we propose a model (Fig. 1) allowing automatically to generate an application dedicated for each support given, while basing ourselves on Model Driven Architecture (MDA) proposed by the OMG [4, 9]. Indeed, it makes it possible to separate the functional specifications from a system of its implementation on a platform (or support). For this purpose, the MDA defines an architecture of specifications structured in models independent of the platforms (IMP) and in specific models (PSM) [4].

The models of approach MDA representing the levels of abstraction of the application and are used to model it and generate its code by successive transformations [10]. MDA defines three levels of models the CIM, the PIM and the PSM:

- CIM (Independent Computational Model) is the model trade which describes the functional requirements of the application independently of the details related to its implementation and gives a vision on the system in the environment.
- PIM (Independent Platform Model) is the model of analysis and design of the application which represents a sight partial of one CIM and makes it possible to give a vision structural and dynamic of the application independently of the technical design of the application.

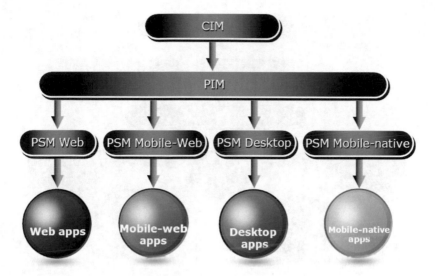

**Fig. 1** Representation of the same functional need in several supports via an approach MDA

- PSM (Platform Specific Model) is the model which approaches more of the final code the application. It depends on the support or the technical platform specified for the application and is used as a basis for the achievable generation of codes this platform.

The source code is the final result of process MDA obtained by automatic generation of the implementation code starting from the model PSM which must be specific for each support (Fig. 1).

The main aim of ZeroCouplage Framework is to make easy the passage of a platform (support) to another without injecting a new code in the existing application nor to have knowledge in the programming Web or mobile because it is based only on the language java. It is a new solution which makes it possible to reduce the number of modules SIM dedicated to each support as it is illustrated in Fig. 2.

## 2 ZeroCouplage Framework

ZeroCouplage is a Framework [5] created for Java developers who waste their time in achieving the same functional need projects on multiple platforms/media: web, desktop and mobile (Android).

ZeroCouplage meets the major needs of the development of IT projects which are: reusability, maintainability and cost reduction. One treatment purely in Java for the three media. The developer does not need any specific skills on JAVA/JEE or Android to develop applications on both web and mobile platforms.

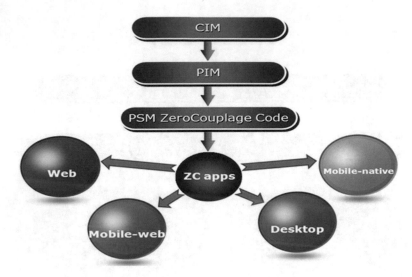

**Fig. 2** Use of the ZeroCouplage framework in an approach MDA

ZeroCouplage is based on the design pattern M2VC and provides a zero cou-
pling between the presentation layer and the business layer hence the name
ZeroCouplage. It also allows easy passage from one media to another just by setting
its principal configuration file.

## 2.1 Presentation of M2VC Design Pattern

### 2.1.1 MVC Structure

The most famous of a programming model example is the Model-View-Controller
(MVC) [11] of Smalltalk developed at Xerox PARC in the late 1970s [12]. MVC is
the fundamental design pattern used in Smalltalk for implementing GUI objects,
and it has been reused and adopted to varying degrees in most other GUI class
libraries and application frameworks [6], since it used in the Architecture Driven
Design and it implementation in Java framework for developing desktop application (Iqbal H. Sarker and K. Apu—2014) [13] (Fig. 3).

### 2.1.2 M2VC Structure

Model-Virtual-View-Controller (M2VC) design pattern is a derivative of the
Model-View-Controller (MVC) design pattern; it is intended for the development of
applications multi support, such as web, mobile and desktop support. M2VC and

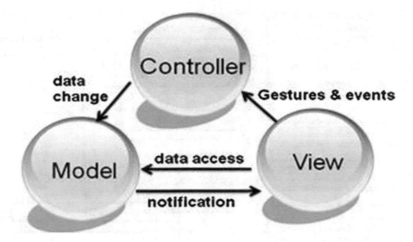

**Fig. 3** Model-View-Controller (MVC) design pattern

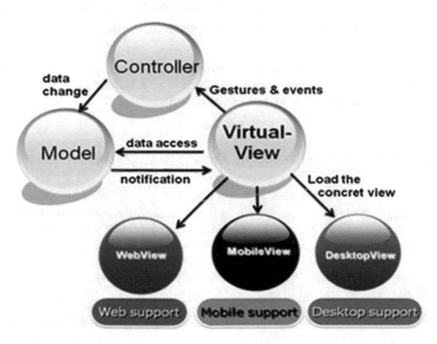

**Fig. 4** Model-Virtual-View-Controller (M2VC) design pattern

MVC share the same concept of the model, but the M2VC controller uses a virtual view to manipulate the concretes views, it loads the concrete view, according to the selected support (Fig. 4).

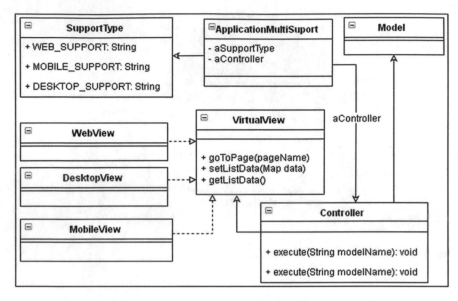

**Fig. 5** Class diagram of the Model-Virtual-View-Controller (M2VC) design pattern

Each concrete view implements the virtual view that manages the navigation between the pages of the same support, and allows for a bidirectional exchange of data between the concrete view and Model (Fig. 5).

## 2.2 Graphical Components of ZeroCouplage Framework

In order to have an independent java application of given media (web, mobile or desktop) graphic interfaces are developed based on ZCComponent components (as ZCPage, ZCLyout, ZCButon...) provided by the framework and represents an abstraction of the layer presentation. The instantiation of this layer is done through the instruction (1); the newComponent method loads the actual implementation (ZCComponentWebImpl, ZCComponentMobileImpl, ZCComponentDesktopImpl) to RunTime:

$$ZCTextField\ text = ZCFactory.newComponent(ZCTextField.class); \quad (1)$$

The display() method of the Fig. 6 returns an object that can be:

- String (HTML) in the case of a web or web-mobile application.
- JComponent (Swing) in the case of a desktop support.
- Native mobile component (Android) in the case of a mobile-native application.

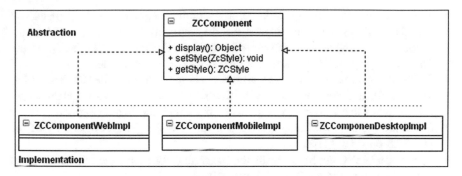

**Fig. 6** Class diagram of the graphical components ZeroCouplage framework

## 2.3    *Zerocouplage Framework Functioning*

In this section we are interested in the diagrams which illustrate the principle and the functioning of Zerocouplage framework, with a detailed description of each diagram, in order to facilitate the understanding of our framework and show the role of M2VC in this kind of framework.

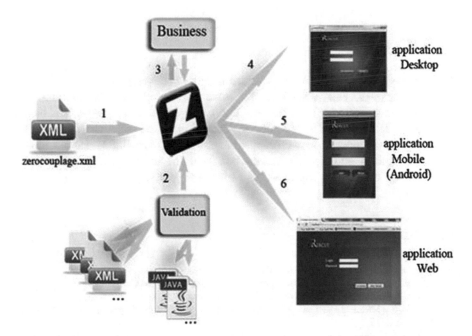

**Fig. 7** Schematic illustration of ZeroCouplage framework

- 1. Zerocouplage framework is based on the configuration of a zerocouplage.xml file. That allows the managing of the different navigation rules and ensures proper configuration to one type of target application independently of the business layer (M2VC model), it's loaded by M2VC controller.
- 2. The Zerocouplage Framework allows developers to use a file ZC_Val.xml to validate form fields, its data is retrieved by the VirtualView.getData().get (fieldsName) that indicated in the class diagram of the figure (Fig. 7).
- 3. Interacting between the controller and the model layer independently of the presentation (view) layer.
- 4, 5, 6. the M2VC controller loads the concrete view switch the selected support that indicated in the zerocouplage.xml file.

## 3 Future Works

To encourage the use of zeroCouplage framework, we will build a development platform ZCProject, integrated as a plugin in eclipse IDE [14] to simplify development, build, and deploy multi-platform applications (web mobile-web, desktop and mobile-native).

## 4 Conclusion

ZeroCouplage is a new framework realized in order to develop one application that can be deployed and visualized on several supports (Web, mobile-web, mobile-native and desktop) without any limits of the Web-Responsive. It is based on a new derivative of MVC design pattern baptized M2VC. The use of ZeroCouplage Framework in MDA approach to generate a multimedia applications, makes possible to optimize the number of the PSM in one, which generates a ZeroCouplage application deployable on several supports.

## References

1. Natda, K.V.: Eduvantage, 2013 Responsive Web Design (2013)
2. Harb, E., Kapellari, P., Luong, S., Spot, N.: Responsive Web Design (2011)
3. Brown, A.: An Introduction to Model Driven Architecture (2004)
4. Mukerji, J., Miller, J.: MDA Guide Version 1.0.1, 2003. http://www.omg.org. Accessed 20 March 2015
5. Ettifouri, E.H., Bouchentouf, T.: Source code of ZeroCouplage framework. http://code.google. com/p/zerocouplage/source/browse/#svn%2Ftrunk%2Fcode%20source%2Fzerocouplage-2.0.0. Accessed 20 March 2015 (2012–2014)

6. JavaZONE, Oslo, JAOO, Århus: The Model-View-Controller (MVC) Its Past and Present (2003)
7. Groves, R.M., Heeringa, S.G.: Responsive design for household surveys: tools for actively controlling survey errors and costs. J. R. Stat. Soc. Ser. A (Stat. Soc.) **169**, 439–457 (2006). doi:10.1111/j.1467-985X.2006.00423.x
8. Bryant, J., Jones, M.: Pro HTML5 Performance, 2012—Responsive Web Design (2012)
9. Arlow, J., Neustadt, I.: Enterprise Patterns and MDA. Addison-Wesley, Boston (2004)
10. Miller, J., Mukerji, J.: Model Driven Architecture (MDA), July 2001. Architecture Board ORMSC (2001)
11. Reenskaug, T.: Xerox PARC technical note (December 1979) defines the MVC terms. http://heim.ifi.uio.no/trygver/1979/mvc-2/1979-12-MVC.pdf. Accessed 20 March 2015
12. Burbeck, S.: Applications Programming in Smalltalk-80: How to Use Model-View-Controller (MVC) (1992)
13. Sarker, I.H., Apu, K.: MVC Architecture Driven Design and Implementation of Java Framework for Developing Desktop Application (2014)
14. Development Guid for eclipse plugin. http://www.eclipsetotale.com/articles/Developpement_de_plugins_Eclipse_partie1.html. Accessed 20 March 2015

# Get the Public Opinion from Content Published on the Web/CSM: New Approach Based on Big Data

**Abdelkader Rhouati, El Hassane Ettifouri, Mohammed Ghaouth Belkasmi and Toumi Bouchentouf**

**Abstract** Public opinion is the engine of the political, economic and social actions. It is vital that leaders of states and countries have a clear, quick and permanent idea of the public opinion for making the right decisions. Nowadays the main way to get public opinion is by surveys, which have limits related to the selected sample of people and the chosen questions. Also the long time needed to prepare questions, and analyze answers is a serious handicap. Our work aims to introduce a new approach to get public opinion based on Big Data and the use of content published on the Web/CMS as an alternative of surveys. Our approach allows getting public opinion on a given topic as a permanent and quick way, and based on a larger number of people.

**Keywords** Big Data · Web · CMS · Content · Data · Public opinion · Hadoop · NoSQL · Web mining · Survey

## 1 Introduction

Public opinion [1–3] is a major factor that directs the most sensitive areas of states and nations. A good and respectful government is one who is consistent with the points of view and the opinions of his own citizens. The most used way to know

A. Rhouati (✉) · E.H. Ettifouri · M.G. Belkasmi · T. Bouchentouf
Team SIQL, Laboratory LSEII, ENSAO, Mohammed First University,
BP 60000, Oujda, Morocco
e-mail: abdelkader.rhouati@gmail.com

E.H. Ettifouri
e-mail: h.ettifouri@gmail.com

M.G. Belkasmi
e-mail: ghaouth@gmail.com

T. Bouchentouf
e-mail: tbouchentouf@gmail.com

© Springer International Publishing Switzerland 2016        413
A. El Oualkadi et al. (eds.), *Proceedings of the Mediterranean Conference on Information & Communication Technologies 2015*, Lecture Notes in Electrical Engineering 381, DOI 10.1007/978-3-319-30298-0_43

about the public opinion is surveys, which are not always efficient. Almost every time surveys are done on insignificant sample of people. They also requires a long time for preparing questions, asking people and finally analyzing all results. So, it's time to think of a new efficient solution to have a clear and precise public opinion.

The number of internet users increases every day around the world as describe on [4]. For example in 2013 over than 56 % of Moroccans became connected to the Internet, it's about 18 million internet users. Those internet users are creating data every second. Tweets, facebook post, articles and comments are all different kinds of content published on the Web. This content reflects the citizen's points of view about topics of their daily: this huge data available on the web represents the public opinion.

In fact, this paper aims to get advantage of the content published on the web, mainly through CMS tools [5], in order to get the public opinion on a given topic. So using Big Data in this case appears interesting, especially that the target content matches exactly to the 5 V's [6] characterizing the Big Data system.

So, how to use Big Data to meet our needs? How data extracted from the web can be stored in a Big Data system? And what method of research and analysis is suitable to get public opinion of a specific topic from a web content?

This work is carried out to answer these questions.

This chapter is organized in different sections. Section 2 presents the concept of Big Data and its tools, especially Hadoop. Section 3 introduces the idea of public opinion and the use of survey as a solution to it. After, on Sect. 4 we detail our approach to take advantage of Big Data features to determine public opinion based on the content extracted from the Web/CMS. Finally, a conclusion and futur works will be presented in Sect. 5.

## 2 The Big Data Systems

### 2.1 Very Large Database Isn't a Big Data

A very large database is a database that contains a very large number of rows [7, 8] (the exact number changes over time, by the evolution of hardware). However, we cannot consider it as Big Data system [9]. In fact most commercial DBMS's are able to manage a large data, which is just only one side of the aspect of Big Data. First the structural systems limiting their ability to handle a different type of data, like streaming data. Second, Almost all DBMS are not adapted to a analyzing large amounts of data, and does not allow to parallelize the management and processing of data.

## 2.2 What Is Big Data?

"Big data technologies describe a new generation of technologies and architectures designed to economically extract value from very large volumes of a wide variety of data, by enabling high-velocity capture, discovery, and/or analysis." as defined by the IDC [10]. A Big Data system is then characterized from a technical point of view by three V's: volume, variety and velocity.

*Volume*: large amounts of data
*Variety*: different forms of data, contains traditional databases, images, documents, and complex records.
*Velocity*: the content of the data is constantly changing.
Other references added two other characteristics that define a Big Data from an economical point of view; value and veracity [6].
*Value*: the usefulness of the results the analyzing process of data in a Big Data system.
*Veracity*: refers to the way we trust data depending on its sources.

From this definition we can conclude that Big Data is just a problem of extracting and storage of data. Which is totally false. Big Data is also a technology that offer a possibility to analyze and process data in the goal to have a small and significant data, that can be used as assistance to decision makers. Like, and for example, when a person goes on Google Map [11] looking for the nearest restaurant to his position, the system proceeds and analyze data, among GPS locations, name of places, images, telephone numbers and other information, and returns a small information; like "10 restaurants in the center of Oujda".

## 2.3 Tools of Big Data

However, the definition of Big Data with the 5 V's necessarily involves issues related to each aspect. Especially, three fundamental issues, around which most of the solutions and technical research is focused, and they are: the data storage issues related to the volume of data and the diversity of data's sources, the data management issues related to variety of structured and unstructured data, and finally the processing data issues as the goal of the whole system [12].

So, many Big Data technology tools have been created in the few last years. The most popular, considered as reference of Big Data, is Hadoop.

Hadoop is composed mainly of two components, as detailed on [13]:

- File System (The Hadoop File System)
- Programming Paradigm (Map Reduce)

# 3 Getting Public Opinion by Surveys

Surveys are a systematic way of asking people volunteer information about their attitudes, behaviors, opinions and beliefs. The success of survey research rests on how closely the answers that people give to survey questions matches reality, how people really think and act. The first problem that a survey researcher has to tackle is how to design the survey in order to get the right information from people. The second problem is how accurate does the survey have to be? Is this a one-time survey or can the researcher repeat the survey on different occasions and with different settings? How will the results be used?

All this problems make us thinking for the necessity of new and alternative way to get public opinion. A way that will be easy, efficient, permanent, significant and a way that will be easy, efficient, permanent, significant and using an area abundant in people and content: The web.

# 4 Using Big Data to Get Public Opinion

Our approach consist in the use of Big Data to get public opinion, based on content derived from the Web/CMS. For that we will use an analytics workflow for Big Data. (Figure 1 illustrates the different steps of our approach.) In fact, our approach is devised on 4 steps:

- Data source: consists in the extraction of data from different web source
- Data management: consists in modeling data for preparing it to be stored on an NoSQL storage (many tools are available to manage NoSQL Database, as example Cassandra[1])
- Modeling: consists in using the Web mining to analyze data.
- Result: consists in visualizing the results by showing clearly the positive or negative opinion of people.

## 4.1 Extracting Data from the Web/CMS

One of the issues of our approach is extracting data from the content of the CMS on the web. This can be done from three different resources.

a. **Web data extraction**: Since the explosion of the World Wide Web, several works have addressed the problem of data extraction from HTML pages available on the web. Then different types of tools have been created to achieve that goal [14–18].

---

[1]Apache database solution—http://cassandra.apache.org/.

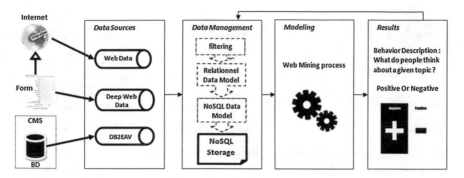

**Fig. 1** Our approach of the analytics workflow for Big Data to define public opinion based on content from the web/CMS

b. **Deep Web data extraction**: Almost research are focused only on one part of the web; it's about pages defined by a unique URL and generally available by search engine using crawling or surfacing method. However huge data on the web are not yet explored, which are accessible through query interfaces. it's called *deep web data*. To extract data from the deep web, we will use a form driven approach [19]. This approach is based on the form fields and the table results to get a structured relational schema of data.

c. **Direct data extraction**: CMS are the most used tools to create and manage data on the web. So it's an important resource of data. Several, even the most of, CMS uses a relational database based on the EAV model [20]. This makes possible to do data extraction directly from databases using API, as the "DB2EAV" API [21]. So we a simple implementation of this API make possible to extract data directly from databases of CMS.

## 4.2 Data Management: To NoSQL Data Model

As our goal is to get public opinion from the content created on the web by people, the first step of data management will the filtering to only keep potential articles and comments. Next, all data will be temporary modeling by a relational data model to ensure that all data retrieved respect a unique structure of three components; article, comment and user (as shown in the Fig. 2) And also to be the input element of the transformation into a NoSQL data model.

NoSQL databases data models are different from relational databases ones. Indeed, we can classify the NoSQL data model in 4 categories: key value stores, document stores, Column Family stores and graphs databases [22].

The column family stores model is more appropriate for storing huge amounts of data and dealing with clustering because the data model can be partitioned very effectively.

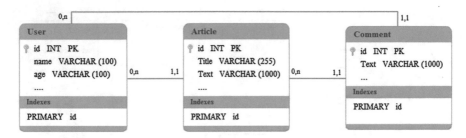

**Fig. 2** Relational data model of data extracted from the Web/CMS

A column family consists of rows that contain many columns. Each column stores a row name and a row value. It's not mandatory that each row store has the same columns. Also, if the number of columns is different, the rest of columns will be completed by null values (see Fig. 3).

As shown on Fig. 3, a column family is constituted of pair of key—value, where the key is associated with a series of columns.

The first step of transition from relational data model to NoSQL data model based on Column family stores, is denormalizing of our model. All relations between tables will be translated to a column integrated on the columns family. This means that the relational aspect is no more handled by the database, it will be handled by the logical application-level [23]. The transformation of the relation data model illustrated on the Fig. 2, produces the definition of three new column family (Fig. 4 details this results).

| Key | Column | Column | Column |
|-----|--------|--------|--------|
|     | Value  | Value  | Value  |

**Fig. 3** Column family data model

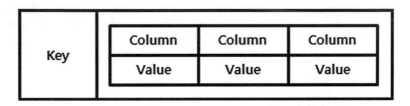

User Column Family                          Article Column Family

Comment Column Family

**Fig. 4** Column family stores data model

**Fig. 5** Public opinion about a
given topic from Big Data
system

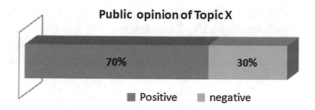

## 4.3 Analyzing Data: Web Mining

The goal of our approach is to define the public opinion of a given topic. To better understand it, Our system must be able to answer the question: What do people think about a given topic?

The current step must give an answer of this question. So we will semantically analyze each rows of data, article and comment, to get the opinion of each Users, by using the Text mining techniques [24]. This processing is known by "Web mining". It's contains the following steps:

- Extract all opinion terms about a given topic after pre-processing each row of data from NoSQL storage.
- Determine if the opinion is positive or negative
- If the number of positive opinion exceeds the number of negative opinion, that mean the public agree whit the given topic. If it's the opposite, that mean the public disagree with the given topic.

## 4.4 Visualization of Results

The final result of the process will be a description of how people on the internet are thinking about a given topic? It can be either a positive or negative opinion which will be presented as shown in Fig. 5.

## 5 Conclusion and Future Works

In this paper we have present a new approach to get public opinion using Big Data system based on data extracted from the Web/CMS. Our approach is an alternative to the limited classic surveys. As a Big Data system, the new approach allows a permanent extraction and analyzing of data from the Web/CMS, which make public opinion always up to date. Furthermore, it makes possible to process a huge quantity of data from the Web/CMS. Therefore the result public opinion is more significant.

We are now working on implementation of our approach through the design and construction of a Big Data platform, which follows the description of the analyzing data workflow described above. So, our future works will be around the validation and evaluation of our approach over the technical solution.

# References

1. Public Opinion Quarterly Journal. http://poq.oxfordjournals.org/ (2015)
2. Mill, J.S.: On Liberty (see Chapter IV). In: The Basic Writings of John Stuart Mill. Modern Library, New York (2002 [1863]); Dewey, J.: The Public and Its Problems. John Dewey: The Later Works, 1925–1953, vol. 2, edited by J. Boydston. Southern Illinois University Press, Carbondale (1988 [1927])
3. Davison, W.P.: The public opinion process. Publ. Opin. Quart. **22**, 91–106; also see Allport, F.: Toward a science of public opinion. Publ. Opin. Quart. **1**, 7–23 (1937)
4. The Word Bank—Internet users (per 100 people). http://data.worldbank.org/indicator/IT.NET. USER.P2
5. Baxter, S., Vogt, L.C.: Content management system. In: Google Patents, US Patent 6,356,903. http://www.google.com/patents/US6356903
6. Assunção, M.D., Calheiros, R.N., Bianchi, S., Netto, M.A.S., Buyya, R.: Big Data computing and clouds: trends and future directions. J. Parallel Distrib. Comput. Available online 27 August 2014. ISSN 0743-7315
7. Ganti, V., Gehrke, J., Ramakrishnan, R.: Mining very large databases. Computer **32**(8), 38–45 (1999)
8. Linstedt, D.E.: What is VLDB? Very Large Databases. In: The Data Administration Newsletter. http://www.tdan.com/view-articles/4967 (2015)
9. Madden, S.: From Databases to Big Data. In: The IEEE Computer Society—1089-7801/12 © 2012 IEEE
10. Gantz, J., Reinsel, E.: Extracting Value from Chaos. In: IDC's Digital Universe Study, sponsored by EMC
11. Google Map. https://www.google.fr/maps (2015)
12. Kaisler, S., Armour, F., Espinosa, J.A., Money, W.: Big Data: issues and challenges moving forward. In: 2013 46th Hawaii International Conference on System Sciences (HICSS), pp. 995, 1004, 7–10 Jan 2013
13. Katal, A., Wazid, M., Goudar, R.H.: Big Data: issues, challenges, tools and Good practices. In: 2013 Sixth International Conference on Contemporary Computing (IC3), pp. 404, 409, 8–10 Aug 2013
14. Laender, A.H.F., Ribeiro-Neto, B.A., da Silva, A.S., Teixeira, J.S.: A brief survey of web data extraction tools. In: SIGMOD Rec. 31, 2 (June 2002), pp. 84–93
15. Zhai, Y., Liu, B.: Web data extraction based on partial tree alignment. In: Proceedings of the 14th International Conference on World Wide Web (WWW '05), pp. 76–85. ACM, New York, NY, USA
16. Crescenzi, V., Mecca, G., Merialdo, P.: RoadRunner: towards automatic data extraction from large web sites. In: Proceedings of the 27th International Conference on Very Large Data Bases (VLDB '01)
17. Myllymaki, J.: Effective Web data extraction with standard XML technologies. In: Proceedings of the 10th International Conference on World Wide Web (WWW '01), pp. 689–696. ACM, New York, NY, USA
18. Ferrara, E., De Meo, P., Fiumara, G., Baumgartner, R.: Web data extraction, applications and techniques: a survey. Knowledge-Based Syst. **70**, 301–323 (2014). ISSN 0950-7051

19. Saissi, Y., Zellou, A., Idri, A.: Extraction of relational schema from deep web sources: a form driven approach. In: The 2nd World Conference on Complex Systems, 10–12 Nov 2014
20. Nadkarni, P.M., Brandt, C.A., Marenco, L.: WebEAV: automatic metadata driven generation of web interfaces to entity–attribute–value databases. J. Am. Med. Inform. Assoc. **7**, 343–356 (2000)
21. Rhouati, A., Ettifouri, H., Belkasmi, M.G., Bouchentouf, T.: The DB2EAV API of mapping database to EAV model as solution of data interoperability between Content Management Systems (CMS). In: The 2nd World Conference on Complex Systems, 10–12 Nov 2014, Agadir, Morocco. ISBN: 978-1-4799-4648-8
22. Hecht, R., Jablonski, S.: NoSQL evaluation: a use case oriented survey. In: Proceedings of the 2011 International Conference on Cloud and Service Computing (CSC '11), pp. 336–341. IEEE Computer Society, Washington, DC, USA (2011)
23. Schram, A., Anderson, K.M.: MySQL to NoSQL: data modeling challenges in supporting scalability. In: Proceedings of the 3rd Annual Conference on Systems, Programming, and Applications: Software for Humanity (SPLASH '12), pp. 191–202. ACM, New York, NY, USA (2012)
24. Khan, K., Baharudin, B., Khan, A., Ullah, A.: Mining opinion components from unstructured reviews: a review. J. King Saud Univ. Comput. Inf. Sci. **26**(3), 258–275 (2014). ISSN 1319-1578

# Part VII
# Online Experimentation and Artificial Intelligence in Education

# STIG: A Generic Intelligent Tutoring System a Multi-agents Based Model

**Mohammed Serrhini and Abdelmajid Dargham**

**Abstract** In this paper is described the design of an intelligent tutoring system model based on multi-agents approach. The purpose is to create a simple—but generic—model that allows us to rapidly derivate an intelligent tutoring system. The model is designed by using the universal language modeling (UML), the platform Jade for multi-agents programming and The Java NetBeans IDE. Our future objective is to use the developed model to create ITSs as supplements for several courses in the Faculty of Sciences (University Mohamed Premier).

**Keywords** Intelligent tutoring system · Multi-agents system · Object-oriented modelling · UML · JADE

## 1 Introduction

Intelligent tutoring systems (ITSs) are in the intersection of education, psychology and Artificial Intelligence (AI). The principal idea of ITSs is to optimize learning and problem solving skills by means of adaptive and individualized instruction because ITSs act as knowledge communication systems [1]. Benefits of ITSs are more:

- Learning by doing: learners are able to learn a lot of concepts by doing exercises and problems.
- Automatic guidance: learners can also obtain more guidance than if they were simply solving a problem in classical instruction mode.

M. Serrhini (✉) · A. Dargham
Departement of Mathematics and Computer Science,
Faculty of Sciences, Oujda, Morocco
e-mail: serrhini@gmail.com

A. Dargham
e-mail: abdelmajid.dargham@gmail.com

© Springer International Publishing Switzerland 2016
A. El Oualkadi et al. (eds.), *Proceedings of the Mediterranean Conference on Information & Communication Technologies 2015*, Lecture Notes in Electrical Engineering 381, DOI 10.1007/978-3-319-30298-0_44

- Large interactivity: ITSs interact with learners at multiple levels [2] (when the learner asks for help, or when he progresses in the learning process or when he makes an error).

In the literature, the general architecture of ITSs is made on four components: (a) the expert model (b) the student model (c) the tutor model and (d) the communication model. The following is a brief description of these four models:

- The **expert model** (or domain model) contains the knowledge base that generates intelligent responses to the student queries during instruction. It consists of sets of rules and procedure that belongs to the domain.
- The **student model** contains personnel data of the student and his learning characteristics. There are several types of student model: overlay, stereotypes, perturbation, machine learning based techniques, cognitive theories, constraint-based model, fuzzy student modeling, Bayesian networks, ontology-based student modeling and open learner model.
- The **tutor model** (or pedagogical model) contains several data about curriculum, cognition, learning process and learning strategies (pedagogies). This model is crucial since it reflects the adaptive interaction between student and the system.
- The **communication model** (or user interface model) is the middle between the first three ITS's components and the real student.

ITSs are more complex systems and their development involves a long time and requires very important costs. We are interesting in the using of ITSs as supplements for several courses in the Faculty of Sciences of the University of Mohamed Premier. In this context, our objective is to propose a generic model of ITSs.

This paper presents a summary of our work. In Sect. 2, we discuss the using of multi-agents approach in order to model a generic ITS. Section 3 presents a detailed description of our model. In Sect. 4, we show some examples of the implementation of developed model.

## 2   ITSs and Multi-agents Approach

### 2.1   Multi-agents Systems

A multi-agents system is a system that involves a certain number of agents. Agents are autonomous software entities that perceive their environment through sensors and perform actions on the environment through actuators, processing information and knowledge [3]. Multi-agents systems are efficient to model complex systems. Agents have some fundamental properties [4]:

- **Autonomy**: they accept high level requests and operate without the direct intervention of humans in deciding on how to meet those requests.
- **Social ability**: they communicate and collaborate with other agents.

- **Reactivity**: they perceive their environment and respond in a timely fashion to changing conditions.
- **Proactivity**: more than just reactive, they are able to exhibit goal-directed behavior by taking the initiative.
- **Mobility**: they are able to travel through computer networks in order to search and retrieve the necessary information to complete the assigned task.

## 2.2 Why Using Multi-agents Systems for ITSs

Multi-agents systems are an evolution of artificial intelligence: the distributed artificial intelligence. According to [5], a multi-agents approach is the best way to design distributed computing systems. The Multi-agent Systems Engineering (MaSE) methodology leads the designer from the initial system specification to the implemented agent system [6]. In the analysis phase, MaSE uses goal hierarchies and role models, whereas the design phase creates agent-class, communication and deployment diagrams. AUML [7], is an extension to UML to represent various aspects of agents by introducing new types of diagrams including agent class diagrams and protocol diagrams. Some raisons to use multi-agents systems in building ITSs are the following:

- Knowledge is distributed into several experts (agents).
- Problems are solved in distributed way.
- Control is done in a distributed method.

## 2.3 How to Use Multi-agents Systems in ITSs

In multi-agents architecture, each component of the ITS can be considered as an request/response interaction between two or more agents. The fundamental properties of multi-agents systems cited above ensure that the global functionality of ITSs is emerging from the interaction of these different agents.

An intuitive approach consist of using one agent to represent each component of ITS:

- A **curriculum agent** to represent the student model.
- An **expert agent** to represent the expert model.
- A **pedagogical agent** to represent the tutor model.
- An **interface agent** to represent the communication model.

Due to the complexity of their roles, their functionalities and their communications, the used agents in ITSs must be cognitive. In [8], an intelligent tutor is definitely a cognitive agent and it has to accomplish the following functions:

- Perceiving the surrounding environment (the actions performed through the GUI, the statements produced by the learner, the learning materials that have been accessed, and the affective status of the user).
- Estimating the student's mental state as regards both cognition and metacognition: some internal representation of such a state has to be available; moreover the agent has to build the representation of the learning domain to enable assessment of the student, and to select suitable learning strategies.
- Acting properly to elicit both cognitive and metacognitive abilities in the learner, thus modifying the surrounding environment.

This architecture can evolve according to the particular objective that the system would manipulate. Thus, we can build ITSs with extra agents:

- **Other interfaces agents** for better studying the interaction between the system and the student.
- **Animated pedagogical agents** for improving the knowledge transmission.
- **Dedicated agents** (for example, the **errors detector/evaluator/qualificator agent** will be integrated in our ITSs).

## 2.4 Overview of ITSs Based on Multi-agents Approach

We now present some ITSs based on multi-agents approach [9].

**LANCA** (Learning Architecture Based on Networked Cognitive Agents)

This intelligent tutoring system is implemented as a Web Application and uses intelligent agents. Its purpose is to provide distance learning via the Internet network by distributing the knowledge and assuming a better quality of learning. Its architecture is founded on four agents: pedagogical agent, dialogue agent, negotiator agent and moderator agent.

**Adele** (Agent for Distance Education—Light Edition)

This is a pedagogical agent on the Web like LANCA. Its role is to interact with student and to trace its learning during a session (doing courses or exercises). The architecture of Adele contains tree components: a GUI for simulation, an expert system and an animated agent in 2D.

**Disciple**

Its architecture is made of two components. The first component is domain independent and contains the following elements: an empty knowledge base, a manager of this base, a module of knowledge acquisition and learning associated with a GUI, and finally a classical (general) problems solver. The second component is domain dependant and contains a specialized problems solver and a GUI.

**Herman the Bug**

This is an animated pedagogical agent implemented in a learning environment for teaching botanic anatomy and physiology in North Carolina State University.

In this environment, the student can draw a plant according to a set of constraints. Herman can observe the state of student progression and then help him to solve the problem by presenting a sequence of hints.

# 3 STIG: A Generic ITS Based on Multi-agents

In this section, we present our approach to building the architecture of our model: STIG (*Système Tuteur Intelligent Générique*). First, we present the context and the theoretical framework that describes the principals that guides our design of STIG. Then we describe the STIG architecture, where we also describe how agents communicate and collaborate. After the description of the architecture, we provide the details of the agents in STIG. Finally, we present some examples of use of STIG.

## 3.1 The Context and the Theoretical Framework

We want to build a general framework for ITSs that can be used after for creating several intelligent tutoring systems as supplements for the different courses in the Faculty of Sciences. This Faculty contains five Departments. First, we want to build the framework and to test it for the Department of Computer Sciences. The theoretical framework of the future system is based on the two following constraints:

- **Constraints 1**: the system must be responsive, distributed and adaptive to individual student behaviors. The multi-agents approach was chosen in accordance to this constraint.
- **Constraints 2**: the system must be used as a supplement of courses delivered by teachers and can be used by student for doing the exercises.

## 3.2 STIG Architecture

We present the architecture of STIG in Fig. 1. Currently, we consider five types of intelligent agents:

- **Curriculum agent**: saves the evolution of the interaction between the system and the student by building the history of events that occurs during learning sessions. Its role consists of building, loading and managing of the student model.
- **Didactic resources manager agent**: holds the knowledge on the addresses of didactic resources (courses, exercises). The role of this agent is to determine the

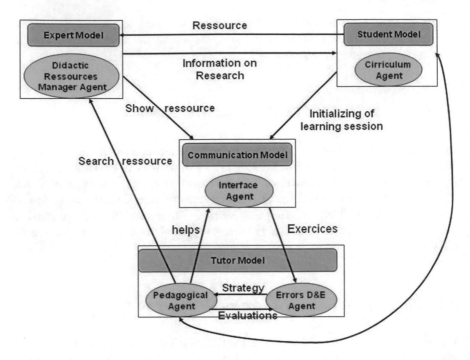

**Fig. 1** Architecture of STIG

address of the resource requested by both the curriculum agent and the pedagogical agent.

- **Pedagogical interface**: ensures the dialog between the student and the system.
  Its role consists of loading the resources delivered by the second agent and
  showing the messages transmitted by the other agents.
- **Pedagogical agent**: helps the student when an error occurs. This agent determines the better learning strategy according to the level of student. Three learning
  strategies are specified: (1) Companion, (2) Advisor, and (3) Perturbator.
- **Errors detector/evaluator agent**: detects and evaluates the student errors. This
  agent is crucial in our model. For example, the pedagogical agent can apply a
  learning strategy according to the information delivered by this agent.

## 3.3 The UML Model

The STIG system contains four actors: (1) the student, (2) the teacher, (3) the
administrator, and (4) the intelligent agent. The last actor is a generalization of the
five intelligent agents described above. Use cases of the system are spliced into
seven sub-use cases: (1) the use case of the student and the interface agent, (2) the

**Use case diagram of student and interface agent.**

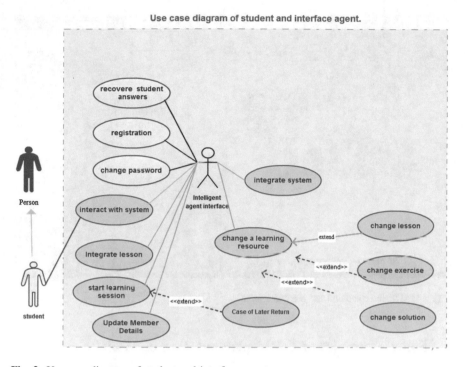

**Fig. 2** Use case diagram of student and interface agent

use case of the curriculum agent, (3) the use case of pedagogical agent, (4) the use case of didactic resources manager agent, (5) the use case of errors detector/ evaluator agent, (6) the use case of the teacher, and (7) the use case of administrator. As example, we present the use case diagram for the student and the interface agent in Fig. 2.

## 3.4 Implementation of the Model

The STIG model was implemented by using: (1) **JADE** (*Java Agent DEvelopment framework*) for building the five intelligent agents, (2) **JDBC** (*Java DataBase Connectivity*) for building the database of the system, (3) **JDOM** (*Java Document Object Model*) for manipulating the didactic resources and student data stored in XML format, (4) **JAVA NetBeans IDE** for coding the entire system. In Fig. 3 is shown the JADE interface and the created agents.

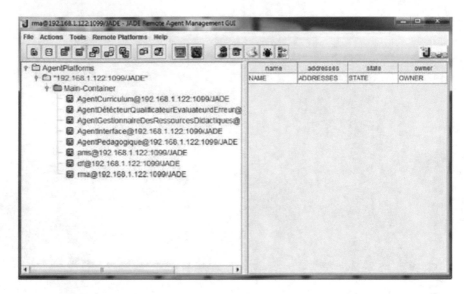

**Fig. 3** JADE interface and the created agents

## 4 Some Examples of the Implementation of Model

We have testing the STIG system for building four ITSs: (1) STIF-Compiler for teaching compiler design, (2) STIG-Math for teaching functions, (3) STIG-CProg for teaching C programming language, and (4) STIG-Algo for teaching algorithms. We present the interface of STIG-Complier in Fig. 4.

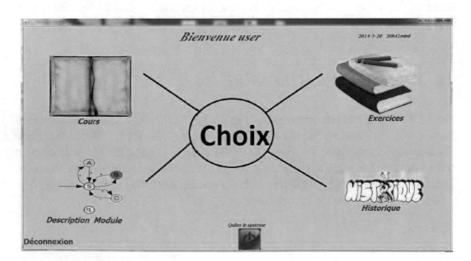

**Fig. 4** GUI for STIG-Compiler

# 5 Conclusion and Future Works

The main objective of this work is to build a generic model for intelligent tutoring systems by using the multi-agents approach. This paper has presented an architecture-based method for MAS development; Agents are mapped to system's goals (agent coordination and agent autonomy). We successfully built such model by using JADE, object oriented modeling (UML) and object oriented programming (Java). The developed model was used to derivate four different intelligent tutoring systems. Future work includes improvements of our model by envisaging other agents.

# References

1. Wenger, E.: Artificial Intelligence and Tutoring Systems: Computational and Cognitive Approaches to the Communication of Knowledge. Morgan Kaufmann, Los Altos (1987)
2. Anderson, J.R., Gluck, K.: What role do cognitive architecture play in intelligent tutoring systems?. In: Klahr, D., Carver, S.M. (eds.) Cognition & Instruction: Twenty-Five Years of Progress, pp. 227–262 (2001)
3. Russell, S., Norving, P.: Artificial Intelligence: A Modern Approach. Prentice Hall (2002)
4. Wang, H., Mylopoulus, J., Liao, S.: Intelligent Agents and Financial Risk Monitoring Systems, pp. 83–88. CACM (2002)
5. Huhns, M.N., Stephens, L.M.: Multiagent systems and societies of agents. In: Weiss, G. (ed.) Mutiagent Systems: A Modern Approach to Distributed Artificial Intelligence (Chapter 2), pp. 79–120 (2000)
6. DeLoach, S.A., Wood, M.F., Sparkman, C.II.: Multiagent systems engineering. Int. J. Softw. Eng. Knowl. Eng. 11(3), 231–258 (2001)
7. Bauer, B., Muller, J.P., Odell, J.: Agent UML: a formalism for specifying multiagent software systems. Int. J. Softw. Eng. Knowl. Eng. 11(3), 207–230 (2001)
8. Pipitone, A., Cannella, V., Pirrone, R.: Cognitive models and their application in intelligent tutoring systems. In: Paviotti, G., Giuseppe Rossi, P., Zakra, D. (eds.) Intelligent Tutoring Systems: An Overview (Chapter 2), pp. 57–83 (2012)
9. Laforcade, P.: Etude et conception des mécanismes d'agents détecteurs, évaluateurs et qualificateurs des erreurs d'un apprenant dans un EAIH. Rapport de DEA (2001)

# Part VIII
# Networking, Cloud Computing and Security

# MDAOrBAC: An MDA Security Framework Based OrBAC Access Control Policies

Aziz Kaddani, Amine Baina and Loubna Echabbi

**Abstract** Securing Critical Infrastructures form unauthorized access to information and system resources became one of the important areas of research last years. In this paper, we present a new approach using the MDA approach to design and generate OrBAC access control policies. The aim of our work is to provide a complete package, a visual model to define a security policy based on OrBAC model, to generate access control rules based XMI files, to use MotOrBAC engine to validate the policy and generate concrete security rules.

**Keywords** Organization based access control · Model driven architecture · Access control · MotOrBAC · Critical infrastructure · Electrical grid

## 1 Introduction

Critical Infrastructure (CI) provides services that are so vital to the social and economic well-being of a society that a disruption or destruction of the infrastructure would have severe consequences. CI sectors include, amongst others, the telecommunications infrastructures, electricity and transport infrastructures. CIs are highly interdependent; attacks on one infrastructure may cascade to another interdependent infrastructure to cause service disruptions [1]. A good example of CIs is the Electrical Grid infrastructure that is a requirement for all others CIs, since everything relies on constant supply of energy to work. Critical infrastructure security has become an important research topic in the last years [2]. In fact, CIs are

A. Kaddani (✉) · A. Baina · L. Echabbi
STRS Laboratory, National Institute of Posts and Telecommunications – INPT,
2, Av. Allal El Fassi, Madinat al Irfane, Rabat, Morocco
e-mail: kaddani@inpt.ac.ma

A. Baina
e-mail: baina@inpt.ac.ma

L. Echabbi
e-mail: echabbi@inpt.ac.ma

© Springer International Publishing Switzerland 2016     437
A. El Oualkadi et al. (eds.), *Proceedings of the Mediterranean Conference on Information & Communication Technologies 2015*, Lecture Notes in Electrical Engineering 381, DOI 10.1007/978-3-319-30298-0_45

becoming more complex and unable to manage sensitive and critical data. Therefore, it is very important to define security policies (by the use of *access control models*) in order to guarantee security properties. Indeed, Access control [3] is a mechanism that is applied extensively to prevent unauthorized access to information and system resources. Many models have been developed and studied to construct and manage access control systems such as RBAC [4] and Organization Based Access Control (OrBAC) [5].

Our new approach combines the Model Driven Architecture (MDA) approach with OrBAC model to define and generate applicable access control policies. In fact, we choose OrBAC model as defined in [5], because it has been used in various research studies, and has already proven its effectiveness in defining and applying security policies [5], in managing collaboration [6, 7], and dealing with integrity issues [8, 9]. We use the Eclipse Modeling Framework (EMF) [10] which is modeling framework and code generation facilities based on the MDA approach, for design and generate security rules. These security rules are expressed in XMI code knowing that there is a specific standard to express access control policy language implemented in XML which is XACML [11].

The remainder of this paper is organized as follows: Sect. 2 discusses the related work. Section 3 gives an overview of the necessary background to understand our approach. Section 4 presents our MDAOrBAC model and its instantiation through a critical infrastructure example. Section 5 details the generated XMI files based abstract security rules. Section 6 shows the manner to deduct concrete security rules using MotOrBAC. Section 7 concludes the paper and gives an overview of our future work.

## 2   Related Work

Related work focus on different research studies combining access control models and the use of MDA approach. SecureUML [12], XIN JIN's Doctoral dissertation [13] and Denisse Muñante's paper [14] are the tree most closely references to our work. The two first references are based on RBAC model. Both of these methods allow users to define access control models using the UML notation and use OCL to specify constraints. Nevertheless, we use in our work the mostly used access control model, OrBAC model [5]. In [14], the approach proposed is almost nearly to our work since we define also a security policy using OrBAC model, we generate security policy based XMI files and finally we use MotOrBAC prototype to validate the given security policy and deduct concrete security rules. Therefore, our approach contribution's is designing and implementing a prototype modeling tool MDAOrBAC with the following functions: (1) Design OrBAC modeling and specification as graphic models with UML notation, (2) OrBAC policy specification validation, (3) Automatic generation of OrBAC XMI policies and (4) when a modification done by the security administrators, the transformations/modifications are applied automatically to the rest of rules reducing thereby the modification time (*ongoing work*).

# 3 Background

## 3.1 Organization Based Access Control—OrBAC

In OrBAC [5], an organization is any entity that is responsible for managing a security policy, in which subjects play specific roles, an activity is a group of one or more actions, a view is a group of one or more objects, and a context is a specific situation that conditions the validity of a rule. Actually, the Role entity is used to structure the link between the subjects and the organizations. The relationship *Empower(org, r, s)* means that the organization "org" employs the subject "s" in the role "r". In the same way, the objects that satisfy a common property are specified through views, and activities are used to abstract actions. Security rules have the following form *Permission(org, r v, a, c)* which means: In the context "c", the organization "org" grants to the role "r" the permission (or the obligation, or the prohibition, or the recommendation) to perform the activity "a" on the view "v" (Obligation, Prohibition and Recommendation are defined similarly).

Two security levels can be distinguished in the OrBAC model [6]. (1) **Abstract level**: the security administrator defines security rules through abstract entities (roles, activities, views) without worrying about how each organization implements these entities. (2) **Concrete level**: when a user requests an access, concrete authorizations are granted (or not) to this user according to the corresponding rule, the belonging organization, the played role, the instantiated view/activity and the current parameters. Concrete security rules are expressed through the rules: *Is_permited, Is_prohibited, Is_Recommended, and Is_Obliged*, containing Subjects, Objects and Actions.

In order to allow security administrators to use OrBAC model, Fabien Autrel et al. proposed in [15], the MotOrBAC prototype (http://motorbac.sourceforge.net) which aims at making easy the use of OrBAC to express security policies.

## 3.2 Model Driven Architecture—MDA

Model Driven Architecture [16] has been proposed as an approach to specify and develop applications where systems are represented as models and transformation functions are used to map between models as well as to automatically generate executable code. MDA makes modeling as the primary focus of the software development process. Actually, MDA approach focuses primarily on the functionality and behavior of the system, not the technology in which it is going to be implemented in. It considers implementation details and business functions at two different levels. Thus, it is not necessary to repeat the process of modeling of an application or system's functionality and behavior each time a new technology comes along [16]. In this paper, we adopt the MDA approach through the use of Eclipse EMF [10].

# 4 MDAOrBAC: Design Model

## 4.1 Modeling OrBAC Model Using EMF Framework

In this subsection, we present our new proposed model named MDAOrBAC which is an enhancement of the first version proposed in our previous work [17]. It is actually an UML modeling of OrBAC entities using EMF framework [10].

We define nine classes in our proposed MDAOrBAC model (Fig. 1); five of them represent the five abstract level entities of OrBAC model: *Role* class represents different roles within an organization. *Activity* class lists all activities in the organization. *Context* class describes the different cases (Normal, Emergency, etc.). *View* class simulates all views (machines, folders, rooms, etc.) on which an engineer, for example, will intervene. *Organization* class represents the organization, institution or department within a company. Each class has an attribute having the same class name, e.g. the *Role* class has the attribute: Role, which represents an enumeration lists (Fig. 2), e.g. the role can be an engineer, a director, a technician, etc. The "Organization" attribute, in the *Organization* class specifies its type: Public, Private or Subcontracting. We define two more attributes in this class: "name" and "address" to assign a name and an address to an organization.

The five classes cited above (*Organization*, *Role*, *Activity*, *View* and *Context*) are related by a composition relationship to the *Rule* class which is the fundamental class of our model because this is where we will define security rules. It has two attributes: (1) "RuleType", to specify the type of rule (Permission, Prohibition, Obligation or Recommendation) and (2) "Id_Rule", to identify the rule by attributing an ID to each new security rule. A rule, or security rule, is characterized by the instantiation of each class related to the *Rule* class. In fact, a new created rule in *Rule* class is composed by one Role, one Activity, one View, one Context and one Organization.

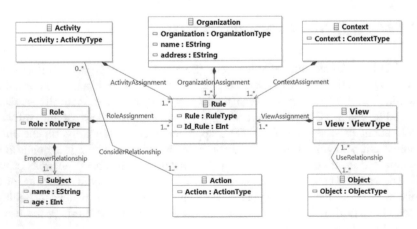

**Fig. 1** MDAOrBAC model

| <<enumeration>> ViewType | <<enumeration>> ActionType | <<enumeration>> ObjectType | <<enumeration>> ActivityType | <<enumeration>> RoleType |
|---|---|---|---|---|
| – Default<br>– Generator_Wind_Farm<br>– Generator_Solar_Farm<br>– Generator_Coal_Plant<br>– Generator_Hydro_Electric_Plant<br>– Transformer_275kv<br>– Transformer_132kv<br>– Transformer_33kv<br>– Transformer_Substation_11kv<br>– Distributor_11kv<br>– Distributor_415_240v | – Default<br>– Click<br>– Read<br>– Delete<br>– Turn<br>– Switch | – Default<br>– Folders<br>– Files<br>– Database<br>– SystemFiles<br>– ConfigurationFiles | – Default<br>– Check<br>– Modify<br>– Execute<br>– Transfer<br>– Change<br>– Repair<br>– Stop | – Default<br>– Director<br>– Engineer<br>– Technician<br>– Supervisor<br>– Project_Manager |
| | <<enumeration>> ContextType | <<enumeration>> OrganizationType | <<enumeration>> RuleType | |
| | – Default<br>– Normal<br>– Warning<br>– Emergency | – Default<br>– Public<br>– Private<br>– Subcontracting | – Default<br>– Permission<br>– Prohibition<br>– Obligation<br>– Recommendation | |

**Fig. 2** Enumeration lists

Three more classes are defined corresponding to the concrete level entities of OrBAC model: *Subject* class to satisfy the relationship Empower, *Action* class to satisfy the relationship Consider, and *Object* class to satisfy the relationship Use. Our future works improve the model with more attributes and methods in the different classes. The next subsection presents our case study, defined in our previous work [17], to illustrate our approach.

## 4.2 Case Study: Defining Security Policies

In this subsection, we present our case study to demonstrate the proof of concept of our approach. For this end, we choose the Electrical Grid as an example of Critical Infrastructure and we divide it in three essential sections (Fig. 3). The first one is for electricity generation (Coal Plant, Solar Farm, Wind Farm, etc.). The second section is for electricity transformation into different voltages, and the last section is for electricity distribution to the end users.

Three scenarios have been proposed in this case study. Each scenario is associated with an Electrical Grid section and a set of views already defined (Fig. 4). Due to space restrictions, we present only the security policy defined for the first scenario. Security policies for the two others scenarios are defined similarly. Defining security policy in OrBAC model means creating abstract security rules by giving for each role, inside an organization, different access rights to do (or not) on a specific or a set of resources.

**Fig. 3** Electrical grid

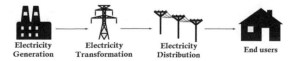

Electricity Generation — Electricity Transformation — Electricity Distribution — End users

| | Scenario 1 | Scenario 2 | Scenario 3 |
|---|---|---|---|
| **Electrical Grid** | Electricity generation | Electricity transformation | Electricity distribution |
| **Organization** | Operator_X | Operator_Y | Operator_Z |
| **Views** | **GeneratorViews:** Generator_Wind_Farm Generator_Solar_Farm Generator_Coal_Plant Generator_Hydro_electric_Plant | **TransformerViews:** Transformer_275kv Transformer_132kv Transformer_33kv Transformer_Substation_11 kv | **DistributorViews:** Distributor_11kv Distributor_415_240v |

**Fig. 4** Case study

**Table 1** Access rights associated to the role *Director*

| Activity | Context | | | | | | | | | | | |
|---|---|---|---|---|---|---|---|---|---|---|---|---|
| | Emergency | | | | Warning | | | | Normal | | | |
| | P | Pr | R | O | P | Pr | R | O | P | Pr | R | O |
| Change | | | 16 | | | | 8 | | | 1 | | |
| Check | 20 | | | | 12 | | | | 5 | | | |
| Execute | 21 | | | | 13 | | | | 6 | | | |
| Modify | 22 | | | | 14 | | | | 7 | | | |
| Repair | | | 17 | | | | 9 | | | 2 | | |
| Stop | | | 18 | 15 | | | 10 | | | 3 | | |
| Transfer | | | 19 | | | | 11 | | | 4 | | |

Table 1 summarizes the different access rights associated to the role *Director* inside the organization *Operator_X* on the set of views *GeneratorViews*. The letters **P**, **Pr**, **R** and **O**, in the table refer to the four types of security rules, P: Permission; Pr: Prohibition; R: Recommendation and O: Obligation. The numbers represent the "Id_Rule" attribute defined in the *Rule* class. Each ID number can be read as follow:

> **Security rule:**
> Id_Rule ⬌ Rule (Organization, Role, Activity, View, Context)

In order to understand to previous schematic rule, we present hereafter a prohibition security rule example:

> Id_Rule = 1 ⬌ Prohibition (Operator_X , Director, Change, GeneratorViews, Normal)

This statement means that in the organization *Operator_X*, the role *Director* is prohibited to do the activity *Change* on the set of views *GeneratorViews* in the context *Normal*. We define in the same way security policies for the two others scenarios for all roles. The next step is to apply these security policies over the our generated EMF application corresponding to our MDAOrBAC model, then XMI files based abstract security rules are generated automatically; the following section tackles this point.

## 5 Code Generation

In order to generate XMI files based abstract security rules, we first generate the corresponding EMF application of the proposed MDAOrBAC model (Fig. 1). We apply all abstract security rules (already defined in the previous subsection) over the generated EMF application. Then, XMI files are generated automatically. Our EMF application generates several XMI files corresponding to each type of entities defined in the proposed model. Indeed, in order to create the following abstract security rule *Permission(Public, Engineer, Check, Transformer_33kv, Normal)*, we must instantiate the fifth classes: *Organization, Role, Activity, View* and *Context*. Due to space restrictions, we present an analysis of one generated XMI file:

```
<?xml version="1.0" encoding="UTF-8"?>
<orbac:Role xmi:version="2.0"
xmlns:xmi="http://www.omg.org/XMI"
xmlns:orbac="http://orbac/1.0" Role="Engineer">
    <RoleAssignment Rule="Permission" Id_Rule="1"/>
    <EmpowerRelationship name="Aziz" age="25"/>
</orbac:Role>
```

It is an instantiation of the *Role* class in which the "Role" attribute is set to the role "Engineer". It is part of one abstract security rule which has the identifier "1". Another information in this XMI file is the fact that the subject "Aziz" (old "25" years) is an "Engineer" (*Empower* relationship).

> **General rule:** To create a new security rule, it is necessary to instantiate each class forming an abstract security rule namely: *Organization, Role, Activity, View* and *Context* classes. Also, an instantiation of the concrete classes namely: Subject, Action and Object, is mandatory in order to satisfy the *Empower, Consider* and *Use* relationships defined in OrBAC model.

After generating the whole XMI files based abstract security rules, we need to deduct concrete security rules using MotOrBAC engine. The next section tacks this point.

# 6   Deducting Concrete Security Rules Using MotOrBAC

In order to complete our approach, we need to deduct concrete security rules from the abstract security rules already defined. To do so, we parse the generated XMI files based abstract security rules using a Java Parser technology to match with the INPUT file of MotOrBAC [15] engine. In fact, we use MotOrBAC engine for validating the policy and deducting concrete security rules. Indeed, our EMF application generates 5 XMI files separately corresponding to one abstract security rule. We use DOM parser (in JAVA) and we develop a module for parsing the generated XMI files. We read and extract the important information that we need to write a new XML file that match with the INPUT file of MotOrBAC engine (XmlOrbacPolicy file (.pof)). Finally, we use MotOrBAC to load the security policies and thus to deduct concrete security rules which are displayed on the MotOrBAC's interface. However, we have no access to XML code based *concrete security rules*.

# 7   Conclusion and Future Work

In this paper, we present a first proof of concept of our global new approach using MDA approach to design and generate access control policies based on OrBAC model. The aim of our work is to provide a complete package, a visual model to define a security policy using OrBAC model, to generate access control rules based XMI files, to use MotOrBAC engine to validate the policy and generate concrete security rules. In other words, we firstly propose the MDAOrBAC model which is a combination between the MDA approach and OrBAC model through the use of Eclipse Modeling Framework (EMF). Secondly, we choose the Electrical Grid as an example of Critical Infrastructure to instantiate our model. In this step, we define a complete scenario of security policies. Thirdly, we generate our EMF application corresponding to the proposed MDAOrBAC. Then, we apply already defined security policies. Fourthly, we develop a module for parsing the generated XMI files using DOM Parser in Java technology to match with the INPUT file of MotOrBAC engine. Finally, we use MotOrBAC engine to deduct concrete security rules from the abstract security rules.

Our MDAOrBAC model had some design problems and limitations that we faced during the development process. We have identified two major points: (1) our EMF application generates a lot of XMI files corresponding to each type of the enumeration lists mentioned in Fig. 2. If we defined several abstract security rules, then the parsing of the generated XMI files becomes more complex. (2) MotOrBAC engine deduct concrete security rules, from the already defined abstract security rules, and displays it on the interface. However, we don't have access to the XML code based *concrete security rules* that we need for the next step of our approach. These issues are currently investigated in an ongoing work. We are also working on the dynamicity of the policy in the sense that the security administrator, for example, can change the parameters of one rule and the rest of the similar rules are transformed/modified automatically using the MDA approach.

# References

1. Ching-Ting, L., et al.: A GIS-based simulator for CIIP interdependency analysis. In: International Computer Symposium (ICS), pp. 901–906 (2010)
2. Schaberreiter, T., et al.: Critical infrastructure security modelling and RESCI-MONITOR: a risk based critical infrastructure model. In: IST-Africa Conference Proceedings, pp. 1–8 (2011)
3. Bai, Q.H., Zheng, Y.: Study on the access control model in information security. In: IEEE Cross Strait Quad-Regional Radio Science and Wireless Technology Conference (CSQRWC), pp. 830–834 (2011)
4. Sandhu, S., et al.: Role-based access control models. Computer 29(2), 38–47 (1996)
5. El Kalam, A., et al.: Organization based access control. In: 4th IEEE International Workshop on Policies for Distributed Systems and Networks (POLICY), pp. 120–131 (2013)
6. El Kalam, A., Deswarte, Y., Baina, A., Kaâniche, M.: Access control for collaborative systems: a web services based approach. In: IEEE International Conference on Web Services (ICWS), pp. 1064–1071 (2007)
7. El Kalam, A., Deswarte, Y., Baina, A., Kaâniche, M.: PolyOrBAC: a security framework for critical infrastructures. Int. J. Critic. Infrastruct. Protect. 2(4), 154–169 (2009)
8. El Kalam, A., et al.: Integrity-OrBAC: an OrBAC enhancement that takes into account integrity. In: the 8th International Conference on Intelligent Systems: Theories and Applications (SITA), pp. 1–7 (2013)
9. Baina, A., Laarouchi, Y.: MultiLevel-OrBAC: multi-level integrity management in organization based access control framework. In: IEEE International Conference on Multimedia Computing and Systems (ICMCS), pp. 933–938 (2012)
10. Steinberg, D., et al.: EMF: Eclipse Modeling Framework, 2nd edn. Addison-Wesley Professional. Part of the Eclipse Series (2008)
11. eXtensible Access Control Markup Language—XACML. http://docs.oasis-open.org/xacml/3.0/xacml-3.0-core-spec-os-en.html
12. Lodderstedt, T., et al.: SecureUML: a UML-based modeling language for model-driven security. In: Proceedings of the 5th International Conference on the Unified Modeling Language, pp. 426–441 (2002)
13. Jin, X.: Applying model driven architecture approach to model role based access control system. Doctoral dissertation, University of Ottawa (2006)
14. Muñante, D., et al.: An approach based on model-driven engineering to define security policies using OrBAC. In: the 8th International Conference on Availability, Reliability and Security (ARES), pp. 324–332 (2013)
15. Autrel, F., et al.: MotOrBAC 2: a security policy tool. In: Conference on Security in Networks Architectures and Security of Information Systems (SARSSI) (2008)
16. Singh, Y., Sood, M.: Model driven architecture: a perspective. In: IEEE International Advance Computing Conference (IACC), pp. 1644–1652 (2009)
17. Kaddani, A., Baina, A., Echabbi, L.: Towards a model driven security for critical infrastructures using OrBAC. In: IEEE International Conference on Multimedia Computing and Systems (ICMCS), pp. 1235–1240 (2014)

# Secure and Flexible RBAC Scheme Using Mobile Agents

Hind Idrissi, Mohammed Ennahbaoui, El Mamoun Souidi,
Said El Hajji and Arnaud Revel

**Abstract** Distributed Computing shows a fast development, that makes sharing
and diffusion of information more easier than before. However, ignoring to adopt a
well defined security policy exposes the information system to serious damages and
a wide variety of attacks. In this paper, we propose a new conception of access
control where subjects and entities are modeled by autonomous mobile agents.
Features of these later are combined with cryptographic mechanisms such as
resistant-MITM Diffie-Hellman Key Exchange and digital signature, symmetric and
elliptic curve encryption, as well as one-way hash functions. We have implemented
the proposed scheme and conducted detailed experiments to evaluate the security
and effectiveness of our scheme.

**Keywords** RBAC · Security · Mobile agent · Cryptography · JADE

H. Idrissi (✉) · M. Ennahbaoui · E.M. Souidi · S.E. Hajji
Laboratory of Mathematics, Computing and Applications,
Faculty of Sciences, University of Mohammed-V, Rabat, Morocco
e-mail: hind.idr@gmail.com

M. Ennahbaoui
e-mail: ennahbaoui.mohamed@gmail.com

E.M. Souidi
e-mail: souidi@fsr.ac.ma

S.E. Hajji
e-mail: elhajji@fsr.ac.ma

H. Idrissi · A. Revel
L3I, University of La Rochelle, La Rochelle, France
e-mail: arnaud.revel@univ-lr.fr

© Springer International Publishing Switzerland 2016
A. El Oualkadi et al. (eds.), *Proceedings of the Mediterranean Conference
on Information & Communication Technologies 2015*, Lecture Notes
in Electrical Engineering 381, DOI 10.1007/978-3-319-30298-0_46

447

# 1 Introduction

Recent days, the organizations become increasingly dependent on the information system (IS), which makes the security issue of IS strongly important, especially in electronic transactions operating sensitive data, where security flaws are evolving through time and may lead to serious harms. Access Control [1] is one of the mechanisms that meet the privacy requirements of an IS, as it monitors permissions and prohibitions to objects (passive entities) by subjects (active entities). In this paper, we propose a security scheme that traces a new direction in access control policies. It takes benefit from the advantages of the mobile agent paradigm [2] to solve communication problems, such as: waiting time, network delay, dependence and server fails. Mobile agents are particular category of intelligent software entities able to freely roam among nodes of the network. Along its mobility, an agent transports all its resources: code, data, state of execution, which allow it to perform independently and autonomously without referring to its native platform neither to the hosting one.

Our paper is organized as follows: Sect. 2 presents our motivation to lead this work compared to some other contributions. Section 3 provides a detailed description of the proposed access control scheme basing mobile agents to model its different entities. In Sect. 4, an evaluation of the implemented proposed scheme is supplied. Finally, further discussion and perspectives are mooted in conclusion.

# 2 Motivation

The intense need to availability of information pushes collaborative organizations to make it more accessible and open, despite the fact that the shared content is not secure, especially that it is used by different users. Thus, it becomes crucial to define flexible access control mechanisms, where data are secure and only accessible to authorized users, in the most efficient and agile forms. Role-Based Access Control (RBAC) [1] is the predominant model for advanced access control, since it reduces the cost of security management. However, there are several anomalies that can be considered when using RBAC for information management systems. The most consistent ones are the security of data and the lack of interoperability, generally due to the complexity and heterogeneity of the relevant systems. Many efforts have been devoted to investigate and explore the role assignment, time constraint, flexibility and security controlled mobility of RBAC. Towards this, Strembeck and Mendling [3] specified a formal meta-model for process-related RBAC models basing UML activity models. TrustBac [4] was proposed to support automatic and dynamic user-role assignment based on credentials and past behaviors of the subject. Nassr et al. [5] proposed a new supervised roles and permissions delegation model for open and dynamic organizations that depends on lines of authority. Santos-Pereira et al. [6] submitted an RBAC based mobile agent model supported by a specially managed PKI for strong authentication and restricted access authorization to allow the exchange of clinical information in health systems.

# 3 Proposed Scheme

This section provides a detailed description of the proposed scheme with a conception of the different execution phases. In order to well define and model our scheme, an RBAC example for a faculty management system is used and illustrated in Fig. 1, which offers administration and hierarchical features. It contains two super roles: User and Staff, from which the other roles are derived. Each role is associated with a set of activities to be performed according to specific constraints.

## 3.1 Platform Architecture

Flexibility and robustness of our RBAC model is ensured through using cooperative mobile agents to simulate its hierarchical architecture. Figure 2 illustrates the structure of our platform, where each role or activity is represented by an agent able to move between the platforms to ask for a task or execute an action. Constraints are left unchanged. Thus, Our platform contains two main agents: an "Authenticator_agent" charged with performing an authentication process based on key exchange, and a "Manager_agent" responsible for managing sensitive data and administering agents communication inside the same platform or with remote ones. Thereby, the architecture also models two containers. The first one assembles a set of mobile agents representing the roles that can be assumed. When a role agent is created and implanted,

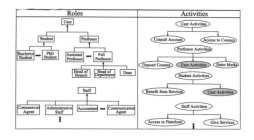

**Fig. 1** RBAC model of a faculty management system

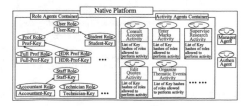

**Fig. 2** The hierarchical architecture for the proposed platform

it is automatically associated with a specific key, that allows an activity to be unlocked and performed if that role appears in its white-list of the authorized ones. The second container includes the activity agents that are able to achieve the requested activity if the role associated is granted the permission to do. For that purpose, each activity agent is designated with a list of hashed keys obtained using the (reduced capacity/free-collision) hash function SHA-3 [7], and which are corresponding to the roles allowed to perform this activity.

## 3.2 Authentication Process and Role Assignment

Our scheme adopts an authentication mechanism where three main entities need to interact: the Native Platform (**NP**), the Remote Platform (**RP**) hosting the subjects requesting access and the Database Server (**DS**) containing the Authentication Credentials (**AC**).

As a first step, we make use of Diffie-Hellman Key Exchange in a new security conception inspired from [8] to avoid the Man-In-The-Middle (MITM) attacks, due to use of a secret ephemeral numbers in the exponential calculations. This protocol illustrated in Fig. 3 creates a shared key named Session Key (T) of 256 bits, and allows to verify if this key was altered through primality. Next steps are described in Fig. 4. When a Subject Agent (**SA**) of a RP needs to execute an activity provided by our NP, it sends a request through the Manager Agent of its platform. This request launches a key exchange process, in which a session key (T) is generated to encrypt the SA, using AES [9], before that it migrates. Once on the NP, the Manager Agent decrypts the SA using the same session key (T). Then, the agent's data need to be verified referring to the DS, where the AC about authorized subjects are stored.

After claiming the public key of the DS, the Manager Agent of the NP constitutes a message formed by the concatenation of the encrypted and signed SA, using

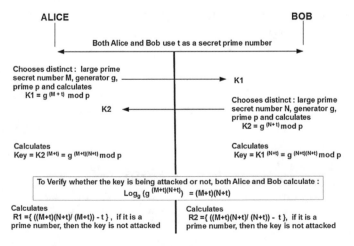

**Fig. 3** DHKE protocol with secret ephemeral to avoid MITM

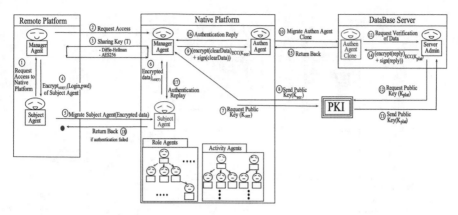

**Fig. 4** The authentication process of the proposed scheme

the elliptic curve encryption (ECC) and the digital signature standard (DSS) [9]. This message is then directed to the Authenticator Agent, that creates a clone to migrate and verify the message on the DS, called "Auhtenticator-Clone Agent" (ACA). The use of ECC in our scheme is mainly due to its robustness, speed, use of shorter key lengths for stronger security, and for its several advantages over public-key cryptosystems, namely RSA. In addition, the employment of DSS is complementary to ensure authenticity and integrity of data, as well as tracking and non-repudiation with minimal costs.

Using its private key, the Database Agent decrypts the first part of the message, signs it and compares the signature it has obtained with the second part of the received message. If the two signatures match, then the Database Agent verifies the AC, decides whether the subject has an authorized access or not, and identifies which role it should be assigned to him. Else, an acquittal of rejected access request is prepared. In both cases, a reply is encrypted and signed using the public key of the NP, and provided to the ACA which returns back. Similarly, the Manager Agent decrypts and verifies the integrity of the reply using its private key. Finally, the SA is terminated to approve a security strategy, that aims to reduce the lifetime of the agent to be unable to adopt any malicious behaviors against the NP. Thus, instead of hosting outer agents inside the NP, the communications are more controlled from far, through creating clones to perform the requested tasks.

## 3.3 Interactions Among RBAC-Agents and Execution of Activities

Once the authentication has passed and the role of the SA has been identified, the Manager Agent sends to the concerned Role Agent a request to activate its services. Figure 5 simulates a scenario where the supposed SA has the role "Student" and wants to perform "Consult Account" as activity. When the "Student Role Agent"

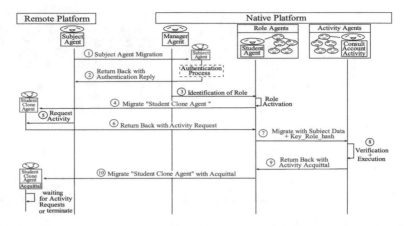

**Fig. 5** Simulation of a "consult account activity" executed by a SRA

(**SRA**) is activated, it creates a clone "Student Clone Agent" (**SCA**) that migrates to the RP and waits for activity requests. The reason behind creating clones of roles agents is to avoid the waiting time that can be added when we have multiple subject agents requesting the same role at the same time.

As it was quoted previously, each Role Agent maintains a specific key allowing it to be identified regarding the activities agents. In parallel, each Activity Agent contains a list of hashes designating the identifier keys of the roles allowed to perform this activity. In our scenario, the SRA charged with the "Consult Account Activity" has to hash its key using SHA-3, and join it to the data requested to execute this activity such as login (name or email) and password. These information are assigned to SCA that migrates to the Activity Agents Container. Once there, it presents the hash to the concerned Activity Agent, that verifies in its own list if that hash exists. If so, the activity requested is then executed, else the request is then rejected. In both cases, an acquittal is given to SCA that returns first to the "Role Container", where it displays this acquittal to the original role agent, then it moves to the RP where it exposes the results of this round-trip, and waits for other activity requests or for termination.

## 4 Evaluation

In this section, we present our experimental investigations to prove the feasibility and the security of the proposed scheme. Thus, heterogeneous machines are used with different OS, Core i7, 4 Go of RAM, and equipped with JADE (4.3.3) [10] for software interoperability among agents. First, we consider the scenario mentioned previously to test the time performance of our scheme, then we evaluate its scalability and efficiency. Our measurements were subdivided into two successive phases: The authentication and the execution of the activities requested.

To compute the time cost of an authentication round-trip (ART), many operations need to be performed, which include: $T_{AR}$ Access request, $T_{DHKE}$ Diffie-Hellman Key Exchange, $T_{M_{SA}}$ Migration of SA, $T_{PKR}$ Requests of public keys from the PKI, $T_{AES_{SA}}$ AES encryption and decryption of SA, $T_{DSS_{AC}}$ Signature of AC, $T_{ECC_{AC}}$ ECC encryption and decryption of AC, $T_{M_{ACA}}$ Migration of ACA to DS, $T_{DSV_{AC}}$ DS's Verification of AC, $T_{DSS_{AR}}$ Signature of AR, $T_{ECC_{AR}}$ ECC encryption and decryption of AR, and $T_{R_{ACA}}$ Return of ACA to NP. Let $T_{ART}$ the total time spent for an ART be calculated as follows:

$$
\begin{aligned}
T_{ART} = {} & T_{AR} + T_{DHKE} + T_{AES_{SA}} + T_{M_{SA}} + T_{PKR} + T_{ECC_{AC}} + T_{DSS_{AC}} \\
& + T_{M_{ACA}} + T_{DSV} + T_{ECC_{AR}} + T_{DSS_{AR}} + T_{R_{ACA}}
\end{aligned}
\tag{1}
$$

Considering that, $T_{ECC_{AC}} = T_{ECC_{AR}}$ and $T_{DSS_{AC}} = T_{DSS_{AR}}$, because the weights of the authentication credentials and the authentication reply are approximately equal. Moreover, $T_{M_{ACA}} = T_{R_{ACA}}$ because the time spent to move the ACA is the same for its return. Equation 2 satisfies the updated constraints and Table 1 shows the time cost of an ART:

$$
\begin{aligned}
T_{ART} = {} & T_{AR} + T_{DHKE} + T_{AES_{SA}} + T_{M_{SA}} + T_{PKR} + 2T_{ECC_{AC}} + 2T_{DSS_{AC}} \\
& + 2T_{M_{ACA}} + T_{DSV}
\end{aligned}
\tag{2}
$$

After the authentication phase, the execution of the requested activity according to the role identified comprises many operations to perform, which include: $T_{A_{SRA}}$ Activation of SRA, $T_{AES^1_{SCA}}$ 1st AES Encryption of SCA by NP/Decryption by RM using the session key, $T_{M^1_{SCA}}$ 1st migration of SCA to RP, $T_{AR}$ activity request, $T_{R_{SCA}}$ return of SCA to NP, $T_{HKey}$ hashing the SRA's secret key, $T_{IM_{SCA}}$ Intra-Migration of SCA from the Role Container to the Activity Container, $T_{EV}$ verification of SRA eligibility for the activity, $T_{ActExec}$ execution of the "CA Activity", $T_{IR_{SCA}}$ Intra-Return of SCA to the Role Container, $T_{AES^2_{SCA}}$ 2nd AES Encryption/Decryption of SCA carrying results, $T_{M^2_{SCA}}$ 2nd migration of SCA to RP to present results and support other activities or end its services. Let $(T_{AERT})$ the total time spent for an Activity-Execution Round-Trip (AERT) be calculated as follow:

$$
\begin{aligned}
T_{AERT} = {} & T_{A_{SRA}} + T_{AES^1_{SCA}} + T_{M^1_{SCA}} + T_{AR} + T_{R_{SCA}} + T_{HKey} + T_{IM_{SCA}} \\
& + T_{EV} + T_{ActExec} + T_{IR_{SCA}} + T_{AES^2_{SCA}} + T_{M^2_{SCA}}
\end{aligned}
\tag{3}
$$

Regarding to the length and type of the results carried, $T_{AES^1_{SCA}} = T_{AES^2_{SCA}}$. Besides, the time spent for the migration is the same for the return, $T_{M^1_{SCA}} = T_{R_{SCA}} = T_{M^2_{SCA}}$.

**Table 1** Time cost of the different operations in an ART

|           | $T_{AR}$ | $T_{DHKE}$ | $T_{AES_{SA}}$ | $T_{M_{SA}}$ | $T_{PKR}$ | $T_{ECC_{AC}}$ | $T_{DSS_{AC}}$ | $T_{M_{ACA}}$ | $T_{DSV}$ | $T_{ART}$ |
|-----------|----------|------------|----------------|--------------|-----------|----------------|----------------|---------------|-----------|-----------|
| Time (ms) | 0.57     | 4.3        | 30.16          | 77.35        | 3.6       | 63.28          | 9.44           | 54.21         | 6.9       | 376.74    |

**Table 2** Time cost of the different operations in an AERT

| | $T_{ASRA}$ | $T_{AES^1_{SCA}}$ | $T_{M^1_{SCA}}$ | $T_{AR}$ | $T_{HKey}$ | $T_{IM_{SCA}}$ | $T_{EV}$ | $T_{ActExec}$ | $T_{AERT}$ |
|---|---|---|---|---|---|---|---|---|---|
| Time (ms) | 0.82 | 32.6 | 74.37 | 0.58 | 0.03 | 21.43 | 5.4 | 18.64 | 356.64 |

Similarly for the mobility inside the same platform which is cheaper, $T_{IM_{SCA}} = T_{IR_{SCA}}$. Equation 4 satisfies the constraints prescribed and Table 2 presents the time cost of an AERT:

$$T_{AERT} = T_{A_{SRA}} + 2T_{AES^1_{SCA}} + 3T_{M^1_{SCA}} + T_{AR} + T_{HKey} + 2T_{IM_{SCA}} + T_{EV} + T_{ActExec} \quad (4)$$

The cost of our scheme is $T_{Total} = T_{ART} + T_{AERT} = 376.74 + 356.64 = 733.38$ ms, compared to 511.87 ms for a traditional-RBAC system. This cost appears admissible, credible and not compromising the performances of the agent or the platform, which benefit from a security feature against any external vulnerabilities. To prove the ability of our scheme to scale in large conditions, we evaluate its behaviors face to the increase of subject agents (S = 2, 4, 8, 16, 30), and particularly when asking for same roles and same activities. The results are presented in Fig. 6, where it is noticed that the time performance is not linear, but becomes better when the tests are reiterated many times, which is mainly due to the Java environment endowed with a code cache memory. The overall difference of non-linearity is between 2 and 9 %.

For comparison purpose, we have conducted experiments to evaluate timing and detecting performances of our scheme versus to the traditional-RBAC one. Figure 7 shows the given results, that illustrate the ability of our scheme to detect 13

**Fig. 6** Scalability of our scheme versus the increase of subjects and requests of similar roles and activities

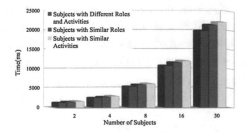

**Fig. 7** Comparison of time and detection performances between traditional RBAC and our scheme

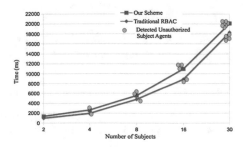

unauthorized subject agents attempting to gain access, while the traditional RBAC comes to pick up only 8 ones. This is achieved through an added security overhead of about 37 % of the overall time cost.

# 5  Conclusion

In this paper, we have proposed a novel contribution based on simulating roles and activities of administrative RBAC models using mobile agents. The interoperability and autonomy of these later are associated to cryptographic mechanisms and infrastructures, to deal with the flexibility and security issues faced in RBAC. In addition, the practical experiments we have conducted show high timing, scalability and detection performances. As perspective, we find it interesting to apply our scheme for extended access control models, namely Attribute-RBAC for spatio-temporal contexts.

# References

1. Samarati, P., Di Vimercati, S.: Access control: policies, models, and mechanisms. In: Foundation of Security Analysis and Design, LNCS, vol. 2171, pp 137–196 (2001)
2. Singh, Y., Gulati, K., Niranjan, S.: Dimensions and issues of mobile agent technology. Int. J. Artif. Intell. Appl. 3(5) (2012)
3. Strembeck, M., Mendling, J.: Modeling process-related RBAC models with extended UML activity models. Inf. Softw. Technol. 53(5), 456–483 (2011)
4. Chakraborty, S., Ray, I.: TrustBAC: integrating trust relationships into the RBAC model for access control in open systems. In: Proceedings of 11th ACM Symposium on Access Control Models and Technologies, New York, USA (2006)
5. Nassr, N., Aboudagga, N., Steegmans, E.: An organizational supervised delegation model for RBAC. In: Information Security, pp. 322–337. Springer (2012)
6. Santos-Pereira, C., Augusto, A.B., Cruz-Correia, R., Correia, M.E.: A secure RBAC mobile agent model for healthcare institutions-preliminary study. In: Information Technology in Bio- and Medical Informatics, pp. 108–111. Springer (2013)
7. Jaffar, A., Martinez, C.J.: Detail power analysis of the SHA-3 Hashing algorithm candidates on Xilinx Spartan-3E. Int. J. Comput. Electr. Eng. 5(4), 410–413 (2013)
8. Biswas, B., Basuli, K., Sarma, S.: On a key exchange technique avoiding man-in-the-middle-attack. J. Glob. Res. Comput. Sci. 3(9) (2012)
9. Mogollon, M.: Cryptography and Security Services: Mechanisms and Applications, pp. 1–488. IGI Global, Hershey, PA (2008)
10. Bellifemine, F., Poggi, A., Rimassa, G.: JADE: a FIPA2000-compliant agent development environment. In: Proceedings of 5th International Conference on Autonomous Agents, pp. 216–217. ACM, Montreal (2001)

# Incorporation and Model Checking of a Quantum Authentication and Key Distribution Scheme in EAP-TLS

Aymen Ghilen, Hafedh Belmabrouk and Mostafa Azizi

**Abstract** The prevalent use of wireless networks has led to noteworthy development in security techniques that aim to protect information. The existing standard for exchanging a secret key of authenticating users and securing data transmission within EAP-TLS protocol relies on Public Key Infrastructure. This technique is secure only in case of limited computational power of eavesdropping. We accordingly propose a quantum extension of EAP-TLS which enables a cryptographic key exchange with the authentication of a remote client peer, with absolute security, ensured by the laws of quantum physics. A model checking approach based on PRISM tool is applied to check some security properties for the proposed scheme.

**Keywords** Quantum key distribution · Quantum authentication · Entanglement · Model checking · EAP-TLS · Quantum cryptography

## 1 Introduction

The wide dissemination of information security techniques is related to the progress in the field of communication networks technology, especially the advent of many critical services such as e-commerce applications and electronic transactions via internet. This kind of applications requires an extremely high security that can be provided by QKD (Quantum Key Distribution) [1, 2]. By connecting to a Wireless

A. Ghilen (✉) · H. Belmabrouk
Laboratory of Electronics and Microelectronics of the Faculty of Sciences of Monastir,
Monastir, Tunisia
e-mail: ghilen06@gmail.com

H. Belmabrouk
e-mail: hafedh.belmabrouk@fsm.rnu.tn

M. Azizi
Department of Computer Engineering ESTO, Mohamed I University, Oujda, Morocco
e-mail: azizi.mos@gmail.com

© Springer International Publishing Switzerland 2016      457
A. El Oualkadi et al. (eds.), *Proceedings of the Mediterranean Conference on Information & Communication Technologies 2015*, Lecture Notes in Electrical Engineering 381, DOI 10.1007/978-3-319-30298-0_47

Local Area Network (WLAN), a legitimate user must be authenticated before any data transfer. So, authentication is an essential security task. EAP-TLS is a powerful solution among a vast range of possible authentication mechanisms available for the EAP standard [3]. According to its guide definition, EAP-TLS is based on PKI (Public Key Infrastructure) which is the basis of the secret keys distribution. The security of public-key cryptography relies basically on various computational problems such as one-way functions and the difficulty of the discrete logarithm which are believed to be intractable. However, PKI can only guarantee computational security, i.e., PKI methods can be cracked by advances in technology and mathematical algorithms. Therefore, the quantum cryptography issued from quantum mechanics provides unconditional security. Moreover, in quantum approach, any states measurement will induce disturbances, hence, eavesdropping can be detected. The application of quantum cryptography in WLAN 802.11 networks has been investigated in many studies [4–7]. However, even the QKD is vulnerable against man-in-the-middle attack. That is why an authentication mechanism must accompany the execution of the quantum key distribution algorithm. To practically implement a running quantum cryptography protocol, it is very important to analyze some security properties and check that the built model complies with the privacy requirement. In the literature, many models for several quantum protocols were developed using automated verification tools [5, 8]. The main objective is to guarantee that the key exchange as well as authentication within EAP-TLS protocol become unconditionally secure. The paper is organized as follows: related works are addressed in Sect. 2. In Sect. 3, we present an overview on EAP-TLS conversation. Our proposed quantum authentication and key distribution protocol is described in the fourth section. In Sect. 5, we introduce a quantum EAP-TLS: a quantum extension of EAP-TLS without all PKI infrastructures but incorporates our quantum authenticated key distribution. Section 6 deals with the security analysis of the proposed scheme. Finally, we conclude the paper in Sect. 7.

## 2  Related Works

Researchers are focusing on exploiting the great potential of quantum cryptography to ensure secure communications. Many approaches were proposed to introduce QKD within existing protocols. Papers [6, 7] present a new scheme to integrate BB84 into EAP-TLS. By removing the classical PKI, such as RSA or Diffie-Hellman, BB84 [9] will generate and distribute a key. The most important flaw of the pre-cited works in [6, 7] is the lack of any authentication mechanism during the BB84 quantum handshake, so the participants may be subject to man-in-the-middle attacks in which the traffic between them is illegitimately compromised and redirected to a third party. To cope with this powerless, an authentication process conjointly with key distribution must be involved instead of BB84. Thereby, our proposed solution will provide not only an authenticated key but also a mutual authentication at the end of the quantum protocol, too.

## 3 EAP-TLS Conversation: The Base Case

EAP-TLS includes certificate-based mutual authentication and key management methods of the TLS protocol. EAP-TLS conversation starts by negotiating EAP. That is, the authenticator sends an EAP-Request/Identity packet to the peer, who will respond with an EAP-Response/Identity packet containing his user-Id. Once having received the peer's Identity, the EAP server (the ultimate endpoint conversing with the peer) must respond with an EAP-TLS/Start packet. The EAP-TLS conversation will then begin, with the peer sending an EAP-Response packet with EAP-Type=EAP-TLS. As depicted in Fig. 1, The EAP server will then respond with a TLS server_hello handshake, possibly followed by TLS certificate, server_key_exchange, certificate_request, server_hello_done and/or finished handshake messages, and/or a TLS change_cipher_spec message. The EAP server must include a TLS server_certificate handshake message for either a public key exchange (such as an RSA or Diffie-Hellman key exchange) or a signature public key (such as an RSA or Digital Signature Standard DSS signature public key), a TLS server_key_exchange handshake message must also be included to allow the key exchange to take place and a server_hello_done handshake message must be the last handshake message encapsulated in this EAP-Request packet. Figure 1 presents the whole EAP-TLS conversation in the base case. To check whether the key exchange and authentication processes were successful, TLS finished messages must be sent and calculated at both sides of the peer and the EAP server using the following formula developed in [3]:

$$
\begin{aligned}
\mathrm{PRF}\ &(\mathrm{master\_secret}, \mathrm{finished\_label}, \mathrm{MD5}(\mathrm{handshake\_messages}) \\
&+ \mathrm{SHA}-1(\mathrm{handshake\_messages}))
\end{aligned} \tag{1}
$$

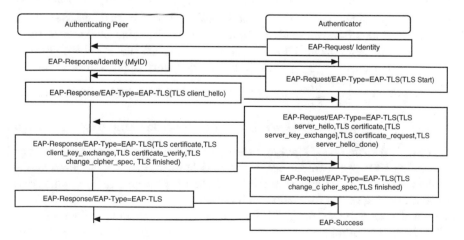

**Fig. 1** EAP-TLS conversation: base case

where PRF is a pseudo random function. For Finished label sent by the client, the string "client finished" is used. For Finished label sent by the server, the string "server finished" is used. MD5 and SHA-1 are examples of one-way hash functions. The value handshake_messages includes all handshake messages starting from client_hello up to, but not including, this TLS finished message. The "+" in the previous formulas indicates concatenation. The master_secret parameter is defined by the expression:

$$\text{master\_secret} = \text{PRF}(\text{pre\_master\_secret}, \text{"master secret"}, \text{ClientHello.random}$$
$$+ \text{ServerHello.random})$$

(2)

ClientHello.random and ServerHello.random are two random values generated and exchanged by the client (the peer) and the authenticator (the server). The pre_master_secret parameter results from the key exchange process and is known to the communicating parties. In this work, we treat only the base case scenario of EAP-TLS conversation. The other scenarios (session resumption, termination, etc.) will be developed in a future work.

## 4 Quantum Authentication with Key Distribution Protocol

Numerous enhancements are applied to our scheme [4]. They are inspired from Shang et al. [10]. Let two parties Alice and Bob who have previously shared $N$ EPR pairs either in $|\Psi^-\rangle$ or $|\Psi^+\rangle$ states, where $|\Psi^-\rangle = \frac{1}{\sqrt{2}}(|01\rangle - |10\rangle)$ and $|\Psi^+\rangle = \frac{1}{\sqrt{2}}(|01\rangle + |10\rangle)$. Considering the $X$ (diagonal) basis, we can express them as $|\Psi^-\rangle = \frac{1}{\sqrt{2}}(|-+\rangle - |+-\rangle)$ and $|\Psi^+\rangle = \frac{1}{\sqrt{2}}(|++\rangle - |--\rangle)$. The first particle is held by Alice however the second is held by Bob. To establish a secret and authenticated key, Bob will send his particle back to Alice after randomly applying one of three unitary operators according to the initial state shared between them. If the pre-shared state is $|\Psi^-\rangle$, then Bob will randomly apply either $I$ (identity operator) or $\sigma_x = \begin{bmatrix} 0 & 1 \\ 1 & 0 \end{bmatrix}$ on his particle. Otherwise, he randomly applies either I or $\sigma_z = \begin{bmatrix} 1 & 0 \\ 0 & -1 \end{bmatrix}$ on it, then sends it back to Alice. The matrices are expressed in the canonical basis $(|0\rangle, |1\rangle)$. Now, Alice performs Bell measurement on the corresponding EPR pair. If the measuring state is in error, then she decides that this particle is not from Bob and consequently aborts the protocol. Table 1 summarizes Alice measuring results in relation with Bob's operator and the pre-shared pair. If the initial state is $|\Psi^+\rangle$, then, Alice's measuring result should be either $|\Psi^+\rangle$ or $|\Psi^-\rangle$ if no eavesdropper exists. Otherwise, an eavesdropper may exist. Once transmission of N particles is completed, the two parties consider $\Psi^-$ and $\Phi^-$ as the

**Table 1** Alice's measuring results versus pre-shared state and Bob's operator

| Initial state | Bob's operator | Alice's measurement | Binary data |
|---|---|---|---|
| $\lvert \Psi^+ \rangle$ | $I$ | $\Psi^+$ | 0 |
| | | $\Psi^-$ | 1 |
| | $\sigma_z$ | $\Phi^-$ or $\Phi^+$ | ERROR |
| $\lvert \Psi^- \rangle$ | $I$ | $\Psi^-$ | 0 |
| | | $\Phi^-$ | 1 |
| | $\sigma_x$ | $\Psi^+$ or $\Phi^+$ | ERROR |

binary data "0" and "1" respectively if initial state is $\Psi^-$ and consider $\Psi^+$ and $\Psi^-$ as "0" and "1" otherwise.

By this way, they will simultaneously share a key and Bob will be authenticated to Alice. Now, we discuss the security of the quantum protocol. Two security requirements must be fulfilled. The first one consists on ensuring that any quantity of information about the key gained by Eve is minimal. In other words, the unconditional privacy of the key must be verified. The protocol must also establish that the presence of an eavesdropper is detected by the legitimate users. We suppose Eve is performing an "intercept and resend" attack. As soon as Bob sends back his particle to Alice, Eve intercepts and measures it, then she resends a fake particle to Alice. To measure the intercepted particle, Eve uses either $X$ or $Z$ basis. The impact of the basis choice by eve is shown on Table 2.

Without knowing the initial state, the probability to detect Eve is globally 1/4. That is, the probability that Eve passes without being detected is 3/4 for each entangled pair. Therefore, for N states, the probability to detect Eve is $\left(1 - \left(\frac{3}{4}\right)^N\right)$. For N large enough, Eve is detected with a probability about 100 %. By this way, we expect that the amount of information gained by Eve about the key is minimal and tends toward zero.

**Table 2** Relation between the probability of detecting Eve and her chosen measurement basis

| Initial state | Possible Alice's measuring state | | Detecting Eve with probability |
|---|---|---|---|
| $\Psi^+$ | $\lvert \Psi^+ \rangle = \frac{1}{\sqrt{2}}(\lvert 01 \rangle + \lvert 10 \rangle) = \frac{1}{\sqrt{2}}(\lvert ++ \rangle - \lvert -- \rangle)$ | | X basis: 50 % Z basis: 0 % |
| | $\lvert \Psi^- \rangle = \frac{1}{\sqrt{2}}(\lvert 01 \rangle - \lvert 10 \rangle) = \frac{1}{\sqrt{2}}(\lvert -+ \rangle - \lvert +- \rangle)$ | | |
| $\Psi^-$ | $\lvert \Psi^- \rangle = \frac{1}{\sqrt{2}}(\lvert 01 \rangle - \lvert 10 \rangle) = \frac{1}{\sqrt{2}}(\lvert -+ \rangle - \lvert +- \rangle)$ | | X basis: 0 % Z basis: 50 % |
| | $\lvert \Phi^- \rangle = \frac{1}{\sqrt{2}}(\lvert 00 \rangle - \lvert 11 \rangle) = \frac{1}{\sqrt{2}}(\lvert -+ \rangle + \lvert +- \rangle)$ | | |

# 5   New Release of EAP-TLS: Quantum EAP-TLS

The proposed extension of EAP-TLS called Quantum EAP-TLS is presented in Fig. 2. Henceforth, the classical key exchange (DH or RSA) is replaced by a quantum key generation process described in Sect. 4. Specifically, the pre_master_key parameter used in the formula (2) of calculation of the master_secret parameter which is needed to compute the TLS finished message, will be generated by the quantum key distribution protocol developed in Sect. 3. To ensure mutual authentication, each side has to calculate its own TLS finished message as mentioned in expression (1).

Since the master_secret is calculated by expression (2), our novel authentication scheme incorporates the modification:

pre_master_key $= K_B$ for the peer side and pre_master_key $= K_A$ for the authenticator side. Once the EAP server receives the peer's TLS finished message inputted by $K_B$, as proposed in the new setting, it subsequently calculates the TLS finished message which is inputted by $K_A$ and verifies if they are equal. If so, the peer is successfully authenticated. It should be mentioned that according to our quantum authentication and key distribution protocol, if $K_A = K_B$, i.e. the peer TLS finished message and the EAP server TLS finished message agree, then the EAP server becomes sure of the identity of the authenticating peer. Conversely, the same approach is performed at the peer after it receives the TLS finished message from the server. Therefore, our new extension of EAP-TLS ensures not only authentication of the peer, but also a mutual authentication for both parties. In the quantum version of EAP-TLS protocol, the messages exchanged during the quantum authentication and key distribution conversation are encapsulated in EAP-Request and EAP-Response packets of EAP-Type = EAP-TLS in both sides of the EAP

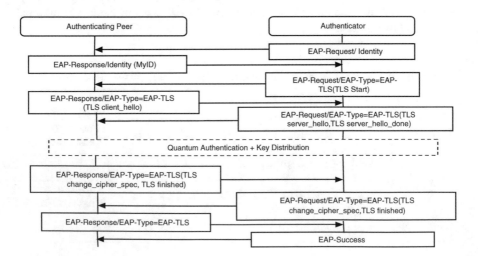

**Fig. 2** Quantum extension of the EAP-TLS conversation

server and the peer respectively. They are part of the value of the handshake_messages.

# 6 Security Analysis of Quantum EAP-TLS

To check the correctness of the proposed protocol, we use PRISM as a formal model checking. It allows modeling and validating systems which have probabilistic behavior, with respect to a temporal specification or a property. The probability that a system M satisfies a property p is written:

$$Pr\{M \models p\} \tag{3}$$

Each model can be parameterized by writing $M = M(x_1, x_2, \ldots, x_n)$. Thus, the probability (3) is computed for numerous values of $x_i$. This is what we call an experiment that enables us to generate relevant curves of probability variation. A PRISM model is composed of components called modules. Each module has its own local variables and its behavior is described by a sequence of actions to be carried out. An action takes the following form:

$$[action]g \rightarrow a_1 : (x'_1 = val_1) + a_2 : (x'_2 = val_2) + \cdots + a_n : (x'_n = val_n) \tag{4}$$

where $x_1, x_2, \ldots, x_N$ are local variables of the current module. The variable $x_i$ is updated by $val_i$ with probability $a_i$. So, we have $\sum_{i=1}^{N} a_i = 1$. A command is enabled if and only if the state satisfies the predicate $g$. In our model, to express the tendency of Eve to choose either $X$ basis or $Z$ basis, we add the action:

$$\begin{aligned} &[eve\_b](e\_st = 0) \rightarrow 0.5 : (e\_st' = 1) \\ &\&(e\_bas' = 0) + 0.5 : (e\_st' = 1)\&(e\_bas' = 1) \end{aligned} \tag{5}$$

In this case, the two choices are equiprobable.

## 6.1  Definition of Our Model $M_{QA}$ Using PRISM

The quantum authentication with key agreement as described in Sect. 3 is modeled using PRISM as a model checker. We built four modules to represent Bob (the peer), Alice as an authenticator, Eve and the quantum channel. Our analysis will focus on an extremely interesting security property: the quantity of valid information that Eve can capture must be minimal. Henceforth, we use our model to calculate

$$Pr\{M_{QA}\vDash p\} \tag{6}$$

where p is a PCTL (Probabilistic Computation Tree Logic) formula. By varying N, the security property is investigated through the calculation of the probability $P_{>\frac{N}{2}}$ that Eve intercepts and measures more than half the photons transmitted between the two parties without being detected:

$$P_{>\frac{N}{2}}(N) = Pr\{M_{QA}\vDash p_{>\frac{N}{2}}\} \tag{7}$$

where $p_{>\frac{N}{2}}$ is a PCTL formula such that its boolean value is true if Eve intercepts and measures more than half the photons between the two parties without being detected.

## 6.2  Expression of $p_{>N/2}$

After intercepting each particle from Bob, Eve selects either X or Z basis to measure its state. In our model $M_{QA}$, we have created a variable called $n_{wd}$ to represent the number of times in which Eve measures Bob's particle without being detected. The corresponding code lines are:

$$
\begin{aligned}
&[eve\_mes](e\_st = 1)\&(e\_bas = 0)\&(ch\_pair = 1)\\
&\&(nwd < N) \rightarrow (e\_st' = 2)\&(nwd' = nwd + 1);\\
&[eve\_mes](e\_st = 1)\&(e\_bas = 1)\&(ch\_pair = 0)\\
&\&(nwd < N) \rightarrow (e\_st' = 2)\&(nwd' = nwd + 1);
\end{aligned} \tag{8}
$$

The parameters e_st, e_bas, ch_pair and N refer to respectively Eve's state, basis used by Eve, the pair chosen by Alice and the number of photons transmitted between Alice and Bob. These code lines deal with the case such that the initial state is $\Psi^+$ and basis is Z as well as the second case where $\Psi^-$ is the initial state and the X basis is used by Eve.

We point out that for the case where the initial state is $\Psi^+$ and basis is X and the other case where $\Psi^-$ is the initial state and the Z basis, Eve has a chance of 50 % to go unnoticed (not detected), therefore, we have to consider these two cases in our model by adding the two code lines:

$$
\begin{aligned}
&[eve\_mes](e\_st = 1)\&(e\_bas = 0)\&(ch\_pair = 0)\&(nwd < N) \rightarrow\\
&lucky : \left(e'_{st} = 2\right)\&\left(nwd' = nwd + 1\right) + (1 - lucky) : \left(e'_{st} = 2\right);\\
&[eve\_mes](e\_st = 1)\&(e\_bas = 1)\&(ch\_pair = 1)\&(nwd < N) \rightarrow\\
&lucky : (e\_st' = 2)\&(nwd' = nwd + 1) + (1 - lucky) : (e\_st' = 2);
\end{aligned} \tag{9}
$$

where lucky = 0.5 and denotes the probability to measure Bob's particle without being detected using inappropriate basis. Consequently, the PCTL formula is expressed in function of $n_{wd}$ and $N$ as:

$$P_{> \frac{N}{2}} = \{\text{true U } (\text{nwd} > \frac{N}{2})\} \tag{10}$$

## 6.3 Impact of Varying the Channel Efficiency on the Information Gained by Eve About the Key

A quantum channel can be a free space or an optical fiber. A specific module was built to manage its behaviour, its interactions with other modules and all parameters relative to the channel. If we consider the case of a perfect channel, PRISM model checker enables us to model it through these following code line:

$$[\text{Bob\_put}](\text{ch\_st} = 1) \rightarrow (\text{ch\_st}' = 2)\&(\text{ch\_op}' = \text{Bob\_choice})$$
$$[\text{eve\_mes}](\text{ch\_st} = 2) \rightarrow (\text{ch\_st}' = 3)\&(\text{ch\_b}' = \text{e\_bas}) \tag{11}$$

The information sent by Bob is unchanged through the channel. The parameters ch_st, ch_op and Bob_choice represent respectively the channel state, the channel operator and the choice of Bob in terms of operators ($I$, $\sigma_z$ or $\sigma_x$) to be applied on its own particle. Within the range $N \in [2\ 30]$, PRISM computes the probability $P_{> N/2}$ and provides the curve $P^{ch(2)}_{> N/2}(N)$ illustrated in Fig. 3. If we consider the second case where the channel is relatively noisy, the lines (11) change as follows:

$$[\text{Bob\_put}](\text{ch\_st} = 1) \rightarrow d0 : (\text{ch\_st}' = 2)\&(\text{ch\_op}' = \text{Bob\_choice}) +$$
$$d1 : (\text{ch\_st}' = 2)\&(\text{ch\_op}' = 2 - \text{Bob\_choice});$$
$$[\text{eve\_mes}](\text{ch\_st} = 2) \rightarrow d0 : (\text{ch\_st}' = 3)\&(\text{ch\_b}' = \text{e\_bas}) +$$
$$d1 : (\text{ch\_st}' = 3)\&(\text{ch\_b}' = 1 - \text{e\_bas}); \tag{12}$$

By setting d0 = 0.7 and d1 = 0.3, a little bit noise arises and the information provided by Bob are slightly parasitized. The relevant curve $P^{ch(1)}_{> N/2}(N)$ is drawn on Fig. 3. To simulate a very noisy channel, in lines (12) we set the values d0 = 0.4 and d1 = 0.6. PRISM accordingly draws the curve $P^{ch(0)}_{> N/2}(N)$. From Fig. 3, we point out that the probability that Eve measures more than half the transmitted photons without being detected decreases exponentially if $N$ increases and we have $\lim_{N \to +\infty} P_{> \frac{N}{2}} = 0$. Thus, the security requirement is verified. More interestingly, as soon as the channel becomes noisy, the quantity of information gained becomes smaller, and we deduce the following inequality:

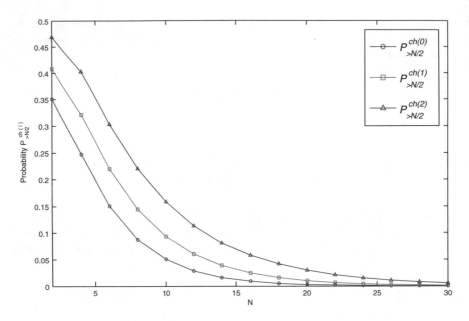

**Fig. 3** Probalities $\{P^{ch(i)}_{>N/2}(N), i = 0, 1, 2\}$ of measuring more than half the transmitted photons without being detected

$$P^{ch(0)}_{>N/2}(N) \leq P^{ch(1)}_{>N/2}(N) \leq P^{ch(2)}_{>N/2}(N) \quad \text{for } 2 \leq N \leq 30 \tag{13}$$

Consequently, we affirm that our proposal satisfies the intended security property.

# 7 Conclusion

In this paper, a quantum extension of EAP-TLS protocol is proposed. More specifically, all the PKI infrastructure was removed and replaced by a quantum authenticated key distribution. Thereby, the key exchange as well as the authentication mechanism is performed with only quantum means. Moreover, the handshake messages become simpler and the security is obviously enhanced. To analyze the security of our proposal, a formal verification of some security properties is performed using PRISM. The results confirm the proof of unconditional security and match with our expectation:

- To minimize the probability to gain information about the private key by Eve, it is mandatory to increase the number of photons transmitted, that is, enlarge the key length.

- Once the channel is noisy, then the probability to collect some information on the key by Eve decreases.

Consequently, our solution takes advantage of the laws of quantum mechanics and the model checking approach to build a new reliable authentication protocol with IEEE 802.11 wireless network and contribute in the evolution of this research field.

# References

1. Payal, P., Soni, D.: An invention of quantum cryptography over the classical cryptography for enhancing security. IJAIEM **2**(2), 243–246 (2013)
2. Sharma, R.D.: Quantum cryptography: a new approach to information security. IJPSOEM **1** (1), 11–13 (2011)
3. Simon, D., et al.: The EAP-TLS authentication protocol. RFC 5216, March 2008
4. Ghilen, A., et al.: Quantum authentication based on entangled states, pp. 75–78. Tunisia, Proc. World Cong. Multimedia Comput. Sci. (2013)
5. Norman, G., et al.: Model checking for probabilistic timed automata. Formal Methods Syst. Design **43**(2), 164–190. Springer (2013)
6. Elboukhari, M., et al.: A new extension of the EAP-TLS protocol based on quantum cryptography. J. Comput. **2**(9), 25–32 (2010)
7. Elboukhari, M., et al.: Integration of Quantum Key Distribution in EAP-TLS. Protocol used for Wireless LAN Authentication. In: Proceedings of ISIVC'10
8. Kwiatkowska, M.: Probabilistic model checking with PRISM. POPL 2015 tutorial, Mumbai, January 2015
9. Bennett, C.H., Brassard, G.: Quantum cryptography: Public key distribution and coin tossing. Theoret. Comput. Sci. **560**, 7–11 (2014)
10. Wei, T.S., et al.: Comment on quantum key distribution and quantum authentication based on entangled state. Int. J. Theor. Phys. **50**, 2703–2707 (2011)

# On the Deployment Quality for Multi-intrusion Detection in Wireless Sensor Networks

Noureddine Assad, Brahim Elbhiri, My Ahmed Faqihi, Mohamed Ouadou and Driss Aboutajdine

**Abstract** The intrusion detection in a Wireless Sensor Network is defined as a mechanism to monitor and detect any intruder in a sensing area. The sensor deployment quality is a critical issue since it reflects the cost and detection capability of a wireless sensor network. When the random deployment is required, which sensor nodes are uniformly randomly distributed over on surface area, determining the deployment quality becomes challenging. In the intrusion detection application, it is necessary to define more precise measures of sensing range and node density that impact overall system performance. To enhance the detection quality for single/multi intrusion, a probabilistic intrusion detection models are adopted, called single and multi sensing probability detection and the deployment quality issue is surveyed and analysed in term of coverage.

**Keywords** Intrusion detection probability · Network coverage · Sensing rang · Node density

N. Assad (✉) · M. Ouadou · D. Aboutajdine
LRIT, Associated Unit to CNRST (URAC 29), FSR, Mohammed V University,
10000 Rabat, Morocco
e-mail: assad.noureddine@gmail.com

M. Ouadou
e-mail: ouadou@fsr.ac.ma

D. Aboutajdine
e-mail: aboutaj@fsr.ac.ma

B. Elbhiri
EMSI, 10014 Rabat, Morocco
e-mail: elbhirij@yahoo.fr

M.A. Faqihi
ENSIAS, Mohammed V University, BP 713 Rabat, Morocco
e-mail: faqihi@ensias.ma

© Springer International Publishing Switzerland 2016
A. El Oualkadi et al. (eds.), *Proceedings of the Mediterranean Conference on Information & Communication Technologies 2015*, Lecture Notes in Electrical Engineering 381, DOI 10.1007/978-3-319-30298-0_48

# 1  Introduction

Recent advances in technology have made possible to develop small low-cost devices that integrate sensors, called sensor nodes, with limited on-board processing and wireless communication capabilities. Sensor nodes may be deployed in large numbers to form a wireless sensor network (WSN) that can be used in many applications. Each sensor senses a field of interest and communicates the collected data to the sink, where the end user can access them. This chapter focuses on WSN surveillance applications like the detection of unauthorized/unusual moving intruders, which requires to characterize some WSN parameters that does not exist in traditional sensor networks. Each field of interest point must be within the sensing range of at least one sensor. The WSN must be able to adapt to changing network because an intruder may be detected by single or multi sensor nodes, that is modelled by single or multi sensing detection.

In literature, most works are targeted at particular applications, but the main idea is still centred on coverage. [1, 2] showed how well an area is monitored by sensor nodes, while [3, 4] derived analytical expressions to enhance the deployment quality, because the network coverage concept is a measure of the quality of service. The deployment quality analysis for multi intrusion detection in WSNs is motivated by the reasons that Random network topology, all sensor nodes are randomly deployed in a sensing area, this implies the efficient deployment of the required coverage. Specifically, given a monitoring region, how we can guarantee that each point of the region is covered by at least one sensor node? In other words, we need to recognize which regions are covered by enough sensor nodes. So, we can enhance the probability of intrusion detection and an intruder not exceed the threshold distance.

The major contributions of this paper can be summarized as follows: a probabilistic approach is developed by deriving some analytical expressions to characterize the topological properties of sensor network, and analyse the intrusion detection model in a wireless sensor network.

The remainder of the paper is organized as follows: In Sect. 2, intrusion detection models and some basic graph theory are presented. Then, multi-intrusion detection model in a homogeneous wireless sensor network is reviewed in Sect. 3. Our simulation results and their analysis are presented in Sect. 4. Finally, we conclude the paper in Sect. 5.

# 2  Intrusion Detection Models and Some Basic Graph Theory

In this section, we define some terms and notations used in theory of graph. Probabilistic intrusion detection models are adopted, and deployment quality issue is surveyed and analysed in terms of network coverage.

## 2.1 Preliminaries and Models

**Definition 1** (*Sensing range*) The sensing range of a node $N_i$ is a disk of radius $R_{SENS}$, centred at $\xi_i$ and defined by:

$$Disk(N_i) = \{\xi_j \in \mathbb{R}^2 : |\xi_i - \xi_j| \leq R_{SENS}\},$$

where $|\xi_i - \xi_j|$ stands for the Euclidean distance between $\xi_i$ and $\xi_j$.

**Definition 2** (*Collaborating sensors*) Consider two nodes $N_i$ and $N_j$ located at $\xi_i$ and $\xi_j$ respectively. Let us note $d_{ij}$ the distance between $N_i$ and $N_j$. The collaborating set of $N_i$ and $N_j$ is defined as the union between $S_{N_i}$ and $S_{N_j}$. Besides, $N_i$ and $N_j$ are said to be collaborating if and only if $d_{ij} = |\xi_i - \xi_j| \leq 2R_{SENS}$. In general, the collaborating sensor set $N_i$ is:

$$S_{col}(N_i) - \bigcup_{\{N_j : |\xi_i - \zeta_j| \leq 2R_{SENS}\}} S_{Nj}.$$

## 2.2 Network Topology and Coverage Degree

Throughout this paper, we consider $N$ sensor nodes randomly distributed over a square region $A$ of edge length $L$ in set $\mathbb{R}^2$ following a uniform distribution, which form a wireless sensor network. We assume that every sensor node has the same sensing range $R_{SENS}$. The region $A$ is said to be covered if every point is at distance at most $R_{SENS}$ from at least one sensor node. We define any convex region $A$ of $\mathbb{R}^2$ with a coverage degree $k$, where every point of the considered region is covered by at least $k$ sensor nodes [5]. Given the surface area $S$ and specified by the application (either before or after deployment), what would be the required value of the sensing range $R_{SENS}$ to achieve a specified coverage degree k, $k > 0$?

Practically, a network with a higher degree of coverage can achieve higher sensing accuracy and be more robust against sensing failure. Therefore, it is important to place or select the effective number of sensor nodes to cover the same monitored area as much as possible, without diminishing the overall field coverage and specified coverage degree $k$, $k > 0$. We can express the efficient number of nodes which can be deployed to cover the sensing area by node availability rate as a variable in our intrusion detection probability model.

## 2.3  Sensing Model Probability

Based on network topology as discussed in previous subsection, we study the sensing model probability that is generally closely coupled with the specific sensor application and the type of sensor device [6]. We adopt a sensing model probability to guarantee the detection of any events happened in the sensing area.

For a uniformly distributed sensor network with node density $\lambda$, we denote the number of sensor nodes covering the surface area $S_{area}$ by $N$. For each sensor node, the probability of that sensor node within the sensing range $R_{SENS}$ distance of a point from sensing area is a Bernoulli trial, where the probability of success is:

$$p = \frac{\pi R_{SENS}^2}{S_{area}}.$$

Hence, the number of sensor nodes within distance $R_{SENS}$ of a point forms a Binomial distribution. Moreover, for large $N$ and small $p$, the Binomial distribution can be represented by a Poisson process. Then, the mean value of the equivalent Poisson process is:

$$Np = \frac{N\pi R_{SENS}^2}{S_{area}}.$$

The sensing model probability that exactly $k$ sensors detects an intruder follows a Poisson distribution is given by the next formula:

$$P_{sensing}(nbrs = k) = \frac{(S\lambda)^k}{k!} e^{(-S\lambda)}.$$

where $\lambda$ is the node density and $S$ is the area swept by intruder following a trajectory $l$, we assume that the intruder starts from a random point in the sensing area and moves in a random fashion as illustrated in Fig. 1. Hence, the previous probability can further be represented as:

**Fig. 1** A simple intrusion scenario. An intruder starts from a random point in the sensing area and moves in a random fashion, and sweeps the surface area $S$ following a trajectory $l$

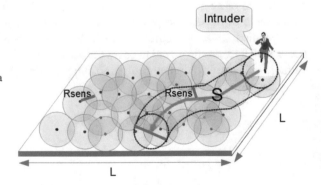

$$P_{sensing}(nbrs = k) = \frac{(S\lambda)^k}{k!} e^{-(S\lambda)}. \tag{1}$$

$$S = 2R_{SENS}l + \pi R_{SENS}^2 \; and \; \lambda = \frac{N}{S_{area}}.$$

# 3   Multi-intrusion Detection Model in a Homogeneous Wireless Sensor Network

Intrusion detection in a WSN is defined as a mechanism to detect unauthorized intrusions. For deterministic deployment of nodes, the quality of deterministic deployment can be determined sufficiently by analysis before the deployment. However, when random deployment is required, it becomes challenging [7]. For this purpose, it is a fundamental issue to characterize WSN parameters like node density and sensing range. In this section, probabilistic multi-intrusion detection models are adopted, and deployment quality issue is surveyed and analysed in terms of network coverage.

## 3.1   Single-Sensing for Multi-intrusion Detection

The probability that no sensor node in a sensing area can detect an intrusion is $\overline{P} = e^{-\lambda S}$. The complement of $\overline{P}$ is the probability that there is at least one sensor node which detects an intrusion, it is determined by:

$$P_{sensing}(nbrs \geq 1) = 1 - \overline{P} = 1 - e^{-\lambda S}.$$

In the multi-intrusion, the probability that $m$ intruders immediately detected is:

$$P_{sensing}(nbrs \geq 1) = \prod_{j=1}^{m}(1 - e^{-\lambda S_j}). \tag{2}$$

According to the intrusion scenario given in the previous subsection, The probability that an intruder does not exceed the threshold distance $l_{THR}$ is:

$$P_{sensing}(nbrs \geq 1, 0 \leq l < l_{THR}) = \prod_{j=1}^{m}(1 - e^{-\lambda(2R_{SENS}l_{THR} + \pi R_{SENS}^2)}).$$
$$P_{sensing}(nbrs \geq 1, 0 \leq l < l_{THR}) = (1 - e^{-\lambda(2R_{SENS}l_{THR} + \pi R_{SENS}^2)})^m. \tag{3}$$

**Theorem 1** (Multi-intrusion detection probability with distance threshold $l_{THR} = 0$ in homogeneous WSNs) *We consider N nodes randomly distributed over a square region A of edge length L in set $\mathbb{R}^2$ following a uniform distribution. If we want to be sure, with a probability of at least $P_{threshold\_sensing}$, that m intruders can be immediately detected, we have:*

$$P_{sensing}(nbrs \geq 1, l_{THR} = 0) \geq P_{threshold\_sensing}.$$

*We can set the sensing range off all node to:*

$$R_{sens} \geq \sqrt{\frac{-\ln(1 - (P_{threshold\_sensing})^{\frac{1}{m}})}{\lambda \pi}}. \tag{4}$$

## 3.2   k-Sensing for Multi-intrusion Detection

In a WSN, the number of required sensor nodes depends on the coverage quality. An area may require that multiple sensor nodes monitor each of its points. This constraint is known as k-coverage, where $k$ is the number of nodes that watch each point. To achieve a coverage degree $k$, the k-sensing detection for multi-intrusion detection model is:

$$P_{sensing}(nbrs \geq k) = \prod_{j=1}^{m}(1 - \sum_{i=0}^{k-1}\frac{(\lambda S_j)^i}{i!}e^{-\lambda S_j}). \tag{5}$$

We derive the probability $P$ that an intruder can be detected within the threshold intrusion distance $l_{THR}$:

$$P_{sensing}(nbrs \geq k, 0 \leq l < l_{THR}) = (1 - \sum_{i=0}^{k-1}\frac{(\lambda S)^i}{i!}e^{-\lambda S})^m. \tag{6}$$

$$S = 2R_{SENS}l_{THR} + \pi R_{SENS}^2 \quad and \quad \lambda = \frac{N}{S_{area}}.$$

## 3.3   Node Availability for Network Coverage

In a dense network, sensing areas of different nodes may be similar to their neighbour nodes, so nodes will transmit redundant information and WSN total energy consumption will increase. Thus, it is important to select the effective sensor number covering the same monitored area without diminishing the overall field coverage. We can efficiently reduce the energy consumption in the WSN, when node power can be put on/off periodically. It is appropriate to take into account the node availability rate $p$ in our analysis. Each sensor can decide whether to become active with probability $p$ or to move to the sleep mode with probability $1 - p$. So, the sensing model probability that exactly $k$ sensor nodes detect an intruder is:

$$P(nbrs = k) = \frac{(\lambda pS)^k}{k!} e^{(-\lambda pS)}. \tag{7}$$

In the multi-intrusion, the probabilities that $m$ intruders can be detected without exceeding a threshold distance $l_{THR}$ in an homogeneous WSN, with node density $\lambda$, sensing range $R_{SENS}$ and node availability $p$, in the single and the k-sensing are:

$$P_{sensing}(nbrs \geq 1, 0 \leq l < l_{THR}) = (1 - e^{-\lambda pS})^m. \tag{8}$$

$$P_{sensing}(nbrs \geq k, 0 \leq l < l_{THR}) = (1 - \sum_{i=0}^{k-1} \frac{(\lambda pS)^i}{i!} e^{-\lambda pS})^m. \tag{9}$$

$$S = 2R_{SENS}l_{THR} + \pi R_{SENS}^2 \quad and \quad \lambda = \frac{N}{S_{area}}.$$

While increasing the sensing range and the number of nodes per unit, the intrusion detection probability in single-sensing and k-sensing increase, all events that happen in the sensing network will be covered. According to the formulas discussed above, the optimal values of sensing range and node density can be determined in advance to guarantee high quality of multi-intrusion detection.

# 4 Discussions and Results

In this section, we investigate our probabilistic model for multi-intrusion in single/multi sensing detection by varying the sensing range, node density, node availability, and intrusion distance. We define by analytical result the sufficient condition to guarantee the network coverage. We consider a random homogeneous WSN composed of static sensors which are independent and distributed uniformly in a square area A.

The results illustrated in Fig. 2 show the intrusion detection probability $P$, that there is at least one sensor node which detects an intruder (single intrusion) in surface area A i.e., $A = 100 \times 100 \, m^2$ versus $A = 200 \times 200 \, m^2$, it is determined by the node number $N$ and the sensing range $R_{SENS}$. Intrusion detection may need a large sensing range or a high node number, thus increasing the WSN deployment cost. We can note that if we increase the node number $N$ or the sensing range $R_{SENS}$, the probability to cover an intrusion happens in the network increases too. This is

**Fig. 2** Intrusion detection
probability that there is at
least one node which detects
an intruder, as a function of
node number $N$ and sensing
range $R_{SENS}$ in surface area A:
$A = 100 \times 100\,\text{m}^2$ versus
$A = 200 \times 200\,\text{m}^2$

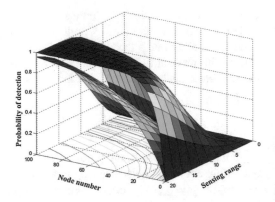

because the increase of sensing range or node number significantly enhances the
network coverage. However, increasing more $N$ or $R_{SENS}$ the probability attend 1
and remains constant, will not affect the robustness of detection. Consequently, for
a given value of sensing range $R_{SENS}$, we can find the optimal node number, which
can be deployed to cover efficiently the controlled region. This node number and
sensing range will be the optimal values, which must be used to totally cover an
area of interest.

The Fig. 3 shows detection probability in k-sensing detection as a function of
intrusion distance. The detection probability increases with the increase of the
intrusion distance. At the same time, the single-sensing detection probability
($k = 1$) is higher than that of multi-sensing detection ($k = 2$ and $k = 3$). This is
because the multi-sensing detection imposes a more strict requirement on detecting
an intruder in the sensing network, at least $k = 2$ ($k = 3$) sensors are required. We
plot in Fig. 3 the intrusion detection probability curves as a function of the intrusion
distance, for different values of node availability rate $p$. It is obvious that if the
intrusion distance $l$ increases, the detection probability $P$ increases too. In the
normal cycle, the node availability $p$ is usually less than 1.0 it is considered to be
satisfied to monitored an area as much as possible without diminishing the overall
field coverage.

**Fig. 3** Single versus multi
sensing detection probability
for different values of node
availability

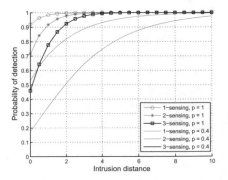

**Fig. 4** Single versus
k-sensing detection
probability in multi-intrusion
($m = 1, 2, 3$)

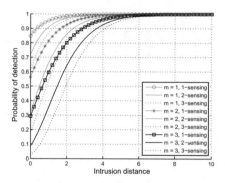

The analytical result of multi-intrusion detection is illustrated in Fig. 4, it is obvious that if the number of intruders increases, the detection probability either single-sensing or k-sensing decrease. This is because the multi-intrusion ($m = 2$ and $m = 3$) imposes a more strict requirement on intrusion detection in sensing network.

## 5 Conclusion

In this work, we developed a probabilistic approach by deriving analytical expressions to characterize the topological properties of network coverage, we analyse the intrusion detection probability in a homogeneous WSN. We investigate our model for multi-intrusion detection in WSN to single-sensing and multi-sensing detection. Our analytical results enable us to design and analyse the homogeneous WSN, and help us to select the critical parameters of network in order to meet the detection quality requirement.

## References

1. Mulligan, R., Ammari, H.M.: Coverage in wireless sensor networks: a survey. Netw. Protoc. Algor. **2**(2), 27–53 (2010)
2. Ravelomanana, V.: Extremal properties of three-dimensional sensor networks with applications. IEEE Trans. Mob. Comput. **3**(3), 246–257 (2004)
3. Onur, E., Ersoy, C., Deliç, H.: How many sensors for an acceptable breach detection probability? Comput. Commun. **29**(2), 173–182 (2006)
4. Gui, C., Mohapatra, P.: Power conservation and quality of surveillance in target tracking sensor networks. In: Proceedings of the 10th Annual International Conference on Mobile Computing and Networking, pp. 129–143. ACM (2004)
5. Kumar, S., Lai, T.H., Balogh, J.: On k-coverage in a mostly sleeping sensor network. In: Proceedings of the 10th Annual International Conference on Mobile Computing and Networking, pp. 144–158. ACM (2004)

6. Chong, C.Y., Kumar, S.P.: Sensor networks: evolution, opportunities, and challenges. Proc. IEEE **91**(8), 1247–1256 (2003)
7. Huang, C.F., Tseng, Y.C.: The coverage problem in a wireless sensor network. Mob. Netw. Appl. **10**(4), 519–528 (2005)

# Implementation an Intelligent Architecture of Intrusion Detection System for MANETs

Sara Chadli, Mohammed Saber, Mohamed Emharraf
and Abdelhak Ziyyat

**Abstract** Mobile Ad Hoc network (MANET) is a self-configuring network composed of mobile devices (PC, laptop, PDA). Data exchange between these devices is done without any dedicated infrastructure, or central component. However, the hostile environment in which they can be deployed, and their dynamic nature make them more vulnerable to attacks, As a result, there is a strong need to for security solutions that protect these types of network. Several Intrusion detection systems (IDS) models have been proposed as a possible solution. However, these IDSs do not tolerate or consider the presence of malicious or faulty nodes. This work presents our new architecture of IDS which is distributed and featuring a cooperation system based on a multi-agent system (SMA). It presents also aspects of practical implementation of agent interaction mechanisms in JADE platform.

**Keywords** Intrusion detection system (IDS) · Mobile ad hoc networks (MANETs) · Multi-agent system (SMA) · JADE

S. Chadli (✉) · A. Ziyyat
Laboratory Electronics and Systems, Faculty of Sciences,
First Mohammed University, Oujda, Morocco
e-mail: chad.saraa@gmail.com

A. Ziyyat
e-mail: abdelhak_ziyyat@hotmail.com
URL: http://www.ump.ma

M. Saber · M. Emharraf
Laboratory LSE2I, National School of Applied Sciences,
First Mohammed University, Oujda, Morocco
e-mail: mosaber@gmail.com

M. Emharraf
e-mail: m.emharraf@gmail.com

© Springer International Publishing Switzerland 2016                         479
A. El Oualkadi et al. (eds.), *Proceedings of the Mediterranean Conference
on Information & Communication Technologies 2015*, Lecture Notes
in Electrical Engineering 381, DOI 10.1007/978-3-319-30298-0_49

# 1 Introduction

Mobile ad hoc network is a distributed network infrastructure consists of several autonomous entities able to communicate them randomly without the existence of a centralized infrastructure. Here, the mobile nodes directly communicate with other nodes without a router and thus the desired features are embedded at each node. Indeed, the mobile nature of mobile ad hoc networks and the lack of access points, make them susceptible to many security problems [1–3], these features make security solutions developed for wired or wireless networks with infrastructure inapplicable in the context of mobile ad hoc networks. Several security solutions for mobile ad hoc networks (MANET) are proposed, but these solutions focus on security of routing protocols such as Ariadne protocols or SRP (Secure Routing Protocol) [4] and the ARAN protocol (Authenticated Routing for Ad Hoc Networks) [5]. However, in MANETs, security mechanisms need a trust model in order to dynamically evaluate the confidence level nodes and ensure network security. This is why an intrusion detection system (IDS) is an effective mechanism to identify when an attack occurs in a MANET. Since the MANET is characterized by its limited resources, it implies many constrains compared to a traditional computer network. Many studies have taken place to propose a new architecture of IDS for MANETs. The existing IDS architectures for MANETs fall under three basic categories [6, 7] (a) stand-alone, (b) cooperative, and (c) hierarchical architecture. In this paper, we have proposed a distributed detection architectures that works by the multi-agents system (SMA), because these architectures are more efficient and can solve the problems of centralized architectures. The paper is organized as follows. Section 2 describes our proposed architecture and reports its implementation. Section 3 presents the results of the experiment. In Sect. 4, conclusion and future work are presented.

# 2 Implementation Our Architecture in JADE Platform

This paper proposes a distributed architecture, based on the proposal [8] in which agents perform the task of detection by communication and collaboration. Our model of security and intrusion detection based on a distributed approach using multi-agent system for receiving intelligence of these agents. It is formed by agents with the capacity to react quickly reactive against different types of attacks. To test the operation of our architecture, the proposed agents have been implemented using the JADE platform. To facility the operation of our proposed model of multi-agent system we are used design based on the UML for describing the interaction between different agents of our architecture (Fig. 1).

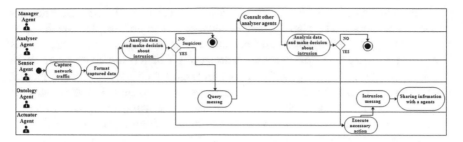

**Fig. 1** Diagram of agents interaction

## 2.1 Sensor Agent (SA)

The sensor agent is the initiator of the detection process. It captures network traffic and records it in a file in the following format: the *Jpcap* method *Packet.toString()* is called and the result string is added to the date and time of the reception of the package, each line of the file representing a captured packet. Then it creates an analyzer agent using the JADE command *createNewAgent* to communicate the formatted results to this agent.

## 2.2 Analyzer Agent (ANA)

The analyzer agent analysis the captured packets and extracts source addresses and destination addresses, source and destination ports, date and time of capturing. It stores the number of packets that have been analyzed. To compare the analysis results we propose to use a trust model associated with detection policies, it is an open and extensible model [9]. The scalability of the model means that new policies can be introduced into the system. This ensures that the model is adaptable to several scenarios. The main step in our detection system is the search for evidence of suspicious behavior that may represent an attack. The method used stochastic detects anomalies by analyzing the distribution of detection parameters with the specified thresholds.

For each target service, the number of packets is counted (in a time period). Two predefined thresholds (*TH_min* and *TH_max*) are used to determine the nature of the traffic. If the number of observed packets is less than *TH_min* then the traffic is normal, in this case the agent will temporarily save this threshold for a case of collaboration with another initiator. Otherwise if the number exceeds *TH_max* then the traffic is abnormal (i.e. malicious activity). The third case is when the number of packets is greater than and less than *TH_min TH_max*, in this case the traffic is deemed suspicious (possibly malicious). Therefore, additional analyzes are required to decide on the type of traffic, so it will be necessary to work with other agents to get more information on the same service.

In case of detection of a number of packets exceeding *TH_max* the analyzer creates an actuator agent, passing as arguments packet characteristics which led to the detection. After the creation of the actuator agent, ontology agent is created, this last will send as arguments its purpose ("intrusion"), which is to include an attack to ontology to make sharing of information. In case of detection of a number of packets than *TH_min* and *TH_max* less than the analyzer agent brand representation as suspicious packets and creates an agent ontology, passing as argument its purpose ("query"), which is to question the other agents about suspicious packages.

After the comparison, the analyzer agent checks if it has received a message about suspects packets. According to receiving a message from a ontology agent we have two cases: if this last said that a suspicion was confirmed, it extracts the characteristics and uses them as parameters to create an actuator agent. If the ontology agent said that it was not possible to confirm a suspicion, it extracts the characteristics of the suspicion and uses them as parameters to create a manager agent. According to receiving information from a other agents of the network generated in another nodes if the information in the message corresponds to a malicious activity is confirmed, it creates a response message with the content we have an intrusion and creates an actuator agent. Then as before an ontology agent is created, this last will send as arguments its purpose ("intrusion"), which is to include an attack to ontology to make sharing of information.

## 2.3   Manager Agent (MA)

To help a analyser agents by requesting to them additional information's about a suspicious packets, it sends a request to have more information's about all active containers. Thus, it uses the JADE command *doMove()* to move container by container. When migrating to a new host, it sends an ACL message to the others analyzer agents informing the characteristics of the suspicion that was passed as an argument in its creation, asking if it is present in its list of suspects. If the others analysers agents show that the suspicion was confirmed, they create an ACL message for the initiator analyzer agent, passing information about the protocol, source and destination addresses and ports which were considered as intrusion.

## 2.4   Actuator Agent (ACA)

The actuator agent extracts the information that has been adopted in its creation and adds additional information like the date of receiving information, the date of execution of the action and time when the alert is generated, save the information in a file.

## 2.5  Ontology Agent (OA)

An ontology agent and checks the arguments passed in its creation by moving to the main container by calling the JADE command *doMove()*. According to its purpose, if the purpose is query the agents goal is to ask the manager agent to help an analyser agent about suspicious packets. If the purpose is intrusion the agents goal is to make a sharing of information about the confirmed intrusion. It sends an ACL message to agents of the network, to inform them about the intrusion. the message content is two instances of the node class (attacker and target) that identify the protocols and network addresses, and two strings that identify the ports numbers.

# 3  Experiment Design and Results

## 3.1  Experiment Design

In order to evaluate our architecture we simulated a MANET. Our experiment model is a network of 3 nodes (Fig. 2a), Node1 and Node2 for the implementation of our architecture, the third node for traffic generator by D-ITG (Distributed Internet Traffic Generator). D-ITG [10] is a platform capable to produce traffic at packet level accurately replicating appropriate stochastic processes for both IDT (Inter Departure Time) and PS (Packet Size) random variables (exponential, uniform, cauchy, normal, pareto, ...). D-ITG supports both IPv4 and IPv6 traffic generation and it is capable to generate traffic at network, transport, and application layer. Secondly, we are we did a study to estimate trust levels TH_min and TH_max in order to configure the agents. Thirdly, we are implement our agents in Jade platform in two nodes, Node1 and Node2 (Figs. 2b and 3).

## 3.2  Experiment Results

The attacks were launched on tow nodes aimed to see if our architecture works well and the interaction between the agents is perfectly.

**Fig. 2  a** Experiment MANET environment; **b** JADE for our architecture

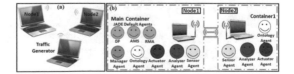

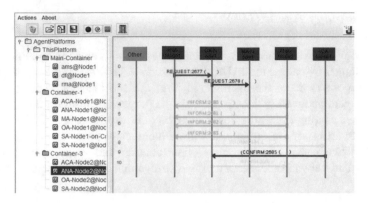

**Fig. 3** The communication between agents in our architecure

During our test we applied many scenarios of attacks, in which the traffic generator was used to send a packet of varying sizes that contain data and other attacks.

We applied several scenario using the confidence level in order to have all three cases (normal, intrusion, suspect ion) after the packets analysis. Table 2 show the results of our experiments. The following aspects in Table 1 were measured in the experiments.

### 3.2.1 Scenario 1

In the first sequence, attacks were made with a number of packets that is greater than *TH_min* and less than *TH_max* (case of suspicious packages). The first scenario is as follows: *Node1* not detect any attack, *Node2* detected the attack through the manager agents. These results are occurred as expected, because *Node1* has detected suspicious activity and requested ontology agents and manager agent for other information. Thus, the agent manager was called, migrated to *Node2* and it also found no information on the attack. That is why Node1 did not detect the attack. *Node2* detects suspicious activity, because when it called the ontology and the manager agents, this latter when migrating to Node1, received information that

**Table 1** Description of packet type

| Packet type | Description |
|---|---|
| Packets tested (**PT**) | The number of packets sent by the traffic generator node |
| Packets received (**PR**) | The number of packets received by Sensor Agent |
| Packets analysed (**PA**) | The number of packets analysed from the total packets Received by Analyser Agent |
| Packets detected (**PD**) | The number of packets detected |

**Table 2** The results of three scenarios

| Node | Scenario 1 | | | | Scenario 2 | | | | Scenario 3 | | | |
|------|----|----|----|----|----|----|----|----|-----|-----|-----|-----|
|      | PT | PR | PA | PD | PT | PR | PA | PD | PT  | PR  | PA  | PD  |
| Node1 | 20 | 20 | 20 | 0 | 20 | 20 | 20 | 20 | 100 | 100 | 100 | 100 |
|       |    |    |    |   | 30 | 30 | 30 | 0  | 30  | 30  | 30  | 30  |
| Node2 | 20 | 20 | 20 | 20 | 20 | 20 | 20 | 20 | 30  | 30  | 30  | 30  |

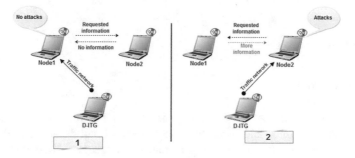

**Fig. 4** Platform of scenario 1

the suspect packages are intrusive, and confirming to an attack. As results the analyser agent calling the actuator agent that generated an alert. Table 2 show the results of our experiment (Fig. 4).

### 3.2.2 Scenario 2

In the second scenario, we generate the same traffic at the first scenario, only at the end we performed another attack against the *Node1* flooding a number of packets greater than TH_min and less than TH_max. As Results, we found exactly as expected: *Node2* generates a single alert and *Node1* has generates one alerts, because this last, when it send request to a manager agent to help it about the traffic, it had no additional information (Figs. 5 and 6).

**Fig. 5** Platform of scenario 2

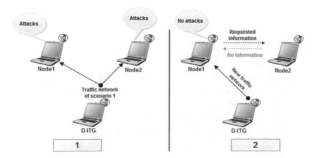

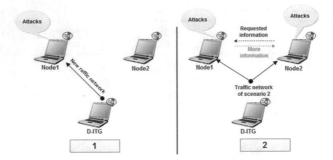

**Fig. 6** Platform of scenario 3

### 3.2.3 Scenario 3

In the third scenario, D-ITG program was to flooding a new traffic to *node1*
exceeding the TH_max, and flooding same traffic as second part of scenario 2 to
*node2* and *Node1*. As Results, we found exactly as expected: *Node1* has generates
two alerts, *Node2* generates a single alert because this last, when it send request to a
manager agent to help it about the traffic, it had additional information from an
analyser agent of *Node1*.

## 4 Conclusion

We propose an IDS based on multi-agent system. To evaluate our solution, we are
implemented it in JADE platform. The test results confirmed that our proposed
architecture has worked well, because nodes detect all attacks flooding by
Distributed Internet Traffic Generator (D-ITG), thanks to the collaboration between
the agents of our solution.

As future work, we figured out improved algorithms for different agents and
especially improve the analyzer agent algorithm with more parameters to characterize the attacks, in order to implement the architecture tests with real network
traffics as data set.

## References

1. Djenouri, D., Khelladi, L., Badache, A.N.: A survey of security issues in mobile ad hoc and
   sensor networks. Commun. Surv. Tutor. IEEE **7**(4), 2, 28, Fourth Quarter (2005). doi:10.1109/
   COMST.2005.1593277
2. Amiri, E., Afshar, E., Naji, H.R., Ardekani, M.M.: Survey on network access control technology
   in MANETs. In: 2012 International Conference on Innovation Management and Technology
   Research (ICIMTR), pp. 367, 372, 21–22 May 2012. doi:10.1109/ICIMTR.2012.6236420

3. Chadli, S., Saber, M., Ziyyat, A.: Defining Categories to Select Representative Attack Test-Cases in MANETs. IEEE Xplore Digital Library (2014), pp. 658, 663. doi:10.1109/CSNT.2014.138
4. Hubaux, J.-P., Buttyn, L., Capkun, S.: The quest for security in mobile ad hoc networks. In: Proceedings of the 2nd ACM International Symposium on Mobile Ad Hoc Networking and Computing, 04–05 Oct 2001, Long Beach, CA, USA. doi:10.1145/501436.501437
5. Sanzgiri, K., Dahill, B., Levine, B.N., Shields, C., Belding-Royer, E.M.: A secure protocol for ad hoc networks. In: Proceedings of ICNP, pp. 78–87 (2002)
6. Anantvalee, T., Wu, J.: A survey on intrusion detection in mobile ad hoc networks, wireless/mobile network security Springer (2006) Chapter 7, p. 170196. Wireless Network Security. doi:10.1007/978-0-387-33112-6_7
7. Xenakis, Christos, Panos, Christoforos, Stavrakakis, Ioannis: A comparative evaluation of intrusion detection architectures for mobile ad hoc networks. Comput. Secur. 30(1), 63–80 (2011). doi:10.1016/j.cose.2010.10.008
8. Chadli, S., Emharraf, M., Saber, M., Ziyyat, A.: Combination of Hierarchical and Cooperative Models of an IDS for MANETs, pp. 230–236. IEEE Computer Society (2014). doi:10.1109/SITIS.2014.32
9. Zaidi, A., Kenaza, T., Agoulmine, N.: IDS adaptation for an efficient detection in high-speed networks. In: 2010 Fifth International Conference on Internet Monitoring and Protection (ICIMP), pp. 11, 15, 9–13 May 2010. doi:10.1109/ICIMP.2010.10
10. Botta, A., Dainotti, A., Pescap, A.: A tool for the generation of realistic network workload for emerging networking scenarios. Comput. Netw. 56(15), 3531–3547 (2012). ISSN 1389-1286. doi:10.1016/j.comnet.2012.02.019

# Intrusion Detection System Using PCA and Kernel PCA Methods

Zyad Elkhadir, Khalid Chougdali and Mohammed Benattou

**Abstract** The network traffic data used to build an intrusion detection system is frequently enormous with important useless information which decreases IDS's efficiency. In order to overcome this problem, we have to reduce as much as possible this meaningless information from the original high dimensional data. In this paper, we compare the performance of two features reduction techniques, the first one is Principal Component Analysis (PCA), and the second is Kernel Principal Component Analysis (KPCA). After the step of dimension reduction, samples are classified using k nearest neighbor (K-NN) algorithm in order to determine whether the test data are normal or anomalous corresponding to attacks against computer networks. Experimental results on KDDcup99 intrusion detection dataset shows that KPCA with the quadratic kernel performs better than many other types of kernels. Furthermore, KPCA outperforms PCA mainly when we vary the number of nearest neighbors from one to four. Finally, we have also noted that KPCA with the quadratic kernel overcomes PCA in detecting denial of service (DOS) and probing attacks.

**Keywords** Network security · Intrusion detection system (IDS) · PCA · KPCA

## 1 Introduction

The security of a computer network is compromised when an intrusion takes place. An Intrusion Detection System (IDS) is an important mechanism that attempts to identify any set of actions or malicious activities which can compromise network security policy.

Z. Elkhadir · M. Benattou
RLCST Research Laboratory, Ibn Tofail University, Kenitra, Morocco

K. Chougdali (✉)
National School of Applied Sciences (ENSA), Ibn Tofail University, Kenitra, Morocco
e-mail: chougdali@gmail.com

© Springer International Publishing Switzerland 2016
A. El Oualkadi et al. (eds.), *Proceedings of the Mediterranean Conference on Information & Communication Technologies 2015*, Lecture Notes in Electrical Engineering 381, DOI 10.1007/978-3-319-30298-0_50

Practically, there are two principal intrusion detection techniques: misuse detection and anomaly detection. Misuse detection recognizes a suspicious behavior by comparing it to a specific attack signature that has been already stored in a database of attacks signatures; unfortunately it can't detect new attacks. Examples of IDS using misuse detection techniques are STAT [1] and Snort [2]. On the other side, anomaly detection defines normal behavior as a model, and tries to check any deviation from the model and thus decides to generate or not the corresponding alert. Anomaly detection was originally introduced by Anderson [3] and Denning [4] and then implemented in some IDS like IDES [5] or EMERALD [6].

Many methods have been developed for anomaly based IDS, such as machine learning, data mining, neural networks and statistical methods. All of them have been applied directly on the rough high dimensional data without any dimension reduction technique. It can be considered as one of the principal factors contributing in IDS inefficiency.

The main idea behind our proposed work is to reduce original features of database connection records by extracting its relevant information. A simple technique to extract the information contained in network connection records is to capture the variance in a collection of connection records. The given information is used to classify these network connection records as normal or attack connection.

Mathematically speaking, we want to find the principal components of the connection records of a dataset. It can be seen as binary TCP/IP network connections records corresponding to a normal connection or to a specified attack. To do this, the approach extracts the relevant information using the eigenvectors of the covariance matrix of all connection records [7]. These eigenvectors can be defined as a set of features used to reduce the variation between record connections. Indeed, each connection is expressed using only the eigenvectors with the largest eigenvalues given by the most variance within the set of connection records. The new space generated is constructed using the Principal Component Analysis (PCA) [8] which has proven to be efficient in intrusion detection [9, 10] and in many application domains including data compression, image analysis, visualization and pattern recognition.

In this paper, we have used two reduction techniques PCA and KPCA (Kernel Principal Component Analysis) which extracts principal components by adopting a non-linear kernel method [11]. The simulation results show that a KPCA with quadratic kernel gives a higher detection rate than PCA when the number of nearest neighbors varies between one and four, particularly for detecting DOS and PROBE attacks.

This paper is organized as follows: Sect. 2 is dedicated to presents briefly the two dimensionality reduction methods PCA and KPCA. Section 3 describes and discusses the obtained results, and Sect. 4 gives the concluding remarks and outlines our future works.

## 2   PCA and Kernel PCA

In this section, we present a modeling concepts and theoretical analysis of PCA and KPCA.

### 2.1   PCA

Principal component analysis (PCA) is a mathematical technique that transforms a number of correlated variables into a number of uncorrelated variables called principal components (PCs). Generally, the number of these principal components is less than or equal to the number of original variables. The main goal of principal component analysis is to reduce dimensionality (number of variables) of the initial dataset, while retaining as much as possible the variance present in this dataset. This is achieved by taking only the first few PCs, sorted in decreasing order, so that they retain most of the variance present in all of the original variables [8].

Suppose we have a training set of M vectors $\omega_1, \omega_2, \ldots, \omega_M$ each vector contain n features. To $n'(n' \ll n)$ get principal components of the training set the procedure is based on the following steps:

1. Compute the average $\sigma$ of this set:

$$\sigma = \left(\frac{1}{M}\right) \sum_{i=1}^{M} \omega_i. \tag{1}$$

2. Subtract the mean $\sigma$ from $\omega_i$ and get $\rho_i$:

$$\rho_i = \omega_i - \sigma. \tag{2}$$

3. Compute the covariance matrix $C$ where:

$$C_{(n \times n)} = \left(\frac{1}{M}\right) \sum_{i=1}^{M} \rho_i \rho_i^T = AA^T. \tag{3}$$

$$A_{(n \times M)} = \left(\frac{1}{\sqrt{M}}\right) \rho_i. \tag{4}$$

4. Let $U_k$ be the $k$th eigenvector of $C$ corresponding to the $\lambda_k$ associated eigenvalue and $U_{(n \times n')} = [U_1 \ldots U_{n'}]$ the matrix of these eigenvectors, so we have:

$$CU_k = \lambda_k U_k. \tag{5}$$

5. Sort the eigenvalues (and the corresponding eigenvectors) in decreasing order and choose the first eigenvectors, those eigenvectors are principal components ($PC_i$). Practically, the number of the principal components chosen depends on the precision we wish to have. The inertia ratio defined by these axes is given by:

$$\tau = \frac{\sum_{i=1}^{n'} \lambda_i}{\sum_{i=1}^{n} \lambda_i}. \tag{6}$$

This ratio defines the information rate kept, from the whole rough input data, by the corresponding $n$ eigenvalues. Finally, the projection of a new column vector sample $x_{new}$ on the space constructed by principal components can be obtained by:

$$t_i = PC_i^T x_{new}. \tag{7}$$

## 2.2 Kernel PCA

PCA allows only a linear dimensionality reduction. However, if the data has more complicated structures, which cannot be well represented in a linear subspace, standard PCA will not be very helpful. Fortunately, kernel PCA (KPCA) allows us to generalize PCA to nonlinear dimensionality reduction. This can be done by a nonlinear mapping function $\varphi$, that transform all samples input into a higher-dimensional feature space F as follows:

$$\varphi : \omega \in R^n \rightarrow \varphi(\omega_i) \in F$$

where $\varphi(\omega_i)$ is sample of F and $\sum_{i=1}^{M} \varphi(\omega_i) = 0$. The mapping of $\omega_i$ is simply noted as $\varphi(\omega_i) = \varphi_i$ and the covariance matrix of this sample in the feature space F can be constructed by:

$$C = \left(\frac{1}{M}\right) \sum_{i=1}^{M} (\varphi_i - mean)(\varphi_i - mean)^T \tag{8}$$

where $mean = \sum_{i=1}^{M} \left(\frac{\varphi_i}{M}\right)$. The covariance matrix $C$ can be diagonalized with nonnegative eigenvalues $\lambda$ satisfying:

$$Cv = \lambda_i v. \tag{9}$$

It's easy to see that every eigenvector $v$ of $C$ can be linearly expanded by:

$$v = \sum_{i=1}^{M} (\alpha_i \varphi_i). \tag{10}$$

To obtain the coefficients $\alpha_i$, a kernel matrix $K$ with size $M \times M$ is defined and its elements are determined as follows:

$$K_{ij} = \varphi_i^T \varphi_j = \varphi_i \cdot \varphi_j = k(\omega_i, \omega_j) \tag{11}$$

where $k(\omega_i, \omega_j) = <\varphi_i, \varphi_j>$ is the inner-product of two vectors in F. If the projected dataset $\{\varphi(\omega_i)\}$ does not have zero mean, we can use the Gram matrix $K'$ to substitute the kernel matrix $K$ using:

$$K' = K - 1_M K - K 1_M + 1_M K 1_M \tag{12}$$

with, $1_M = (1/M)_{M \times M}$.

In order to solve the eigenvalue problem in (9), we can reformulate this equation as [11]:

$$K'\alpha = M\lambda\alpha. \tag{13}$$

Let column vectors $\alpha_i$ be the orthonormal eigenvectors of $K'$ corresponding to the $p$ largest positive eigenvalues $\lambda_1 \geq \lambda_2 \geq \cdots \geq \lambda_p$, hence the orthonormal eigenvectors $v_i$ of $C$ can be expressed as:

$$v_i = \left( \frac{1}{\sqrt{\lambda_i}} \varphi_i \alpha_i \right). \tag{14}$$

For a new column vector sample $x_{new}$, the mapping to the feature space F is $\varphi(x_{new})$ and then the projection of $x_{new}$ onto eigenvectors $v_i$ is:

$$t = (v_1, v_2, \ldots, v_p)^T \varphi(x_{new}). \tag{15}$$

The $i$th KPCA transformed feature $t_i$ can be obtained by:

$$t_i = v_i^T \varphi(x_{new}) = \left( \frac{1}{\sqrt{\lambda_i}} \right) \alpha_i^T k(\omega_i, x_{new}). \tag{16}$$

It can be noted that the kernel matrix can directly constructed from the training dataset. The commonly used kernels are:

- Gaussian kernel

$$k(x, y) = e^{\left(-\frac{\|x-y\|^2}{2 \times sigma^2}\right)} \tag{17}$$

- Polynomial kernel

$$k(x, y) = (x^T y + 1)^d \quad where \ d \in \mathbb{N} \tag{18}$$

- Power kernel

$$k(x, y) = \|x - y\|^d \quad where \ d \geq 1 \tag{19}$$

- Rational Power kernel

$$k(x, y) = \|x - y\|^d \quad where \ 0 < d < 1. \tag{20}$$

## 3   Experiments and Discussion

This section is dedicated to present and evaluate the results obtained when applying the two dimensionality reduction methods PCA and KPCA with K-NN. In our simulation experiments, we have chosen from the 10 % of KDDcup99 [12] training dataset, a subset composed of 1000 normal data, 100 DOS data, 50 U2R data, 100 R2L data, and 100 PROBE data. Likewise, the test dataset is composed of 100 normal data, 100 DOS data, 50 U2R data, and 100 R2L data, all randomly selected. To evaluate the performance of the proposed system we have used detection rate (DR) defined as follows:

$$DR = \left(\frac{TP}{TP + FN}\right) * 100 \tag{21}$$

where TP are true positives (intrusions correctly classified), FN are false negatives (intrusions wrongly classified), FP are false positive (normal instances wrongly classified), and TN are true negatives (normal instances correctly classified).

After applying PCA on training dataset, we obtained principal components (PC) then we have computed the projection of the test dataset on different number of (PC).

Figure 1 shows that, only the first three principal components give a highest detection rate (%) with inertia ratio $\tau > 0.99$ (Eq. 6).

After that we have fixed the number of PCs at 3 and tried to find the optimal detection rate for every type of attacks (DOS, U2R, R2L, and PROBE). Table 1 shows this manipulation. It's clear that, the two categories of attacks DOS and

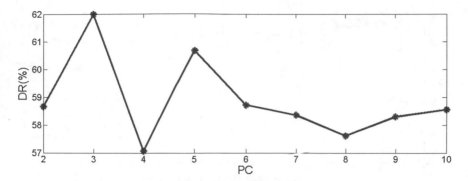

**Fig. 1** Detection rate (%) versus number of principal components (PC)

| **Table 1** Attack's detection rate for PCA | DOS | U2R | R2L | PROBE |
|---|---|---|---|---|
| | 95,13 (%) | 8,13 (%) | 3,5 (%) | 69,8 (%) |

PROBE are detected with a rate of 95,13 % for DOS and 69,8 % for PROBE. In the other hand, U2R and R2L are not well detected with only 8,13 % for U2R and 3,5 % for R2L.

In the second part of our experiments, we have evaluated the effectiveness of Kernel PCA in intrusion detection system. Firstly, we implement four kernels, described by Eqs. (17)–(20). Secondly, we try to pick up the maximum detection rate varying the different kernels parameters. Indeed, to achieve this goal we set the number of principal components at three, and then we have changed kernel's parameters.

A next experience seeks to identify the best kernel for KPCA. For this reason, in Fig. 2 we have compared detection rates obtained by using the four kernels with the adequate parameters. We can observe that the power kernel gives higher detection

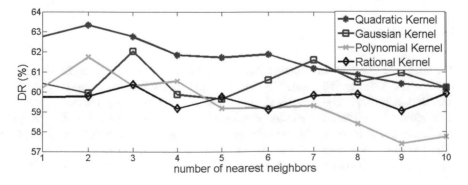

**Fig. 2** Performance comparison of different kernels

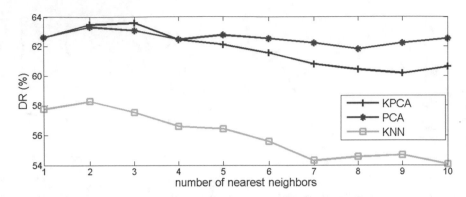

**Fig. 3** Detection rate (%) of KPCA, PCA and KNN versus number of nearest neighbors

| Table 2 Attack's detection rate for KPCA | DOS | U2R | R2L | PROBE |
|---|---|---|---|---|
| | 96,13 (%) | 9.93 (%) | 3,46 (%) | 73,8 (%) |

rate than the others kernels especially when K nearest neighbors is less than seven. Hereafter, we will call the power kernel with d = 2 as a quadratic kernel.

The results of comparison described in Fig. 3 of PCA and KPCA with quadratic kernel, show that KPCA outperforms PCA when number of nearest neighbors is between 1 and 4.

Finally, we visualize detection rate for every type of attacks using kernel PCA in Table 2. Indeed, the two categories of attacks DOS and PROBE are well detected with a rate of 96.13 % for DOS and 73.8 % for PROBE attacks. Furthermore, we can conclude that these detection rates are better than those founded with PCA (95.13 % and 69.8 %). In the other hand, U2R and R2L attacks are not well detected with only 9.93 % for U2R (slightly better than PCA which gives 8.13 %) and just 3.46 % for R2L.

# 4 Conclusion

The main idea behind the work presented in this paper is to reduce the original features that represent all connection records stored in a dataset for the purpose of intrusion detection. The proposed work shows, how we can extract relevant information using PCA and KPCA in order to build a robust network IDS with the maximum detection rate. The experimental results show that a KPCA with quadratic kernel gives a higher detection rate than PCA when the number of nearest neighbors varies between one and four, particularly for detecting DOS and PROBE attacks. Our future works will be oriented towards advanced dimension reduction techniques in order to improve the performance of an IDS particularly for the CPU time consuming.

# References

1. Kumar, S., Spafford, E.H.: A Software architecture to support misuse intrusion detection. In: Proceedings of the 18th National Information Security Conference, pp. 194–204 (1995)
2. Beale, J.: Snort 2.1 Intrusion Detection. Syngress (2004)
3. Anderson, J. P.: Computer Security Threat Monitoring and Surveillance. Technical report, James. P. Anderson Co., Fort Washington, Pennsylvania (1980)
4. Denning, D.: An intrusion detection model. IEEE Trans. Software Eng. **13**(2), 222–232 (1987)
5. Lunt, T., Tamaru, A., Gilham, F.: A Real-time Intrusion Detection Expert System (IDES) Final Technical Report. Computer Science Laboratory. SRI International, Menlo Park, California (1992)
6. Porras, P.A., Neumann, P.G., EMERALD: Event monitoring enabling responses to anomalous live disturbances. In: Proceedings of National Information Systems Security Conference, Baltimore MD (1997)
7. Jolliffe, I.T.: Principal Component Analysis. Springer, New York, NY (2002)
8. Kirby, M., Sirovich, L.: Application of the Karhunen Loeve procedure for the characterization of human faces. IEEE Trans. Pattern Anal. Mach. Intell. 103–107 (1990)
9. Bouzida, Y., Cuppens, F., Cuppens-Boulahia, N., Gombault, S.: Efficient intrusion detection using principal component analysis. In: 3éme Conférence sur la Sécurité et Architectures Réseaux (SAR), La Londe, France (2004)
10. Hashem, S.H.: Efficiency of Svm and Pca to enhance intrusion detection system. J. Asian Sci. Res. 381–395 (2013)
11. Scholkopf, B., Smola, A., Muller., K.R.: Nonlinear component analysis as a kernel eigenvalue problem. Neural Comput. 1299–1319 (1998)
12. KDD 99 Task. Available at: http://kdd.ics.uci.edu/databases/kddcup99/task.html (1999)

# Predicting System Failures on Mobile Devices

Stanislav Marcek and Martin Drozda

**Abstract** It is rare that mobile applications run without any failures over an extended period of time. Such failures can result in user data loss and/or user dissatisfaction. Herein we apply machine learning approaches to predict when a system failure is likely, so that any necessary steps including application rejuvenation can be applied in order to improve user experience. This study is based on the Device Analyzer data set that is a result of a multi-year project run by the University of Cambridge. We investigate several classification algorithms to predict failures. In order to increase their performance we apply several feature selection algorithms. We also investigate whether cascading classification is an effective approach for failure detection.

**Keywords** Device Analyzer data set · Machine learning · System failure detection

## 1 Introduction

Our goal is to detect abnormal behavior of mobile devices. Abnormal behavior is a state, when mobile device is either unresponsive or its behavior is unpredictable. Such behavior can have severe implications such as data loss leading to user/customer dissatisfaction. Herein we investigate whether it is possible to predict OS freezing, which is one of the leading causes of abnormal behavior. OS freezing can result in the necessity to re-start mobile device and can cause irreversible damage. In some severe cases it might be necessary to take out and re-insert device battery, which may require expert skills.

S. Marcek (✉) · M. Drozda
Faculty of Electrical Engineering and Information Technology, Slovak University of
Technology, Ilkovičova 3, 812 19 Bratislava, Slovakia
e-mail: stanislav.marcek@stuba.sk

M. Drozda
e-mail: martin.drozda@stuba.sk

© Springer International Publishing Switzerland 2016
A. El Oualkadi et al. (eds.), *Proceedings of the Mediterranean Conference
on Information & Communication Technologies 2015*, Lecture Notes
in Electrical Engineering 381, DOI 10.1007/978-3-319-30298-0_51

Mobile devices can be interpreted as event systems. Events are user actions such as initiating a phone call, starting an application, receiving a text message, changing a base station, locking a GPS position etc. To model event driven systems we can use Petri nets. Petri net is a modeling approach for representing distributed systems and process analysis. Petri nets thus also allow for detecting and evaluating deadlocks. User perceives deadlock as a state when application is unresponsive to user actions. This can be accompanied with data loss.

Our evaluation is based on a massive data set compiled by the University of Cambridge within their Device Analyzer project [10]. Thousands of users installed their logger application that recorded a large number of events and states in mobile devices. This resulted in a terabyte data set that can be obtained upon agreeing to licensing requirements.

Given the data set size and the limitations of current Petri net model synthesis approaches, we decided to apply standard machine learning approaches (including feature selection) to OS freezing detection. Petri net approaches will be investigated in the future. Our goal is to identify features that could predict such abnormal behavior and that allow for high detection accuracy.

The impact of our results can be viewed in terms of software aging and rejuvenation, when applications with a high failure probability get re-initiated in order to protect user data and improve the over-all user experience. In Sect. 2 we discuss related work on software rejuvenation. In Sect. 3 we describe the Device Analyzer data set. In Sect. 4 we present our evaluation approach. Finally, in Sect. 5 we present our results, which are followed by conclusions.

## 2   Related Work

Software aging and rejuvenation is a topic widely analyzed since 1990s at AT&T Bell Laboratories [5].

Controneo et al. [2] survey software aging and rejuvenation as well as discuss numerous open issues. Software aging is described as the state of software which leads to a component or an entire system degradation, and can result in a system crash. Software aging degrades the system by accumulation of needed resources and failures that cannot be correctly restored. Possible causes are for example erroneous physical memory management leading to memory leaks, unterminated CPU threads or unreleased file locks. Controneo et al. [2] divide approaches of rejuvenation analysis to model-based, measurement-based and hybrid. Our approach based on a log data set falls within measurement-based approaches.

Software rejuvenation is a cost-effective solution to counteract software aging that concerns servers, virtual machines, communication devices etc. Rejuvenation focus is on reliability, availability and serviceability of software in order to obtain the best possible performance. Controneo et al. apply statistical approaches to forecast software aging and plan rejuvenation.

Huang et al. [5] implement a module for UNIX system that periodically and preemptively re-starts an application. The disadvantage of this action is increased downtime of the mentioned client-server application. A re-start depends on time, program purpose and several performance measures.

Software rejuvenation has been also applied in managing virtual machines and servers. Garg et al. [3] describe an approach based on Markov Regenerative Stochastic Petri Nets for client-server applications. In this work, the authors present efforts to minimize downtime of server for clients requests. Their concern is both on software fault as well as on software aging, because not all executed code is non-intrusive.

Guo et al. [4] introduce rejuvenation based on regenerating applications that need to be re-booted due to software aging. They show that partial rejuvenation is better for stability in dependence to downtime cost. Kim et al. [6] propose a software rejuvenation framework for WSNs (Wireless Sensor Networks). They investigate an approach for repairing the results of intrusions, attacks, accidents and failures. Their methodology is based on semi-Markov processes and on the discrete-time Markov chains.

# 3 Device Analyzer Data Set

## 3.1 Data Set Description

The data set is described on the project web site [10]. The complete data set log has about one terabyte and contains of 17,103 devices with operating system Android 2.1 or above.

We focused on the 172 file logs from the most active users. The biggest log contains over 10 millions rows and the smallest over 1.1 mil. rows. The files store data in csv format, each line corresponds to a single update (log) of system.

## 3.2 Improper System Shutdown

We are looking for early warnings, which could indicate that system is going to shut down. Each application variously depletes battery energy. As improper shutdown we consider battery depletion, hardware shutdown or battery removal that results to device shutdown. We are looking for different sequences of events, that could have the above effect. The causes can be for example faulty application, virus or device misuse.

We are looking for attributes in logs that indicate improper shutdown. Since there is no explicit *crash* attribute, that would hold information on failures, we consider system crash to be a sequence of *startup* or *time|bootup* not followed by an explicit user initiated *shutdown*. These border conditions enclose events that we investigate in feature selection and crash prediction.

In addition to battery level information, we also consider CPU usage, system settings, application parameters and information on available memory. We also consider attributes that might be important due to secondary effects such as the impact of GSM signal strength on battery level. Given our focus, we do not evaluate information related to networking, sensor, ring tone and similar events.

## 3.3   Information Extraction—Attribute Description

We transform log events to states in conformity with the last known state before new boot-up of device. We considered the following attributes divided into several groups. The attributes with prefix $A$ include the class *app*, which contains information about currently running processes (note: the $A\_i$ attributes below are sorted by priority of importance. Each process has one of the levels):

- $A\_iF$: the number of processes in foreground level,
- $A\_iV$: the number of processes in visible level,
- $A\_iS$: the number of processes in service level,
- $A\_iB$: the number of processes in background level,
- $A\_iE$: the number of processes in empty level,
- $A\_ii$: the number of processes in priority level other than five above mentioned,
- $A\_num$: the number of processes that are logged in the last known state,
- $A\_priv$: the sum of total private dirty memory usage of processes,
- $A\_pss$: the sum of total proportional set size of memory usage of processes,
- $A\_share$: the sum of total shared dirty memory usage of processes,
- $A\_sMax$: the maximum elapsed time of running processes,
- $A\_sMin$: the minimum elapsed time of running processes.

The attributes with prefix $B$ capture information focused on battery:

- $B\_health$: the nominal attribute that corresponds to status of battery health,
- $B\_lvl$: the battery level from range [0;1],
- $B\_status$: the nominal value reflecting whether battery is charging, discharging or full,
- $B\_temp$: the temperature of battery.

The attributes with prefix $C$ capture static system information about device:

- $C\_api$: the highest api version on the device,
- $C\_name$: the name/type of CPU,
- $C\_num$: the number of available cores in CPU on the device.

The attributes with prefix $S$ are attributes with storage information:

- $S\_eMem$: the free memory on external storage (in percentage),

- *S_iMem*: the free memory on internal storage (in percentage),
- *S_lMem*: low internal memory indicator,
- *S_tMem*: the system information about threshold for free main memory below which OS starts terminating processes,
- *S_fMem*: the amount of free main memory available to the JVM (java virtual machine).

The attributes corresponding to phone GSM signal information:

- *GSM_ASU*: the signal strength information in ASU level,
- *GSM_srvS*: the service state of telephony subsystem with four nominal values (inservice, emergencyonly, outofservice, poweroff).

The grouped attributes of other system information:

- *cpu*: the information about time the first core CPU has spent in different sleep state,
- *display*: the nominal value that holds the last state of display. So-called screen power status with four states,
- *uptime*: the time elapsed from the last known log, the time in milliseconds between two boot-ups,
- *shutdown*: device powering down indicator.

Summed up, including the ID attribute, we get a data set with 31 columns (attributes) and 38,462 rows. This data set is much shorter than the original data set, since we removed information unrelated to failure detection.


# 4 Evaluation

We validate any applied machine learning algorithm (MLA) with standard 9-fold cross-validation with stratified sampling. To compute and visualize our results, we use Rapidminer software [9] in version 5.3.013 with Weka machine learning library extension.We apply the following MLAs:

- RBFN (Radial Basis Function Network),
- JRip proposed by Cohen in [11], a rule induction MLA,
- DTNB (Decision Table/Naive Bayes), a hybrid classifier applying rule induction and Naive Bayes,
- NaB (Naive Bayes),
- NBT (Naive Bayes Tree), a tree induction MLA; this is a classifier for generating a decision tree with naive Bayes classifiers at the leaves,
- J48, an improvement of the classic C4.5 decision tree algorithm,

- E2LSH (Exact Euclidean Locality Sensitive Hashing [1]), which is an approximate nearest neighbor algorithm. The E2LSH algorithm allows only for numeric attributes, therefore the number of attributes is increased to 54 as a result of feature space transformation [7].

Except for E2LSH, all other MLAs are implemented in Rapidminer. We also apply cascading classification discussed in [8]. This approach showed promising results when classifying the KDD99 network intrusion data set. It is based on several classifiers in cascade, where a simpler classifier gets applied before a more complex classifier (in terms of learning complexity). We use the following notation for two consecutive classifiers $K_1 \odot K_2$, where $K_1$ classifier precedes $K_2$ classifiers.

We evaluate detection rate in terms of accuracy $acc$, precision $prec$ and false positive rate $FPR$:

$$acc = \frac{TP+TN}{TP+FP+FN+TN}, \quad prec = \frac{TP}{TP+FP}, \quad FPR = \frac{FP}{FP+TN},$$

where TP is the number of true positives, i.e. correctly classified shutdown sequences, TN is the number of true negatives. FP is the number of false positives i.e. incorrectly predicted shutdown sequences and FN is the number of false negative.

## 5 Results

In the case when single classifier gets applied, the best result is obtained with NBT algorithm with almost 83 % accuracy and FPR about 27 %. J48 delivers comparable results with respect to standard deviation. The best FPR is obtained with the E2LSH algorithm giving 80 % accuracy but only about 25 % FPR; see Table 1. We also investigate performance when several standard feature selection algorithms get applied. The highest accuracy is achieved with J48 with DTNB and NBT providing comparable results with respect to standard deviation.

Since the E2LSH algorithm can also return zero neighbors (zero neighbors within a predefined radius), classification is only done by the next classifier in cascade. A similar situation can arise when the number of neighbors equals two and each belongs to a different class. In all other cases, classification is done according to the two nearest neighbors. We applied several radii r = {0.1, 0.3, 0.9, 2.5}.

Table 2 shows the results when E2LSH and other classifiers are applied in cascade. The best results are obtained with Weka J48 with 83.75 % accuracy and 25.28 % FPR. For comparison, the third and fourth column shows the result when mRMR and correlation weights feature selection gets applied, respectively. We can

**Table 1** Performance of different MLAs with/without FS algorithms

|  |  | acc [%] | prec [%] | TP | FP | FN | TN |
|---|---|---|---|---|---|---|---|
| All attributes | RBFN | 69.36 ± 0.61 | 71.94 ± 0.46 | 19071 | 7441 | 4344 | 7606 |
|  | JRip | 81.09 ± 0.57 | 80.94 ± 0.56 | **21118** | 4976 | **2297** | 10071 |
|  | DTNB | 79.17 ± 0.67 | 79.95 ± 0.78 | 20563 | 5159 | 2852 | 9888 |
|  | NaB | 63.01 ± 0.91 | 75.01 ± 0.59 | 13778 | 4591 | 9637 | 10456 |
|  | NBT | **82.67 ± 0.39** | 83.63 ± 0.46 | 20829 | 4079 | 2586 | 10968 |
|  | J48 | 82.19 ± 0.79 | **84.01 ± 0.74** | 20461 | 3897 | 2954 | 11150 |
|  | E2LSH r = 2.5 | 80.24 ± 0.65 | 83.59 ± 0.58 | 19675 | **3862** | 3740 | **11185** |
| Fwd. FS | RBFN | 76.16 ± 0.58 | 75.52 ± 0.67 | 21087 | 6840 | 2328 | 8207 |
|  | JRip | 81.39 ± 0.47 | 81.55 ± 0.55 | 21010 | 4754 | 2405 | 10293 |
|  | DTNB | 82.27 ± 0.64 | 82.44 ± 0.79 | 21091 | 4496 | 2324 | 10551 |
|  | NaB | 76.09 ± 0.57 | 77.27 ± 0.48 | 20144 | 5927 | 3271 | 9120 |
|  | NBT | 83.14 ± 0.49 | 83.41 ± 0.62 | **21135** | 4205 | **2280** | 10842 |
|  | J48 | **83.75 ± 0.41** | **84.65 ± 0.52** | 20970 | **3805** | 2445 | **11242** |
| Correlation weight FS | RBFN | 73.55 ± 0.99 | 73.57 ± 1.03 | 20676 | 7434 | 2739 | 7613 |
|  | JRip | 77.25 ± 0.72 | 77.62 ± 0.73 | 20617 | 5951 | 2798 | 9096 |
|  | DTNB | 79.04 ± 0.49 | 79.20 ± 0.51 | 20821 | 5468 | 2594 | 9579 |
|  | NaB | 73.25 ± 0.62 | 75.35 ± 0.32 | 19510 | 6385 | 3905 | 8662 |
|  | NBT | 77.59 ± 0.72 | 77.12 ± 0.60 | **21041** | 6244 | **2374** | 8803 |
|  | J48 | **79.44 ± 0.61** | **79.95 ± 0.75** | 20702 | **5195** | 2713 | **9852** |
| mRMR FS | RBFN | 70.65 ± 0.35 | 79.85 ± 1.00 | 16230 | 4104 | 7185 | 10943 |
|  | JRip | 78.44 ± 0.51 | 80.05 ± 0.89 | 20151 | 5028 | 3264 | 10019 |
|  | DTNB | 78.78 + 0.56 | 81.41 ± 0.81 | 19773 | 4520 | 3642 | 10527 |
|  | NaB | 66.74 ± 0.39 | 78.63 ± 0.52 | 14588 | **3966** | 8827 | **11081** |
|  | NBT | 76.30 ± 2.06 | **81.48 ± 0.71** | 18502 | 4204 | 4913 | 10843 |
|  | J48 | **80.59 ± 1.02** | 80.73 ± 0.93 | **20952** | 5003 | **2463** | 10044 |

observe that cascading classification gives comparable results. Figure 1 shows the feature selection progress for each MLA. It can be observed that the performance of several MLAs could be negatively impacted if feature selection does not get applied. RBFN and Naive Bayes require a small number of attributes, further increase of attributes degrades their ability to learn. DTNB degrades when forward feature selection is applied, however it occurs at a later stage. Other MLAs after the 12th iteration show only a negligible increase or decrease of accuracy performance.

As a result of feature selection, the following three attributes, namely *display*, *GSM_srvS*, *B_lvl* remain relevant for all MLAs.

**Table 2** Failure detection accuracy with standard deviation of E2LSH with different MLAs as a second classifier

| MLAs | | all attributes acc [%] | Fwd. Sel. acc [%] | mRMR Sel. acc [%] | corr. weights acc [%] |
|---|---|---|---|---|---|
| E2LSH r = 0.1 | ⊙ RBFN | 73.92 ± 0.45 | 78.51 ± 0.54 | 76.90 ± 0.44 | 77.15 ± 0.79 |
| E2LSH-2 r = 0.1 | | 73.53 ± 0.46 | 78.67 ± 0.57 | 75.92 ± 0.50 | 77.24 ± 0.83 |
| E2LSH-2 r = 2.5 | | 81.57 ± 0.51 | 82.41 ± 0.49 | 81.38 ± 0.77 | 82.19 ± 0.53 |
| E2LSH r = 0.1 | ⊙ JRip | 80.88 ± 0.51 | 80.94 ± 0.58 | 79.98 ± 0.48 | 78.90 ± 0.67 |
| E2LSH-2 r = 0.1 | | 81.36 ± 0.51 | 81.43 ± 0.58 | 80.14 ± 0.43 | 79.20 ± 0.86 |
| E2LSH-2 r = 2.5 | | 83.18 ± 0.54 | 83.19 ± 0.67 | 82.73 ± 0.64 | 82.71 ± 0.64 |
| E2LSH r = 0.1 | ⊙ DTNB | 79.86 ± 0.62 | 82.04 ± 0.71 | 81.10 ± 0.70 | 80.01 ± 0.57 |
| E2LSH-2 r = 0.1 | | 80.10 ± 0.70 | 82.46 ± 0.70 | 81.11 ± 0.75 | 80.35 ± 0.66 |
| E2LSH-2 r = 2.5 | | 82.91 ± 0.59 | 83.41 ± 0.67 | 82.85 ± 0.65 | 82.99 ± 0.61 |
| E2LSH r = 0.1 | ⊙ NaB | 73.09 ± 0.88 | 78.12 ± 0.51 | 75.74 ± 0.30 | 76.59 ± 0.68 |
| E2LSH-2 r = 0.1 | | 71.28 ± 1.10 | 78.16 ± 0.56 | 74.18 ± 0.48 | 76.38 ± 0.80 |
| E2LSH-2 r = 2.5 | | 80.66 ± 0.52 | 82.31 ± 0.63 | 80.98 ± 0.62 | 82.13 ± 0.61 |
| E2LSH r = 0.1 | ⊙ NBT | 81.61 ± 0.67 | 82.37 ± 0.52 | 80.52 ± 0.65 | 80.05 ± 0.59 |
| E2LSH-2 r = 0.1 | | 82.19 ± 0.70 | 82.92 ± 0.49 | 80.30 ± 0.73 | 80.41 ± 0.69 |
| E2LSH-2 r = 2.5 | | **83.36 ± 0.64** | 83.50 ± 0.66 | 82.29 ± 0.66 | 82.89 ± 0.51 |
| E2LSH r = 0.1 | ⊙ J48 | 81.86 ± 0.44 | 82.66 ± 0.39 | 81.36 ± 0.62 | 80.01 ± 0.46 |
| E2LSH-2 r = 0.1 | | 82.34 ± 0.41 | 83.31 ± 0.37 | 81.72 ± 0.63 | 80.41 ± 0.56 |
| E2LSH-2 r = 2.5 | | 83.25 ± 0.55 | **83.78 ± 0.42** | **83.09 ± 0.70** | **83.07 ± 0.56** |

**Fig. 1** Accuracy and precision (*dotted*) for each iteration of forward feature selection

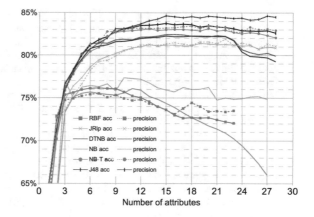

## 6 Conclusion

We present our preliminary results on how to efficiently predict system failures. As discussed, such failures can result in user data loss as well as user dissatisfaction. Our results show that about 84 % accuracy can be achieved. We applied a multitude of classification approaches including cascading classification. We also applied feature selection in order to decrease the number of applied feature and thus possibly increase classification performance. Our results indicate that failure detection approaches that do not take event causality into consideration might be insufficient for our purpose. Missing information on causality is a clear disadvantage of the Device Analyzer data set, however, this data set is the most comprehensive data set available.

**Acknowledgment** This publication was created with the support of the OP Research and Development for project: *Establishment, Development and Scientific Management of a Research Center for the Analysis and Protection of Data*, ITMS: 26240120037, co-funded by the EU.

## References

1. Andoni, A., Indyk, P.: Near-optimal hashing algorithms for approximate nearest neighbor in high dimensions. In: Proceedings of 47th Annual IEEE Symposium on Foundations of Computer Science (FOCS), pp. 459–468 (2006)
2. Cotroneo, D., Natella, R., Pietrantuono, R., Russo, S.: A Survey of Software Aging and Rejuvenation Studies. J. Emerg. Technol. Comput. Syst. 8:1–8:34 (2014)
3. Garg, S., Puliafito, A., Telek, M., Trivedi, K.: Analysis of software rejuvenation using markov regenerative stochastic petri net. In: Sixth International Symposium on Software Reliability Engineering. pp. 180–187. IEEE (Oct 1995)
4. Guo, J., Li, W., Song, X., Zhang, B., Wang, Y.: Software rejuvenation strategy based on components. In: Second World Congress on Software Engineering (WCSE). vol. 2, pp. 80–83. IEEE (2010)

5. Huang, Y., Kintala, C., Kolettis, N., Fulton, N.: Software rejuvenation: Analysis, module and applications. In: Twenty-Fifth International Symposium on Fault-Tolerant Computing (FTCS-25), pp. 381–390 (June 1995)
6. Kim, D., Shazzad, K., Park, J.: A framework of survivability model for wireless sensor network. In: The First International Conference on Availability, Reliability and Security (ARES 2006), pp. 8–pp. IEEE (2006)
7. Marcek, S., Drozda, M.: Network intrusion detection with cascading classification. In: Proceedings of 5th International Conference on Intelligent Systems, Modelling, and Simulation (ISMS) (2014)
8. Marcek, S., Drozda, M., Juhas, G., Lehocki, F.: Network intrusion detection in high dimensional space. In: Proceedings of 2nd International Symposium on Applied Sciences in Biomedical and Communication Technologies (ISABEL), pp. 1–7 (2009)
9. Rapidminer: Rapidminer tool. Website (November 2013), http://www.rapidminer.com
10. Wagner, D., Rice, A., Beresford, A.: Device Analyzer: Large-scale Mobile Data Collection. SIGMETRICS Perform. Eval. Rev. 41(4), 53–56 (april 2014)
11. William, C. et al.: Fast effective rule induction. In: Twelfth International Conference on Machine Learning. pp. 115–123 (1995)

# Performance Evaluation of an Intrusion Detection System

Mohammed Saber, Sara Chadli, Mohamed Emharraf
and Ilhame El Farissi

**Abstract** Intrusions detection systems (IDSs) are systems that try to detect attacks as they occur or after the attacks take place. IDSs collect network traffic information from some point on the network or computer system and then use this information to secure the network. Research in intrusion detection systems aims to reduce the impact of these attacks. In the recent years, research in intrusion detection systems aims to reduce the impact of attacks, and to evaluate the system. The evaluation of an IDS is a difficult task. We can make the difference between evaluating the effectiveness of an entire system and characteristics of the system components. In this sheet of paper, we present an approach for IDS evaluating based on measurement of its components performance. In this context, we have proposed a hardware platform based on embedded systems for the implementation of an IDS (SNORT) components. After, we tested a system for generating traffics and attacks based on Linux KALI (Backtrack) and Metasploite 3 Framework. The obtained results show the IDS performance is linked to the characteristics of these components. The obtained results show that the performance characteristics of an IDS depends on the performance of its components.

M. Saber (✉) · M. Emharraf · I.E. Farissi
Laboratory LSE2I, National School of Applied Sciences,
First Mohammed University, Oujda, Morocco
e-mail: mosaber@gmail.com
URL: http://wwwensa.ump.ma

M. Emharraf
e-mail: m.emharraf@gmail.com

I.E. Farissi
e-mail: ilhame.elfarissi@gmail.com

S. Chadli
Laboratory Electronics and Systems, Faculty of Sciences,
First Mohammed University, Oujda, Morocco
e-mail: chad.saraa@gmail.com

© Springer International Publishing Switzerland 2016                    509
A. El Oualkadi et al. (eds.), *Proceedings of the Mediterranean Conference
on Information & Communication Technologies 2015*, Lecture Notes
in Electrical Engineering 381, DOI 10.1007/978-3-319-30298-0_52

**Keywords** Evaluation · Intrusion detection system (IDS) · Network security · Performance · Embedded system · Field-Programmable Gate Array (FPGA) · SNORT · Traffic generator

# 1    Introduction

The evaluation of the intrusion detection systems is a difficult task, demanding a thorough knowledge of techniques relating to different disciplines, especially intrusion detection, methods of attack, networks and systems, technical testing and evaluation [1]. What makes the evaluation more difficult is the fact that assorted intrusion detection systems have different operational environments and can use a variety of techniques for producing alerts corresponding to attacks. The task is even more difficult as the IDS must be evaluated not only under normal conditions. However, especially in a malicious environment, taking particular account of unexpected and sometimes even unknown patterns of use (this is true of almost all the tools dedicated to security such as firewalls, IPS (Intrusion Prevention Systems) and anti-viruses. All these considerations make it difficult to build representative data for evaluation.

Therefore, normally before beginning any experimental test, it is extremely important to identify clearly the objectives of the evaluation. First, it is important to distinguish between the tests of evaluating the entire system effectiveness, and the whole characteristics of IDS [2], which are interested in testing components of IDS. In this case, we estimate the performance of these components, which allows subsequently the evaluating of its characteristics.

In practice, most IDS suffer from several problems, taking into consideration the large number of false positives and false negatives, and the evolution of attacks. All these obstacles increase the need for establishing an evaluation system of the IDS. In this context many attempts to assessment took place [3–10]. They are based on the classification of attacks, which aims to classify the attacks to simplify detection, either by technology or detection range, or by the generation of attack scenarios to understand the behavior of attacks and by other criteria. Whereas, the great weakness of these assessments is that they do not cover all the characteristics of an IDS as cited in [2].

In this sheet, we propose a new approach EIDS (Embedded Intrusion Detection System) for evaluating IDS mounted on a hardware platform based on embedded systems to test the performance of the various components of an IDS, which can allow later to evaluate its characteristics. For this we set up an EIDS platform based on a comprehensive LYRTECH SFF SDR, and we set up the IDS SNORT in this platform.

In the remaining sections of this article. We present the proposed approach to evaluate performance IDS, and we describe the implementation of SNORT with our platform in Sect. 2. In Sect. 3, we present the experiment to evaluate performance of IDS components. In Sect. 4, we present the results and discussions of our evaluation. In Sect. 5, we cease our document with a conclusion.

## 2 The Proposed Approach to Evaluate Performance IDS

### 2.1 SNORT Software Architecture

The SNORT Architecture, presented in Fig. 1a, consists of four phases: Sniffer Component, Preprocessor Component (14 predefined plugins in Fig. 1b), Detection Engine Component, Alert/Logging Component.

### 2.2 Presentation of Our Prototype

In this section, we describe the components of our prototype through the implementation of the IDS platform based on an embedded system (Embedded Intrusion Detection System). Our platform EIDS uses an advanced system, based onto the caracteristics of (TMS ARM). The architecture of our hardware platform is shown in Fig. 2b. It uses LYRTECH SFF SDR in Fig. 2a.

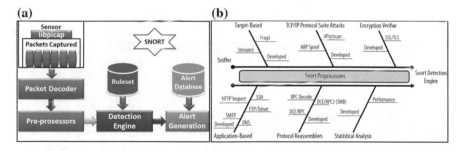

**Fig. 1 a** SNORT basic software components; **b** Snort Preprocessor plugin categories

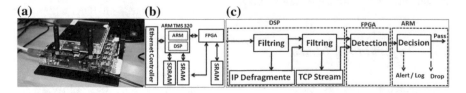

**Fig. 2 a** Platform hard LYRTECH SFF SDR; **b** EIDS based on a ARM TMS; **c** EIDS pipeline detection

## 2.3 Implementation of Components SNORT on Our Platform

We briefly describe implementation of the main components of SNORT Fig. 2c. The first operation, the capture and the filtering are based on Libpcap (SNORT uses the libpcap packet capture library, to access network packets). The second operation, the packets are transmitted to the decoder to process their headers, are realized by the DSP, we have developed its running by C (Embedded C). The third operation, each packet is then passed through a series of preprocessors, including IP fragmentation and reassembly of TCP and UDP flows. The last operation, the packets are controlled by the detection engine (ARM) (Fig. 2c).

Deep packet inspection forms the backbone of any Network Intrusion Detection (NID) system [11]. It involves matching known malicious patterns against the incoming traffic payload. Pattern matching in software is prohibitively slow in comparison to current network speeds. Due to the high complexity of matching, only FPGA (Field-Programmable Gate Array) platforms can provide efficient solutions. FPGAs facilitate target architecture specialization due to their field programmability. Our FPGA-based solution performs high-speed pattern-matching while permitting pattern updates without resource reconfiguration. An off-line optimization method first finds common sub-patterns across all the patterns in the SNORT database of signatures. For that, we based on work [12, 13] to make implement the part pattern-matching. To manage the pattern-matching, we have developed an FPGA design by VHDL (VHSIC hardware description language; VHSIC, very-high-speed integrated circuits).

## 3 Experiment Design Our Platform and SNORT

### 3.1 Network Design

To verify that our system produces the correct results, we compared it with the standard SNORT software distribution and EIDS. We have created a network in which a computer connected to our platform for the supervision operations and recovery results. The network (Fig. 3) is composed of 9 computers, depending on our need of generating (running both open source tools and commercial tools) smaller packet size on high traffic speeds.

### 3.2 Performance Metrics

Performance metrics are used in the experiments to measure the ability of the SNORT and our platform EIDS to perform a particular task and to fit within the performance constraints. These metrics measure and evaluate the parameters that

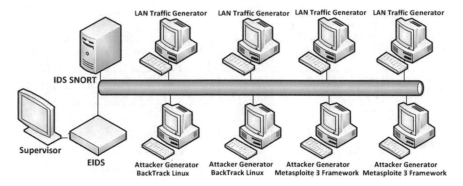

**Fig. 3** Network test

**Table 1** Description of performance metrics

| Performance metrics | Description |
|---|---|
| Packets received (**PRE**) | The percentage of packets received from the total packets transmitted |
| Packets analyzed (**PAN**) | The percentage of packets analyzed from the total packets received |
| Packets dropped (**PDR**) | The percentage of packets dropped from the total packets received |
| Packets detected (**PDE**) | The percentage of packets detected from the total packets abnormal |

impact SNORT and our platform EIDS performance. The following aspects were measured in the experiments. The metrics are the percentages of the total packets processed by Snort. The specific metrics used are shown in Table 1.

# 4 Results and Discussions

Test scenarios were designed to test the performance of our platform EIDS and SNORT. Both IDS were subject to the same tests and under the exact same conditions. The following subsections will give more view of the test senarios. The purposes of the senarios are: High speed traffic (Senario 1), Heavy traffic (Senario 2), Large data traffic (Senario 3), and Attack detection rate (Alerts) (Senario 4).

## 4.1 Senario 1. EIDS and SNORT Reactions to High-Speed Network Traffic

We used LAN Traffic Genarator tools, for we sent 21,000 packets (14,000 TCP, and 7000 UDP) at different transmission time intervals (16, 8, 4 ms). Table 2 show the EIDS and SNORT output and results of our experiments.

**Table 2** Same number of packets but different transmission time intervals

| Packets type | Metrics | SNORT reaction | | | EIDS reaction | | |
|---|---|---|---|---|---|---|---|
| | | 16 ms (%) | 8 ms (%) | 4 ms (%) | 16 ms (%) | 8 ms (%) | 4 ms (%) |
| TCP (14,000) | PRE | 100 | 100 | 100 | 100 | 100 | 100 |
| | PAN | 100 | 71.28 | 43.66 | 99.50 | 63.83 | 24.48 |
| | PDR | 0 | 28.72 | 56.34 | 0.50 | 36.17 | 75.52 |
| UDP (7000) | PRE | 100 | 100 | 100 | 100 | 100 | 100 |
| | PAN | 100 | 74.01 | 62.17 | 99.34 | 62.34 | 30.74 |
| | PDR | 0 | 25.99 | 37.83 | 0.66 | 37.66 | 69.26 |

Our experiment demonstrated that both systems analysis performance was decreased as the speed of transmission increased. We deduce that the capacity of analysis to the components have the difficulty in order to increase the speed transmission.

## 4.2 Senario 2. EIDS and SNORT Reactions to Heavy-Traffic Networks

Here, the transmission rate of packets was kept to the same speed (16 ms intervals) to obtain a fair analysis of different numbers of packets (each packet carried 1024). We sent 100, 500 and 1000 packets batches at 16 ms intervals. Table 3 show the SNORT and EIDS results. Our experiment shows that as the number of packets increases, more packets are dropped.

## 4.3 Senario 3. EIDS and SNORT Reactions to Large Packets

For this experiment, the number of packets was kept to the same value (18,000) and the same speed (16 ms) to obtain a fair analysis of different sizes (lengths) of

**Table 3** Same speed limit and different numbers of packets

| Metrics | SNORT reaction to heavy traffic | | | EIDS reaction to heavy traffic | | |
|---|---|---|---|---|---|---|
| | 100 (%) | 500 (%) | 1000 (%) | 100 (%) | 500 (%) | 1000 (%) |
| PRE | 100 | 100 | 100 | 100 | 100 | 100 |
| PAN | 100 | 69.68 | 30.98 | 92.66 | 50.29 | 26.64 |
| PDR | 0 | 30.32 | 69.02 | 7.34 | 49.71 | 73.36 |

**Table 4** Same speed and value but different packet size

| Metrics | SNORT reaction to heavy traffic | | | EIDS reaction to heavy traffic | | |
|---------|----------|----------|-----------|----------|----------|-----------|
|         | 100 (%)  | 500 (%)  | 1000 (%)  | 100 (%)  | 500 (%)  | 1000 (%)  |
| PRE     | 100      | 100      | 100       | 100      | 100      | 100       |
| PAN     | 100      | 68.10    | 45.45     | 99.55    | 57.40    | 37.69     |
| PDR     | 0        | 21.90    | 54.55     | 0.45     | 42.60    | 62.31     |

packets. We increased the size of each packet sent started from 256 byte, to 512 bytes, and to 1024 bytes. Table 4 show the performance detection results.

Our experiment demonstrated that more packets will be dropped as packet size increases.

## 4.4 Senario 4. EIDS and SNORT Reactions to Generate the Number of Alerts (Attack Detection Rate)

During the evaluation, attacks have been generated to evaluate the performance of both IDSs in a heavy and mixed traffic. The initial test was perfomed with background traffic only. This was done to confirm that both EIDS and SNORT are configured to generate the number of alerts. We then went on generating the same attacks for both EIDS and SNORT in high speeds network. The results are presented in Table 5.

Our experiment demonstrated that the component of detection will be well to configure to give significant results.

## 4.5 Discussion of Results

Critical analyses were done for experiments senario1, senario2, senario3 and senario4. The obtained results show that both systems performance analysis throughput is affected by high-speed and heavy traffic, and more packets are dropped as the number and size of packets and the speed of traffic increases. Both systems had a limited time to process and analyse any traffic successfully and if a network's traffic speed limit is higher than both systems limit. This problem due to the limited characteristics of the components.

**Table 5** Same number of alerts but different speed

| Speed | SNORT rate of attacks detection | | | | EIDS rate of attacks detection | | | |
|-------|-----------------|-----------------|----------------|----------------|-----------------|-----------------|----------------|----------------|
|       | 250 Mbps (%)    | 500 Mbps (%)    | 1.0 Gpbs (%)   | 2.0 Gbps (%)   | 250 Mbps (%)    | 500 Mbps (%)    | 1.0 Gpbs (%)   | 2.0 Gbps (%)   |
| PDE   | 100             | 100             | 100            | 99,3           | 100             | 89,97           | 65.12          | 41.28          |

# 5 Conclusion

An IDS is considered to be one of the best technologies to detect threats and attacks. IDSs have attracted the interest of many organizations and governments, and any Internet user can deploy them. An IDS usually features four stages to secure a computer system network: scanning, analysing, detecting, and correcting. In this paper we proposed an approach for evaluating an IDS with these characteristics. This approach based on tests to measure performance indicators of the components of an IDS. For this we chose a hardware solution based on embedded systems, that we implemented the components of SNORT. Our solution is a hardware platform that gives the hand to measure the performance indicators of the components of an IDS. As a result of our approach, systems can be configured such that attacks can be thwarted more easily. The obtained results show that the performance characteristics of an IDS depends on the performance of its components.

# References

1. Khorkov, D.A. Methods for testing network-intrusion detection systems. Sci. Tech. Inform. Process. **39**(2), 120–126. doi:10.3103/S0147688212020128
2. Peter, M., Vincent, H., Richard, L., Josh, H., Marc, Z.: An overview of issues in testing intrusion detection systems. Technical report, National Institute of Standard and Technology (2003)
3. Lippmann, R., Haines, J.W., Fried, D.J., Korba, J., Das, K.: Analysis and results of the 1999 DARPA off-line intrusion detection evaluation. In: Recent Advances in Intrusion Detection, pp. 162–182. Springer Berlin Heidelberg. doi:10.1007/3-540-39945-3_11 (2000)
4. Mohammed, S., Toumi, B., Abdelhamid, B., Mostafa, A.: Amelioration of attack classifications for evaluating and testing intrusion detection system. J. Comput. Sci. **6**(7), 716–722. doi:10.3844/jcssp.2010.716.722 (2010)
5. Akhlaq, M., Alserhani, F., Awan, I., Mellor, J., Cullen, A. J., Al-Dhelaan, A.: Implementation and evaluation of network intrusion detection systems. In: Network performance engineering, pp. 988–1016. Springer Berlin Heidelberg. doi:10.1007/978-3-642-02742-0_42 (2011)
6. Mohammed, S., Toumi, B., Abdelhamid, B.: Generation of attack scenarios by modeling algorithms for evaluating IDS. IEEE xplore digital library 1–5. doi:10.1109/ICMCS.2011. 5945730 (2011)
7. Muhammad, A.J., Jihyung, L., Sangwoo, M., Insu, Y., Deokjin, K., Sungryoul, L., Yung, Y., KyoungSoo, P.: Kargus: a highly-scalable software-based intrusion detection system. In: Proceedings of the 2012 ACM Conference on Computer and Communications Security (CCS'12), ACM, New York, NY, USA, pp. 317–328. doi:10.1145/2382196.2382232
8. Mohammed, S., Emharref, M., Toumi, B., Abdelhamid, B.: Platform based on an embedded system to evaluate the intrusion detection system. IEEE xplore Digital library 894–899. doi:10.1109/ICMCS.2012.6320253 (2012)
9. Albin, E., Rowe, N.C.: A Realistic experimental comparison of the suricata and snort intrusion-detection systems. In: 26th International Conference on Advanced Information Networking and Applications Workshops (WAINA), 2012, pp. 122, 127, 26–29 March 2012. doi:10.1109/WAINA.2012.29 (2012)

10. Xinli, W., Alex, K., Lihui, H., Matt, G., Derrick, S.: Administrative evaluation of intrusion detection system. In: Proceedings of the 2nd annual conference on Research in information technology (RIIT'13). ACM, New York, NY, USA, pp. 47–52. doi:10.1145/2512209. 2512216 (2013)
11. Young, H.C., William, H.M.: Deep network packet filter design for reconfigurable devices. ACM Trans. Embed. Comput. Syst. **7**(2), Article 21 (January 2008), 26 p. doi:10.1145/ 1331331.1331345 (2008)
12. Abhishek, M., Walid, N., Laxmi, B.: Compiling PCRE to FPGA for accelerating SNORT IDS. In: Proceedings of the 3rd ACM/IEEE Symposium on Architecture for Networking and Nommunications Systems (ANCS'07), ACM, New York, NY, USA, 127–136. doi:10.1145/ 1323548.1323571 (2007)
13. Guinde, N.B., Sotirios, G.Z.: Efficient hardware support for pattern matching in network intrusion detection. Comput. Secur. **29**(7), 756–769. doi:10.1016/j.cose.2010.05.001 (2010)

# Enhanced Algorithm for Type II Firewall Policy Deployment

**Bezzazi Fadwa, Mohammed El Marraki and Ali Kartit**

**Abstract** Firewall is a core element in any network architecture. However, the large size of modern networks makes the firewall policy management more complex and error-prone. In fact, firewall policy is a list of filtering rules written by an administrator or management tool in specific order to block or permit traffic across the network. Therefore, the order of those rules is a primary for the efficiency of firewall performance. Several researchers have proposed works about firewall policy deployment, focusing on how to pass from a policy to another one while respecting the constraints of security. In this paper we will present our algorithm named ROSO (Rule Ordering using Swap Operation) based on type II editing language. A comparison with existing strategies will validate the efficiency of our algorithm which seems correct and safe since it gives a good result and can be applied to a large size of firewall policies.

**Keywords** Firewall policy · Network security · ROSO algorithm

## 1 Introduction

Modern network topology is become more complex due to their large size which makes the configuration of firewall policies complex and error-prone. Some firewall policies may contain more than 10 K rules, which makes configuration very hard

B. Fadwa (✉) · M.E. Marraki
LRIT Associated Unit to CNRST (URAC 29), Faculty of Sciences,
Mohammed V-Agdal University, BP 1014 Rabat, Morocco
e-mail: bezzazi.fadwa@gmail.com

M.E. Marraki
e-mail: elmarrakimohamed@gmail.com

A. Kartit
Laboratoire de Technologie de l'Information (LTI),
ENSA Chouaib Doukkali University, El Jadida, Morocco
e-mail: alikartit@gmail.com

© Springer International Publishing Switzerland 2016
A. El Oualkadi et al. (eds.), *Proceedings of the Mediterranean Conference
on Information & Communication Technologies 2015*, Lecture Notes
in Electrical Engineering 381, DOI 10.1007/978-3-319-30298-0_53

stuff and actually need smart management tools to solve this issue [1]. We remind that the aim of firewall is to control the access to the network from incoming or outgoing traffic according to a list of rules called firewall policy [2].

These rules are accompanied by an action of permission or rejection of connection. Firewall policy deployment is a big concern that deserves the attention of researchers in order to provide safe strategies enabling enterprises or private networks to deploy their policy efficiently. Four characteristics must be provided by the management tool [3]: correctness, safety, confidentiality and speed.

- Correctness: When the deployment successfully implements the target on the firewall so that the target policy becomes the current used policy. It is a primary criteria for efficiency.
- Safety: The deployment must deny all illegal packets without exception and accept only legal ones.
- Confidentiality: Is provided when the communication is secured between the firewall and the management tool. For this we use the encryption communication protocols such as SSH, SSL or SMTP.
- Speed: The time of execution of the strategy of deployment must be reduced to the maximum so we can apply it easily even for large policies.

The reminder of this paper is organized as follow: Sect. 2 will present the definition of deployment policy, the policy editing language and also define the deployment efficiency. A literature review is important where several works with the same objective but treated differently are exposed in Sect. 3. Section 4 is reserved for explaining the aim of our algorithm named Rule Ordering using Swap Operation (ROSO). We will compare our method to existing ones (SanitizeIT (SI) and Efficient Deployment (ED)) in Sect. 5 and finally conclude by the aim of near future work in Sect. 6.

## 2 Deployment Policy

Many security requirements lead administrators to update their firewall policies. Firewall policy deployment is the process that allows us to pass from a policy I to another one T. Efficient deployment policy should satisfy four major characteristics previously explained: Correctness, Safeness, Confidentiality and Speed.

### 2.1 Firewall Policy Editing Language

In order to deploy a firewall policy an administrator should use a set of commands called policy editing language: We use (app r) to append a rule r at the end of the current policy R, (del r) to delete a rule r from R, while (del i) deletes a rule from a certain position, (ins i r): inserts r at position i in R and (mov i j): moves the *i*th rule to the *j*th position in R.

Some firewalls don't support all these commands. Several types of editing language exist but we will expose the most representative editing languages:

Type I: In type I we are allowed to append or delete a rule r from a policy R. The use of (app r) means that we append r at the end of current policy R but if r exist already in R the command failed then. In same way (del r) deletes a rule r from R but if it doesn't exist the command failed. Some recent firewalls such as FSWM 2.x and JUNOSe 7.x only support Type I editing language

Type II: To get over the incompetence of type I editing, new firewalls have introduced type II editing by adding other commands as: (ins i r), (del i) and (mov i j) which support rule position numbers. (ins i r) means that the rule r must be inserted at position i in the current policy R while (del i) refers to delete *i*th rule from R. The effect of (mov i j) can be achieved by (del i) followed by (ins j r) but in this case can represent a lack of security caused by the absence of rule r between deletion and insertion. (mov i j) allow us to mov a rule from one position to another one

## 2.2 Deployment Efficiency

In order to minimize the communication cost and CLI processing time it is recommended to minimize the number of editing commands sent by a management tool. A deployment is most-efficient if it utilizes the minimum number of editing commands in a given language to correctly deploy a target policy on a firewall by transforming initial policy I into T [4]. Efficient deployment should be correct, safe, confidential and speed.

## 3 Literature Review

In this section we summarized some of the significant methods used to deploy firewall policy. Authors In [4] present a type II algorithm called (Efficient Deployment) it aims to deploy a firewall policy using an intermediate policy that contains the longest list of common rules of I, T and all rules that have to move up or down. Authors assumed that the algorithm provides an efficient results but its complexity is not negligible. In [1] authors propose a greedy2phase representing type II algorithm. This algorithm can deploy various firewall policy but it can not be considered as a correct algorithm since it doesn't respect the order of rules in some cases. In [5–8] authors give much importance in their work to the relation between rules assuming that it's the key for an efficient deployment of firewall policies, the important is not to update the firewall policy by passing from a policy to another but the most relevant point is how to do this operation while providing safeness and security to the network. they highlights the anomalies discovered in the policies and define all relations that may connect two rules from the same policy. Based on this study they classify the anomalies and propose solution to generate free-anomaly policy using (FPA) management tool.

# 4  Our Approach

To deal with the problem of rule ordering and satisfaction of criteria of efficiency especially correctness, we provide a new efficient algorithm. The proposed algorithm is called ROSO (see Algorithm 1).

---

**Algorithm 1** Rule Ordering using Swap Algorithm

---

```
 1: SwapOrderingRuleDeployment(I, T){
 2: /* Phase 1: Balancing lists */
 3: LA ← emptystack
 4: LD ← emptystack
 5: for  t ← 1  to SizeOf(T)
 6: if  T[t] ∉ I  then
 7: LA.append(T[t])
    I.append(T[t])
    endif
      endfor
 8: for t ← 1 to SizeOf(I)
 9: if  I[t]  ∉ T then
10: LD.append(I[t])
11: T.append(I[t]) endif
      endfor
12: /* Phase 2: Calculation of movement values and rule ordering*/
13: Vd ← hashtable   tmp ← 0   val ← 0
14: for i ← 1 to Sizeof(I)
15: Vd[i] ← Indexof(Rᵢ, I) − Indexof(Rᵢ, T)
    end for
16: for t ← 1 to Sizeof(Vd)
17: if  Vd[i] < 0 ∧ Vd[i + 1] >= 0
    then
18: swap(Rᵢ, Rᵢ₊₁)
19: tmp ← Vd[i]
20: Vd[i] ← Vd[i + 1] − 1
21: Vd[i + 1] ← tmp + 1
    endif
      endfor
22: /* Phase 3: Deletion of extra rules */
23: for i ← Sizeof(LD) downto 1
24: del(I[Sizeof(I) - i + 1])
    endfor
25: fin
```

---

**Fig. 1** Calcule of Vd

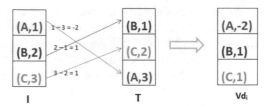

The main phases of the ROSO algorithm can be summarized in the three following points:

1. Balancing the policies I and T
2. Rule ordering
3. Deletion of supplementary rules

In order to pass from initial policy to a target one while respecting the order of the rules in the target list, we first balance the two policies by creating a list LD (*Line8–10*) of rules that exists in I but not in T, and a list LA (*Line5–7*) of rules contained in T but not in I. Once we defined LA and LD we append the LD to the end of the initial policy I and LA to the end of the target policy T such that we obtain two policies with the same size.

**Definition** 1 I and T are balanced: $Sizeof(I) = Sizeof(T)$ and $\forall R_i \in \{I, T\} : R_i \subseteq I \Leftrightarrow R_i \subseteq T$.

In the first phase (*Line5–11*) we obtain two balanced policies with the same set of rules but in a different order. We will discuss in the second phase (*Line14–21*) how to reorder those rules to get the same as target policy order. For this reason we calculate the movement value ($Vd_i$) such as $Vd_i$ is the result of the difference between index of order of the same rule in the two policies as shown in Fig. 1.

Based on these values saved in a hash table Vd, we do our movements using swap operation which means the permutation of two successive rules. Since each rule has only one position in the policy, and each position in the policy can contain only one rule. We can show that the sum of Vd values is null:

$$\sum_{i=1}^{n} Vd_i = 0 \quad \text{with} \quad n = Sizeof(Vd)$$

At the end of phase II we have the current policy equivalent to T, so the next step is to delete the unnecessary rules from current policy to get the target one. This operation is the aim of phase III (*Line23–25*), we use the LD stack to delete all supplementary rules from the current policy so we can obtain the final one (T).

At any moment in ROSO algorithm, the current policy R is a shuffle of I and T which means that R is safe according to the shuffling theorem [1]. In the first phase we scan through I and T only once to create LA and LD which takes a constant time $O(n)$. The IndexOf and $\in$ operators used in the algorithm will take constant time also. In the last phase we run over LD to delete extra elements. The algorithm takes $O(n)$ time and space where $n = max(I + T)$. The complexity of swap $(R_i, R_j)$ operation is negligible.

## 5 Experimentation Results

As shown in the previous section, ROSO algorithm takes $O(n)$ time and space and also we assume that throughout the execution the algorithm is safe.

In this section we expose the performance of this algorithm using four policies; small size (2000), medium size (5000), large size (10,000) and extra large size policy (25,000). We did same test five times as in [1, 4] which requires respectively 10, 500, 1000, 60, 90 % for test 1–5. We mention that the percentages are taken from the initial policy. The algorithm is implemented in Java and all tests are performed on Lenovo 20236 with intel(R) core(TM)i3, 2.4 GHz processors and 4G RAM.

The Table 1 below represents the results of tests, the time taken by ROSO algorithm is specified in the column ROSO, while the column SI and ED specifies the total time taken by respectively SanitizeIt algorithm given in [1] and Efficient Deployment given in [4]. all times are represented in seconds.

From Table 1 it's clear that ROSO algorithm gives a better result than ED or SI in all cases for the five tests. We assume that during the execution of the ROSO the results were satisfying and respect the criteria of correctness, in the same way the algorithm respects the criteria of safeness. When the order of the rules is respected and the deployment is done in a short time, the safeness is ensured as well. Indeed, the complexity of our approach is lower than the Efficient Deployment. Our algorithm is shown to be simple and can be applied to a large size of policies in a short time.

The curve illustrated in Fig. 2, resumed that ROSO algorithm is more efficient than both EfficientDeployment and SanitizeIt. The running time for ROSO algorithm is almost linear. Furthermore, SanitizeIt seems to have a polynomial running time. We notice this effect especially in case of test 5 and Policy 4, where SanitizeIt compute a deployment sequence in almost 27 s.

**Table 1** Experimentation results

| Tests | Small policy (2000) | | | Medium policy (5000) | | | Large policy (10,000) | | | Extra large policy (25,000) | | |
|---|---|---|---|---|---|---|---|---|---|---|---|---|
| | ED | SI | ROSO | ED | SI | ROSO | ED | SI | ROSO | ED | SI | ROSO |
| Test1 | 0.00783 | 0.007 | 0.002 | 0.01646 | 0.018 | 0.010 | 0.03585 | 0.031 | 0.025 | 0.08027 | 0.169 | 0.050 |
| Test2 | 0.00704 | 0.002 | 0.0018 | 0.01813 | 0.018 | 0.015 | 0.04721 | 0.036 | 0.031 | 0.08116 | 0.122 | 0.0651 |
| Test3 | 0.00684 | 0.015 | 0.001 | 0.01859 | 0.023 | 0.012 | 0.03826 | 0.039 | 0.027 | 0.08409 | 0.122 | 0.060 |
| Test4 | 0.00684 | 0.015 | 0.004 | 0.01837 | 0.026 | 0.018 | 0.03713 | 0.127 | 0.035 | 0.08247 | 0.627 | 0.043 |
| Test5 | 0.00696 | 0.008 | 0.004 | 0.01687 | 0.031 | 0.020 | 0.03454 | 0.234 | 0.040 | 0.08761 | 1.031 | 0.0892 |

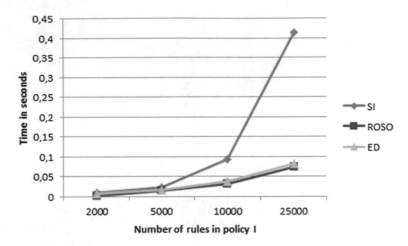

**Fig. 2** Comparison of experimentation's average results

## 6 Conclusion

In modern networks firewall policy manage large size of rules. These policies have
to be updated frequently to deal with the security constraints. The problem of
firewall policy deployment is a big concern that needs more interest from
researchers. Indeed if a firewall policy is not well deployed the system of network
security will be threatened. In this paper we proposed an algorithm with complexity
in order of n (O(n)) that allow us to easily deploy any firewall policy while
respecting the criteria of safety and correctness. ROSO algorithm is safe and close
to linear. A simple comparison between our algorithm and exiting shows that our
approach is more efficient than SanitizeIT and give results as good as Efficient
Deployment and better in case of extra large policy size. In the near future work we
will focus on the optimization of rule ordering using metaheuristic algorithms.

## References

1. Zhang, C.C., Winslett, M., Gunter, C.A.: On the safety and efficiency of firewall policy
   deployment. In: Proceedings of the 2007 IEEE Symposium on Security and Privacy,
   Washington, DC, USA. IEEE Computer Society. 3350 (2007)
2. Karen, S., Paul, H.: Guidelines on firewalls and firewall policy. NIST Recommendations, SP
   800–41 (2008)
3. Bezzazi, F., Kartit, A., El Marraki, M., Aboutajdine, D.: Optimized strategy of deployment
   firewall policies. In: Second International Conference on Innovative Computing Technology
   (INTECH), IEEE, 46–50 (2012)

4. Zeeshan, A., Imine, A., Rusinowitch, M.: Safe and efficient strategies for updating firewall policies. In: INRIA Nancy Grand Est. INRIA ARC 2010 Access and FP7-ICT-2007-1 Project No.216471 Avantssar
5. Abedin, M., Nessa, S., Khan, L., Thuraisingham, B.: Detection and resolution of anomalies in firewall policy rules. In: Data and Applications Security XX Lecture Notes in Computer Science, 4127, pp. 15–29 (2006)
6. Alshaer, E., Hamed, H.: Dynamic rule-ordering optimization for high-speed firewall filtering. In: Proceeding ASIACCS'06 Proceedings of the 2006 ACM symposium on Information, computer and communications security, pp. 332–342 (2006)
7. Hu, H., Ahn, G.J.: Detecting and resolving firewall policy anomalies. IEEE Trans. Dependable Secure Comput. 9(3) (2012)
8. Bezzazi, F., Kartit, A., El Marraki, M., Aboutajdine, D.: Enhaced strategy of deployment firewall policies. J. Theor. Appl. Inform. Technol. 54(1) (2013)

# Innovative Solution for an eID Secure Management in a Multi-platform Cloud Environment

Primož Cigoj and Borka Jerman Blažič

**Abstract** The security provision within multi-platform cloud computing environment in all levels is still considered not to be properly solved due to different problems with technical and human-based origin. This paper presents an attempt to provide an authentication and authorization solution based on the single sign-on (SSO) approach for cloud service users and administrators in a multi-platform environment.

**Keywords** Identity management · Cloud computing · Security · Cloud · Management · Cloud provisioning · Infrastructure-as-a-Service

## 1 Introduction

Cloud computing has revolutionized the provision of computing services, transferring them from local to unspecified remote environments being controlled by third party service providers. This new data processing model has contributed several challenges to emerge in the data management and in the security. The main cloud computing challenges are still in the area of data protection within the applied cloud technology its services and the user data protection requiring adequate security measures. One of the major problem appears from the fact that the individuals have no control over their data, nor influence on used applications nor the data identity stored in the cloud. In addition, cloud technology is faced also with other security and privacy issues that require additional solutions. Installation management procedures in the cloud environment are complex, no adequate and no complete security solution has been developed that ensures a safe deployment. Consequently,

P. Cigoj (✉)
Jožef Stefan International Postgraduate School, Ljubljana, Slovenia
e-mail: primoz@e5.ijs.si

P. Cigoj · B.J. Blažič
Laboratory for Open Systems and Networks, Jozef Stefan Institute, Ljubljana, Slovenia
e-mail: borka@e5.ijs.si

© Springer International Publishing Switzerland 2016       529
A. El Oualkadi et al. (eds.), *Proceedings of the Mediterranean Conference on Information & Communication Technologies 2015*, Lecture Notes in Electrical Engineering 381, DOI 10.1007/978-3-319-30298-0_54

some important security issues remains a topic of concern for both the vendors and the users of the clouds. According to the survey conducted by Microsoft and the U.S. National Institute of Standards and Technology (NIST), security in the cloud computing model is still the ICT executives' main concern [1, 2]. Consequently, many business entities all over the world are not very keen on adopting the cloud computing processing model. The recent disclosure of the Open Technology Institute report about the CIA/NSA controlled surveillance and espionage on the telecommunication networks contributes further to the lack of trust in the cloud computing model especially when privacy and security are questioned.

## 2 Problem Description

Despite of several year of existence of cloud computing services this model is still considered as a new term in the mainstream of the ICT field, first popularized in 2006 by Amazon's EC2 application. The need of establishing trust, on a sufficient level of trust in the cloud, services is very important subject receiving a growing attention from both academia and industry [3]. Besides the user initiated and stored data that need to be protected, the cloud computing provider operate with other sensitive data disclosure and their protection is also recognized as very important [4]. In 2010, the European Network and Information Security Agency (ENISA), a security incident response agency of the European Union, reported on the cloud computing benefits but also on the growing risks that appeared to grow steadily. Recommendations about the need for improved Information Security measures were reported as well [5]. The customer-related security risks that have been analyzed are summarized in the top list of identified risks and this includes: loss of governance over the data, data lock in, isolation failure(s), growing compliance risks, use of management interface compromises without ensured protection, lack of sufficient data protection, insecure and incomplete data deletion, and malicious insider. A search for solutions of the security problems seems to be one of the most important task for the further deployment of the cloud computing services. However, the well-established security requirements and the applied security measures for ICT services may be used into the cloud computing as well. Among those identification, authentication and authorization services are listed as very important. Users and cloud administrators simultaneously access multiple systems; and as each of these systems has a different login dialog users are forced to memorize a vast amount of usernames and passwords. In addition, the strong authentication at the user level (3) requires management and creation of multilevel authentication mechanisms for several cloud services and platforms. This introduces additional complexity and potentially lower reliability in the system. One of the approach to solve this problem is to provide authentication and authorization services with just single user credential that can be used for access to several services offered in the cloud. In this paper, we presented the design of a system for authentication and authorization service applied in a multi-platform cloud computing environment. The designed solution is based on the Single Sign On (SSO) principle and is implemented on two different cloud computing platforms enabling smooth but secure and trustable transition for the users that are passing from use of one service to another. The designed solution enables an efficient management of the electronic identities (eID) service by the cloud administrator on an easy and friendly way. The developed implementation of the designed service was evaluated and tested within a large National Competence Centre project that has developed cloud assisted services for different fields of application. The paper presents the solution through several sections. After the introduction, short overview of the related and previous works is provided, followed with the description of the basic design principles. The section describing the implementation and the testing is followed with conclusion and thoughts for future work.

## 3 Multi-platform Cloud Infrastructure and the Authentication Services

### 3.1 Overview of the Requirements

According to the CSA research report [6] one of the largest identified cloud security problem has origin in the shared technology used for building the cloud infrastructure. For that reason some cloud service providers do not build their own infrastructure but hire infrastructure services from someone else—Elastic Compute Cloud (EC2) [7]. This kind of service is provided by elastically allocating physical or virtual resources on-demand, delivering storage, networking, or computational capabilities in the form of wrapped services by corroborating the utility computing side of the clouds. This service is known as Infrastructure as a Service (IaaS) and only basic system security is provided such as the firewalls. The presence of the Virtual Machine Monitors (VMMs) are critical components in this type of cloud computing services. They are expected to provide complete isolation throughout all Virtual Machine (VM) instances in the cloud which is not a simple task.

Infrastructure resource sharing can potentially enable one consumer to peek into another consumer's data in cases when the cloud service does not provide strong system of authorization and authentication. The problems of account hijacking or user credential theft is also possible if not sufficient protection techniques system are in-built. For comparison, the traditional identity management (IdM) approach used in the enterprise networks [8] is centralized compared to the current solution used in the cloud systems. These IdMs are usually based on knowledge about the user personal data, such as real name, user name, e-mail address, identification number, access permissions, etc. The use of the separate IdM system within an enterprise in a cloud environment, demands an inter-connection with the cloud infrastructure to be allowed and set up. This is very complicated task as currently no simple way to extend the IdM use into the cloud networks has been proposed and deployed [9]. The design of such system must include the presence of the following services:

- Registration of the identities: Verification of a user account must be done before the user registration. The registration system should be based on implemented security standards principles and standard procedures. Organizations or enterprises that decide to transfer their user accounts to a cloud service must be assured that the provider account management system is up-to-date and safe.
- Authentication: Management and implementation of the user authentication system by the provider must be performed in a trustworthy way. The IdM system must allow a multi-level configuration of the in-built authentication procedures.
- Federation of identities: Federation of identities allows users to use the same set of credentials to obtain access to different cloud resources. The user's electronic identity and its attributes are securely shared across multiple systems and the access is granted with the authenticated eID. Federation of identities can be set up in several ways; but the most popular is by use of SAML protocol or the OpenID solution. The federated eIdM system must be capable to allow establishing of a secure connection between the different cloud services based on this single user credential.
- Authorization: Authorization specifies the rights of an individual to have a user account. The cloud account management procedures for authorization of particular account rights must be verified by the highest system authority according to the agreed rules and contracts. In addition, the granted right should be consistent with the policy for access distribution to the privileged applications.
- Access control: Access control requirements vary widely depending on the type of the end-user (an individual user or an organization). The implementation of the access control system must follow the access control policy. The implementation should allow the performed actions to be traceable for non-repudiation purposes in case of dispute or complains.

The SSO principle is the best approach the service requirements listed above to be met in solution designed for a multi-platform network system. SSO is "a mechanism whereby a single action of user authentication and authorization can permit a user to access all computers and systems where that user has access permission, without the need to enter multiple passwords".

## 3.2   Design of the Authentication System—MUPASS

The design of the MUPASS (Multi-platform cloud authentication system) system was part of the development of the general authentication service for the National cloud computing infrastructure with an intention to offer several cloud based services such as e-learning, e-health, e-government etc. As the infrastructure was designed to be built up from the OpenStack and VMware cloud platforms the envisaged authentication system had to incorporate the platform specific properties. The selection of the platforms for the cloud infrastructure was due to their origin. OpenStack is an open source platform and VMware is a proprietary owned platform, so combination of both was expected to conform to the requirements.

OpenStack platform is a platform frequently used for the construction of private and public clouds. The management identity service in OpenStack is known under the name Keystone. Keystone authenticates the users and provides them with authorization tokens that allow the access to the OpenStack services. The current version of Keystone is centralized, and all its users must be registered in the Keystone database.

The other platform, VMware is a very popular platform due to its complete and rich set of features and functionalities. VMware enables enterprises to build secure, multi-tenant private clouds by pooling infrastructural resources into virtual data centers. The users access the service through web-based interface (REST) which is a fully automated, catalogue-based service. The VMware vCloud Director for ID management supports the RBAC features (RoleBasedAccessControl). The authentication method used in VMware vCenter uses a single set of credentials.

By studying the requirements for an eID secure management system and both platforms features which are sufficiently open and flexible, the development of a centralized authentication system based on the SSO principle seemed to be possible. The task has taken few months and the developed system MUPASS appeared to be very useful and secure.

# 4 Implementation

## 4.1 Functional Description

The main objectives followed during MUPASS design were to make possible the cloud administrators and the users to access the heterogeneous cloud's infrastructure services by use of only one single credential.

At the time of preparing this paper, in the investigated literature of cloud computing the authors did not found any report about similar solution, so the solution presented here is original in its design and implementation. The developed code was already copied and used by the Open Source community and can be found on the URL (https://github.com/primozc/sso). In addition to the main objective some more goals were set up and followed during the design:

- provide nice and friendly user experience;
- enable the cloud users to move between services securely and without interruption,
- provide communication to the users with only one dashboard for an interaction
- reduce the processing costs, (follows from the reduction of the number of calls),

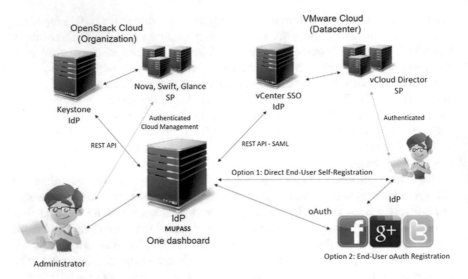

**Fig. 1** The MUPASS architectural design

- assure significantly less time to be needed for the management of the multiple cloud accounts,
- provide higher level of the information security,
- enable provision of an audit log of the user operations and actions.

The operational principles implemented in the system design are shown on Fig. 1. The system MUPASS is composed from two parts—front end and back end. The front end is a web interface, coded with PHP, HTML, and JavaScript. The interface is responsible for communication with the users enabling them to sign and register into MUPASS. The system administrator is responsible to provide user permissions and to manage the access to different cloud platforms by exchanging messages between both parts. The back end part of MUPASS is responsible for mapping, synchronizing and removing the end users accounts from the remote cloud platforms. The back end is designed in a way that uses the REST access calls (REST over HTTP protocol) for performing the needed communication with the remote terminals. The format of the exchanged data depends on the end cloud platform accessed (it can be JSON or XML). MUPASS provides two ways of end-user registration; first option is a self-registration, and the second one is registration through a social network account (oAuth2). The oAuth2 solution, which is a popular open standard for user authorization within the social networks is supported by most known social networking systems such as those offered by Google, Facebook, Twitter and Yahoo. This feature is provided as it enables simple usable access that is liked by most of the users.

The instances of the remote cloud platform is created by the MUPASS administrator before the end user is re-directed to the cloud platform and the service respectively. The administrator within MUPASS selects the type of the platform requested and triggers the generation of an extra data fields for the selected platform. In case when the platform is OpenStack, the required fields by the remote platform administrator, such as password, token and tenant, are filled with the user/account related data. Once the cloud platform is initiated, the administrator connects the user to the remote cloud platform and store his/her data in the MUPASS administrator dashboard. The back end part of the system automatically map each new user data by using the help commands of the REST calls. When the platform is OpenStack then these REST calls insert the user's data into the Keystone application. The users perform the log-in action with the MUPASS initiated login credentials. After a successful login, the user is presented with the availability of the authorized remote platforms and with a single user click on the login button the service is opened to the user. The end user then can create, run and power off the cloud virtual instances.

## 4.2  System Implementation

The implementation of the MUPASS was carried out by use of the available APIs of each of the platforms and with known application tools such as CURL, REST, and HTTPS. Python and PHP programming languages were used for the code development.

PHP scripts with the support of the tool CURL (a command line tool for transferring data) were used for ensuring transfer of the users request from MUPASS to the remote OpenStack platform. The same tools were used for an automated processing of the common user and administrative tasks. Keystone's API interface features were used for communication and implementation of the user authentication process between the MUPASS and the remote OpenStack platform. The authentication and operation requests directed to the Keystone API are performed with the SSL protocol over the HTTPS protocol as both protocols ensure the required data protection. The Keystone SQL identity back end facility is used together with the MUPASS functionality for user credential mapping. The Keystone API supports both JSON and XML data serialization formats and this feature was used in the MUPASS implementation. When the X-Auth-Token is send from the OpenStack, it is accompanied with the other service API endpoints URLs of some other service in the OpenStack cloud.

The vCloud of VMware platform connection to MUPASS is similar to the solution that was developed for the OpenStack platform. During the development of MUPASS a special attention was paid to the migration of the users from the single platform authentication systems to the SSO federated system. With MUPASS an end user who is authenticated by MUPASS is able to access the administrator assigned cloud platform because a trusted relationship is set up between the remote cloud platform Identity service and the MUPASS identity provider. For all users registered and authenticated by MUPASS the additional management of their eIDs in the remote platform identity service is dropped out.

The identity administrative operations are performed for mapping the user's credentials from the local MUPASS database to the OpenStack Keystone and the vCloud user database. These operations provided by MUPASS enable the cloud administrators to obtain and validate access tokens, manage users, tenants, roles, and service endpoints. The user mapping enabled through the integration of the OpenStack with MUPASS require an administration token. The administrative token is generated by the Keystone Identity API which verifies the token and defines the administration role. Generated administration tokens are stored in the local MUPASS database system always when a new cloud platform to the cloud infrastructure is added.

The connection between the vCloud platform component and the CSS is similar to one described above. CSS connects a user to vCloud on the user request by an API call where the user credentials are passed as parameters. Since vCloud supports SAML, the open source SimpleSAMLphp library, which implements the SAML 2.0 standard was used as the main protocol for exchange of the user messages from the CSS and the remote vCloud platform. CSS SAML IdP was defined according to the specification in the vCloud Organization Federation Settings. End users, user groups' data and their roles in vCloud are required to be mapped from the organization's database and from the SAML provider. This restriction required additional functionality to be developed in CSS that enables the data mapping stored in the vCloud Director database to be available on a request. As Simple SAMLphp application does not support dynamic generation of metadata in a SP's remote configuration file an additional upgrading code was developed for overcoming the problem. The XML metadata file from the SAML IdP contains several certificates and information (e.g., SingleLogutService and AssertionConsumerService) which are necessary the vCloud Director to be capable for to communicate with the SSO solution and to validate that the SSO solution is sufficiently trustable. The generated XML metadata file from our SAML/SSO IdP needs to be uploaded into the vCloud federation metadata XML form (13). Since SAML users and groups cannot be found by use of a search function, user's data have to be mapped into vCloud with an automated API call. With the addition of this feature into the SSO solution the vCloud Director integration became complete.

## 5   Conclusion

The developed solution connects different platforms in a cloud infrastructure enabling secure authentication service based on a SSO principle and friendly remote user management. MUPASS is integrating the user authentication of two different platforms, one is proprietary and the other is an open source platform but its concept is sufficiently general to be applied for other platforms and in other environments. The current application and the referenced usage by the national cloud infrastructure confirm the usability of the system and the required level of the service security provided.

The Cloud computing area needs to be improved in the area of development and deployment regarding security and trustable services to offer to users. The reputation and re-building of the user trust become especially relevant after the disclosure of NSA surveillance over the US based cloud computing infrastructure. For achieving these goals, the future development should be oriented towards deployment of more flexible but secure interfaces that will allow accommodation and interconnections of many well-known cloud providers such as Amazon, Digital Ocean, Slicehost, Rackspace that will follow the recommendations for user secure authentication, provision of privacy and data protection.

# References

1. Jansen, W., Grance, T.: Guidelines on security and privacy in public cloud computing. NIST Spec. Publ. **800**, 144 (2011)
2. Microsoft: Microsoft Urges Government and Industry to Work Together to Build Confidence in the Cloud 2010. http://www.microsoft.com/en-us/news/press/2010/jan10/1-20brookingspr.aspx.3
3. Abbadi, I.M., Martin, A.: Trust in the Cloud. Inf. Secur. Tech. Rep. **16**(3), 108–114 (2011)
4. Behl, A.: Emerging security challenges in cloud computing: an insight to cloud security challenges and their mitigation. In: 2011 World Congress on Information and Communication Technologies (WICT), vol. 5 (pp. 217–222), IEEE (2011)
5. Catteddu, D.: Cloud computing: benefits, risks and recommendations for information security. In: Web Application Security (pp. 17–17). Springer, Berlin, Heidelberg (2010)
6. Alliance, C.S.: Top threats to cloud computing v1. 0. Cloud Security Alliance, USA. https://cloudsecurityalliance.org/topthreats/csathreats.v1.0.pdf (2010)
7. Fernandes, D.A., Soares, L.F., Gomes, J.V., Freire, M.M., Inácio, P.R.: Security issues in cloud environments: a survey. Int. J. Inf. Secur. **13**(2), 113–170 (2014)
8. Chow, R., Golle, P., Jakobsson, M., Shi, E., Staddon, J., Masuoka, R., Molina, J.: Controlling data in the cloud: outsourcing computation without outsourcing control. In: Proceedings of the 2009 ACM workshop on Cloud computing security vol. 9, pp. 85–90, ACM (2009)
9. Gonzalez, N., Miers, C., Redígolo, F., Simplicio, M., Carvalho, T., Näslund, M., Pourzandi, M.: A quantitative analysis of current security concerns and solutions for cloud computing. J. Cloud Comput. **1**(1), 1–18 (2012)

# Protecting Co-resident VMs from Side-Channel Attack in Cloud Environment: SAFEPERIMETER System

**Zakaria Igarramen and Mustapha Hedabou**

**Abstract** Today, the use of Cloud Computing is constrained by its vulnerabilities, because sharing the same resources with potential attackers may breaks the isolation between VMs. Also, some studies demonstrated that cache-based side channel attacks can break full encryption keys of RSA, DES and AES. In this paper we investigate side channel attacks via shared memory caches that can break the isolation between Vms, and present our new system, which has been called SAFEPERIMETER, and which aims to mitigate such discussed attacks, in a special thread, like AES encryption. It consists on securing access to L3 cache by locking it line by line, for a special confidential thread, without assigning any locked cache line to any VM. We also present the idea of implementing this service on a particular Image Management System.

**Keywords** Cloud computing · Security · Virtual network · VM · Isolation · BERNSTEIN's attack · Cache misses · Cache hits · SAFEPERIMETER · AES · MIRAGE

## 1 Introduction

Cloud computing has become a fashionable and important technology; both for its economic side and also transparency and simplicity that can bring this new concept.

For companies, the most significant reason that prevents them from moving to cloud computing is security. It is less secure than the traditional network infrastructure.

One of main fundamental concern is Isolation. A VM can monitor another one or access the underlying network interfaces. According to an analysis of vulnerabilities

Z. Igarramen (✉) · M. Hedabou
MTI Laboratory Ensa School, Cadi Ayyad University, Safi, Morocco
e-mail: z.igarramen@gmail.com

M. Hedabou
e-mail: mhedabou@gmail.com

© Springer International Publishing Switzerland 2016
A. El Oualkadi et al. (eds.), *Proceedings of the Mediterranean Conference on Information & Communication Technologies 2015*, Lecture Notes in Electrical Engineering 381, DOI 10.1007/978-3-319-30298-0_55

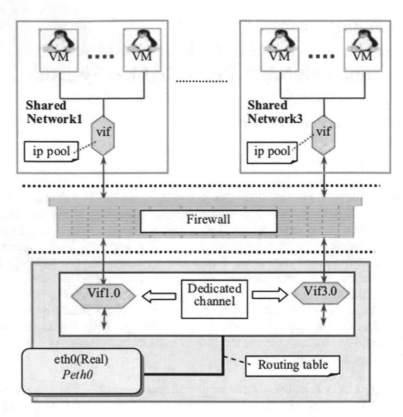

**Fig. 1** Virtual Network Model (*source* Article Network Security for virtual machine in Cloud Computing, P. 20)

existed in VMs, Wu et al. [1], proposed a novel virtual network model to make the communication among VMs more secure. This model is composed of three layers: routing layer, firewall and shared network, as shown in Fig. 1.

Unfortunately, this model has its limits: Thanks to routing and firewall layers, the security is established only between shared networks, but we cannot secure communication among VMs within the same shared network and the same physical host, and therefore, it's difficult to efficiently prevent VMs from attacks such as side-channel which breaks the isolation.

The idea consists on securing access to L3 cache by locking it line by line, for a special thread. We can understand this, by viewing it like a shared parking slot; only one car at a time can park in slot.

Otherwise, this will allow us to prevent retrieving or overwriting of some or all of a microprocessor instruction or data cache.

The remainder of this paper is organized as follows: Sect. 2 presents Cache-Based Side Channel Attack and describes BERNSTEIN'S attack, the new challenges to AES cache attacks. In Sect. 3, we provide a brief description of

SAFEPERIMETER system that can operate positively a side-channel to overcome discussed attack. It's a novel approach to mitigate cache-based side channel attacks. We then conclude in Sect. 4 with an outlook of the future.

## 2 Cache-Based Side Channel Attack

Side Channels, include execution time, power consumption, heat, electromagnetic radiation, or even sound level emanating from a device. Early, such side channel attacks were used to break specialized systems such as smart cards. However, side channel attacks that exploit the shared resources in conventional microprocessors have been recently demonstrated. Cache-based attacks are extremely powerful because they do not require the attacker's physical presence to observe the side-channel and can therefore be launched remotely using only legal non-privileged operations. Consequently, hostile Vms can potentially extract sensitive data, such as passwords and cryptographic keys, from other Vms resident on the same physical machine by using memory caches as side channels. This attack consists on:

1. Measuring cache load
2. Analyzing cache load
3. Extracting private information.

### 2.1 Memory Cache

A CPU cache [2] is a cache used by the central processing unit (CPU) of a computer to reduce the average time to access data from the main memory. The cache is a smaller, faster memory which stores copies of the data from frequently used main memory locations. Most CPUs have different independent caches, including instructions and data caches, where the data cache is usually organized as a hierarchy of more cache levels (L1, L2 etc.). In this paper, our approach is based on L3 cache as it is shared by all the Cores in the case of CPU Multicore processors.

### 2.2 BERNSTEIN'S Attack

BERNSTEIN's attack [3, 4] exploits the memory latency against AES encryption process. Through the timing behavior of memory systems, an attacker can easily retrieve the full AES key.

During encryption, the AES tables used do not all fit, or stay in cache. In theory if the cache size was large enough or the AES tables small enough they would fit, but other processes would inevitably compete for cache. This leads to cache-misses during encryption. These cache-misses cause encryption to take a variable amount of time.

The attack is based on measuring the time variances for encryption of various inputs under a known key, and comparing these to time variances under a secret/unknown key. With enough timing information the full key can be recovered.

# 3  Related Work

## 3.1  STEALTHMEM System

Kim et al. [5], propose a system-level protection mechanism against cache-based side channel attacks in the cloud. It manages a set of locked cache lines per core, which are never evicted from the cache, and efficiently multiplexes them so that each VM can load its own sensitive data into the locked cache lines.

This system is very efficient, but is limited over the number of VMs that a physical host can lodge. That's because each VM has its own locked cache line.

The idea of our SAFEPERIMETER System is to assign to each VM a private locked plage only during the execution of a confidential thread, like AES encryption.

## 3.2  MIRAGE System

Microsoft's Azure [6], IBM's Smart Business cloud offerings [7], Google's AppEngine [8] and Amazon's EC2 [9] are the most used cloud services. They share large-scale computing resources among a large number of users. However, such systems expose multiple vulnerabilities. That's why Jinpeng et al. [10], propose an image management system called Mirage that addresses the following security management features:

1. Access control framework that regulates the sharing of VM images. This reduces the publisher's risk of unauthorized accesses to her images.
2. Image filters that are applied to an image at publish and retrieve time to remove unwanted information in the image.
3. Provenance tracking mechanism that tracks the derivation history of an image and the associated operations that have been performed on the image.
4. Set of repository maintenance services, such as periodic virus scanning of the entire repository, that detect and fix vulnerabilities discovered after images are published.

This approach to manage image security could introduce huge performance. Unfortunately, it is not enough, mostly it don't secure VMs from some attacks like cache based side channel attacks.

# 4 Our Contribution: SAFEPERIMETER System

## 4.1 Problem: A Confidential Information Can Be Recovered Exploiting Cache Hits and Cache Misses

For the example of BERNSTEIN's attack mentioned above, a malicious VM can recover the full key by measuring the time variances for encryption of various inputs under a known key, and comparing these to time variances under a secret/unknown key.

## 4.2 Solution: Cache Locking

In order to mitigate such discussed attack above, the idea around our cache locking consists on securing access to L3 cache by locking it line by line, for a special thread. We can understand this, by viewing it like a shared parking slot; only one car at a time can park in slot.

Otherwise, cache locking is the ability to prevent retrieving or overwriting of some or all of a microprocessor instruction or data cache. Cache locking can occur for either an entire cache or for individual ways within the cache.

Figure 2 shows the interaction model between CPU, Cache and Main Memory, when a Cache line is locked, and in two different cases:

1. Case 1: the thread is authorized to access the locked line.
2. Case 2: the thread is not authorized to access the locked line.

In case 1, the CPU is looking for the content of "X" in the Cache (1).

Finding the content in the cache, the CPU is trying to retrieve the content of "X" from the cache (2).

The CPU is holding the content of "X" to execute the instruction (3).

In case 2, The CPU is looking for the content of "Y" in the Cache (I).

This thread is not authorized to retrieve the content of "Y", even if it is effectively present in the cache (II).

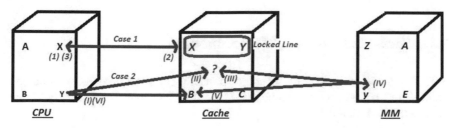

**Fig. 2** Interaction model between CPU, Cache and MM, when a Cache line is locked. Two different cases

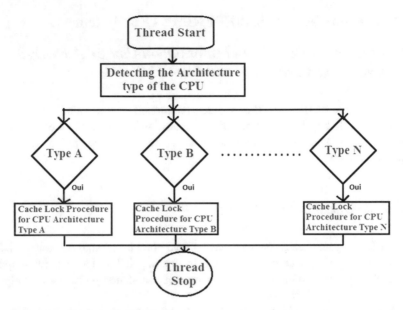

**Fig. 3** Algorithm: Execution of Cache Lock Procedure according to the architecture of the CPU

The CPU is trying to find its content in the Main Memory (III).

Finding the content in the MM, the CPU is trying to hold its content in the cache for a future reuse (IV). In this case, the thread cannot store the content in the locked line of the cache, and instead of that, the content of "Y" will be stored by overwriting any other content, for example the content of "B" (V). The CPU is holding the content of "Y" to execute the instruction (VI).

The procedure for locking down a line in the Instruction Cache and the Data Cache are slightly different, depending on the processor architecture.

In this paper, we experience the ARM [11] architecture and explain the corresponding cache lock procedure. Compared to other architectures, the idea of our SAFEPERIMETER System is being a short software routine that can automatically detect the processor architecture and launch the corresponding program who will handle the Cache Lock procedure.

Figure 3 presents the algorithm of this SAFEPERIMETER Service.

### 4.3 Cache Locking in ARM Architecture

ARM is a family of instruction set architectures for computer processors based on a reduced instruction set computing (RISC [15]) architecture developed by British company ARM Holdings.

In this architecture, and in both cases of locking down a line in the Instruction Cache and the Data Cache:

1. The cache must be put into lock down mode by programming register 9.
2. A line fill must be forced.
3. The corresponding data must be locked in the cache.

If more than one line is to be locked, a software loop must repeat this procedure.

### 4.3.1 Data Cache Lock Down

For the Data Cache, the procedure is as follows:

1. Writing to `CP15 register 9, setting DL = 1 and Dindex = 0.`
2. Initializing the pointer to the first of the 16 words to be locked.
3. Executing an `LDR` [12, 13] from that location. This forces a line fill from that location, and the resulting four words are captured by the cache.
4. Incrementing the pointer by 16 to select cache bank 1.
5. Executing an `LDR` from that location. The resulting line fill is captured in cache bank 2.
6. Repeating steps 1 to 5 for cache banks 3 and 4.
7. Writing to `CP15` [14] `register 9, setting DL = 0 and Dindex = 1`.

If there were more data to lock down, at the final step, step 7, the DL bit should be left HIGH, Dindex incremented by 1 line, and the process repeated. The DL bit should only be set LOW when all the lock down data has been loaded.

### 4.3.2 Instruction Cache Lock Down

For the I Cache, this procedure is as follows:

1. Writing to `CP15 register 9, setting IL = 1 and Iindex = 0.`
2. Initializing the pointer to the first of the sixteen words to lock down.
3. Forcing a line fill from that location by writing to `CP15 register 7`.
4. Incrementing the pointer by 16 to select cache segment 1.
5. Forcing a line fill from that location by writing to `CP15 register 7`. The resulting line fill is captured in segment 1.
6. Repeating for cache segments 3 and 4.
7. Writing to `CP15 register 9, setting IL = 0 and Iindex = 1`.

If there were more data to lock down, at the final step, step 7, the `IL bit` should be left HIGH, `Iindex` increment by 1 line and the process repeated. The `IL bit` should be set LOW when all the lock down data had been loaded.

Performing lock down in the I Cache involves a similar sequence of operations, except that the `IL` and `Iindex` of `CP15 register 9` are accessed.

The only significant difference in the sequence of operations is that an `MCR` instruction must be used to force the line fill in the I Cache, instead of an `LDR`, this is due to the Harvard nature of the processor. During the `MCR`, the value set up in the pointer register is output on the instruction address bus, and a memory access is forced. As this misses in the cache (due to earlier flushing), a line fill occurs.

The rest of the sequence of operations is exactly the same as for D Cache lock down.

## 4.4 Implementing SAFEPERIMETER Service on the Mirage System

So, the idea is to implement our SAFEPERIMETER Service into the Mirage system, especially that Jinpeng, Xiaolan, Glenn, Vasanth and Peng expect that many more such services can be implemented efficiently in their image management system.

Finally we get an image management system more secure and more reliable.

Figure 4 presents the global Mirage system after implementing SAFEPERIMETER service.

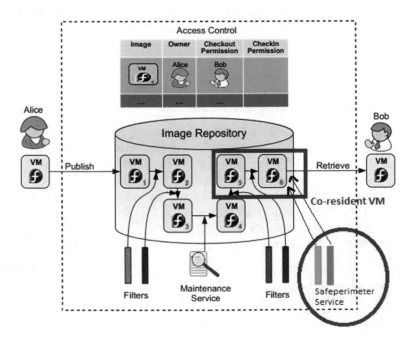

**Fig. 4** Security Features of the Mirage System after implementing SAFEPERIMETER Service

# 5 Conclusion and Future Work

Cloud computing offers great potential to improve productivity and reduce costs. It also poses many new security risks. In this paper, we proposed our novel approach called SAFEPERIMETER System for mitigating cache based side-channel attacks. It consists on securing access to L3 cache by locking it line by line, for a special confidential thread, without assigning any locked cache line to any VM.

We also, proposed to implement our service into the Mirage system, so we get finally an image management system more secure and more reliable.

Our research can be extended in several directions. First, we are going to implement this model to validate its security, and then we can find out how to further improve security of the system. Second, further study should be conducted to evaluate the performance of the model in the other CPU architectures. Third, we plan to test this service on Mirage system, in collaboration with their authors.

# References

1. Wu, H., Ding, Y, Winer, C., Yao, L.: Network security for virtual machine in cloud computing. In: 5th International Conference on Computer Sciences and Convergence Information Technology (ICCIT), pp. 18–21. IEEE Computer Society Washington, DC, USA (2010)
2. O'Hanlon, M., Tonge, A.: Investigation of Cache-Timing Attacks on AES. School of Computing, Dublin City University, Singapore (2005)
3. Bernstein, J.: Cache-Timing Attacks on AES. Department of Mathematics, Statistics, and Computer Science, The University of Illinois at Chicago (2005)
4. Weiß, M., Heinz, B., Stumpf, F.: A Cache Timing Attack on AES in Virtualization Environments, pp. 3–13. Fraunhofer Research Institution AISEC, Garching, Germany (2011)
5. Kim, T., Peinado, M., Mainar-Ruiz, G.: STEALTHMEM: System-Level Protection Against Cache-Based Side Channel Attacks in the Cloud (2012)
6. Microsoft. Azure Services Platform. http://www.microsoft.com/azure/defalt.mspx
7. IBM. IBM Cloud Computing. http://www.ibm.com/ibm/cloud
8. Google. Google App Engine. http://code.google.com/appengine
9. Amazon. Amazon EC2. http://aws.amazon.com/ec2
10. Wei, J., Zhang, X., Ammons, G., Bala, V., Ning, P.: Managing Security of Virtual Machine Images in a Cloud Environment, pp. 3–4 (2009)
11. ARM Architecture Reference Manuel (2005)
12. Maranget, L., Sarkar, S., Sewell, P.: A Tutorial Introduction to the ARM and POWER Relaxed Memory Models, pp. 21–25 (2012)
13. ARMv8 Instruction Set Overview. Architecture Group (2011)
14. JTAG debug interface for GNU Debugger, p. 49 (2014)
15. The Haverford Educational RISC Architecture (2014)

# Part IX
# ICT Based Education and Services ICT Challenges and Applications

# Improving the Session Identification Using the Ratio of Auxiliary Pages Estimate

**Michal Munk and Ľubomír Benko**

**Abstract** Data pre-processing is an important part of web log mining. This paper focuses on one of the phases of data pre-processing—on session identification. Cutoff time is an important part of the session identification using the Reference Length method. The aim of this paper is to compare the influence of subjective and sitemap estimation of the auxiliary pages ratio on calculation of cutoff time. Based on the sitemap and subjective estimation the calculation of auxiliary pages ratio was compared. The ratio of auxiliary pages has only impact on the quantity of extracted rules in the files with path completion.

**Keywords** Web usage mining · Data pre-processing · Session identification · Auxiliary pages · Reference length · Log files

## 1 Introduction

Log files are files that store anonymous information about user actions that occurred on a web portal. We can capture important information from the log files, such as understanding of users' interests and habits. We need to distinguish the web site visitors from each other on the ground of storing anonymous information. There are many methods helping us with the user identification based on the log file. In this paper, we will focus on the method called Reference Length, particularly on the ratio of auxiliary pages and its influence on session identification.

The rest of the paper is structured subsequently: in the Sect. 2 we summarize the related work of other authors dealing with data pre-processing issues in the field of web usage mining. Subsequently, we particularize the experiment in the Sect. 3.

M. Munk (✉)
Constantine the Philosopher University in Nitra, Nitra, Slovakia
e-mail: mmunk@ukf.sk

Ľ. Benko
University of Pardubice, Pardubice, Czech Republic
e-mail: lubomir.benko@gmail.com

© Springer International Publishing Switzerland 2016
A. El Oualkadi et al. (eds.), *Proceedings of the Mediterranean Conference on Information & Communication Technologies 2015*, Lecture Notes in Electrical Engineering 381, DOI 10.1007/978-3-319-30298-0_56

This section describes the research methodology and summary of experiment results. Finally, we discuss the obtained results in the Sect. 4.

## 2 Related Work

Data preparation is the most important part of Web Usage Mining and various techniques for data pre-processing were proposed. In [1], researchers have proposed a novel approach using Longest Common Subsequence algorithm for classifying the user navigation patterns. In the preparation phase the algorithm needs a current active session which represents the user's route on portal. Compared to existing approaches the results in [2] show the performance of proposed algorithm and give good results for web usage mining. The work with log files is essential for data pre-processing. Researches in [3] focused on data preparation of Web usage mining where they introduce an algorithm called "USIA" which can exactly identify user and session. In [4], an analysis of modified log files with the use of petri nets helped to eliminate the interfering elements from processes of e-course modelling. In [5], an offline data preparation and clustering approach are used to determine groups of users with similar browsing patterns. Path completion, described in [6], is an important part of data preparation.

Each user when browsing the web visits some web pages and spends some time on the web portal. In the process of session identification it is important to divide user's visits into sessions. According to [7] the session is characterized as an activity of one user in a certain time on the web portal. Solution for session identification is offered by time oriented heuristics, structure oriented heuristics or navigation oriented. Cooley et al. [8] introduced the time oriented heuristic called heuristic h1, which creates sessions based on a time window of 30 min. Spiliopoulou et al. [9] advised to identify sessions based on a time window of 10 min, called heuristic h2.

This paper is focused on a navigation oriented session identification method, namely the Reference Length method. The Reference Length approach is based on the assumption that the time which a user spends on a page correlates whether the page is classified as auxiliary or content page [8, 10]. Based on previous research [10] we assume that the variance of times spent on the auxiliary pages is small, because the user 'only' passes through the pages to his/her search target. If we have an estimation of the cutoff time $C$, then the session (visit) is a sequence $k$ of visited pages with the time mark, for which is valid: the first $k - 1$ pages are classified as the auxiliary pages, the time spent on these pages is less or equal to the cutoff time and the last $k$th page is classified as a content one and the time spent on this page is higher than cutoff time.

# 3 Experiment

We used a log file of university web site to compare the methods of calculating the cutoff time $C$. Records were cleaned using conventional log file pre-processing methods. Consequently we focused on session identification using the Reference Length technique and also on the reconstruction of activities of web users using the sitemap. In the experiment, we compared four files which were prepared on various levels. Each file was cleaned from unnecessary data with the same algorithm. In the phase of session identification we used the Reference Length method with the different calculation the cutoff time. We compared the influence of ratio of auxiliary pages on the calculation based on the sitemap and subjective estimate. In the subjective estimate the ratio of auxiliary pages was defined by the web portal creator or administrator (what he/she defines as auxiliary page). On the other hand, calculation of ratio of auxiliary pages from the sitemap offers more accurate information about the ratio. The calculation was made using $p = \frac{a}{n}$, where $a$ is a number of auxiliary pages and $n$ is a number of all pages. The number of auxiliary pages was obtained from the sitemap and it corresponds to the number of unique referring pages on the web portal.

## 3.1 Research Methodology

Experiment was realized in several steps as in [6]. In File A1, sessions were identified using the Reference Length method and the ratio of auxiliary pages was calculated from the sitemap (12.3 %). In File A2, sessions were identified using the Reference Length, the ratio of auxiliary pages was calculated from the sitemap (12.3 %) and we reconstructed activities of the web users. In File B1, sessions were identified using the Reference Length and the ratio of auxiliary pages was the subjective estimate (30 %). In File B2, sessions were identified using the Reference Length and the ratio of auxiliary pages was the subjective estimate (30 %) and we reconstructed activities of the web users. In the next steps we applied sequence analysis on these files to extract sequence rules for each file. Finally, we joined these rules into one data matrix where each rule can be occurred once.

We articulated the following assumptions: (a) we expect that the identification of sessions using the Reference Length method, calculated from a sitemap, will have a significant impact on the quantity of extracted rules; (b) we expect that identification of sessions using the Reference Length method, calculated from the sitemap, will have a significant impact on decreasing the portion of trivial and inexplicable rules.

## 3.2    Results

With sequence analysis we can get actionable (useful), trivial and inexplicable rules (Table 1). For deciding on the type of rule, there is no algorithm. Useful rules contain high quality, actionable information. Trivial results are already known by anyone at all familiar with the business. Inexplicable results seem to have no explanation and do not suggest a course of action [11].

The analysis (Table 1) resulted in sequence rules which we obtained from frequent sequences fulfilling their minimum support (in our case, min $s = 0.01$). Frequent sequences were obtained from identified sequences, i.e. visits of individual users during 1 week. We used STATISTICA Sequence, Association and Link Analysis, for sequence rules extraction. It is an implementation of algorithm using the powerful a priori algorithm together with a tree structured procedure that only requires one pass through data [12].

There is a high coincidence between the results (Table 1) of sequence rule analysis in terms of the portion of found rules in case of files with the identification of sessions based on sitemap estimate and subjective estimate without the path completion (A1, B1). The most rules were extracted from files with the path completion; be specific, 62 were extracted from the file A2, which represents over 79 % and 77 were extracted from the file B2, which represents over 98 % of the total number of found rules. Generally, more rules were found in the observed files with the completion of paths.

Based on the results of Q test (Table 1), the zero hypothesis, which reasons that the incidence of rules does not depend on individual levels of data preparation for web usage mining, is rejected at the 1 % significance level.

Kendall's coefficient of concordance represents the degree of concordance in the number of found rules among examined files. The value of coefficient (Table 2) is approximately 0.37, while 1 means a perfect concordance and 0 represents a discordance. Low values of coefficient confirm the Q test results.

**Table 1**  Incidence of discovered sequence rules in particular files

| Body | ==> | Head | A1 | A2 | B1 | B2 | Type of rule |
|---|---|---|---|---|---|---|---|
| (http://www.ukf.sk) | ==> | (http://www.ukf.sk/ university/directory ) | 1 | 1 | 1 | 1 | Useful |
| ⋮ | ==> | ⋮ | ⋮ | ⋮ | ⋮ | ⋮ | ⋮ |
| (http://www.ukf.sk/ university-parts), (http://www.ukf.sk) | ==> | (http://www.ukf.sk/ university-parts/ faculty-of-arts) | 0 | 0 | 0 | 1 | Inexplicable |
| Count of derived sequence rules | | | 35 | 62 | 39 | 77 | |
| Percent of derived sequence rules (Percent 1's) | | | 44.9 | 79.5 | 50.0 | 98.7 | |
| Percent 0's | | | 55.1 | 20.5 | 50.0 | 1.3 | |
| Cochran Q test | | | $Q = 86.63190$; df = 3; $p < 0.001$ | | | | |

**Table 2** Homogeneous groups for incidence of derived rules in examined files

| File | Incidence mean | 1 | 2 | 3 |
|------|----------------|---|---|---|
| A1 | 0.449 | *** | | |
| B1 | 0.500 | *** | | |
| A2 | 0.795 | | *** | |
| B2 | 0.987 | | | *** |
| Kendall Coefficient of Concordance | | | | 0.37022 |

From the multiple comparisons [13] (Scheffe test) one homogenous group (Table 2) was identified in terms of the average incidence of found rules (A1, B1). Statistically significant differences were proved in the average incidence of found rules between files A2 and B2 as well as X2 and X1, on the 5 % significance level.

The ratio of auxiliary pages has an important impact on the quantity of extracted rules only in case of the path completion (A2, B2).

# 4 Conclusion

Both assumptions concerning the quantity of extracted rules in terms of decreasing the portion of trivial and inexplicable rules of sessions identified using the Reference Length method, calculated from the sitemap were only proven partially. They were fully proven after the path completion. On the contrary, path completion is dependent on the accuracy of session identification. The ratio of auxiliary pages has the impact on quantity of extracted rules only in the case of files with path completion (A2 vs. B2). However, making provisions for the identification of sessions, based on the estimation of the ratio of auxiliary pages, has no significant impact on the quantity of extracted rules in the case of files without the path completion (A1 vs. A2).

**Acknowledgments** This paper is supported by the project VEGA 1/0392/13 Modelling of Stakeholders' Behaviour in Commercial Bank during the Recent Financial Crisis and Expectations of Basel Regulations under Pillar 3- Market Discipline.

# References

1. Jalali, M., Mustapha, N., Mamat, A., Sulaiman N.: A recommender system for online personalization in the WUM applications. In: Proceedings of the World Congress on Engineering and Computer Science 2009, vol, II, pp. 741–746, San Francisco, USA (2009)
2. Vellingiri, J., Pandian, S.: A Novel technique for web log mining with better data cleaning and transaction identification. J. Comput. Sci. **7**, 683–689 (2011) (Science Publications)
3. Huiying, Z., Wei, L.: An intelligent algorithm of data pre-processing in Web usage mining. In: Intelligent Control and Automation, vol. 4, pp. 3119–3123, IEEE (2004)
4. Balogh, Z., Turcani, M., Magdin, M.: Design and creation of a universal model of educational process with the support of petri nets. Lect. Notes Electr. Eng. **269**, 1049–1060 (2014)

5. Sumathi, C.P., Padmaja Valli, R., Santhanam, T.: Automatic recommendation of web pages in web usage mining. Int. J. Comput. Sci. Eng. **2**, 3046–3052 (2010)
6. Munk, M., Kapusta, J., Švec, P.: Data preprocessing evaluation for web log mining: reconstruction of activities of a web visitor. Procedia Comput. Sci. **1**, 2273–2280 (2010)
7. Patil, P., Patil, U.: Preprocessing of web server log file for web mining. World J. Sci. Technol. **2**, 14–18 (2012)
8. Cooley, R., Mobasher, B., Srivastava, J.: Data Preparation for Mining World Wide Web Browsing Patterns. Knowl. Inf. Syst. **1**, 5–32 (1999)
9. Spiliopoulou, M., Mobasher, B., Berendt, B., Nakagawa, M.: A framework for the evaluation of session reconstruction heuristics in web-usage analysis. In: Journal on Computing, vol. 15, pp. 171–190, INFORMS, Linthicum (2003)
10. Kapusta, J., Munk, M., Drlik, M.: Cut-off time calculation for user session identification by reference length. In: 2012 6th International Conference Application of Information and Communication Technologies (AICT), pp. 1–6, Tbilisi (2012)
11. Berry, M.J., Linoff, G.S.: Data Mining Techniques: For Marketing, Sales, and Customer Relationship Management. Wiley Publishing Inc, Chichester (2004)
12. Klocokova, D.: Integration of heuristics elements in the web-based environment: Experimental evaluation and usage analysis. Procedia Soc. Behav. Sci. **15**, 1010–1014 (2011)
13. Pilkova, A., Volna, J., Papula, J., Holienka, M.: The influence of intellectual capital on firm performance among Slovak SMEs. In: Proceedings of the 10th International Conference on Intellectual Capital, Knowledge Management and Organisational Learning (ICICKM-2013), pp. 329–338 (2013)

# Modeling an Intelligent Architecture of Intrusion Detection System for MANETs

Sara Chadli, Mohammed Saber, Mohamed Emharraf and Abdelhak Ziyyat

**Abstract** Mobile Ad hoc Network consists of some nodes that are stand randomly in operational environment. Because nodes are without any predefined infrastructure and mobility then there are susceptible to intrusions and attacks. Securing is an important field in this type of network. Intrusion Detection Systems (IDSs) may act as defensive mechanisms, since they monitor network activities in order to detect malicious actions performed by intruders, and then initiate the appropriate countermeasures. IDS for MANETs have attracted much attention recently and thus, there are many publications that propose new IDS solutions or improvements to the existing. In this paper, we propose a new IDS architecture for MANETs, this architecture is a combination model hierarchical based on clusters and cooperation model based on a multi-agent system (SMA). In this paper, we are used AUML language to describe the operation of different agents of our architecture.

**Keywords** Intrusion Detection System (IDS) · IDS architectures · Mobile ad hoc networks (MANETs) · AUML · Multi-Agent System (SMA) · MANETs security · Security attacks

S. Chadli (✉) · A. Ziyyat
Laboratory Electronics and Systems, Faculty of Sciences, First Mohammed University, Oujda, Morocco
e-mail: chad.saraa@gmail.com
URL: http://www.ump.ma

A. Ziyyat
e-mail: abdelhak_ziyyat@hotmail.com

M. Saber · M. Emharraf
Laboratory LSE2I, National School of Applied Sciences, First Mohammed University, Oujda, Morocco
e-mail: mosaber@gmail.com

M. Emharraf
e-mail: m.emharraf@gmail.com

© Springer International Publishing Switzerland 2016
A. El Oualkadi et al. (eds.), *Proceedings of the Mediterranean Conference on Information & Communication Technologies 2015*, Lecture Notes in Electrical Engineering 381, DOI 10.1007/978-3-319-30298-0_57

# 1   Introduction

A mobile ad hoc network (MANET) is a collection of autonomous nodes that form
a dynamic, purpose-specific, multi-hop radio network in a decentralized fashion.
The wireless mobile natures of MANETs in conjunction with the absence of access
points, providing access to a centralized authority, make them susceptible to a
variety of attacks [1]. An effective way to identify when an attack occurs in a
MANET is the deployment of an Intrusion Detection System (IDS).

The existing IDS architectures for MANETs fall under three basic categories
[2, 3] (a) stand-alone, (b) cooperative and (c) hierarchical. The employed intrusion
detection engines are also classified into three main categories [4]: (i) signa-
ture-based; (ii) anomaly-based engines and (iii) specification based engines.

In this paper, we have proposed a hybrid model of intrusion detection system
which combines between models hierarchical based on of clusters and cooperation
model based on a multi-agent system (SMA. The rest of this chapter is organized as
follows. Section 2 we present in detail the design of proposed architecture.
Section 3, we present in detail the operating of proposed architecture. Finally,
Sect. 4 contains the conclusions.

# 2   Proposed IDS Architecture for MANETs

In this section we present our IDS architecture for MANETs. First, we include the
objectives of our classification, and then we present the design of our architecture.

## 2.1   Design of Our IDS Architecture for MANETs Based on Multi-agent

Our architecture is a combination of hierarchical model based of clusters and
cooperative model based on a multi-agent system. The Clustering in MANETs is an
effective way to structure the network. Its purpose is to identify a subset of nodes
(Fig. 1a) in the network and assigned him a the designated node (DN).

The node is elected by applying the clustering algorithm HEED (Hybrid,
Energy-Efficient, Distributed) [5], because this algorithm has four primary objectives: (i) prolonging network lifetime by distributing energy consumption, (ii) terminating the clustering process within a constant number of iterations,
(iii) minimizing control overhead (to be linear in the number of nodes), and
(iv) producing well-distributed cluster heads.

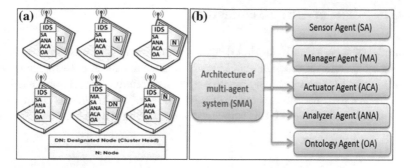

**Fig. 1** **a** Architecture of Cluster in MANET; **b** architecture of multi-agent system

In our proposed architecture, the agents realizing detection tasks by communication and collaboration between them. The architecture consists of a multi-agent detection system that uses five classes of agents (Fig. 1b): Sensor Agent (*SA*), Manager Agent (*MA*), Ontology Agent (*OA*), Agent actuator (*ACA*) Agent Analyzer (*ANA*).

## 2.2 Design of a MANET Node in Our Architecture

In our architecture, the agents (SA), (ANA) (ACA) and (MA) are installed in the various Member nodes of MANET cluster. The designing a node based agents as shown in (Fig. 2).

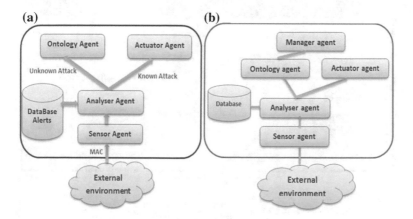

**Fig. 2** **a** IDS Architecture node in SMA; **b** IDS architecture of cluster head in SMA

# 3   Operation of Our Architecture

Our model of security and intrusion detection based on a distributed approach using multi-agent system for receiving intelligence of these agents. It is formed by agents with the capacity to react quickly reactive against different types of attacks (Fig. 3).

To facility the operation of our proposed model of multi-agent system we are used design based on the AUML language for describing the operation of agents as follows.

## 3.1   Sensor Agent(SA)

The sensor agent (SA) *Sensor()*, is the initiator of the detection process. It is down the chain of operation, it captures network raw traffic by using functions *get_info()* and *get_origin()*, and formate in a predefined format by using *set_format()* function that will be sent to the agent analyzer (ANA) with *Send_analyser* function. The

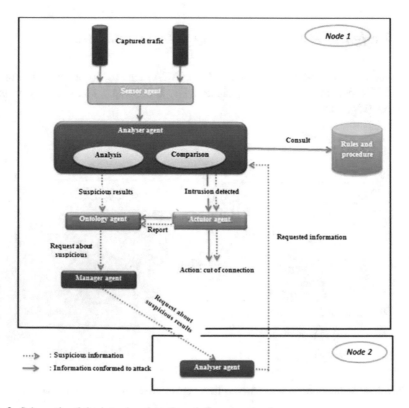

**Fig. 3**   Schematic of the intrusion detection platform proposed

**(a)** **(b)** **(c)** **(d)**

**Fig. 4** Implementation level of the: **a** Sensor Agent; **b** Actuator Agent; **c** Analyzer Agent; **d** Manager Agent

following figure (Fig. 4a) shows a portion of the class agent diagram for the agents sensor.

## 3.2 Analyzer Agent(ANA) and Actuator Agent (ACA)

The analyzer agent *analyser()* analyse the formatted data with *app_analysis()* function and compare the analysis result by applies rules of detection recorded in his database with *comparison()* and *set_decision()* functions (Fig. 4c). According to the analysis result, this agent will initiate an external treatment. Two cases are possible:

- If a malicious activity is confirmed an attack signature, then the analyzer agent use *call_action()* function and *call_actuator()* functions to communicates the result to the actuator agent to perform the necessary actions by using *action()* and *cut_connection()* functions (Fig. 4b).
- In second case if the activity is judged suspicious by comparison of detection thresholds, the agent analyzer (ANA) request additional information to confirm the nature of the activity we say that it is initiating agent collaboration. The analyser agent send request to ontology agent by using *call_ontology()* function to offers semantic verification service of the knowledge to facility the operation for other agent the following figures (Fig. 4b) show a portion of the class agent diagram for the Analyser agent and actuator agent.

## 3.3 Manager Agent(MA)

The manager agent can ask other agents for local information related to suspicious activity with *help()* and *get_request()* functions, in this case one or more agents analyzer located in different nodes in cluster can provide local information to the initiator. In this case one or more agents analyzer located in different nodes in cluster can provide local information to the initiator. This last will repeat the same operation as we see in first case. For its part the manager agent can invoke a data sharing with *set_sharing()* and *send_all()* functions to enrich local data at the initiator. The following figures (Fig. 4d) show a portion of the class agent diagram for Manger agent.

# 4 Conclusion

In this paper we did a study the IDS architectures for MANETs and we have proposed a hybrid model of intrusion detection system which combine between model hierarchical based on of clusters and cooperation model based on a multi-agent system (SMA). Our new architecture is distributed architecture based on the intelligence of the multi-agent system allowing both to reacting rapidly against complex attacks and evaluating the state of flows relative to predefined rules and procedures and other hand enhances the level of security provided to the target monitored.

In the future, after the design phase of the proposed model, we think to use an open source platforms (JADE and MADKIT) to develop the new system intrusion detection and we conducted a simulation platform that reflects the goals already set.

# References

1. Chadli, S., Saber, M., Ziyyat, A.: Defining categories to select representative attack test-cases in MANETs. In: IEEE Xplore DIGITAL LIBRARY, pp. 658, 663 (2014). doi:10.1109/CSNT. 2014.138
2. Anantvalee, T., Wu, J.: A survey on intrusion detection in mobile ad hoc networks, wireless/mobile network security Springer (2006) Chapter 7, pp. 170–196. Wireless Network Security. doi:10.1007/978-0-387-33112-6_7
3. Xenakis, C., Panos, C., Stavrakakis, I.: A comparative evaluation of intrusion detection architectures for mobile ad hoc networks. Comput. Secur. **30**(1), 63–80 (2011). doi:10.1016/j. cose.2010.10.008
4. Li, Y., Qian, Z.: Mobile agents-based intrusion detection system for mobile ad hoc networks. In: 2010 Intl Conf on and Information Technology & Ocean Engineering Innovative Computing & Communication, 2010 Asia-Pacific Conf on (CICC-ITOE), pp. 145, 148, 30–31 Jan 2010. doi:10.1109/CICC-ITOE.2010.45
5. Younis, O.; Fahmy, Sonia, HEED: a hybrid, energy-efficient, distributed clustering approach for ad hoc sensor networks. IEEE Trans. Mob. Comput. **3**(4), 366, 379. doi:10.1109/TMC.2004.41

# A Fault Tolerance Mechanism in Distributed and Complex System on a LAN

Farid Lassoued and Ridha Bouallegue

**Abstract** The fuzzy cognitive maps [1–6] are qualitative tools which can capture the extent from cause to effect in the links that exist within a complex system such as the information system. These cognitive maps are a simple way of representing knowledge with a huge capacity of interpreting the information. Indeed, they are exploited for the decision-making, the prediction of future states and the explanation of past actions. Added to these capacities, when the information is applied and propagated through the model, the topology of the map itself can be used in the diagnosis of breakdowns by identifying the causes of the nodes of interest [7]. The main objective of this paper is the conception of impact analysis of engine of rules in an environment object in which objects and their links of impact (CIs: elementary components of the information system contributing to the delivery of a service), are neither defined nor ordered and the real time restoration of analysis of impact results and presentation according to the various orchestrated processes. Thus our role is to conceive a design of impact analysis and its development guaranteeing in times of answer by using the inference of the fuzzy cognitive maps.

**Keywords** Fuzzy cognitive maps · Fuzzy inference · Complex system · Analysis of impacts · CI (configuration item)

## 1 Introduction

The principal objective of this paper is the conception of impact analysis of engine of rules in an environment object in which objects and their links of impact are neither defined nor ordered and the real time restoration of the results of the

F. Lassoued (✉) · R. Bouallegue
Innov'COM Laboratory, Higher School of Communications of Tunis,
Sup'Com, Carthage University, Tunis, Tunisia
e-mail: lassoued.ferid@gmail.com

R. Bouallegue
e-mail: ridha.bouallegue@gmail.com

© Springer International Publishing Switzerland 2016          563
A. El Oualkadi et al. (eds.), *Proceedings of the Mediterranean Conference
on Information & Communication Technologies 2015*, Lecture Notes
in Electrical Engineering 381, DOI 10.1007/978-3-319-30298-0_58

calculation of analysis of impact and presentation according to the various orchestrated processes. The goal of our spot is conceive a design of impact analysis and its development by guaranteeing performances in times of answers.

Thus, the researches within laboratory focus essentially on the study of the priority problems and optimization of an information system by integrating the various problems which arise from an industrial operator (phone, network networks IP, waiters' networks/software/IT applications/services). Machines, software and IT services are resources users that we can represent by nodes of a graph (or a map) interconnected or by autonomous agents. This graph forms a tree of dependence which connects the resources in a complex weft [8–10]. The arcs of this graph are then colored by a value which indicates the intensity of the relations of dependence inter nodes. The graph is directed; certain nodes can be entry points which then spread in the entire graph. The network manager has to know, every time, for every node (or agent): Its availability, its characteristics of daily activity, of debit, of memory capacity and its position within the graph of dependence.

It is easy to understand intuitively that the more a node is taken away from entry points, the more its weight is easy in the arborescence, but it is not necessarily the reality. The problem arises at two levels including:

- To determine the affectations of links between entrances and other resources minimizing the number of switching links.
- To analyze the impacts relations between the various nodes (CIs) of the project by using the most adequate method among those envisaged methods.

## 2 Study of CIs's Plan for the System

### 2.1 A Plan of CIs

After we note as follows (Fig. 1):

C1: Service; C2: Physical Machine; C3: Controller raid; C4: Multicore; C5: HDA; C6: HDB; C7: CPU0 et C8: CPU1

### 2.2 Study by Inference of the Fuzzy Cognitive Maps

In a first place we are going to determine the weights of the relations between various CIs. For this fact, we have to use the various formulae and the functions of thresholds proposed by the expert. Several cases which appear from this plan:

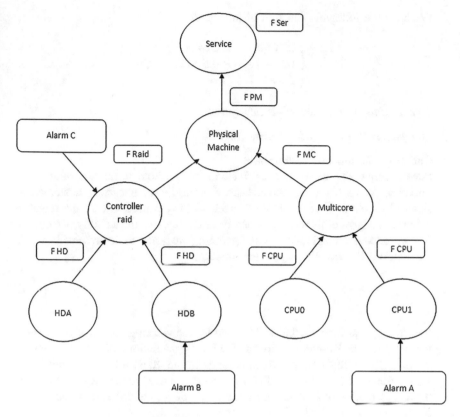

**Fig. 1** A plan of the relations of impact between CIs

- Case of the alarm B

$FHD = alarm\ value\quad with \begin{cases} HDB\ Ko : k1 = 1 \\ HDA\ Ok : k0 = 0 \end{cases}$

$FRaid = 0.6 \times k0 + 0.6 \times k1$

*If F Raid* $> 0.7$ *then F Raid* $= 1$ *else FRaid*

*So F Raid* $= 0.6$

$FPM = 1\ if\ C3 = 1\ or\ C4 = 1$

$We\ have \begin{cases} C3 = 0.6 \\ C4 = 0 \end{cases}$

$Fmph = 0.5 \times C3 + 0 \times C4 = 0.5 \times 0.60 + 0 = 0.3$

$Fser = 1 \times C2 = 1 \times 0.3 = 0.3$

We have the following function of threshold:

$$\begin{cases} 0 \prec Fser \prec 0.2 : Ok \\ 0.2 \prec Fser \prec 0.5 : Degraded \\ Fser \succ 0.5 : Ko \end{cases}$$

*Thus in that case the system is deg raded*

### Analysis of Impact by the Fuzzy Inference

- Choice of the function of threshold:
  Several functions can be used as function of threshold in the algorithm of the inference of the fuzzy cognitive maps. Among these functions we studied the sigmoid function. This function of threshold [11] is used in case the representation of a degree of increase, one degree of decrease or the stability of a concept is required and strategic scenarios of planning are going to be presented. Among the sigmoid functions we quote the function

$$F(x) = \frac{e^{2x} + 1}{e^{2x} - 1} \tag{1}$$

  It is the sigmoid function that will be held for the working continuation to satisfy the need of the laboratory experts if who have relations of impacts between various CIs having weights which belongs to the interval [−1, 1]. Thus in our application of the algorithm of fuzzy inference we need to have a scenario where the representation of a degree of increase, one degree of decrease or the stability of a concept is required.
- Algorithm of fuzzy inference:
  We call back here the algorithm of the fuzzy inference [12, 13] which was detailed in the first part of the paper.
  Thus the stages to be followed for the causal inference and the interpretation of the results are:

  - Create the adjacency-matrix E.
  - Opt for a vector of entrance Ci.
  - Do repetitive multiplications of vectors (E × Ci) up to a point of balance.
  - Every time, afterward of multiplication, we apply a function of threshold.
  - The vector-result becomes Ci + 1 = F [Ci × E] or F is the function of threshold.
  - Until affect a point of balance (vector which repeats several times successive).

- Analysis by fuzzy inference by using the sigmoid function:
  After the choice of the threshold function which satisfies best our needs, which is the sigmoid function, we apply the algorithm of causal inference to the previous plan after we determined the various weights of links between nodes. This choice of function of threshold justifies itself by making that the sigmoid

function more relevant than the other functions. It is of use to a decision-making with a good management of contrast. Several scenarios can appear while considering the failure of every alarm of the plan. We study after the impact of the excitements, with various percentages of increase (small, average or big), on the alarms A, B and C try everything to handle all the possible scenarios.

- Excitement of the alarm B:

In that case we are going to study the effect of an average increase of 0.8 on the node HDB (C6).

- Creation of the adjacency-matrix E:

$$
E = \begin{pmatrix}
0 & 0 & 0 & 0 & 0 & 0 & 0 & 0 \\
0.3 & 0 & 0 & 0 & 0 & 0 & 0 & 0 \\
0 & 0.6 & 0 & 0 & 0 & 0 & 0 & 0 \\
0 & 0 & 0 & 0 & 0 & 0 & 0 & 0 \\
0 & 0 & 0 & 0 & 0 & 0 & 0 & 0 \\
0 & 0 & 1 & 0 & 0 & 0 & 0 & 0 \\
0 & 0 & 0 & 0 & 0 & 0 & 0 & 0 \\
0 & 0 & 0 & 0 & 0 & 0 & 0 & 0
\end{pmatrix}
$$

- Affectation of the initial vector:

$$
C1 = (0 \quad 0 \quad 0 \quad 0 \quad 0 \quad 0.8 \quad 0 \quad 0)
$$

- Repetitive multiplication of the vectors with threshold:

$$
\begin{aligned}
C1 &= (0 \quad 0 \quad 0 \quad 0 \quad 0 \quad 0.8 \quad 0 \quad 0) \\
C1 \times E &= (0 \quad 0 \quad 0.8 \quad 0 \quad 0 \quad 0 \quad 0 \quad 0) \\
C2 &= (0 \quad 0 \quad 0.608 \quad 0 \quad 0 \quad 0.8 \quad 0 \quad 0) \\
C2 \times E &= (0 \quad 0.364 \quad 0.8 \quad 0 \quad 0 \quad 0 \quad 0 \quad 0) \\
C3 &= (0 \quad 0.311 \quad 0.608 \quad 0 \quad 0 \quad 0.8 \quad 0 \quad 0) \\
C3 \times E &= (0.093 \quad 0.364 \quad 0.8 \quad 0 \quad 0 \quad 0 \quad 0 \quad 0) \\
C4 &= (0.081 \quad 0.311 \quad 0.608 \quad 0 \quad 0 \quad 0.8 \quad 0 \quad 0) \\
C4 \times E &= (0.093 \quad 0.364 \quad 0.8 \quad 0 \quad 0 \quad 0 \quad 0 \quad 0) \\
C5 &= (0.081 \quad 0.311 \quad 0.608 \quad 0 \quad 0 \quad 0.8 \quad 0 \quad 0) : \textit{The point of balance}
\end{aligned}
$$

- Result and interpretation:

The vector C5 = (0.081 0.311 0.608 0 0 0.8 0 0) is a point of balance for the algorithm of inference and shows the enchainment provoked by an important excitement on a node of the plan of the relations between CIs. Then an important increase of 0.8 on the node HDB (C6) provokes an important increase of 0.608 on the node C3, a small increase of 0.311 on the node C2 and a small increase of 0.081 on the node Service (C1).

# 3 Conclusion

In this work, we were interested, more particularly, in analysis of impacts strictly speaking on explanatory examples of the relations of impacts between CIs by using algorithms of inference of the fuzzy cognitive maps in particular the inference by using a function of threshold of type sigmoid in a LAN. The working objective in this part is essentially to allow managing the impacts in a plan of CIs. This will allow add or modify CIs and their direct impacts (fault tolerance) without having to worry about verifying the impact on the final service.

Naturally, this work is far from being ended and there is certain number of points to be examined. The present work can follow up studies on the function of threshold used during the algorithm of the fuzzy inference. Why not to use an innovative function of threshold in the place the function thus sigmoid the project will be 100 % a project of innovation. Furthermore, we can enrich this work by the extension of software package of classic cognitive inference of maps exploited previously in a software package which manipulates in addition to that the inference of the fuzzy cognitive maps.

# References

1. Julie, A., Kosko, B.: Virtual worlds as fuzzy cognitive maps. In: Virtual Reality Annual International Symposium, vol. 12, pp. 471–477 (1993)
2. Tsadiras, A.K.: Comparing the inference capabilities of binary, trivalent and sigmoid fuzzy cognitive maps. Inf. Sci. **178**, 3880–3894 (2008)
3. Kosko, B.: Fuzzy cognitive maps. Int. J. Man Mach. Stud. **24**, 65–75 (1986)
4. Stylios, C.D., Georgopoulos, V.C., Malandraki, G.A., Chouliara, S.: Fuzzy cognitive map architecture for medical decision support systems. Appl. Soft Comput. **8**, 1243–1251 (2008)
5. Pelaez, C.E., Bowles, J.B.: Using fuzzy cognitive maps as a system model for failure modes and effects analysis. Inf. Sci. **88**, 177–199 (1996)
6. Yaman, D., Polat, S.: A fuzzy cognitive map approach for effect-based operations: An illustrative case. Inf. Sci. **179**, 382–403 (2009)
7. Pagageorgiou, E.I., Spyridonos, P.P., Glotsos, D.T., Stylios, C.D., Ravazoula, P., Nikiforidis, G.N., Groumpos, P.P.: Brain tumor characterization using the soft computing technique of fuzzy cognitive maps. Appl. Soft Comput. **8**, 820–828 (2008)
8. Perusich, K.: Using fuzzy cognitive maps to identify multiple causes in troubleshooting systems. Integr. Comput. Aided Eng **15**, 197–206 (2008)
9. Perusich, K.: Using fuzzy maps to assess multi-operator situation awareness. In: Collaborative Crew Performance in Complex Operational Systems, pp. 20–22 (1998)
10. Lee, K.C., Kim, J.S., Chung, N.H., Kwon, S.J.: Fuzzy cognitive map approach to web-mining inference amplification. Expert Syst. Appl. **22**, 197–211 (2002)
11. Lee, K.C., Lee, S.: Causal knowledge-based design of EDI controls: an explorative study. Comput. Hum. Behav. **23**, 628–663 (2007)
12. Chunmei, L.: Combination study of fuzzy cognitive map. Int. J. Energ. Environ. **1**, 65–69 (2007)
13. Rodriguez-Repiso, L., Setchi, R., Salmeron, J.L.: Modelling IT project success with fuzzy cognitive maps. Expert Syst. Appl. **32**, 543–559 (2007)

# A New Approach to Detect WEB Attacks Senario in Intrusion Detection System

Noureddine Rahmoun, Mohammed Saber, Elhassane Ettifouri,
Adil Zeaaraoui and Toumi Bouchentouf

**Abstract** The tremendous growth of the web-based applications has increased information security vulnerabilities over the Internet. Security administrators use Intrusion-Detection System (IDS) to monitor network traffic and host activities to detect attacks against hosts and network resources. The solutions proposed in the literature actually achieved good results for the detection rate, while there is still room for reducing the false positive rate. To this end, in this paper we propose a model of an IDS based on combination of Markov chain and Naïve Bayes, to reduce the rate of false positive.

**Keywords** WEB attacks · Intrusion detection system (IDS) · Naïve bayes classifier · Markov chain · False positive

## 1 Introduction

The protection of Web applications is challenging, because they are in general large, complex, highly customized and often created by programmers with poor security background. On the other hand, a requirement that a tool to protect Web

N. Rahmoun (✉) · M. Saber · E. Ettifouri · A. Zeaaraoui · T. Bouchentouf
Laboratory LSE2I, National School of Applied Sciences,
First Mohammed University, Oujda, Morocco
e-mail: rahmoun.noureddine@gmail.com

M. Saber
e-mail: mosaber@gmail.com

E. Ettifouri
e-mail: h.ettifouri@gmail.com

A. Zeaaraoui
e-mail: adilzeaaraoui@yahoo.fr

T. Bouchentouf
e-mail: tbouchentouf@gmail.com
URL: http://www.ensa.ump.ma

© Springer International Publishing Switzerland 2016                   569
A. El Oualkadi et al. (eds.), *Proceedings of the Mediterranean Conference on Information & Communication Technologies 2015*, Lecture Notes in Electrical Engineering 381, DOI 10.1007/978-3-319-30298-0_59

applications is desired to meet is being as autonomous as possible, i.e., it should not require extensive administration overhead. Several hardware and software solutions have been developed, and are available on the market.

Several works that focused on anomaly-based high speed classification, proposed the use of simple statistics on the application-layer payload to characterize the normal behavior of Web applications [1–3]. We share the definition of "normal behavior" provided by [4]: the term normal behavior generally refers to a set of characteristics (e.g., the distribution of the characters of string parameters, the mean and standard deviation of the values of integer parameters) extracted from HTTP messages that are observed during normal operation. Initially, the main obstacle to the large scale deployment of anomaly based IDS solutions has been the too high false positive rate, as not all the detected anomalies are actually related to attack attempts.

The payload is the data portion of a network packet, that is the portion of the network packet which carries the HTTP message. According to RFC 2616, the HTTP payload contains the Request-Line plus the Request-Header fields used by the client to pass additional information to the server RFC 2616.

In last years, several statistical models for IDS based on solutions anomaly. Among they there is the work [5] on IDS, Naïve Bayes classifier [6, 7] is used. This technique identifies the most important HTTP traffic features that can be used to detect HTTP attacks. The highlight is to enhance IDS performance through preparing the training data set allowing to detect malicious connections that exploit the HTTP service. The weakness of this work has not decreased the false positive rate.

In this work, we are interested in the analysis of the HTTP. The HTTP analysis offers the great advantage that the security of a Web application can be guaranteed by simply analyzing the network traffic incoming to the Web server, without a detailed knowledge of how the Web application is implemented and works. To this end, in this paper we propose a model of an IDS based on combination of Markov chain and Naïve Bayes, to reduce the rate of false positive.

In the remaining sections of this article, we describe the proposed our approach based on mixed Naïve Bayes and Markov chain in Sect. 2.

## 2  New Approach Based on Mixed Naïve Bayes and Markov Chain for Securing HTTP Services

The new approach that we present is based on the combination of the use of Naïve Bayes networks and Markov chains. The diagram in Fig. 1 explains the presented algorithm and describes our approach. Now, we explain in detail our approach.

The advantage of Naïve Bayes classification is it requires relatively little training data (database) to estimate the parameters necessary for classification, namely the means and variances of the variables. The drawback of this approach is it does not

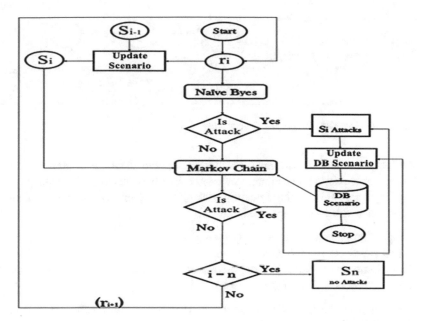

**Fig. 1** The proposed IDS model based on combination of Naïve Bayes and Markov Chain

take in consideration that web attacks are based on HTTP service which is essentially a transmission protocol data over the web (TCP), that is to say it is a client–server communication protocol. Therefore, the classification Naïve Bayes does not take in consideration the order of running queries as it handles them one after another. A set of queries with a specific order can be an attack scenario. So, this classifier is inefficient to identify an attack of type scenario.

We will model a scenario $S_n$ which is composed by requests $r_1$, $r_2$, $r_3$, ..., $r_n$ by the following notation $S_n = (r_1, r_2, r_3, ..., r_n)$. We then apply the algorithm of Fig. 1 to $S_n$ as follows:

1. If $n = 1$, then $S_1 = (r_1)$. In this case the Naïve Bayes classifier is effective for determining whether $S_n$ is an attack or not.
2. If $n > 1$ in this case we will apply Naïve Bayes $r_i$ with $1 \leq i \leq n$.
3. If there is a $1 \leq i \leq n$ such that $r_i$ is an attack by then the $S_n$ scenario is considered an attack.
4. If not, for all $i = 1$ we check for each $r_i, i \in 1, ..., n$, if the generated scenario is an attack or not using Markov chains as follows:
   if the scenario is an attack, so we will update the knowledge database that contains a set of queries $r_i$ of the different scenarios either attack or not, which are already covered by the presented scenario. Otherwise we will first see if the last query $r_n(i = n)$, if that is the case we store the scenario in the database as normal one. If not, we move the next request $r_1 + 1$.

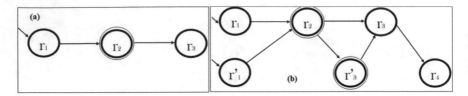

**Fig. 2** **a** Attack scenario $S$; **b** Attack scenario $S'$

As we mentioned in the previous paragraph, the database contains a set of different scenarios $r_i$ queries already processed by the scenario. This database plays a very important role in the detection and gives a graph visualization of attack scenarios in which each graph contains:

1. Graphs Summits are $r_i$ of each $S_n$ scenario.
2. Initial states are $r_1$ of each scenario.
3. Final State are $r_j$ with $j$ is the iteration that caused the stop (Stop) of the proposed model. If normal scenario $j = n$.

We will give the following example to understand the database mechanism: Consider the following scenario $S$:
$S = (r_1,\ r_2,\ r_3)$ with $r_2$ an attack, therefore as graph (Fig. 2a would be:
Lets take another scenario $S'$:
$S' = (r'_1,\ r_2,\ r'_3,\ R_3,\ R'_4)$ with $S'r'_3$ an attack and $r_2$, $r_3$ are of the $S$ scenario; so the scenario $S$ of Fig. 2b becomes as follows.

In this example, we see that according to this model we can deduce a dynamic graph that can help us detect attacks submitted from each scenario.

For the presented model it adds more detection efficiency and accuracy in attack using the algorithm based on Markov chains which is based on a scenario database detect Naïve Bayes classifier and the $S_i$ scenario as described in Fig. 1.


# 3 Conclusion

In this paper, we proposed an IDS designed to detect attacks against Web applications through the analysis of the HTTP by combination of chain Markov and Naïve Bayes classifier. With this work we provided for several innovative contribution. In this work, we are interested in the analysis of the HTTP. The HTTP analysis offers the great advantage that the security of a Web application can be guaranteed by simply analyzing the network traffic incoming to the Web server, without a detailed knowledge of how the Web application is implemented and works. In this work the proposed model of an IDS aims to reduce the rate of false positive. We plan to generalize this model on different attacks to improve the detection of attack scenarios.

# References

1. Krgel, C., Toth, T., Kirda, E.: Service specific anomaly detection for network intrusion detection. In: Proceedings of the 2002 ACM Symposium on Applied Computing (SAC '02). ACM, New York, NY, USA, pp. 201–208 (2002). doi:10.1145/508791.508835
2. Mahoney, M.V.: Network traffic anomaly detection based on packet bytes. In: Proceedings of the 2003 ACM Symposium on Applied Computing (SAC '03), pp. 346–350. ACM, New York, NY, USA (2003). doi:10.1145/952532.952601
3. Perdisci, R., Ariu, D., Fogla, P., Giacinto, G., Lee, W.: McPAD: a multiple classifier system for accurate payload-based anomaly detection. Comput. Netw. **53**(6), 864–881 (2009). doi:10.1016/j.comnet.2008.11.011
4. Maggi, F., Robertson, V., Kruegel, C., Vigna, G.: Protecting a moving target: addressing web application concept drift. In: Kirda, E., Jha, S., Balzarotti, D. (eds.) Proceedings of the 12th International Symposium on Recent Advances in Intrusion Detection (RAID '09), pp. 21–40. Springer, Berlin (2009). doi:10.1007/978-3-642-04342-0_2
5. Abd-Eldayem, M.M.: A proposed HTTP service based IDS. Egypt. Inf. J. **15**(1), 13–24 (2014). doi:10.1016/j.eij.2014.01.001. ISSN 1110-8665
6. Luo, Y.-X.: The research of Bayesian classifier algorithms in intrusion detection system. In: International Conference on E-Business and E-Government (ICEE), 2010, pp. 2174, 2178, 7–9 May 2010. doi:10.1109/ICEE.2010.550
7. Chen, J., Huang, H., Tian, S., Qu, Y.: Feature selection for text classification with Naïve Bayes. Expert Syst. Appl. **36**(3), 5432–5435 (2009). doi:10.1016/j.eswa.2008.06.054. ISSN 0957-4174

# GEANT4 Simulation of $^{192}Ir$ Source to Study Voxelization and Number of Event Effect on the Dose Distribution

Mustapha Zerfaoui, Abdeslem Rrhioua, Abdelilah Moussa, Samir Didi, Yahya Tayalati and Mohamed Hamal

**Abstract** The GEANT4 is a C++ computational tool that samples from known probability distributions to determine the average behavior of a system. It is used in brachytherapy to improve our understanding of the production, transport and the ultimate fate of therapeutic radiation. Here we present the effect of the voxelization and the number of events on the dose distribution. We calculate the dose distribution of GammaMedplus HDR $^{192}Ir$ in a water phantom and compare it with the measured dose by Gafchromic film EBT2. Our Monte Carlo results are validated by a comparison with dose data measured.

**Keywords** GEANT4 · C++ · Monte carlo simulation · Gafchromic film EBT2 · Brachytherapy

## 1 Introduction

GEANT4 is a software toolkit for the simulation of particles and their interactions with matter. Used in different application domains such high energy physics, radiation protection and medical physics [1–5]. Recently, this platform has been intensively explored in brachytherapy [6–9] which is the subject of the present paper. Brachytherapy is an internal radiotherapy treatment modality that involves

M. Zerfaoui (✉) · A. Rrhioua (✉) · A. Moussa · S. Didi · Y. Tayalati · M. Hamal
LPRM, Faculté des Sciences, Université Mohamed1er, Oujda, Morocco
e-mail: zerfaouim@yahoo.fr

A. Rrhioua
e-mail: rrhioua_ab@hotmail.com

M. Zerfaoui · A. Moussa
Ecole Nationale des Sciences Appliquées, Al Hoceima, Morocco

S. Didi
Centre Régional d'Oncologie Hassane II, Oujda, Morocco

© Springer International Publishing Switzerland 2016
A. El Oualkadi et al. (eds.), *Proceedings of the Mediterranean Conference on Information & Communication Technologies 2015*, Lecture Notes in Electrical Engineering 381, DOI 10.1007/978-3-319-30298-0_60

the embedded radioactive source close or within a tumor volume. Usually this technique complements and boosts the external beam therapy to bring better health care outcomes [10].

The success of the brachytherapy depends deeply on the quality of the Treatment Planning. Among much available software, GEANT4 provides the most accurate predictions [5, 8]. However many parameters should be tuned in order to reach high quality precision.

Here we will present the effect of both the voxelization granularity and the number of primary events on the brachytherapy dose distribution profiles. For this propose, we will use the GEANT4 simulation tools to calculate the dose distribution of GammaMedplus HDR $^{192}Ir$ sources in a water phantom. Our Monte Carlo results will be validated by a comparison with the dose data measured.

## 2   Materials and Methods

### 2.1   Brachytherapy Source

Details of the $^{192}Ir$ HDR encapsulated source used in this study are obtained from the manufacturer [11]. The active core contains a pure $^{192}Ir$ with an effective density of $22.42\,\mathrm{g\,cm^{-3}}$. It has 3.5 mm active length with active diameter of 0.60 mm. The core remains covered by the stainless steel type AISI 316L capsule of density $8.02\,\mathrm{g\,cm^{-3}}$. The lateral and top sides of the active core are hollow. The inner and outer diameters of stainless steel encapsulation are 0.07 cm and 0.09 cm, respectively.

### 2.2   Radiochromic Film

The radiochromic film is used widely in the dosimetry of brachytherapy treatments process [12, 13]. Its chemistry composition and density mimic the human tissue. We used the EBT2 Gafchromic Film produced by the International Speciality Products (ISP Technologies, Lot Number F020609), it has a high spatial resolution and a high sensitive dosimetry covering (0.0150 Gy) dose range. The film is also dose rate independent and has the property to cumulate dose fractions [14]. First of all, we calibrate the film according to the method described in [15] using the Elekta accelerator (6 MV) instead of Cobalt source. Then the film is digitalized by the powerlook 2100XL scanner. Later, the output scanner Tiff image is analyzed by the Multidata RTD4 software.

## 2.3   Monte Carlo

Radiation transport calculations were performed on a Linux based personal computer (intel core I5, Ram 4GBY) running the GEANT4 (version 4. 9.1) Monte Carlo computer code. The code is used to model $^{192}Ir$ wires with surrounding geometry, the catheter, the film and the water phantom. We used the monochromatic spectrum at 356 keV in their simulation with GEANT4 for brachytherapy treatment [16].

## 3   Results and Discussion

We represent in (Fig. 1a, b), the source and the experimental device simulated by GEANT4. The time required for a given representation is such about 6 h with 100 events.

We performed dose measurements in the plane (x, y) for a fixed Z. We chose 3 position $Z = 0$, $Z = 1$ cm and $Z = 1.5$ cm. For phantom size equal to $Z = 10 \times 10 \times 10 \, cm^3$, the computation time consumed depends on the voxel size and the number of events. This dependence is shown in (Table 1).

We note that the calculus time increases with the number of event but it decreases when voxel size increases. Figure 2 represents the MC curves calculated and measured by the EBT2 film for a number of event equal to 100000 and a voxel size equal to 2.5 $mm^3$, for $Z = 0$ and $Z = 1$ cm. The results show that we have a very good agreement between calculated profiles and the measured one when it is close to the source center. The deviation increases when away from the plane ($Z = 0$).

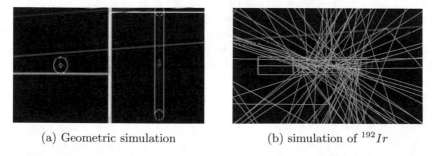

(a) Geometric simulation                    (b) simulation of $^{192}Ir$

**Fig. 1**   Geometric simulation of the catheter + film + source. **a** Geometric simulation **b** simulation of $^{192}Ir$

**Table 1**   Dependence of computing time on the voxel size and number of event

| Event number | 10,000 | 100,000 | 100,000 | 1,000,000 | 1,000,000 |
|---|---|---|---|---|---|
| Voxel size ($mm^3$) | 5 | 2.5 | 5 | 2.5 | 5 |
| Computation time (h) | 48 | 480 | 240 | 720 | 480 |

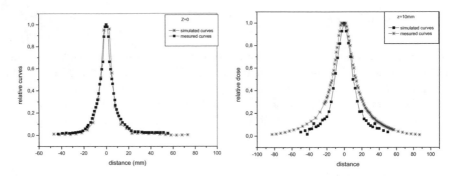

**Fig. 2** MC curves calculated and measured by the EBT2 film

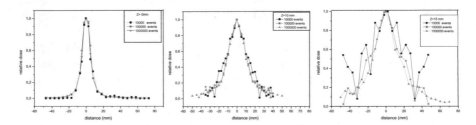

**Fig. 3** Effect of event number

The effect of the number of events is also studied. Figure 3 presents the profiles curves calculated for different event numbers. This simulation is done for the three considered planes ($Z = 0$ cm, $Z = 1$ cm, $Z = 1.5$ cm). Note that the simulated curves converge to the measured curves where the number of event increases. Fluctuations noticed far from $x = 0$, can be explained by the low statistics in these areas.

On (Fig. 4) we draw the dose distribution versus the distance for two values of voxel 2.5 and 5. The analysis of these curves shows that the shape of the curves is the same. However there is slight differences in certain areas ($Z = 1$ cm, $Z = 1.5$ cm). Indeed, when increasing the size of the voxel, the calculation time decreases, but we lose information in some areas. Thus, for the case $Z = 1$ cm)

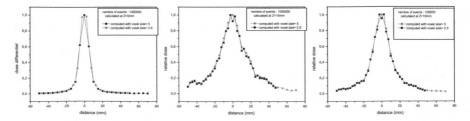

**Fig. 4** Effect of voxel size

a significant drop in the dose is observed (in the cap of the capsule) for the calculations made with a voxel size of 2.5. This drop disappears when we double the voxel size.

# 4 Conclusion

In this work we show that the simulation of the dose distribution by changing the number of events and the size of the voxel gives good results on the axis. Thus we can save time if we increase the size of the voxel but we will lose precision away from the center. On the other hand if we rise the number of events we obtain a good precision on the calculation of the dose but this requires a longer computing time.

# References

1. Paschalis, P., Mavromichalaki, H., Dorman, L.I., Plainaki, C., Tsirigkas, D.: Geant4 software application for the simulation of cosmic ray showers in the Earths atmosphere. New Astron. **33**, 26–37 (2014)
2. Singh, V.P., Medhat, M.E., Badiger, N.M.: Photon energy absorption coefficients for nuclear track detectors using Geant4 Monte Carlo simulation. Radiat. Phys. Chem. **106**, 83–87 (2015)
3. Tajik, M., Rozatian, A.S.H., Semsarha, F.: Calculation of direct effects of $^{60}Co$ gamma rays on the different DNA structural levels: a simulation study using the Geant4-DNA. Nucl. Instrum. Methods Phys. Res. Sect. B: Beam Interact. Mater. Atoms, **346**(1), 53–60 (2015)
4. André, T., et al.: Comparison of Geant4-DNA simulation of S-values with other Monte Carlo codes. Nucl. Instrum. Methods Phys. Res. Sect. B: Beam Interact. Mater. Atoms, **319**, 87–94 (2014)
5. Taylor, R.E., Rogers, D.W.O.: EGSnrc Monte Carlo calculated dosimetry parameters for $^{192}Ir$ and $^{169}Yb$ brachytherapy sources. Med. Phys. **35**, 4933 (2008)
6. Rivard, M.J., et al.: Update of AAPM task group no. 43 report: a revised AAPM protocol for brachytherapy dose calculations. Med. Phys. **31**, 633–674 (2004)
7. Daskalov, G.M., Lffler, E., Williamson, J.F.: Monte Carlo-aided dosimetry of a new high dose-rate brachytherapy source. Med. Phys. **25**, 2200 (1998)
8. Ababneh, E., Dababneh, S., Qatarneh, S., Wadi-Ramahi, S.: Enhancement and validation of Geant4 Brachytherapy application on clinical HDR $^{192}Ir$ source. Radiat. Phys. Chem. **103**, 57–66 (2014)
9. Gaudreault, M., Reniers, B., Landry, G., Verhaegen, F., Beaulieu, L.: Dose perturbation due to catheter materials in high-dose-rate interstitial $^{192}Ir$ brachytherapy, **13**(6), 627–631 (2014)
10. Sureka, C.S., Aruna, P., Ganesan, S., et al.: Computation of relative dose distribution and effective transmission around a shielded vaginal cylinder with $^{192}Ir$ HDR source using MCNP4B. Med. Phys. **33**, 1552–1561 (2006)
11. Casado, F.J., Garca-Pareja, S., Cenizo, E., Mateo, B., Bodineau, C., Galn, P.: Dosimetric characterization of an $^{192}Ir$ brachytherapy source with the Monte Carlo code PENELOPE. Phys Med. **26**(3), 132–9 (2010)
12. Uniyal, S.C., Naithani, U.C., Sharma, S.D.: Evaluation of Gafchromic EBT2 film for the measurement of anisotropy function for high-dose-rate $^{192}Ir$ brachytherapy source with respect to thermoluminescent dosimetry. Rep. Pract. Oncol. Radiother. **16**(1), 14–20 (2011)

13. Schumer, W., Fernando, W., Carolan, M., Wong, T., Wallace, S., Quong, G., F.R.A.C.P, F.R. A.C.R and Geso M. Verification of brachytherapy dosimetry with radiochromic film. Med. Dosim. **24**(3), 197–203 (1999)
14. Ballester, F., et al.: Technical note: Monte-Carlo dosimetry of the HDR 12i and Plus $^{192}Ir$ sources. Med. Phys. **28**, 258691 (2001)
15. Varian Medical Systems, BrachyVisionAcuros algorithm reference guide (2010)
16. Pia, M.G., Foppiano, F., Agostinelli, S., Garelli, S., Paoli, G., Nieminen, P.: The application of GEANT4 simulation code for brachytherapy treatment. In: Proceedings of the 9th International Conference on Calorimetry in High Energy Physics, Annecy, pp. 1–15 (2001)

# Automatic Mobile Diagnostic System for Collaborative Patients Based on Data Migration and Users Experiences

Achraf Taitai, Mostafa Ezziyani, Loubna Cherrat
and El Mamoune Soumaya

**Abstract** The knowledge extraction is an interactive and iterative process of analyzing a mass of raw data with the aim of extracting distributed knowledge, usable and adaptable to a given situation and profile. In the perspectives of implementing a medical knowledge extraction system, we present in this paper the study, design and development of an incremental knowledge base for medical diagnosis based on decision trees. The knowledge results from shared experiences, by connected users to the system, produce a base of incremental knowledge improved gradually for every new experience of various users. We are interested in the construction process of knowledge allowing the exchange and the adaptation of the shared results according to the user profile. Our solution includes an automatic verification system to the knowledge base in order to validate user experiences. It also helps to promote the most profitable and reliable experiences, giving users more choices and options for a safe operating procedures.

**Keywords** Knowledge extraction · Distributed knowledge · Incremental knowledge base · User experiences

## 1 Introduction

The implementation of any solution that supports not only the operation of medical data, but also performs processing and generates dynamic knowledge models to provide suitable and adaptable outcomes for every patient seems to be difficult to adapt. It is mainly due to the constraints required by the sensitivity of medical data

A. Taitai (✉) · M. Ezziyani · L. Cherrat · E.M. Soumaya
Laboratory of Mathematics and Applications, Abdelmalek Essaadi University, Tangier, Morocco
e-mail: achraf.taitai1@gmail.com

M. Ezziyani
e-mail: ezziyani@gmail.com

© Springer International Publishing Switzerland 2016
A. El Oualkadi et al. (eds.), *Proceedings of the Mediterranean Conference on Information & Communication Technologies 2015*, Lecture Notes in Electrical Engineering 381, DOI 10.1007/978-3-319-30298-0_61

and the lack of knowledge to produce the best results, judging by the scarcity of existing programs that usually exploit a predefined database.

Our solution, called MDS-CP: Mobile Diagnostic System for Collaborative Patients, is based on a new methodology for adaptive systems modeling dynamic decision support. It is based on three criteria: The interaction with the environment, the Auto-improvement and the Evaluation.

This study presents a new approach of modeling and visualization of system behaviors with its different advantages, and also services synchronization and resources sharing.

On the other hand, as it is necessary to produce classification procedures understandable by the user, we have chosen to work through decision trees that fit this requirement and represent graphically a set of rules and knowledge, that make it more easily interpreted.

## 1.1 General Presentation of MDS-CP

MDS-CP is a medical diagnostic tool by accumulative knowledge, which allows the user to generate treatments, diagnosis… via a decisional system, through the consultation of all the solutions collected from the experiences of other users who have a close profile and a similar situation to the use case.

The application will also allow the user to treatment and even suggest solutions that will be added to the system, the system will perform updates and generate decision trees after each knowledge validation.

## 2 Description of the Project

### 2.1 Decisional Phase

The decision-making system that implements our tool is an expert system defined as an artificial intelligence system using facts, knowledge and reasoning techniques to solve problems that require some kind of human skills [1].

Our expert system differs from other conventional systems by its ability to acquire new accumulative knowledge from application users. The users, who are in the same network, can trigger the change of the knowledge base that fits with the number of views of a given data. Thus, the automatic updates of the knowledge base, ensures the integrity and reliability of data by fostering, in the decision tree, the most profitable branches and removing system vulnerabilities.

By default, each new update generate a new knowledge tree, more reliable and much richer. Taking into account the views of the users give more choices and options for a safer data exploitation.

### 2.1.1 The Knowledge Base Construction

The central idea in our approach is to split, recursively and as efficiently as possible, the knowledge, set through defined tests, by using attributes until we get subsets of examples containing almost only the examples that belong to the same class.

Thus a verification procedure will take place for each node, rejecting the knowledge causing disturbances, and keeping the validated and approved ones. The resulting tree will still be awaiting for shared knowledge to perform data validation procedure.

## 2.2 Contextual Research Phase

The purpose of the contextual research is to meet the needs of the user by providing adapted information to their specific research context [2].

This phase is related to the treatment of the data flows and the information exchanges, be it those transiting between users (Ex. Smartphones) via the server, or those traveling between smartphones and decisional system. We are also interested in the knowledge construction process to adapt the shared results according to the profile of their target user.

The figure above shows the transmission and reception of user data to the system base, the data will then be shared to all users.

The data flows in the system as a social network, except that a shared data goes through a validation process to detect anomalies and ensure the reliability of data flowing in the network (Fig. 1).

**Fig. 1** The data flow in the system; it is shared in the entire network via internet

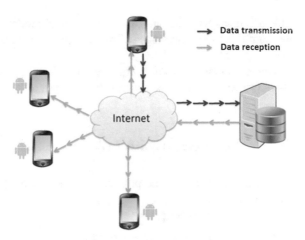

### 2.2.1   User Profile Adaptation

The traditional search engines do not place enough emphasis on the context of research, thus, it was necessary to integrate an engine that will adapt the results to the user's profile. In our case, the fundamental need is to give patients [3, 4] access to their medical record (treatment, medication…) as well as other patients sharing knowledge with similar target user. The Profile here means all the information attached to a patient: gender, location, weight, height, diet…

The relevance of the information provided, its intelligibility and its adaptation to user preferences and usage, are key factors for success or rejection of such information systems. For this, we focused on the modeling of user preferences in information research area in several heterogeneous resources integrated via mediators. The architecture of a user profile [5] is defined through:

- The focus dimension based on historical interrogation
- The integration of this profile into an information search process to customize the results returned by the mediation system.

The use of profile ontologies [6, 7] allows us to:

- Facilitate access to multiple information sources.
- Exchange information to improve, and participate in a collaborative process of distributed resolution.
- Distribute and balance the workload.

## 3   Functional and Conceptual Study

This part deals with the design and the general modeling of the features that define the solution to be developed. This design is carried out with the language of modeling UML.

Our main class diagram brings together all the classes that compose our mobile client (see Fig. 2), and the relationships between these different classes.

## 4   The Solution Architecture

The figure below shows the physical architecture of the applications that compose the solution (Fig. 3).

We have three main parts:

- Client part: Representing the HMI (Human Machine Interface) front office and all visible features. Our HMI will be the Android mobile.

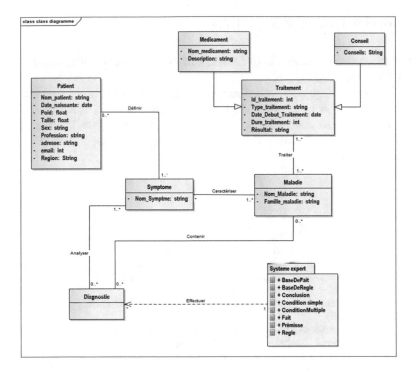

**Fig. 2** Client-side class diagram

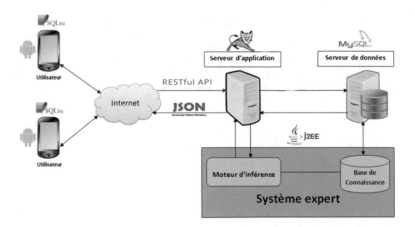

**Fig. 3** The different components of the system

- Business part: In charge of processing mechanisms and the implementation of the project business processes. Our decisional system is developed on this part.
- Data part: Deployed on a data server and accessed by client/business parts.

## 5 Conclusion and Perspectives

In this study, we propose a fully functional powerful solution that will make life easier for many people.

The project is of great technical interest; an extension of this work would be the realization of robustness tests and deployment of the solution. The solution can also functionally evolve by responding to new needs thanks to its flexible architecture, carefully designed, and enable the introduction of good programming practices to encourage re-use.

## References

1. Castillo, J., Silvescu, A., Caragea, D., Pathak, J., Honavar,V.: Information extraction and integration from heterogeneous, distributed, autonomous information sources—a federated ontology-driven querycentric approach. In: IEEE International Conference on Information Reuse and Integration IRI03, pp.183–191 (2003)
2. Agrawal, R., Rantzau, R., Terzi, E.: Context-sensitive ranking. In: SIGMOD Conference ACM, pp. 383–394 (2006)
3. Calvanese, D., Lembo, D., Lenzerini, M.: Survey on methods for query rewriting and query answering using views. Technical Report D2I, (2001)
4. Ullman, V.: Information integration using logical views. In: Proceedings of the 6th International Conference on Database Theory, (ICDT'97), pp. 19–40 (1997)
5. Cali, A., Calvanese, D., Giacomo, G., Lenzerini, M.: On the expressive power of data integration systems. In: Proceedings of the 21st International Conference on Conceptual Modeling, pp.338–350 (2003)
6. Beneventano, D., Bergamaschi, S., Castano, S., Corni, A., Guidetti, R., Malvezzi, G., Melchiori, M., Vincini, M.: Information integration: the MOMIS project demonstration. In: Proceedings. of the 26th International Conference on Very Large Data Bases (VLDB), (2000)
7. Rousset, M.-Ch., Bidault, A., Froidevaux, C., Gagliardi, H., Goasdoue, F., Reynaud, Ch., Safar, B.: Construction de médiateurs pour intégrer des sources d'information multiples et hétérogènes: le projet PICSEL. Revue 2(1), 5–59 (2002)

# Methods and Systems for the Linked Data

**Fouad Nafis and Dalila Chiadmi**

**Abstract** The amount of data published in the Web of data is growing continuously. To comply with standards and principles of linked data, especially the fourth principle, we must always find links to the data of other people. Several tools that use different techniques and operate under several stages provide this task of link. The purpose of this paper is to give a comparative study of several existing tools according to various criteria not supported by other studies.

**Keywords** Linked data · Alignment · Ontology · Entity · Semantic Web

## 1 Introduction

In recent years, the Web has transformed from a traditional web in which users have a set of pages (often using the standard HTML format) interconnected via hyperlinks to a web of data where each data is the central point through which each data producer or user must pass. These data are structured using the RDF (Resource Description Framework) representation language, and connected by explicit links. These links can be created manually or using a tool to make this task automated. The manual method quickly finds its limits in the fact that the number of published data are increasing. Several tools have emerged in recent years to solve the problem of data linking. Each tool works on one or more axes. The data sources are heterogeneous and can share the same ontology or operate each on a different ontology. The purpose of this article is to make a comparison of some data binding tools depending on a number of criteria that can make the difference when choosing a method of publishing data in the web of data among all methods exist.

F. Nafis (✉) · D. Chiadmi
Computer Science Department, EMI School,
Mohammed V th University-Agdal, Rabat, Morocco
e-mail: fouadnafis@research.emi.ac.ma

D. Chiadmi
e-mail: chiadmi@emi.ac.ma

© Springer International Publishing Switzerland 2016
A. El Oualkadi et al. (eds.), *Proceedings of the Mediterranean Conference on Information & Communication Technologies 2015*, Lecture Notes in Electrical Engineering 381, DOI 10.1007/978-3-319-30298-0_62

In this paper, the second part is an introduction to the principle of linked data and related technologies. The third part presents the related works to our research area. The fourth section focuses on our search space and addresses the issue of criteria considered for the comparison of tools. It also presents the different systems studied and concludes with a comparative table highlighting the strengths and weaknesses of each link tool. The final section provides a conclusion and present potential prospects.

## 2 The Web of Data Principles

### 2.1 Linked Data

The linked data [1] is a movement that emerged in 2006 following the initiatives of Tim Berners-Lee [2, 3] and his staff, to refer to best practices of publication and linking of structured data on the web. A significant number of organizations and data producers use these best practices to the publication of their data and the result is a large graph called Linked Open Data Cloud (LODC) containing a large number of RDF triples. In January 2015, the number of RDF triples is over than 74 trillion triples, linked with over 700 million links.[1]

### 2.2 RDF/RDFS

RDF [2](Resource Description Framework), the basic language of the Semantic Web is a knowledge representation model as a graph called RDF graph.

RDFS[3] (RDF Schema) provides basic elements for the definition of ontologies and vocabularies intended to structure RDF resources (triples).

### 2.3 OWL

OWL[4] (Web Ontology Language) is a knowledge representation language based on RDF structure. It allows the representation of classes through the properties of the instances and the types of properties. This language offers more possibilities for data description, compared to RDF and RDFS.

---

[1]http://stats.lod2.eu.
[2]http://www.w3.org/RDF/.
[3]http://www.w3.org/TR/rdf-schema/.
[4]http://www.w3.org/2001/sw/wiki/OWL.

# 3 Related Works

Comparing the link tools is not a new task. Several studies have tried to draw a comparison of systems that exist, according to a number of criteria. The works that stood out were, firstly [4] who made a comparison of the link tools that responded to a call for expressions of interest. The owners of these tools specify the comparison criteria, which puts night objectivity of the study, since every team will give the criteria they want. In addition, the comparison area is very small, since we have compared the systems that responded to the request, which disqualifies other tools and invalidates the comparison task of his major interest, which is to find the most optimal and most efficient solution to satisfy a specific need.

The authors in [5] made a detailed comparison with a number of tools more important and criteria, but they have missed the criteria we considered important in our work. Then they included tools that work on a particular ontology as LD Mapper [6] in which the authors were able to show that the link operation provides relevant results if the data sources share the same ontology (MusicOntology). Finally, several criteria were ignored as: conviviality of the interface of the tool, the type of the proposed tool (Open Source, free, etc. ...) the type of output result (Links owl: sameAs, linkset or other).

# 4 Link Tools Analysis

## 4.1 Comparison Criteria

For each study tool, we have taken into account the following:
Type of treatment (Automated/Semi-automated), Ontologies (Same ontology/different ontologies), Mapping techniques, the proposed interface (GUI, web service...), Input, Output, and Domain of work.

## 4.2 Link Tools

**SILK** [7]

SILK is a platform developed in Python. This is a link tool, using a language called LSL (Link Specification Language). The user must specify the entities to link, in a separate file written in LSL. It must also specify some parameters for similarity measures used by the tool (Alignment of strings, dates alignment, digital similarity measures, measure the distance between concepts in a taxonomy (levenshteinDistance, jaroWinkler, jaro, softjaccard)). Operators (MAX, MIN, AVG) between these alignment algorithms can be used to improve the results of the link process.

## LIMES

LIMES [8] requires some user involvement, particularly in the configuration step (Choice of thresholds, choice of metrics, comparison operators...). It uses the mathematical properties of metric spaces (triangle inequality, identity...) to optimize the alignment techniques, and thus reduce the number of comparison to make.

## KnoFuss

KnoFuss [9] is developed in Java and runs using data sources described with several ontologies. It allows merging multiple sources of heterogeneous knowledge. Several algorithms similarity measure are used. The tool allows the fusion of the original data sources and merging ontologies used, and works with local data sources.

## RDF-AI

RDF-AI [10] is developed in Java, and uses local copies, and XML configuration files specifying the dataset structure and alignment techniques and other parameters (threshold for the link generation, properties to align...). The structure of the data sources and the resources to be linked are provided in two different files.

## SERIMI

Serimi [11] operates in two steps: we first search in the dataset source the candidates to link, and then we use the names of the entities found to look up the corresponding entities in the dataset. We then construct, classes from instances found, these classes are linked to the corresponding classes in the target dataset.

**Table 1** Comparison of the different tools according to the criteria considered (1/2)

| The tool | Type (automated/semi) | Ontologies | Mapping techniques | Domain |
|----------|----------------------|-----------|-------------------|--------|
| LIMES | Semi automated | Multiple + alignment | String similarity | Multiple |
| SILK | A configuration file is required + semi automated | Multiple + implicit alignment | String similarity + adaptive learning | Multiple |
| SERIMI | Automated | Multiple ontologies | String similarity + classes of interest | – |
| Okkam | Semi automated | One entity at time | String similarity + similarity combination techniques | – |
| KnoFuss | Semi automated | Multiple + explicit alignment | Aggregated attribute − based similarity | Multiple |
| | | | Unsupervised attribute − based similarity | |
| RDF-AI | Semi automated | One ontology | String similarity + wordnet | Multiple |

**Table 2** Comparison of the different tools according to the criteria considered (2/2)

| The studied tool | The proposed interface | Input | Output | Type |
|---|---|---|---|---|
| Limes | GUI | SPARQL local copies | Owl :sameas | Free + open source |
| Silk | GUI + command line | Silk LSL (XML) SPARQL | Linkset + Owl : sameas | Free + open source |
| SERIMI | Command line | – | Owl :sameas | – |
| Okkam | – | – | Owl :sameas | – |
| KnoFuss | – | Local copies | Merged data + linkset | Free + open source |
| RDF-AI | – | XML + RDF dump | Linkset + merged data | Free + open source |

## OKKAM

OKKAM[5] [12] proposes the use of Entity Name Servers component (ENS) having the role of resource directories. Entities descriptions are stored as pairs: Key-property. For a new entity, an alignment algorithm is executed based on measurements of similarity between the entity and the entities of the server to decide whether a new entity to be added or not. The similarity measurement algorithm uses Strings similarity.

We summarize, in Tables 1 and 2, the major differences between these tools according to the criteria mentioned above, that caught our attention during this study.

## 5 Conclusion and Future Works

This comparative study has allowed us through the overview of the various link tools, to know more about these tools and the components they contain. The e-gov is a field for us to explore and develop a link tool for this area will be for us the best combination of exploitation of the web of data technologies and the open data movement to produce solutions helping to improve the quality of services provided by public agencies. In our future work, we intend to present an ontology concerning Members, in the Moroccan context. We then use this ontology, in addition to data sources that already exist to produce a file of data that represent the same information about the MPs and to generate a solution, decision support in this direction.

---

[5]http://api.okkam.org/.

# References

1. Berners-Lee, T.: Linked data. W3C design issues (2006)
2. Bizer, C., Heath, T., Berners-Lee, T.: Linked data-the story so far. Int. J. Seman. web Inf. Syst. **5**(3), 1–22 (2009)
3. Bizer, C.: Evolving the web into a global data space. In: BNCOD, vol. 7051, p. 1 (2011)
4. Scharffe, F., Euzenat, J.: Méthodes et outils pour lier le web des données. In: Actes 17e conférence AFIA-AFRIF sur reconnaissance des formes et intelligence artificielle (RFIA), pp. 678–685 (2010)
5. Ferraram, A., Nikolov, A., Scharffe, F.: Data linking for the semantic web. Semantic Web: ontology and Knowledge Base Enabled Tools, Services, and Applications, **169** (2013)
6. Raimond, Y., Sutton, C., Sandler, M. B.: Automatic interlinking of music datasets on the semantic web. In: LDOW (2008)
7. Volz, J., Bizer, C., Gaedke, M., Kobilarov, G.: Silk-a link discovery framework for the web of data. LDOW, **538** (2009)
8. Ngomo, A. C. N., Auer, S.: LIMES: a time-efficient approach for large-scale link discovery on the web of data. In: Proceedings of the Twenty-Second international joint conference on Artificial Intelligence-Volume Volume Three, pp. 2312–2317. AAAI Press (2011)
9. Nikolov, A., Uren, V., Motta, E., De Roeck, A.: Integration of semantically annotated data by the KnoFuss architecture. In: Knowledge Engineering: Practice and Patterns, pp. 265–274. Springer, Heidelberg (2008)
10. Scharffe, F., Liu, Y., Zhou, C.: Rdf-ai: an architecture for rdf datasets matching, fusion and interlink. In: proceedings of the IJCAI 2009 workshop on Identity, reference, and knowledge representation (IR-KR), Pasadena (CA US) (2009)
11. Araujo, S., De Vries, A., Schwabe, D.: Serimi results for OAEI 2011. In: Proceedings of the Sixth International Workshop on Ontology Matching, pp. 212–219 (2011)
12. Bouquet, P., Stoermer, H., Cordioli, D., Tummarello, G.: An entity name system for linking semantic web data. In: *LDOW* (2008)

# Toward a Big Data-as-a-Service for Social Networks Graphs Analysis

Siham Yousfi, Dalila Chiadmi and Fouad Nafis

**Abstract** Big Data analytic and Cloud Computing are the new trends that have submerged the IT industry. While big data environment requires powerful cluster infrastructure, new ideas about combining this two paradigms were born to enhance business agility and productivity and enable greater efficiency and reduce costs. The objective of our research is to introduce a Big data as-a-service (BDAAS) solution based on Hadoop ecosystem that analyses a graph built from heterogeneous data sources such as social media, database tables and so on. We created a prototype of our solution that determines the most popular deputy based on internet users' tweets.

**Keywords** Big Data · Cloud Computing · Big data-as-a-service · GraphBuilder · GraphLab

## 1 Introduction

Recently the amount of data generated by individuals, organizations and academy has been increasing. The environment of Big Data needs a high-level infrastructure of clusters that may handle large volume, high velocity and different formats. Besides, data collected by companies are generally not ready for analysis. Therefore, they need to perform a set of commons steps, such as information extraction, data cleaning and modeling [1]. Cloud computing is a new technology that allows clients to use a set of services related to infrastructure, platform or software. Users can create scalable virtual machines quickly and easily while

S. Yousfi (✉) · D. Chiadmi · F. Nafis
Computer Science Department, EMI School,
Mohammed Vth University-Agdal, Rabat, Morocco
e-mail: sihamyousfi@research.emi.ac.ma

D. Chiadmi
e-mail: chaidmi@emi.ac.ma

F. Nafis
e-mail: fouadnafis@research.emi.ac.ma

© Springer International Publishing Switzerland 2016           593
A. El Oualkadi et al. (eds.), *Proceedings of the Mediterranean Conference on Information & Communication Technologies 2015*, Lecture Notes in Electrical Engineering 381, DOI 10.1007/978-3-319-30298-0_63

maintenance and configuration tasks are performed by service providers. In order to deal with big data issues, a new cloud computing service was proposed: big data as-a-service (BDAAS) that offers services for managing large scale of data. The objective of the present paper is to describe and provide a prototype of BDAAS service that extracts data from "Twitter". Before storage, raw data are filtered and processed semantically and the results are provided as a graph that could be used later for further analysis. In order to illustrate our proposition, we will consider a use case that builds a graph representing deputy reputation from tweets. The remainder of this paper is organized as follows: First, the Sect. 2 presents an overview of big data and cloud computing paradigms. Then, we describe our use case in Sect. 3. Section 4 introduces our proposition and the final two sections will present the related works, conclude and introduce our perspectives.

## 2   Overview of Big Data and Cloud Computing Concepts

With the increasing amount of stored information, the percentage of processed data is rapidly decreasing [2]. Big Data represents the intensive data centric technologies [3]. It indicates the continuity of the technology advancement on various individual activities including production, collaboration and consumption. In a 2001, MetaGroup publication, Gartner analyst Doug Laney described Big data using the 3 V's characteristics Volume (data size) Velocity (how fast data are arriving and stored) and Variety (different types of data) [4]. Big data environment requires a powerful infrastructure of clusters and well configured tools. The implementation of such environment induces a risk of wasting money without guaranteeing a positive result. Moreover, Big Data analysis involves preliminary common steps like data extraction, formatting storage and processing. These operations are performed separately by many companies. Thereby, using a cloud computing platform will help companies to delegate the management of these steps to a specialized organization, ensure efficiency and reduce costs.

The primary objective of "Cloud computing" is to be flexible and responsive to customers' needs by providing IT services in a scalable way via internet to a number of clients, at low costs. In fact, users will not care about managing various hardware and software installation, configuration and updates. All these operations will be performed by services providers [5]. With the increase of big data, cloud computing paradigm is facing many challenges. According to [6], Big Data computational paradigms and tools such as Hadoop and MapReduce aren't optimal in the cloud and cause an increase of cost. Also, big data environment is dynamic and cloud computing needs to gradually converge to support big data by creating adequate services: Big Data-as-a-service. This service refers to common big data services provided as cloud hosted service. According to [7] BDAAS encapsulates big data techniques into three layers Big data infrastructure-as-a-service, Big data platform-as-a-service BDPAAS and big data software-as-a-service BDSAAS. The paper [1] describes these architectures and challenges.

# 3 Case of Study

Analyzing publicly available opinions shared using social media, has become an increasingly popular method for studying socio-political issues, and can help government managers to improve the quality of their services and activities. Our case of study is about analyzing the reputation of Deputies. In fact, on arrival of each new election, the members are selected based on the choice of the citizens. Analyzing the popularity of deputies may be interesting for both parties and citizens:

- For parties: It could be interesting to know in advance their most popular members in social networks so as they would present them to the court of elections and have more chance to win these elections.
- For citizens: They may prefer compare their opinions with the other citizen's. And will be more likely to choose an honest deputy to represent them in parliament.

To collect information from citizens we are using Twitter a microblogging service that allows user to communicate by publishing small texts called "tweets" [8]. However, the reliable personal data of deputy may be stored in another data source type like an e-government database. Consequently, the solution that we are describing offers a BDAAS that allows data extraction from a set of heterogeneous sources like relational databases and social network. Our research will be in the first step on explicit patterns for example "I like *" "I appreciate *" where "*" is the deputy name already stored in the database. Data will be displayed as a directed graph where MPs and Internet users are nodes, and edges model the feeling of appreciation expressed by a user in twitter. The analysis of such a graph will allow us later to identify which is the most popular deputy by analyzing the most significant node which will be the one with the largest number of incoming edges.

# 4 Description of BDAAS Solution

The aim of our paper is to describe a BDAAS solution that analyses data extracted from twitter. The output data of the BDAAS could be provided in different formats raw data, Text/XML files, graphs and so on. We chose to provide a graph data for the following reason (1) Graphs are an intuitive way to represent the relationship between entities. (2) Graph analytics provides answers to business problems like shortest path, centrality. etc. (3) Graph analytics helps machine learning solutions by adding new parameters about relationships between objects. "Figure 1" is illustrating the business workflow process of our solution in the context of our use case described in Sect. 3 it contains the following steps:

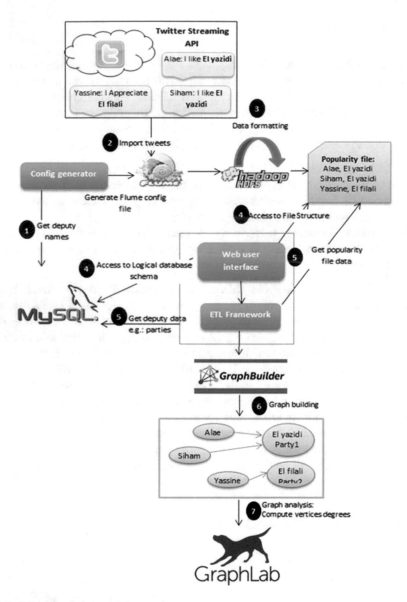

**Fig. 1** BDAAS solution workflow

**Generate Flume configuration file (1)**: In our context we have two sources: MySQL database that contains deputies' data and tweets that contain citizens' feeling. We need to use flume framework to collect tweets talking about an existing member where the deputy name is stored in our database. "Config generator" is a program that generates the flume configuration file automatically where keywords are filled from deputies' data table stored in MySQL database. **Data gathering & storage (2)**: In order to allow developers accessing twitter's data "Twitter" company provides the "Twitter Streaming API" that outputs a constant stream of tweets in a JSON format. Flume will collect every tweet and store the results in HDFS. **Data formatting (3):** This step aims to format the retrieved data from twitter so as to keep only plain text. Then we proceed to a semantic analysis based on simple patterns such as "I like *" and "I appreciate *". The result of this step is a text file named "popularity.txt" where each line mentions the user name and the deputy that he likes e.g. {user,deputy}. **Filling graph structure (4)**: we built a web application that allows users to customize at each time the properties of the graph nodes and edges on the basis of the logical database schema and the structure of the "popularity.txt" file. This application returns two tables of objects called "NodeProperties" and "EdgeProperties". **Graph Data extraction (5)**: based on "NodeProperties" and "EdgeProperties" we created a solution that selects data relating to nodes and edges from MySQL database, and the file popularity.txt. **Graph formation (6)**: We are using the GraphBuilder library to form a graph and perform other operations such as transformation, normalization and partitioning. **Graph Analysis (7)**: the analysis is then performed by GraphLab that will read the stored graph file from GraphBuilder and computes the vertex degree which represents the number of edges that point to each vertex. The result is a list of deputies sorted by popularity.

# 5 Related Articles

Many authors especially bloggers tried to extract data from social media for example: [9] Shows the technical method of tweets gathering and analysis using flume and hive, our study proposes a solution that allows extending unstructured results by structured data stored in RDBMS. Furthermore, [10] proposes a method for collecting data such as comments, posts and status from a Facebook pages. However, the article didn't talk about real time data extraction. Since social networks are sources of big data that are updated very quickly, they require a real-time collection and analysis.

# 6 Conclusion and Perspectives

This paper introduced a solution built on top of Hadoop ecosystem that collects data from social networks and database tables and constructs graph using GraphBuilder Library. The use case that we presented is about creating a graph for deputy reputation based on tweets where nodes are Twitter's users and deputies, and edges are the appreciation feeling. Using GraphLab, we analyzed the reputation of the deputies by computing vertices degrees. In the next step, we will try to improve the semantic analysis of text tweets by considering other ways of expressing opinions like smiley, and we will also improve the resulting graph data to perform other analysis tasks (e.g. determine the region where the deputy is popular etc).

# References

1. Zheng, Z., Zhu, J., Lyu, M.R.: Service-generated big data and big data-as-a-service: an overview. In: 2013 IEEE International Congress on Big Data (BigData Congress), IEEE, pp. 403–410 (2013)
2. Zikopoulos, P., Eaton, C., et al.: Understanding Big Data: Analytics for Enterprise Class Hadoop and Streaming Data. McGraw-Hill Osborne Media, New York (2011)
3. Demchenko, Y., Grosso, P., de Laat, C., Membrey, P.: Addressing big data issues in scientific data infrastructure. In: 2013 International Conference on Collaboration Technologies and Systems (CTS), IEEE, pp. 48–55 (2013)
4. Laney, D.: 3D data management: controlling data volume, velocity and variety. META Group Research Note 6 (2001)
5. Wang, L., Von Laszewski, G., Younge, A., He, X., Kunze, M., Tao, J., Fu, C.: Cloud computing: a perspective study. New Gener. Comput. **28**, 137–146 (2010)
6. Kasson, P.M.: Computational biology in the cloud: methods and new insights from computing at scale. In: Pacific Symposium on Biocomputing, World Scientific, p. 451 (2013)
7. EMC Solution Group.: Big data-as-a-service: a market and technology perspective. Tech. Rep (2012)
8. Java, A., Song, X., Finin, T., Tseng, B.: Why we twitter: understanding microblogging usage and communities. In: Proceedings of the 9th WebKDD and 1st SNA-KDD 2007 Workshop on Web Mining and Social Network Analysis, ACM, pp. 56–65 (2007)
9. Natkins, J.: How-to: Analyze Twitter Data with Apache Hadoop. http://blog.cloudera.com/blog/2012/09/analyzing-twitter-data-with-hadoop/ (2012)
10. Shah, C., et al.: Politics 2.0 with Facebook–Collecting and Analyzing Public Comments on Facebook for Studying Political Discourses (2011)

# Sender Assisted Receiver Initiated MAC with Predictive-Wakeup for Wireless Sensor Networks

Shagufta Henna and Bilal Saleeemi

**Abstract** In this paper, we present PW-SA-MAC, an asynchronous energy efficient duty cycle MAC protocol. Similar to RI-MAC, PW-SA-MAC employs receiver-initiated transmissions with a sender-assisted contention resolution. PW-SA-MAC enables senders to wake up at predicted times according to the receiver's schedule. We evaluate PW-SA-MAC using ns-2 and compare it to RI-MAC and SA-RI-MAC, under clique, grid, and random networks. In random network scenarios, the average duty cycle of PW-SA-MAC is 66 % less compared to RI-MAC and SA-RI-MAC.

**Keywords** PW-SA-MAC · Predictive Wake-up MAC · Asynchronous MAC

## 1 Introduction

Idle listening is one of the main sources of energy consumption. Duty cycle MAC protocols can be classified as synchronous or asynchronous. Synchronous MAC protocols conserves energy by synchronizing sensor node's sleep and wake-up intervals. These protocols require synchronization of sender and receiver clocks. In contrast, asynchronous duty cycle MAC protocols allow sensor nodes to operate in an independent way. With the sender-initiated approach, sender informs the receiver for an incoming packet by sending a preamble prior to data transmission [1, 2]. In this paper, we present the design and evaluation of a new asynchronous energy-efficient MAC protocol called PW-SA-MAC. PW-SA-MAC outperforms other duty cycle MAC protocols. PW-SA-MAC reduces delivery latency experi-

S. Henna (✉)
Department of Computer Science, Bahria University, Islamabad, Pakistan
e-mail: shaguftahenna@gmail.com

B. Saleeemi
Department of Computer Science, Capital University of Science and Technology,
Islamabad, Pakistan
e-mail: bilalsaleemi@gmail.com

© Springer International Publishing Switzerland 2016
A. El Oualkadi et al. (eds.), *Proceedings of the Mediterranean Conference
on Information & Communication Technologies 2015*, Lecture Notes
in Electrical Engineering 381, DOI 10.1007/978-3-319-30298-0_64

enced by the nodes to send data by allowing sender to wake up right before the receiver is awake. In particular PW-SA-MAC inherits the receiver initiated approach of RI-MAC [3] with a sender assisted contention resolution mechanism of SA-RI-MAC. In contrast to SA-RI-MAC and RI-MAC, PW-SA-MAC utilizes a pseudo-random number to schedule sender and receiver's wake-up times. The use of a pseudo-random helps the sender to predict the wake-up time of a receiver. We have evaluated PW-SA-MAC using ns-2. In all experiments, PW-SA-MAC maintained 100 % packet delivery ratio (Drat), with lowest latency, and duty cycle.

Section 2 discusses related work. In Sect.3, we present the design of PW-SA-MAC including its contention resolution mechanism and predictive wake-up mechanism. Section 4 presents the operation of PW-SA-MAC. Section 5 presents the results of our evaluation. Finally, in Sect. 6, we present conclusions.

## 2   Related Work

A synchronous duty cycle MAC protocol requires synchronization between the sender and receiver [4, 9]. Synchronous duty cycle MAC protocols require time synchronization which incurs extra overhead. On the other hand, in asynchronous duty cycle MAC protocols each node independently sleeps and wakes up [1, 2]. WiseMAC [1] and X-MAC [2] use preambles and are called sender-initiated MAC techniques. In receiver-initiated duty cycle MAC protocols, it is the receiver who initiates the transmission, such as RI-MAC [5]. In RI-MAC, receiver uses a beacon to announce the time when it will be ready to receive the DATA. However, PW-SA-MAC enables senders to wake up according to the predicted time of the receiver. SA-RI-MAC [6] is based on RI-MAC [5] and utilizes sender-assisted contention resolution to resolve the contention of SA-RI-MAC. WiseMAC enables the sender to predict the next wake-up time of the receiver based on the repeated schedule of the receiver and therefore wakes up shortly before the wake-up time of the receiver. However, preamble transmissions from multiple transceivers may create contention at the receiver side. Unlike WiseMAC, nodes with PW-SA-MAC utilize a pseudo-random schedule to ensure that nodes will not share same time interval which reduces number of collisions. PW-SA-MAC uses an on-demand wake-up prediction mechanism coupled with the benefits of receiver initiated transmissions and sender-assisted contention resolution.

## 3   PW-SA-MAC Design

PW-SA-MAC allows the sender to predict the wake-up time and turn its radio on to receive a beacon right before the receiver is active. Similar to PW-MAC [7], we use a pseudo-random wake-up interval to schedule nodes. However, we have modified the linear congruential generator (LCG) [8] according to the traffic condition. LCG equation used by PW-MAC is given in Eq. 1.

$$X_{n+1} = (aX_n + c) \bmod m \tag{1}$$

where $m$ is modulus, $a$ is the multiplier, and $c$ is an increment. $X_n$ indicates current seed. $X_{n+1}$ is a pseudo-random number generator. Sender in PW-MAC predicts the wake-up time of the receiver with the help of this new seed and the time difference between the sender and the receiver clocks. We have modified the Eq. 1 as in Eq. 2.

$$X_{n+1} = (aX_n + c) \bmod N \tag{2}$$

$$Y_{n+1} = Z_w + X_{n+1} \tag{3}$$

The value of $N$ is less than $m$. In Eq. 3, $Z_w$ is selected according to the current traffic load and vary from $Z_{low}$ to $Z_{high}$, where $Z_{low}$ indicates low traffic load and $Z_{high}$ denotes high traffic load. Nodes use a random back-off interval to avoid collisions. After broadcasting the beacon, node waits for the incoming packets from the sender. If no packet is received for a period of time, node turns its radio off. In order to measure the current traffic load on the network, we have considered the $R_{busy}$ [8]. In a MAC protocol with RTS/CTS enabled, given $T_{suc}$ and $T_{col}$ be the time slots for successful transmission and transmissions resulting in collision as below:

$$T_{suc} = rts + ccts + crn + data + ack + 3sifs + difs \tag{4}$$

$$T_{col} = rts + cts\_timeout + difs = rts + eifs \tag{5}$$

$rts$, $cts$, and $data$ are the time taken to transmit a RTS and CTS and data respectively. The $R_{busy}$ is computed as follows:

$$R_{busy} = \frac{T_{suc} + T_{col}}{T_{tot}} \tag{6}$$

where $R_{busy}$ is defined as the total number of slots due to successful transmissions and collisions during $T_{tot}$ time. It can be compared with a predefined threshold $Th_{busy}$. Similar to SA-RI-MAC, if a receiver is under high contention, sender evaluate if any of its neighbours have CHANNEL ACCESS FAILURE higher, it turns its radio off. For details refer to [6].

## 4   PW-SA-MAC Protocol Operation

Based on the traffic load each node adjusts its duty cycle according to the following rules:

S. Henna and B. Saleeemi

- If $R_{busy} > Th_{busy}$, PW-SA-MAC considers it high traffic load, and adjusts the duty cycle by selecting $Z_{low}$ in the Eq. 3. This will increase the duty cycle and reduces the number of collisions and delay.
- If $R_{busy} < Th_{busy}$, PW-SA-MAC considers it low traffic load in the network, and node adjusts the duty cycle by selecting $Z_{high}$ in the Eq. 3.

## 5 Evaluation

We have evaluated RI-MAC, SA-RI-MAC, and PW-SA-MAC using ns-2 simulator under random, clique, and $(7 \times 7)$ grid networks. $T_x$ range for each node is 250 m, slot time is 320 μs, and SIFs duration is 192 μs. Duty cycle for each node is 1 %. $T_x$ and $R_x$ power for each node 31.2, and 22.2 mW respectively.

In a clique network, all nodes are within each other's transmission range and are twice the number of flows. Time between generation of two packets is uniformly distributed between 0.5 and 1.5 s. Each simulation lasts for 100 s for three random clique networks. Figure 1 shows, PW-SA-MAC shows the competitive delivery ratio for flows greater than 15, since it adjusts the duty cycle according to the network contention. Figure 2 shows the average per-node duty cycle of PW-SA-MAC. Each sender with RI-MAC and SA-RI-MAC stays in the awake state to receive any incoming beacon from the receiver, and therefore wastes energy. PW-SA-MAC uses predictive wake-up mechanism dynamically according to the traffic conditions, therefore reduces idle listening. PW-SA-MAC can reduce energy consumption more than 98 % compared to RI-MAC. Figure 3 shows PW-SA-MAC has the lowest delivery latency due to predictive wake-up mechanism coupled with the sender

**Fig. 1** Drat versus flows

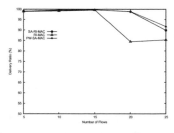

**Fig. 2** Duty cycle versus flows

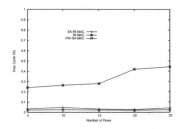

**Fig. 3** Avg. delay versus
flows

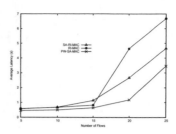

assisted contention resolution. In a 7 × 7 grid, 49 nodes are deployed where two
nodes are 100 m apart with a sink node at the centre. Traffic is generated according to
RCE model [5]. An average over 3 random runs each for 10,000 has been taken with
a total of 48 events. Figure 4 shows that as the sensing range increases above 400 m
the delivery ratio of all the protocols start to decrease due to increase in traffic load.
PW-SA-MAC shows the competitive performance for all the sensing ranges larger
than 400 m due to its adaptive predictive wake up mechanism. Figure 5 shows
PW-SA-MAC experiences lowest energy consumption due to its predictive wake-up
mechanism. Figure 6 shows that as the sensing range increases, PW-SA-MAC
shows the lowest minimum delivery latency.We evaluate performance of
PW-SA-MAC in 3 random network scenarios with 40 nodes in an area of
1000 × 1000. Each experiment lasts for 100 s. Figure 7 shows that as flows exceed
25, PW-SA-MAC shows the competitive delivery ratio compared to RI-MAC and
SA-RI-MAC. Figure 8 shows that average duty cycle of PW-SA-MAC is significantly better for varying number of flows. PW-SA-MAC uses predictive wake-up

**Fig. 4** Drat versus sensing
Rx

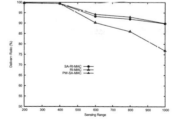

**Fig. 5** Duty cycle versus
sensing Rx

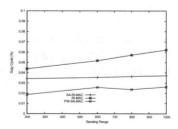

**Fig. 6** Min. delay versus
sensing Rx

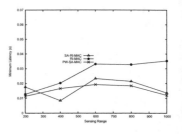

**Fig. 7** Drat versus flows

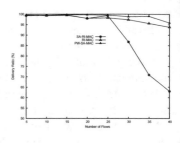

**Fig. 8** Duty cycle versus
flow

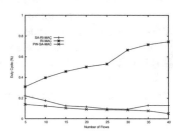

**Fig. 9** Avg. delay versus
flows

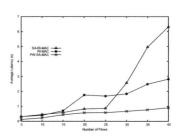

mechanism coupled with the sender-assisted contention resolution, therefore it
reduces idle listening. As shown in Fig. 9. PW-SA-MAC experiences lowest latency
due to predictive wake-up mechanism with the sender-assisted contention resolution.

# 6 Conclusion

With a simple predictive wake-up mechanism based on pseudo-random number coupled with the contention resolution, PW-SA-MAC achieves higher delivery ratio and lower energy consumption. With a detailed packet-level simulations we evaluate and verify that PW-SA-MAC achieves better energy efficiency with improved delivery latency and delivery ratio compared to previous MAC solutions.

# References

1. El-Hoiydi, A., Decotignie, J., WiseMAC: An ultra low power MAC protocol for multi-hop WSNs. In: Proceedings of Algorithmic Aspects of Wireless Sensor Networks (2004)
2. Buettner, M., Yee, G.V., Anderson, E.: X-MAC: a short preamble MAC protocol for duty-cycled WSNs. In: Proceedings of Embedded Networked Sensor Systems (2006)
3. Sun, Y., Du, S., Gurewitz, O., Johnson, D.B.: DW-MAC: a low latency, energy efficient demand-wakeup MAC protocol for wireless sensor networks. In: Proceedings of Symposium on Mobile Ad Hoc Networking and Computing (2008)
4. Ye, W., Silva, F., Heidemann, J.: Ultra-low: duty cycle MAC with scheduled channel polling. In: Proceedings of Conference On Embedded Networked Sensor Systems (2006)
5. Sun, Y., Gurewitz, O., Johnson, D.B.: RI-MAC: a receiver initiated asynchronous duty cycle MAC protocol for dynamic traffic loads in wireless sensor networks. In: Proceedings of Conference On Embedded Networked Sensor Systems (2008)
6. Henna, S.: SA-RI-MAC: sender-assisted receiver-initiated asynchronous duty cycle MAC protocol for dynamic traffic loads in WSNs. In: Proceedings of Conference on Mobile Lightweight Wireless Systems (2011)
7. Tang, L., Sun, Y., Gurewitz, O., Johnson, D.B.: PW-MAC: an energy-efficient predictive-wakeup MAC protocol for wireless sensor networks. In: Proceedings of Conference on Computer Communications (2011)
8. Knuth, D.E.: The Art of Computer Programming, 3rd edn, vol. 2: Seminumerical Algorithms (1997)
9. Sun, Y., Du, S., Gurewitz, O., Johnson, D.B.: DW-MAC: a low latency, energy efficient demand-wakeup MAC protocol for wireless sensor networks. In: Proceedings of Symposium on Mobile Ad Hoc Networking and Computing (2008)

# How Big Open Data Can Improve Public Services

Fouad Nafis, Siham Yousfi and Dalila Chiadmi

**Abstract** During the last few years, IT companies and academies are focusing more researches on data. Indeed, data are important since the success of companies in a competitive market place depends on their ability to identify a trend, a problem or opportunities only seconds or microseconds before others. Public services are also experiencing a great need for the use of data held by different government agencies. Especially since the birth of the movement of open data. The objective of this work is to propose a global architecture for using technologies of big data, in accordance with the standards of open data, to improve the performance of public services.

**Keywords** Open data · Big data · E-gov · Public service · Open Government Data

## 1 Introduction

During the last few years, IT companies and academies are focusing more researches on data. Indeed, data are important since the success of companies in a competitive market place depends on their ability to identify a trend, a problem or opportunities only seconds or microseconds before others. While business sector has been the leader in Big Data development for many years, many other sectors, like public sector, become more interested to this emerging concept. In Sect. 2 of this paper, we make a quick discussion on the work in relation with our research

F. Nafis (✉) · S. Yousfi · D. Chiadmi
Computer Science Department, EMI School, Mohammed Vth University-Agdal,
Rabat, Morocco
e-mail: fouadnafis@research.emi.ac.ma

S. Yousfi
e-mail: sihamyousfi@research.emi.ac.ma

D. Chiadmi
e-mail: chiadmi@emi.ac.ma

© Springer International Publishing Switzerland 2016      607
A. El Oualkadi et al. (eds.), *Proceedings of the Mediterranean Conference
on Information & Communication Technologies 2015*, Lecture Notes
in Electrical Engineering 381, DOI 10.1007/978-3-319-30298-0_65

focus. In Sect. 3, we present some definitions of the concepts involved in our work. In Sect. 4, we show the components of the Moroccan Open Data Platform. In Sect. 5, we develop and detail our overall architecture. The last section provides a conclusion and some perspectives.

## 2 Related Works

Several works address the problematics of big data. Others have chosen to focus on the emergence of the movement of open data and principles and its relationship with government data. In [1] the authors have tried to give an overview on the use of open data in the field of finance, but without focusing on the side of big data. Our work takes its originality, is that it emphasizes the point that government data fulfill all the conditions of the BIG DATA and therefore they can be exploited in this direction to improve public services.

## 3 Open and Big Data

### 3.1 Big DATA

**Characteristics of BIG DATA**

**Volume**

The first characteristic is volume that refers to data size. As already mentioned, it is obvious that the volume of data is the most important challenge that requires specific optimization for traditional technologies. As mentioned in [2] the example of Low Frequency Array a radio telescope that collects five PB every hour.

**Velocity**

Velocity means how fast data are arriving and stored. In fact, effective management of big data, forces organization to process data while it has a value i.e. in near "real-time" for two reasons [3]: The first one is to gain a competitive advantage. The second reason is to filter data.

**Variety**

Variety represents different types of data. With the explosion of internet, several sources are providing data which are most of time unstructured (images, videos) and semi structured (web pages, logs).

**Veracity**

IBM declared Veracity as the fourth big data V that refers to the uncertainty of data wondering, "How can you act upon information if you don't trust it?" [4]. In fact, the uncertainty of data may be caused by data inconsistency (statistics are not reliable) or data trustworthiness (subjective data like feelings and opinions).

**Value**

Most of the existing data may be useless and companies need to filter these data in order to benefit from the true value that the extracted data could add to the intended activity.

## 3.2   OPEN GOVERNMENT DATA

**OPEN DATA**

Open data[1] is a new movement that emerged, encouraging any person or organization to publish their data in any format and make them available, modifiable, redistributable, and usable under a free license or limited.

**GOVERNMENT DATA**

The Governmental data is the data held by any public body or agency. Generally, when we talk about open data is quickly designed government data. OPEN GOVERNMENT DATA is a set that takes the benefits of both subsets OPEN DATA and DATA GOV.

To improve their services, organizations are encouraged to publish their data on the principle of open data. This will generate immediate benefits on several levels of government activities [5, 6, 7]:

*Increase transparency* [8]
Transparency is a fundamental aspect of democratic practice. Giving citizens the right of access to public information has become a necessity for modern governments. With the OPEN DATA, citizens can get an idea about how the taxes they pay are spent, without having to explore documents of hundreds of pages. They can also see how the state budgets are distributed and have an overview on state spending.

*Increase participation*
The OPEN DATA encourage citizens to participate[2] in the various decisions of governments. Some initiatives have emerged to allow citizens to propose law projects or just ask questions to deputies.[3]

---

[1] http://opendefinition.org/.
[2] FIKRA.EGOV.MA.
[3] Project nouabook in Morocco: http://www.nouabook.ma.

*Increase Collaboration*

OPEN DATA also improve collaboration among state agencies on the one hand, encouraging the exchange of data held by these organizations. In addition, between citizens and the government on the other hand through the development of social networks and other interactive solutions.

More benefits can be gained from the adoption of the principles of open data are cited in [5].

# 4 Open Data Platform in Morocco

In Morocco, on 31 March 2012, the Economic Council General Meeting, Social and Environmental instructed the Committee on Cultural Affairs and new technologies to study the question of the OPEN DATA as a site of application of the right of citizen's access to public information. The Ministry of Industry, Trade and New Technologies, as part of the Digital Morocco plan, launched "Data.gov.ma" platform in March 2011. This platform identifies just over 105 datasets (DATASET) published in most cases in Excel format (XLS or XLSX). On the website there is no application, this shows that this field is to explore, and considerable efforts remain to be deployed to firstly encourage organizations and governments to publish their data and on the other hand encourage associations and individuals to produce applications using this data.

# 5 The Proposed Solution

## 5.1 Proposed Architecture

The figure below show the proposed architecture (Fig. 1).

The data sources are supplied by users, which can be private citizens, public or private agencies. They are also involved in the selection of e-government services, which are driven by the data. The data because of their diversity and quantity are stored in a big data environment. Using MapReduce algorithms, this data are formatted to make them in easily readable formats, for the chosen e-gov services. After executing adequate services, several results are obtained as output, or as usable data directly or through gross outcomes for which data mining techniques [9] can be used (Machine Learning, Artificial Intelligence algorithms, …).

The operation of choice and improvement of e-government services is a task that also remains to detail. Several studies have tried to go in this direction as [10].

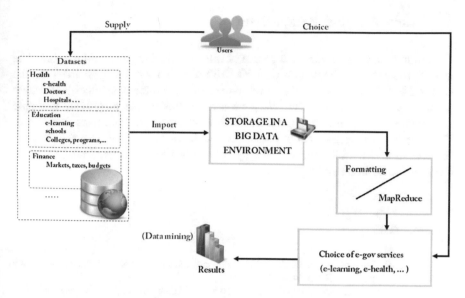

**Fig. 1** The proposed architecture

## 5.2 BIG DATA to Improve E-GOV Services

In this section, we will describe how we can use open big data in order to improve government services:

**Health sector**

In health sector, saving historical medical data allows keeps a record for each medical information related to patients like the list of medical consultation, symptoms, doctors' diagnostics and drugs to which the patient is allergic.

**Cartography**

Using big data sent by the electronic systems of cars, the government can identify bottlenecks and propose other itinerary, track car services used by government employees and identify whether it is used for personal purposes or track criminal cars know their slightest movement and catch them in the act.

**Weather**

Weather sensors are generating and sending a huge volume of logs quickly. For example, the South Korea Meteorological Administration's collects more than 1.6 terabytes of data every day [5].

**Finance**

In finance sector, the publication of financial data allows citizens to report misuse of a public budget.

## 6    Conclusion and Perspectives

Government data are a very rich source of information that we could explore. The use of tools and technologies of BIG DATA based on the principles of the OPEN DATA may lead to the development of government services with more performance and more quality. We have tried in this article to provide a simple architecture to help improving public services using the technologies mentioned. The next step is to implement this architecture with real data present in the platform "data.gov.ma" or ministries websites by following the steps mentioned above to produce the desired results.

## References

1. O'Riain, S., Edward, E., Harth, A.: XBRL and open data for global financial ecosystems: a linked data approach. IJAIS 141–162 (2012)
2. Demchenko, Y., Grosso, P., de Laat, C., Membrey, P.: Addressing big data issues in scientific data infrastructure. In: 2013 International Conference on Collaboration Technologies and Systems (CTS), pp. 48–55 (2013)
3. Wood, B.: "Info-Graphic: How Big is Big Data?". http://www.americanis.net/2013/info-graphic-how-big-is-big-data/ (2013)
4. Zikopoulos, P., Eaton, C., et al.: Understanding big data: analytics for enterprise class hadoop and streaming data. McGraw-Hill Osborne Media, New york (2011)
5. 4-Vs-of-big-data. http://www-01.ibm.com/software/data/bigdata/images/4-Vs-of-big-data.jpg
6. Kucera, J., Chlapek, D.: Benefits and risks of open government data. J. Syst. Integr. 5(1), 30–41 (2014)
7. Conradie, P., Choenni, S.: On the barriers for local government releasing open data (2014)
8. Zuiderwijk, A., Janssen, M.: Open data policies, their implementation and impact: a framework for comparison (2013)
9. Davies, T.G.: Transparency and open data (2013)
10. Lausch, A., Schmidt, A., Tischendorf, L.: Data mining and linked open data–New perspectives for data analysis in environmental research. Ecol. Model. 295, 5–17 (2015)

# Resources Scheduling in Virtual Environment of Cloud Computing

Yassine El Mahoti, Noura Aknin, Souad Amjad
and Kamal Eddine El Kadiri

**Abstract** Cloud Computing is a technique that allows companies to benefit from many features as a service (i.e. Resources as a service, network as a service, etc. …). Despite the great success proved by the cloud in terms of diversity and flexibility of applications given for costumers, it began to have a serious problem in terms of computing overloading due to massive use of devices connected to the internet (e.g. computers, mobile devices). In this paper, we propose a smart scheduling system implemented in the cloud architecture to minimize the hosts overloading, inducing a best power saving of energy consumption by introducing a smart multiplexing and migration of virtual machines.

**Keywords** Cloud computing · Scheduling · Load balancing · Virtualization

## 1  Introduction

The goal of this paper is provide a solution for the cloud architecture that might help to offer a best performance despite the growing number of customers. We propose to introduce a built-in system scheduler that enables a very efficient management of

Y. El Mahoti (✉) · N. Aknin · S. Amjad · K.E. El Kadiri
Information Technology and Modeling Systems Research Unit,
Faculty of Sciences, Abdelmalek Essaadi University, 93030 Tétouan, Morocco
e-mail: yassinemahoti@yahoo.fr

N. Aknin
e-mail: aknin@ieee.org

S. Amjad
e-mail: souad@uae.ma

K.E. El Kadiri
e-mail: elkadiri@uae.ma

Y. El Mahoti · N. Aknin · S. Amjad
Computer Sciences, Operational Research and Applied Statistics Laboratory,
Faculty of Sciences, Abdelmalek Essaadi University, 93030 Tétouan, Morocco

© Springer International Publishing Switzerland 2016
A. El Oualkadi et al. (eds.), *Proceedings of the Mediterranean Conference
on Information & Communication Technologies 2015*, Lecture Notes
in Electrical Engineering 381, DOI 10.1007/978-3-319-30298-0_66

equipment by implementing a smart multiplexing and migration. This will allow a minimization of the interaction time between the server and the clients and therefore a minimization of energy consumption in both directions to reach in the end an ecological IT system (Green computing). The paper is organized as follow: Sect. 1 presents the introduction, Sect. 2 describes the Cloud computing and the Sect. 3 proposes the scheduled system and its results.

## 2 Cloud Computing

Faced with the ever-increasing costs of implementation and maintenance of IT systems, companies outsource more often their systems management [1].

Indeed, this outsourcing of resources and software is cloud computing. Cloud computing allows the use of multiple services on demand, and the company pays only what it consumes [2]. This new model of IT management has given a great flexibility for companies to manage their data, and a significant economization at material and human resources. Cloud computing has proven to be a very important contributor at the ecological and economical levels by [3]:

- Reducing the energy consumption by resources polling (sharing the same resources with several customers).
- Reducing cost: the companies pays only what they consume.
- Reducing waste: infrastructure managed internally are often underutilized, while a cloud infrastructure is pooling all resources for a large number of companies.
- Making resources flexible: the company can increase the capacity of its infrastructure without major investment [4].

## 3 Related Work

We implemented a Cloud infrastructure by the VMWARE technology that comprises a cluster with three physical hosts (ESXI). Each physical server hosts a several virtual machines (VMs) with the following specifications: 128 KB of VRAM, 1 VCPU and the Windows XP operating system. A software layer has been established to a better host scheduling rather the case of VMware DRS (Dynamic Scheduler resources). The results obtained in this paper show the effectiveness of our proposed solution instead of the technology that VMware use to schedule their ESXi hosts.

## 3.1  Description

This software layer permits an efficient scheduling resources in Cloud architecture by achieving a smart multiplexing and migration of VMs for saving energy consumption and resources allocation. The proposed system is composed of these elements as shown in Fig. 1:

- Global Agent: a server that centralize all the information coming from local agents and contacts the VMM to indicate its job (multiplexing or migration).
- Local agent: a local VMs implemented in each host to return in real time by the SNMP protocol all information about this host and other VMs: (resources allocations, version of OS, etc....) and send this information to the global agent.
- VMM: virtual machine manager, a controller of VMs, which allows adding resources and turn off/on the VMs.

  This system interacts on two levels:

- Multiplexing: When the VM is overloaded, a local agent searches on the network VMs that have the same profile (OS, application) and chooses the most suitable to perform the multiplexing.
- Migration: A global agent searches on the network the ideal host witch have the VMs with the same profile as a source host to predict in the future a multiplexing of VMs if is necessary.

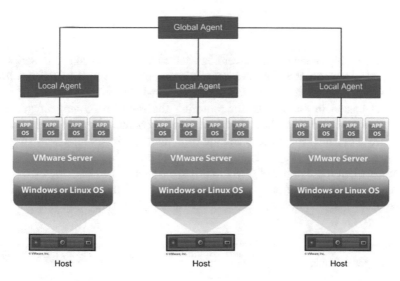

**Fig. 1**  The proposed scheduled system

## 3.2  Functioning

For the system to be effective, it must communicate with the following components:

- Local Agent executes a script for return information about host and VMs resources.
- Local agent collects all this information and send them to the global agent.
- Global agent centralizes all of this data coming from different agent and sort them a way that for each host and VM choose the best elected for migration or multiplexing when is it necessary.
- VMM executes the request coming from the global agent with a specific information (which VMs can be multiplexed and which Host where VMs will be migrated).

This middleware architecture can be describe as below: While the host is running, the Local agent sends VMs and Hosts information to the global Agent, next when the global agent detects an overload host, it selects the best elected host for migration based on a matrix of parameter (CPU, RAM, network channel, profile of virtual machines, etc.) and transmits the information to the VMM for executing the request. This proposed software layer interacts also in a second level; the multiplexing of virtual machines if needed, when a VM is overload the local agent selects the best elected for multiplexing and sending the information to the global agent for contacting the VMM to do the request.

## 3.3  Result

To simulate the behavior of servers in the case of overloading, a script was developed to increase the resources occupations such as the use of CPU and RAM.

This script, which developed in language C, installed in the different of virtual machines hosted in this cluster for returning the state functioning of this cloud environment when the number of customer and resources allocation increase.

As we said, this architecture middleware interacts in multiplexing of virtual machines when is needed, as shown in Table 1, the execution time of this script decreases when the number of VMs multiplexed increase.

This reduction time of process treatment allows also the economization in term of resources consumption by servers because the processing time for this script

**Table 1** Processing time in function of multiplexing

| Number of multiplexing of VMs | Processing time (ms) |
| --- | --- |
| 1 | 122,656 |
| 2 | 7671 |
| 4 | 6468 |

**Table 2** CPU % utilization by physical host

| Mode | Min | Max | Average |
|---|---|---|---|
| DRS | 0 | 89 | 27.8 |
| Scheduled system | 9.91 | 39.55 | 14.7 |

**Fig. 2** The proposed scheduled system

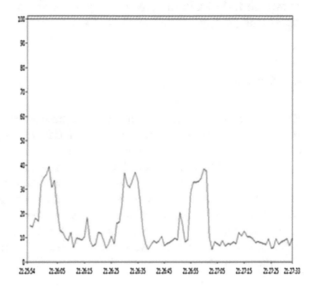

decreases. To verify that the proposed system allows also a reduction of consumption in resources, the same script was executed 3 times in an interval of 1 min and 40 s on the different virtual machine and we have the following results of CPU utilization of hosted server as shown in Table 2 and Figs. 2 and 3:

**Fig. 3** The DRS system

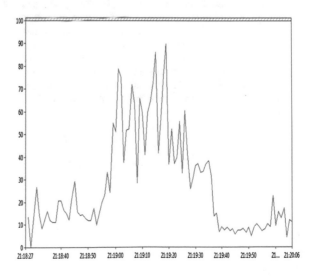

The maximum CPU utilization is 39 whereas in the DRS case is 89, the increasing value of CPU usage is justified that the virtual machine becomes much overloaded when processing a query and takes a longer time to finalize it, which consequently affects the resources of the physical server. This proposed system allows the rate of CPU average utilization to decrease from 27 to 14 % by using a smart multiplexing and migration of VMs.

# 4  Conclusion

The cloud computing had shown that it can provide multiple services to consumers, but against a huge energy consumption which can cause a serious impact on the environment (e.g. electric power, air conditioning).

The scheduling system studied in this paper will be able in the same time to have a powerful and robust system that handles the query's clients effectively by allocating resources dynamically (smart multiplexing) and performing best scheduling resources by migration (smart migration) when the system becomes overloaded.

This system will provide a significant Green computing system capable of handling multiple services with minimal resources and very short response to customer requests.

# References

1. Dong, W.E., Nan, W.; Xu, L.: In: Fifth International Conference on Shiyang Computational and Information Sciences (ICCIS), Shiyang. 21–23 June 2013
2. Shaikh, F.B., Haider, S.: Security threats in cloud computing. In: IEEE conference on Internet Technology and Secured Transactions (ICITST), Abudahbi (2011)
3. Sahu, Y., Pateriya, R.K., Gupta, R.K.: Cloud server optimization with load balancing and green computing techniques using dynamic compare and balance algorithm. In: IEEE Conference on Computational Intelligence and Communication Networks (CICN), Mathura (2013)
4. AbdElminaam, D.S., Abdul Kader, H.M., Hadhoud, M.M., El-Sayed, S.M.: Elastic framework for augmenting the performance of mobile applications using cloud computing. In: 9th International IEEE Conference on Computer Engineering (ICENCO), 2013, 28–29 Dec. 2013, Egypt. Online Library (www.wileyonlinelibrary.com), 11 October 2011. doi:10.1002/wcm. 1203

# Transformations from UML to Android Using MOF 2.0 QVT

Hanane Benouda, Mostafa Azizi and Mimoun Moussaoui

**Abstract** The diversity of mobile operating systems is pushing companies to offer more services to users of smartphones and tablets. This makes developers face a challenge to develop cross-platform mobile applications, knowing that this involves an operating system (OS), a programming language, development tools and a download store. For example, Android uses the Android SDK tools, JAVA and XML, and Google Play blind. That is why code generation based on UML Models becomes very important. This paper presents a technique based on MDA approach (Model Driven Architecture) to generate from UML models codes for Android models while using the standard MOF 2.0 QVT (Meta-Object Facility 2.0 Query-View-Transformation) as a language for transforming. This technique allows converging different software development process environment depending on the end use into a single driven process model sharing resources and reducing the workload.

**Keywords** MDA · Android · UML · MOF QVT · OCL

## 1 Introduction

Model Driven Engineering is a discipline for software engineering, allowing the use of modeling languages to automate the generation of all or a part of computer applications from the models. The OMG had made public its initiative MDA in 2001 as a restriction of the MDE. MDA allows separating the functional specification of an application from its implementation on given platform, and doing

H. Benouda (✉) · M. Azizi · M. Moussaoui
MATSI Laboratory, EST Oujda, Mohammed First University, Oujda, Morocco
e-mail: benouda.89@gmail.com

M. Azizi
e-mail: azizi.mos@gmail.com; azizi.mos@ump.ma

M. Moussaoui
e-mail: m.moussaoui@ump.ma

© Springer International Publishing Switzerland 2016
A. El Oualkadi et al. (eds.), *Proceedings of the Mediterranean Conference on Information & Communication Technologies 2015*, Lecture Notes in Electrical Engineering 381, DOI 10.1007/978-3-319-30298-0_67

interoperability between applications. It makes transformation of Computation Independent Models (CIM) to Platform Independent Models (PIM), and then transforms this PIM to Platform Specific Models (PSM) which will be used for the code generation task. Android as a mobile operating system is mostly used for mobile devices with touch screens as smartphones, and tablets. Android is being developed under open source licenses, and based on the Linux Kernel, using Java programming language to help developers creating mobile applications.

This paper is organized as follows: the second section defines the MDA [1] approach, while the third presents the Android system. MOF 2.0 QVT [2] and the OCL [3] constraints language are the main topics of the fourth section. In the fifth section, we present the UML and Android meta-models. Before conclusion, in the sixth section, we present the transformation rules using MOF 2.0 QVT from source to the Android target model.

## 2 Model Driven Architecture (MDA)

The MDA approach considers the models to transform as a productive element to be used to automatically generate the application source code [1]. MDA is composed of several models that express every aspect. The passage of a contemplative view models a productive view is due to the vehicle information necessary for generating the source code of the application made by these models. The MDA defines three levels of models, CIM, PIM, and PSM, for modeling the application and then by successive transformations to generate source code.

## 3 Android

The android operating system was created in November 2007, while gathering around him constructors, software editors and companies specializing in the development of mobile applications (Open Handset Alliance "OHA": Motorola, Samsung, Toshiba, LG, Huawei, Asus, HTC). This mobile OS which is hardware independent is open source [4].

## 4 Transformations of MDA Models

The MDA approach defines three levels of models: CIM, PIM, and PSM. These models are used to model the application and then by successive transformations to generate source code. There are three approaches in MDA model transformations:

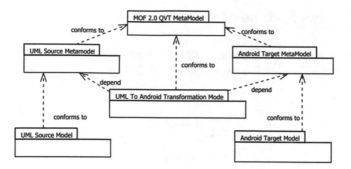

**Fig. 1** Modeling approach

The approach by Modeling, the approach by programming and the approach by Template. The approach by modeling (Fig. 1) is the one used in this paper. The objective of the modeling approach is to make productive and perennial transformation models, and express their independence towards the platforms of execution.

## 4.1 MOF 2.0 QVT

The standard MOF 2.0 defined by OMG, defines the meta-model for the development of transformation models. The Fig. 2 shows the architecture of the QVT standard. In this study we use the QVT-Operational mappings language included in the Eclipse Model to Model Transformation Project [2, 5].

## 4.2 Object Constraint Language (OCL)

The Object Constraint Language (OCL) [3] is a pure specification language and a formal language based on the common core of UML [6] and MOF. OCL can be used as a query language, to describe pre and post conditions on Operations and Methods.

**Fig. 2** QVT architecture

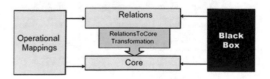

# 5   UML and Android Meta-Models

We present here the different meta-classes forming the simplified UML source
meta-model [5] (Fig. 3) and the simplified Android target meta-model (Fig. 4).

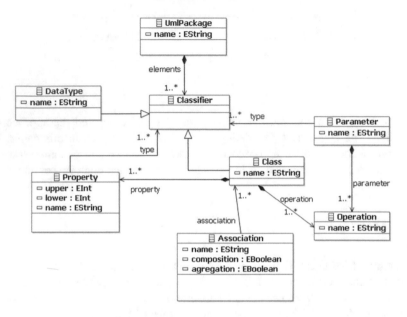

**Fig. 3** Simplified UML source meta-model

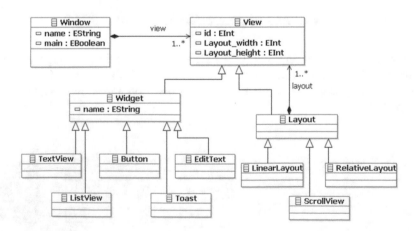

**Fig. 4** Simplified android target meta-model

**Fig. 5** Instance of UML
model

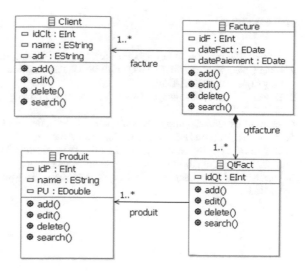

# 6 The Transformation Process of UML Model into Android Model

Our study targets to implement CRUD operations to generate an Android model from a UML model composed by the class Customer (Client), Invoice (Facture), Product (Produit), and Quantity Invoiced (QtFacture) (see Fig. 5). First we developed the corresponding Ecore models to our UML source and Android target meta-model, and then we implemented the algorithm for M2M transformation, using QVT Operational Mappings.

# 7 Conclusion

In this paper, we presented a technique that involves MDA approach to generate Android applications based on UML class diagrams. The transformation rules were developed using the approach by modeling and MOF 2.0 QVT to browse the class diagram and then generate XML file that handles all CRUD operations that can be used for producing the necessary code of the target application. Others aspects could be covered as perspective issues, such as extending our study to the other mobile OS.

# References

1. MDA—The Architecture Of Choice For Changing World. http://www.omg.org/mda/
2. Meta Object Facility (MOF) 2.0 Query/View/Transformation, V1.1, January 2011. http://www.omg.org/spec/QVT/1.1/PDF/
3. Object Constraint Language (OCL), Version 2.4, February 2014. http://www.omg.org/spec/OCL/2.4/PDF
4. Android Developers. http://developer.android.com/index.html
5. Esbai, R., Erramdani, M., Mbarki, S., Arrassen, I., Meziane, A., Moussaoui, M.: Transformation by modeling MOF 2.0 QVT: from UML to MVC2 web model. In: INFOCOMP, vol. 10, no. 3, pp. 01–11, Sept 2011
6. Unified Modeling Language (UML), Version 2.4.1, August 2011. http://www.omg.org/spec/UML/2.4.1/

# Taxonomy and Unified Access
# of E-Learning Platforms

**Karima Aissaoui and Mostafa Azizi**

**Abstract** In great majority of educational institutions, different e-learning platforms are simultaneously used for several reasons. As a consequence of this situation, learners must connect separately in every platform, which makes it a repetitive, redundant and time-consuming task. To address this problem, we propose in this paper to sweep most known existing platforms and extract theirs important features, and then to make an unified access to the e-learning platforms being used by learners, which means here that the learner will be invited to make a single authentication and then will be automatically connected in all platforms under use.

**Keywords** E-learning · Platforms · Taxonomy · Unified access

## 1 Introduction

Nowadays, e-learning has become one of the most interesting components that must be included in the pedagogical process. Its goal is to use technology in order to create a synchronous and asynchronous communication independently of time and location of both teacher and learner. Consequently, many e-learning platforms were created which offers a large choice to users of e-learning platforms. This diversity makes the choice of one platform so specific and personal. In an institution where many platforms are used, learners must be connected independently to these platforms used by their teachers. Whereas, learners often do not have links to switch from one platform to another. For these reasons, we suggest in this paper to unify access to e-learning platforms. This paper is structured as follows: first, we will

K. Aissaoui (✉) · M. Azizi
MATSI Lab, University Mohammed First, Oujda, Morocco
e-mail: aissaoui.karima@gmail.com

M. Azizi
e-mail: azizi.mos@gmail.com

© Springer International Publishing Switzerland 2016
A. El Oualkadi et al. (eds.), *Proceedings of the Mediterranean Conference on Information & Communication Technologies 2015*, Lecture Notes in Electrical Engineering 381, DOI 10.1007/978-3-319-30298-0_68

introduce e-learning, than we will propose a classification of eight popular systems. After that, we will present issues of multiple accesses and finally we present our technique to unify the access to different platforms.

## 2  E-Learning and Platforms Classification

Many definitions were given to e-learning. It refers to the use of electronic tools in delivering instructions including: CD-ROM, Internet, Intranet, Extranet, audio, video, etc. [1]. This way of learning allows learners to access information independently of their locations which decreases the cost of learning and makes this operation easier, faster and more efficient. Many e-learning platforms were developed and commercialized. In a previous work [2], we have classified the most used e-learning platforms using the Classification Tree Method (CTM) [3] and its software tool Classification-Tree Editor. In this paper, we include new criteria for classification and new platforms. E-learning platforms allow their users to have environment to work collaboratively in order to share knowledge and manage the pedagogic content. Developers and searchers are more and more interested by this field. Basically, we find two main families of e-learning platforms: Open source systems and appropriate solutions. Also, architectures of these platforms are various and different. Behind the large choice of platforms offered, the user finds difficulty in choosing the most suitable system for him, and consequently, a clear classification is necessary in order to make the choice of a platform easier. The Fig. 1 illustrates a classification of eight e-learning platforms: Atutor [4], Blackboard Learn [5], Chamilo [6], Claroline Connect [7], Dokeos [8], e-front learning [9], Moodle [4, 9] and Olat [10]. Our classification is based on Classification Tree Method and on many essential criteria on which every user should be informed about in order to choose the most suitable platform for him.

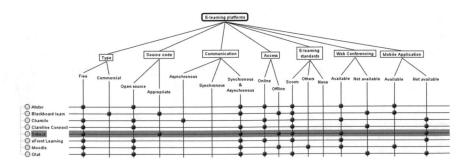

**Fig. 1** Taxonomy of some e-learning platforms

# 3   Issues of Multiple Accesses

Due to the very large choice of e-learning platforms, teachers choose different systems according to their necessities. In an institution where the e-learning platform is not unified, student is obliged to connect in all platforms used by all his teachers. This means that the task of authenticating will be done in all platforms used by all teachers of the same student, which makes them less encouraged to discover all those platforms. In order to resolve these problems, we propose in this paper an architecture of a system serving to unify access to multi e-learning platforms. Using our system, students will avoid authenticating in all platforms used by their teachers. Our solution will be a real benefit for all actors of the e-learning process including: the institution (or the administration), teachers and students.

## 3.1   Case of Administration

Having more than one platform used by teachers of the institution makes the management of data difficult. It must be collected from every database independently, because there is no communication or interconnection between all systems used by teachers in the same institution. Our architecture will facilitate the management of data concerning students. Using our system, the manager will not need to extract data from every platform separately one by one due to the centralized system management of all platforms used by teachers of the same institution.

## 3.2   Case of Teacher

The teacher will have the possibility to use more than one platform or even to change the usual one. Using our proposed architecture, he/she will be able to change his/her usual platform without any fear of lose and without need to recreate courses, learning resources or users in the new platform.

## 3.3   Case of Student

Using our system, students will have a common interface of all platforms used by their teachers. They will be able to connect only one time to be automatically connected on all platforms included in our system and to switch between those platforms easily. First, the student must be connected via our interface. The system checks if connection elements are correct and according to special access rights already predefined in each platform, the student is switched to convenient platforms.

## 4 Our Technique of Unified Access

Our proposed architecture comes to help all actors in e-learning process to improve their participation in this process. It comes as a radical solution to institutions where more than one platform is used. First, it will allow students to avoid signing in all platforms connected with the proposed system. The administration should add easily all platforms installed in the institution in order to give students the ability to sign in just one time in the common interface and automatically he/she will be connected in all platforms independently. This specification will encourage students to use e-learning platforms. Consequently, the diversity of platforms in an institution will not constitute a real problem to the administration. In addition to this, our architecture will offer to the institution a centralized system which makes all management tasks relative to e-learning platforms installed easier. Furthermore, our proposed architecture is easy to use and does not require special knowledge in computer science. As shows the Fig. 2, the administrator uses a web browser to choose the platforms to add to the system. Users of different platforms will be stocked in a Lightweight Directory Access Protocol (LDAP) to allow students to have access to all platforms added to the system. After this, students become able to use one unified interface using web browser to switch between platforms without need to login in all platforms independently. The performance of the proposed architecture resides in reassembling all existing platforms in the same institution in

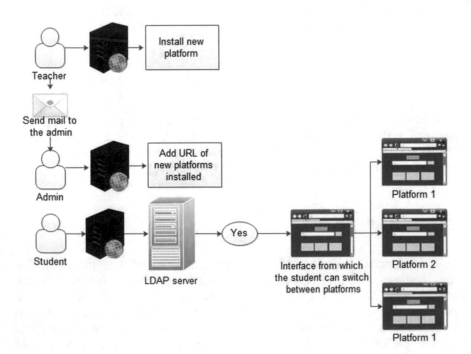

**Fig. 2** Process of use of our proposed system

order to have a centralized architecture which can be more efficient for managers to administer data. Furthermore, this solution supports the rights of access of each platform, which means that only students having right to access in a platform 'A' are able to connect to the same platform 'A' using our solution. This specificity allows this system to take into account the security which stills a very essential element in such system.

## 5  Conclusion and Future Works

Many researches were done in the field of e-learning and so many e-learning platforms were created. Users face challenges to choose one of them from a rich list. As a consequence, institutions find difficulties to impose the use of a specific unified system and allow thus teachers to use their preferred systems independently of others. Consequently, students must be connected in every platform, which makes them less motivated. Our system comes as a radical solution to this problem. It helps institutions to control access to different platforms using a simple web interface that helps to install those platforms in our proposed system. Furthermore, using our system, students will be able to sign in just one time and automatically they will be connected on all platforms installed in the system. However, our actual system includes two platforms which are Moodle and Chamilo, in the perspective to generalize our idea to most locally used platforms.

## References

1. Rosenberg, M.J.: E-learning: strategies for delivering knowledge in the digital age. McGraw-Hill, New York (2001)
2. Karima, A., Mostafa, A.: Classification of m-learning platforms. In: 8th IEEE International Conference On Intelligent Systems: Theories and Applications, pp. 239–243 (2013)
3. Grochtmann, M., Grimm, K.: Classification trees for partition testing. Software Testing, Verification and Reliability 3(2), 63–82 (1993)
4. Lengyel, P., Herdon, M., Szilgyi, R.: Comparison of moodle and ATutor LMSs (2006)
5. Foster, I., Kesselman, C.: The Grid: Blueprint for a New Computing Infrastructure. Morgan Kaufmann, San Francisco (1999)
6. Pishva, D., Nishantha, G.G.D., DANG, H.A.: A survey on how blackboard is assisting educational institutions around the world and the future trends. In: Advanced Communication Technology (ICACT), 2010 The 12th International Conference on. IEEE, pp. 1539–1543 (2010)
7. Maes, J.M.: Chamilo 2.0: a second generation open source e-learning and collaboration platform. Int. J. Adv. Corp. Learn. (iJAC), 3(3), 26–31 (2010)
8. Graf, S., List, B.: An evaluation of open source e-learning platforms stressing adaptation issues. IEEE 163–165 (2005)
9. Fariha, Z., Zuriyati, A.: Comparing Moodle and eFront software for learning management system. Aust. Basic Appl. Sci. 8(4), 158–162 (2014)
10. Consortium Claroline, http://www.claroline.net/

# Automatic Metrics for Machine Translation Evaluation and Minority Languages

Daša Munková and Michal Munk

**Abstract** Translation quality and its evaluation play a crucial role in the field of machine translation (MT). This paper focuses on the quality assessment of automatic metrics for MT evaluation. In our study we assess the reliability and validity of the following automatic metrics: Position-independent Error Rate (PER), Word Error Rate (WER) and Cover Disjoint Error Rate (CDER). These metrics define an error rate of MT output and also of MT system itself, in our case it is an on-line statistical MT system. The results of the reliability analysis showed that these automatic metrics for MT evaluation are reliable and valid, whereby the validity and reliability were verified for one translation direction: from the minority language (Slovak) into English.

**Keywords** Machine translation · Quality · Evaluation · Minority language

## 1 Introduction

Translation quality assessment (TQA) has significant influence not only in the academic sphere [1, 2] but also in the translation industry. Academic and industrial interests in TQA are mainly the criteria for translation quality (TQ), approaches to translation quality or parameter setting in TQA models. Machine translation (MT) evaluation is a highly evitable and challenging task.

An approach to the evaluation of MT output can be divided into a manual approach (human evaluation) and an automated approach (automatic evaluation), both with a wide range of proposed techniques. Papineni et al. [3] stated that methods and metrics of manual evaluation are too slow and time consuming for MT system development, for which fast feedback on TQ is extremely important. Vilar

D. Munková · M. Munk (✉)
Constantine the Philosopher University in Nitra, Nitra, Slovakia
e-mail: mmunk@ukf.sk

D. Munková
e-mail: dmunkova@ukf.sk

© Springer International Publishing Switzerland 2016
A. El Oualkadi et al. (eds.), *Proceedings of the Mediterranean Conference on Information & Communication Technologies 2015*, Lecture Notes in Electrical Engineering 381, DOI 10.1007/978-3-319-30298-0_69

et al. [4] pointed out that subjectivity, which is characterized for manual evaluation causes a problem in terms of biased judgments towards MT, as well as having no clear definition of a numerical scale by TQA. By "fluency and adequacy" metric evaluators have a difficult task in separating these two criteria of translation without any instructions how to quantify meaning (adequacy), or how many grammatical errors separate different levels of fluency [5–8]. In an effort to make evaluation more effective and objective, i.e. without the intervention of human evaluators, automatic techniques have been further applied to evaluation. Several methodologies and metrics were proposed on how to assess TQ, which are mostly based on the similarity between MT output and human translation (HT). They are fast, cheap and re-usable, and comprise speech, cost and usability.

Many experiments showed that automatic metrics for MT evaluation correlate with human judgment [4, 5, 9–12] but only for majority languages.

## 2   Automatic Metrics for MT Evaluation

*PER* metric (*Position-independent Error Rate*) is very similar to metric *Recall*. Both have the same denominator—a reference length, i.e. a number of words in reference sentence but *PER* metric calculate the discrepancy not the concurrence. It takes into account extra words which are considered as errors and they need to be removal from the long translation (sentence).

Formally:

$$PER = 1 - \frac{correct - \max(0, L_{hypothesis} - L_{reference})}{L_{reference}}, \tag{1}$$

where the *correct* means a number of matches, $L_{reference}$ means a number of words in the reference sentence and $L_{hypothesis}$ means a number of words in translation.

*WER* metric (*Word Error Rate*) belongs to the automatic metrics of the first generation for MT evaluation. It comes from the field of speech recognition. It is based on the edit distance and it considers the word order. The edit distance is represented by the Levenshtein distance, which is defined as a minimum number of edit operations (insertion, deletion and substitution) that are necessary to achieve a concurrency between two sequences (sentences).

Formally:

$$WER(h, r) = \frac{\min_{e \in E(h,r)}(insertion(e) + deletion(e) + substitution(e))}{|r|}, \tag{2}$$

where the *insertion* (*e*) means a number of insert words, *deletion* (*e*) means a number of delete words, *substitution*(*e*) means a number of substitutions in a sequence or in a string *e*, *r* is a reference for the hypothesis *h* and $e \in E(h, r)$ is a sequence of the least number of edit operations to modify hypothesis to achieve reference.

Difference between metrics *PER* and *WER* is that the *PER* metric ignores a word position in a sentence.

*CDER* metric (*Cover Disjoint Error Rate*) is an extension of the *WER* metric. It is based on the Levenshtein distance. It utilizes the fact that a number of blocks in a sentence is the same as a number of gaps among them plus 1. The *long jump* is equal to block movement. The *long jump* is a jump over the gap between two blocks, whereby it does not penalize the movement of the whole blocks.

Formally:

$$CDER(h,\ r) = \frac{\min_{e \in E(h,r)}(insertion(e) + deletion(e) + substitution(e) + long\_jump(e))}{|r|},$$

(3)

where the *insertion* (*e*) means a number of insert words, *deletion* (*e*) means a number of delete words, *substitution* (*e*) means a number of substitutions, *long_jump*(*e*) means a number of long jumps in a sequence or in a string *e*, *r* is a reference for the hypothesis *h* and $e \in E(h,\ r)$ is a sequence of the least number of edit operations to modify hypothesis to achieve reference.

# 3 Reliability Analysis of Automatic Metrics for MT Evaluation

The aim of the research was to assess the quality—reliability and validity—of the automatic evaluation of MT; in our case on-line statistical MT system was used. The first part of our research consists of reliability verification—reliability of automatic metrics for MT evaluation (*PER*, *WER* and *CDER*). These metrics are called metrics of error rates, i.e. the higher values of these metrics, the lower the translation quality. Consequently, it consists of the identification of metrics which decrease the total score of reliability of automatic evaluation. The second part of the research rests on the verification of the validity of the automatic metrics for MT evaluation with respect to reference translation and using the manual metrics for MT evaluation, i.e. through the *fluency* (*F*) and *adequacy* (*A*).

We used our own design tool for evaluation which computes automatic metrics based on the comparison of hypothesis (MT output) with reference (reference HT).

***Reliability***. The value of *coefficient of reliability* (Table 1) is approximately 0.95 (95 %). It represents a portion of variability of the sum of metrics score to the whole variability of automatic metrics of MT evaluation (MT error rate). Both estimations (*Crombach's alpha* and *Standardized alpha*) are very similar (Table 1), i.e. the individual metrics have the same variability.

**Table 1** Results of the reliability of the automatic metrics—*PER*, *WER* and *CDER*

| Valid N | Mean | Std dev | Avg inter-metrics correlation | Cronbach's alpha | Standardized alpha |
|---|---|---|---|---|---|
| 74 | 160.94 | 56.36 | 0.885 | 0.950 | 0.953 |

The automatic metrics defining MT error rate and representing the automatic evaluation of MT are considered highly reliable based on the direct estimation of reliability.

All items (Table 2) correlate with the total score of evaluation and after their eliminating the coefficient of reliability has not increased except the *WER* metric. After elimination of the metric *WER*, the coefficient of reliability—*Crombach's alpha* increased from 0.950 to 0.966, but that it is negligible.

Even though the reliability analysis of automatic metrics characterizing the error rate of MT evaluation showed that the *WER* metric is the most deviated from the other automatic metrics of automatic MT evaluation (*PER* or *CDER*) in TQA. It is understandable seeing that, the *WER* metric is very strict to syntax errors (word order).

*Validity*. Criterion for validity is the scores of manual metrics for MT evaluation, i.e. the *fluency* (*F*) and *adequacy* (*A*) score. Both scores are measured on the scale from 1 to 5, where 1 means none (adequacy)/incomprehensible (fluency) and 5 means all meaning (adequacy)/flawless (fluency).

Between automatic and manual metrics for MT evaluation (Table 3) is a medium to large degree of linear dependency.

Validity is affected by reliability of MT evaluation. The lower the reliability, the lower the correlation among automatic and manual metrics for MT evaluation. Therefore we have to compute a corrected coefficient of correlation [13]. The values of corrected correlation coefficients among automatic and manual metrics for MT evaluation achieve values higher than 0.6 in cases of the *fluency* and also *adequacy* of the machine translation.

The automatic metrics, characterizing error rate for MT evaluation can be regarded as valid.

**Table 2** Statistics of automatic metrics, characterizing error rate, for MT evaluation

|  | Mean if deleted | Std dev if deleted | Metrics-total correlation | Alpha if deleted |
|---|---|---|---|---|
| *PER* | 115.48 | 39.34 | 0.882 | 0.939 |
| *WER* | 100.23 | 36.76 | 0.853 | 0.966 |
| *CDER* | 106.16 | 37.20 | 0.964 | 0.874 |

**Table 3** Correlations among automatic and manual metrics for MT evaluation

|  | PER | WER | CDER |
|---|---|---|---|
| *Fluency* (F) | −0.505 | −0.525 | −0.573 |
| *Adequacy* (A) | −0.453 | −0.455 | −0.503 |

# 4 Conclusions and Future Direction

Results of presented research are an analysis, design and implementation of algorithms of the automatic MT evaluation based on metrics defining an error rate such as *Position-independent Error Rate* (*PER*), *Word Error Rate* (*WER*) and *Cover Disjoint Error Rate* (*CDER*). By the text representation we arose from the transaction-sequence model which is described in [14–16]. The results of reliability analysis showed that the automatic metrics characterizing an error rate are reliable and valid, whereby the validity and reliability were verified for the minority language such as Slovak language. We examined the quality of MT output from Slovak into English or German. We verified the validity of automatic metrics for MT evaluation by the metrics of manual MT evaluation. We defined a concurrent validity as a corrected multiple correlation coefficient between the manual metrics for MT evaluation (*adequacy*, *fluency*) and automatic metrics for MT evaluation (*PER*, *WER*, *CDER*). In addition to validity, we also verified reliability—reliability of the automatic metrics for MT evaluation. We used two coefficients of reliability —*Crombach's alpha* and *Standardized alpha*—to estimate reliability. Both estimations were very similar, i.e. individual automatic metrics for MT evaluation have the same variability. Based on the results of the reliability analysis of the automatic metrics for MT evaluation we can say that the automatic metrics highly negative correlate with the manual metrics for MT evaluation. For example, if the judges assessed low the fluency and adequacy of translation, it also depicted in metrics of automatic evaluation, namely with the high values of *PER*, *WER* and *CDER*. In future research we would focus on the metrics of automatic evaluation of the second generation (*IITER* or *METEOR*).

**Acknowledgments** This work was supported by the Slovak Research and Development Agency under the contract No. APVV-14-0336 and Scientific Grant Agency of the Ministry of Education of the Slovak Republic (ME SR) and of Slovak Academy of Sciences (SAS) under the contracts No. VEGA-1/0559/14 and No. VEGA-1/0392/13.

# References

1. House, J.: Translation Quality Assessment: a Model Revisited. Tubingen, Narr (1997)
2. Nord, Ch.: Text Analysis in Translation. Theory, Methodology, and Didactic Application of a Model for Translation-Oriented Text Analysis. Amsterdam, Rodopi (2005)
3. Papineni, K., Roukos, S., Ward, T., Zhu, W.: BLEU: a method for automatic evaluation of machine translation. In: ACL '02, pp. 311–318 (2002)
4. Vilar, D., Leusch, G., Ney, H., Banchs, R.E: Human evaluation of machine translation through binary system comparisons. In: StatMT '07, pp. 96–103 (2007)
5. Banerjee, S., Lavie, A.: METEOR: an automatic metric for MT evaluation with improved correlation with human judgments. In: ACL '05 pp. 65–72 (2005)
6. Callison-Burch, Ch., Fordyce, C., Koehn, P., Monz, Ch., Schroeder, J.: (Meta-) evaluation of machine translation. In: StatMT '07, pp. 136–158 (2007)

7. Popovic M., Ney, H.: Word error rates: decomposition over POS classes and applications for error analysis. In: ACL '07 WSMT (2007)
8. Gornostay, T.: Machine translation evaluation. http://www.ida.liu.se/labs/nlplab/gslt/mt-course/info/TatianaG_assignment1.pdf (2008). Accessed 22 June 2014
9. Doddington, G.: Automatic evaluation of machine translation quality using n-gram co-occurrence statistics. In: HLT-02, pp. 138–145 (2002)
10. Snover, M., Dorr, B., Schwartz, R., Micciulla, L., Makhoul, J.: A study of translation edit rate with targeted human annotation. In: ASMTA (2006)
11. Koehn, P.: Statistical Machine Translation. Cambridge University Press (2010)
12. Lavie, A., Agarwal, A.: Meteor: an automatic metric for mt evaluation with high levels of correlation with human judgments. In: ACL '07 WSMT, pp. 228–231 (2007)
13. Pilkova, A., Volna, J., Papula, J., Holienka, M.: The influence of intellectual capital on firm performance among slovak smes. In: ICICKM-2013, pp. 329–338 (2013)
14. Munková, D., Munk, M., Adamová, Ľ.: Modelling of language processing dependence on morphological features. In: Advances in Intelligent Systems and Computing (2013) 77-86
15. Munková, D., Munk, M., Vozár, M.: Data pre-processing evaluation for text mining: transaction/sequence model. In: ICCS 2013, pp. 1198–1207 (2013)
16. Munková, D., Munk, M., Vozár, M.: Influence of stop-words removal on sequence patterns identification within comparable corpora. In: Advances in Intelligent Systems and Computing, pp. 67–76 (2013)

# Part X
# Advances in ICT Modeling and Design ICT Developments and Recent Progress

# Comparing HMM, LDA, SVM and Smote-SVM Algorithms in Classifying Human Activities

**M'hamed Bilal Abidine and Belkacem Fergani**

**Abstract** Accurately recognizing the rare activities from sensor network based smart homes for monitoring the elderly person is a challenging task. Typically a probabilistic models such as the Hidden Markov Model (HMM) and Linear Discriminant Analysis (LDA) are used to classify the activities. In this work, we demonstrate that discriminative model named Support Vector Machines (SVM) based on the Synthetic Minority Over-sampling Technique (Smote) outperforms HMM, LDA and standard SVM and it can lead to a significant increase in recognition performance. Our experiments carried out on multiple real world activity recognition datasets, consisting of several weeks of data.

**Keywords** Activity recognition · HMM · LDA · SVM · Imbalanced classification

## 1 Introduction

As the number of elderly people in our society increases and the households will include someone who needs help performing basic activities of daily living such as cooking, dressing, toileting, bathing and so on [1, 2]. For their comfort and because the healthcare infrastructure will not be able to handle this growth, it is suggested to assist sick or elderly people at home. Sensor based technologies in the home is the key of this problem. Sensor data collected often needs to be analysed using data mining and machine learning techniques [3] to determine which activities took place. State of the Art methods used for recognizing activities can be divided in two main categories: generative models and discriminative models [3].

M.B. Abidine (✉) · B. Fergani
Laboratoire d'Ingénierie des Systèmes Intelligents et Communicants,
Faculty of Electronics and Computer Sciences, University of Science and Technology
Houari Boumediene (USTHB), Algiers, Algeria
e-mail: abidineb@hotmail.com

B. Fergani
e-mail: bfergani@gmail.com

© Springer International Publishing Switzerland 2016

A. El Oualkadi et al. (eds.), *Proceedings of the Mediterranean Conference on Information & Communication Technologies 2015*, Lecture Notes in Electrical Engineering 381, DOI 10.1007/978-3-319-30298-0_70

639

However, activity recognition datasets are generally imbalanced, meaning certain activities occur more frequently than others. However, the learning system may have difficulties to learn the concept related to the minority class. Many popular machine learning algorithms have been tried to see how well they can cope with the imbalanced situation [4], e.g. Weighted Support Vector Machine (WSVM) [5], $k$-Nearest Neighbors $k$-NN [5], random forests [6] and CS-SVM [7].

The main contribution of our work is twofold. Firstly, we demonstrate the efficiency of the standard discriminative method named Support Vector Machines (SVM) [3] combined with the Synthetic Minority Over-sampling Technique [8] in order to avoid the overfitting caused by imbalanced activity samples in smart homes. Secondly, this method is compared with the standard SVM, Linear Discriminant Analysis (LDA) [9] and Hidden Markov Model (HMM) [2].

## 2 Discriminative Models for Activity Recognition

### 2.1 Linear Discriminant Analysis (LDA)

Given a set of observations in $n$-dimensional space: $D_i = \left\{ x_1^i, \ldots, x_{m_i}^i \right\}$ $(x_j^i \in R^n)$ from class $C_i (i = 1, \ldots, N, N$ is the number of classes), we assume that each of the class probability density functions can be modeled as normal distribution. Define the prior probabilities $p(C_i)$, means $\bar{m}_i$, and covariance matrices $\Sigma_i$ of each class:

$$\Sigma_i = \frac{1}{m_i} \sum_{i=1}^{m_i} (x_i - \bar{m})(x_i - \bar{m})^{\mathrm{T}} \tag{1}$$

where $m_i$ is the number of patterns in class $C_i$. With LDA all classes are assumed to have the same covariance matrices $\Sigma_i, \ldots, \Sigma_N$, on (1). We assign the new feature vector that is to be classified $x$ to $C_i$ with the linear discriminant function $d_i$. This function is obtained by simplification the quadratic discriminant rule [9]

$$d_i(x) = \log(p(C_i)) - \frac{1}{2} m_i^T S_W^{-1} \bar{m}_i + x^T S_W^{-1} \bar{m}_i \tag{2}$$

in which $S_w$ is the common covariance matrix

$$S_W = \sum_{i=1}^{N} \frac{m_i}{m - N} \Sigma_i \tag{3}$$

The classification rule is given in Eq. 4.

$$f(x) = i^* :\Leftrightarrow i^* = \arg \max_i d_i(x) \tag{4}$$

## 2.2 Proposed Approach for Activity Recognition (Smote-SVM)

Smote-SVM approach is shown in Fig. 1. In the training phase, we perform the necessary pre-processing on the activity data represented in a feature space. We need only to correct the class imbalance using the pre-classification named Smote strategy. The balanced data is then used to learn the SVM classifier. It will then be used to process a new observation during the testing phase where the associated ADL class will be predicted.

a. *The Synthetic Minority Over-sampling Technique (SMOTE)*

The SMOTE algorithm generates artificial data based on the feature space similarities between existing minority examples in the training set. Synthetic examples are introduced along the line segment between each minority class example and one of its $k$ minority class nearest neighbors. The $k$-nearest neighbors ($k$-NN) are defined as the $k$ elements of subset $S_{min} \in S$ whose Euclidian distance between itself and $x_i \in S_{min}$ under consideration exhibits the smallest magnitude along the $n$-dimensions of feature space $X$. To create a synthetic sample, the $k$-nearest neighbors are randomly chosen, then multiply the corresponding feature vector difference with a random number $\delta \in [0, 1]$, and finally, add it to $x_i$

$$x_{new} = x_i + (\hat{x}_i - x_i) \times \delta, \tag{5}$$

where $x_i \in S_{min}$ is the minority instance under consideration, $\hat{x}_i$ is one of the $k$-NN for $x_i$: $\hat{x}_i \in S_{min}$.

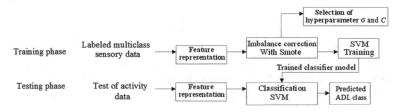

**Fig. 1** Diagram of Smote-SVM approach

b. *Support Vector Machines (SVM)*

We assume that we have a training set $\{(x_i, y_i)\}_{i=1}^{m}$ where $x_i \in R^n$ are the observations and $y_i$ are class labels either 1 or $-1$. The dual formulation of the SVM can be solved by representing it as a Lagrangian optimization problem as follows [3]:

$$\max_{\alpha_i} \quad \sum_{i=1}^{m} \alpha_i - \frac{1}{2} \sum_{i=1}^{m} \sum_{j=1}^{m} \alpha_i \alpha_j y_i y_j K(x_i, x_j)$$

$$\text{Subject to} \quad \sum_{i=1}^{m} \alpha_i y_i = 0 \; and \; 0 \leq \alpha_i \leq C \tag{6}$$

where $K(x_i, x_j)$ is the kernel, the radial basis kernel function (RBF) is used in the study: $K(x, y) = \exp\left(\frac{-1}{2\sigma^2} \|x_i - x_j\|^2\right)$. $\alpha_i > 0$ are Lagrange multipliers. The regularization parameter $C$ is used to control the trade-off between maximization of the margin width and minimizing the number of training error.

Solving (6) for $\alpha$ gives a decision function in the original space for classifying a test point $x \in R^n$ [3]

$$f(x) = \text{sgn}\left(\sum_{i=1}^{m_{sv}} \alpha_i y_i K(x, x_i) + b\right) \tag{7}$$

with $m_{sv}$ is the number of support vectors $x_i \in R^n$.

In this study, a software package LIBSVM [10] was used to implement the multiclass classifier algorithm. It uses the one-versus-one method [3].

## 3 Experimental Results

We use an openly datasets [11] gathered from three houses KasterenA, KasterenB, KasterenC, having different layouts and different number of sensors, thus providing a diverse testbed. The activities performed using a wireless sensor network with a single man occupant. Data are collected using binary sensors, such as reed switches and float sensors. The sensor data were labelled using different annotation methods using Bluetooth headset or Handwritten diary. We separate the data into a test and training set using a "Leave one day out cross validation" approach [2].

As the activity instances were imbalanced between classes, we evaluate the performance of our models by two measures, the accuracy and the class accuracy. The accuracy shows the percentage of correctly classified instances, the class accuracy shows the average percentage of correctly classified instances per classes. They are defined as follows:

$$Accuracy = \frac{\sum_{i=1}^{m} [\inf erred(i) = true(i)]}{m} \tag{8}$$

$$Class = \frac{1}{N} \sum_{c=1}^{N} \left[ \frac{\sum_{i=1}^{m_c} [\inf erred_c(i) = true_c(i)]}{m_c} \right] \tag{9}$$

in which [a = b] is a binary indicator giving 1 when true and 0 when false. $m$ is the total number of samples, $N$ is the number of classes and $m_c$ the total number of samples for class $c$. A problem with the accuracy measure is that it does not take differences in the frequency of activities into account. Therefore, the class accuracy should be the primary way to evaluate an activity classifier's performance.

In our experiments, for the Smote-SVM method, the minority class examples were over-sampled using $k = 4$ nearest neighbors for Smote. We utilize the leave-one-day-out cross validation technique for the selection of width parameter for the SVM classifier. We found $\sigma_{opt} = 1$, $\sigma_{opt} = 1$ and $\sigma_{opt} = 2$ for these datasets respectively. The summary of the accuracy and the class accuracy obtained, for HMM, LDA, SVM and Smote-SVM methods performed using various real world datasets are shown in Table 1. This table shows that Smote-SVM performs better in terms of class accuracy.

Our results give us early experimental evidence that Smote-SVM works better for model classification; it consistently outperforms the other methods in terms of the class accuracy for all datasets. In the rest of section, we explain the difference in terms of performance between HMM and our method. HMM is trained by splitting the training data in which a separate model $P(x|y)$ is learned for each class, as

**Table 1** Class accuracy and Accuracy for HMM, LDA, SVM and Smote-SVM

| Houses | Models | Class (%) | Accuracy (%) |
|---|---|---|---|
| KasterenA | HMM [11] | 66 | 92 |
| | LDA | 60.9 | 89.8 |
| | Standard SVM | 50.3 | 92.1 |
| | Smote-SVM | 75 | 88.6 |
| KasterenB | HMM [11] | 55 | 80 |
| | LDA | 48.7 | 80.6 |
| | Standard SVM | 39.3 | 85.5 |
| | Smote-SVM | 67.3 | 64.6 |
| KasterenC | HMM [11] | 40 | 76 |
| | LDA | 40.1 | 79.7 |
| | Standard SVM | 35.6 | 80.7 |
| | Smote-SVM | 56.6 | 78.3 |

parameters are learned for each class separately. This is why HMM performs better for the minority activities. Our method shows that SVM becomes more robust for classifying the minority class.

# 4 Conclusion and Perspectives

Our experiments on real world datasets show that the choice of Smote-SVM approach can significantly increase the recognition performance to classify multi-class sensory data, and are less prone to overfitting caused by imbalanced datasets. It significantly outperforms HMM, LDA and SVM. Developing Classifiers which are robust and skew insensitive or hybrid algorithms can be point of interest for the future research in activity recognition. It would be interesting to compare Smote-SVM and Smote-CS-SVM [7] and then deciding which gives the best results.

# References

1. Abidine, M.B., Fergani, L., Fergani, B., Fleury, A.: Improving human activity recognition in smart homes. Int. J. E-Health Med. Commun. (IJEHMC) 6(3), 19–37 (2015)
2. Van Kasteren, T., Noulas, A., Englebienne, G., Kröse, B.: Accurate activity recognition in a home setting. In: Proceedings of UbiComp'08, pp. 1–9. ACM, New York, USA (2008)
3. Vapnik, V.N.: The Nature of Statistical Learning Theory. Springer, New York (2000)
4. Chawla, N.V.: Data mining for imbalanced datasets: an overview. In: Data Mining and Knowledge Discovery Handbook, (pp. 875–886). Springer, US (2010)
5. Abidine, M.B., Fergani, B.: A new multi-class WSVM classification to imbalanced human activity dataset. J. Comput. 9(7), 1560–1565 (2014)
6. Chen C., Liaw, A., Breiman, L.: Using random forest to learn unbalanced data. Technical Report 666, Statistics Department, University of California at Berkeley (2004)
7. Abidine, M.B., Fergani, B., Oussalah, M., Fergani, L.: A new classification strategy for human activity recognition using cost sensitive support vector machines for imbalanced data. J. Kybernetes 43(8), 1150–1164 (2014)
8. Chawla, N.V., Bowyer, K.W., Hall, L.O., Kegelmeyer, W.P.: SMOTE: synthetic minority over-sampling technique. J. Artif. Intell. Res. 16, 321–357 (2002)
9. Croux, C., Filzmoser, P., Joossens, K.: Classification efficiencies for robust linear discriminant analysis. Statistica Sinica 18(2), 581–599 (2008)
10. Chang, C.C., Lin, C.J.: LIBSVM: a library for support vector machines. ACM Trans. Intell. Syst. Technol. 2:1–27 (2013). http://www.csie.ntu.edu.tw/~cjlin-/libsvm/
11. Van Kasteren, T., et al.: Effective performance metrics for evaluating activity recognition methods. In: Proceedings of ARCS 2011 Workshop on Context-Systems Design, Evaluation and Optimisation, pp. 22–23. Italy (2011)

# Agent Based Fuzzy Clustering Algorithm

Hanane Barrah and Abdeljabbar Cherkaoui

**Abstract** Recently, the agents and multi-agent systems (MAS) field has known a significant breakthrough and covered a wide range of application domains. In order to keep up with this technological evolution, this paper concentrates on developing a MAS based on a fuzzy clustering algorithm, which is a widely useful clustering method in several fields, in the aim of making it easy to be extended and incorporated into other MASs. In the demonstration section, we show the effectiveness of our algorithm against the standard fuzzy c-means algorithm, and demonstrate theoretically how this system can reduce the running time of the applications based on it.

**Keywords** Agents · Multi-agent system · Fuzzy clustering · c-means algorithm · Mahalanobis distance

## 1 Introduction

Agents and MASs are a very large and active research field situated at the cross-roads of several areas such as computer science and industry. It is centered on developing theories and methods of designing and implementing agent-based and MASs. In this context, a wealth of works have been done so far. Indeed, an overview of research and a historical context to the field were presented by Jennings et al. [1].

Our intention in this work is to provide a MAS able to cluster data using a modified fuzzy clustering algorithm [2]. To show the importance of this system, we extended it to an application, of extracting ellipses from a series of scattered

H. Barrah (✉) · A. Cherkaoui
Laboratory of Innovative Technologies, National School of Applied Sciences,
Tangier, Morocco
e-mail: Hananbarah@gmail.com

A. Cherkaoui
e-mail: Cherkaoui.lti@gmail.com

© Springer International Publishing Switzerland 2016
A. El Oualkadi et al. (eds.), *Proceedings of the Mediterranean Conference on Information & Communication Technologies 2015*, Lecture Notes in Electrical Engineering 381, DOI 10.1007/978-3-319-30298-0_71

two-dimensional points, which is going to be incorporated, in a future work, into another MAS to segment the MRI (Magnetic resonance Imaging) images.

The remainder of this chapter is structured as follows: In Sect. 2, we present the key concepts of the field. In Sect. 3, we outline our modified fuzzy clustering algorithm. We describe our MAS in Sect. 4. Section 5 is dedicated for some experimental results and we conclude in Sect. 6.

## 2 Multi-agent Systems

According to literature, there is no unique definition of the term 'agent'. In this work, we have adopted the Wooldridge and Jennings's definition [3]: An agent is a hardware or a software-based computer system located in an environment and works autonomously and flexibly to achieve the objective for which it was designed.

Based on this definition, a MAS can be defined as a network of autonomous agents that interact with each other and with their environment to achieve a common goal.

Designing agents comes back to designing architectures, which has become an interesting research subject of several researchers. Agent architectures can be divided into three types: *Deliberative* architectures; contain a basic knowledge about the environment and can make decisions; *reactive* architectures; don't have any basic knowledge of the environment and don't reason; and *hybrid* architectures; try to combine the best aspects of the previous ones [1] in order to design more complex and sophisticated architectures.

In order to meet the purposes for which a MAS is designed, their elements must be able to interact with each other. Generally, we can distinguish between three types of interactions: *cooperation*; corresponds to a collective work in order to achieve a common goal; *coordination*; aims to keep the coherence in the system; and *negotiation*; which seeks to find an agreement that satisfied all the involved agents.

## 3 C-Means Based on a Modified Mahalanobis Distance

The c-means algorithm [4] is a fuzzy clustering method that creates fuzzy clusters by minimizing an objective function. In order to extract clusters of different shapes, this algorithm was extended to the Gustafson-Kessel algorithm [5] which is based on the Mahalanobis distance. From a practical standpoint, this latter algorithm has two major problems that prevent it from clustering data [2]. In order to overcome these drawbacks, we extended this algorithm to another one based on a modified Mahalanobis distance [2] (Table 1).

**Table 1** The algorithm steps

| Algorithm |
| --- |
| 1. Initialize the algorithm with a few iterations of the standard c-means |
| 2. Update the cluster centers |
| 3. Update the fuzzy partition matrix |
| 4. Calculate the norm of two consecutive fuzzy partition matrices, if it is smaller than a fixed threshold, stop executing, otherwise, repeat the process from the second step |

## 4 Our Multi-agent System

In this work, we are concentrated on developing a MAS for our fuzzy clustering algorithm [2] in the aim of getting benefits from the agent technology and making the algorithm easy to be extended and incorporated into other systems from other application domains. Based on the algorithm depicted in Sect. 3, we designed the following system (Fig. 1) that takes as input an unlabeled dataset and returns through its output fuzzy partitions of it.

The interactions within this system are carried out using a communication via a shared memory and message passing between agents.

**Initializer**. This agent gets the dataset and the desired number of clusters from a user, initializes the clustering parameters and puts them in the shared. When it accomplishes its task correctly, it sends a stimulus to the CenterUpdater agent.

**CenterUpdater**. This agent updates the clusters centers and when it accomplishes its task, it sends a stimulus to the MatrixUpdater agent.

**MatrixUpdater**. This agent has to update the partition matrix and when it does well, it sends a stimulus to the Decider.

**Decider**. If the stop condition is verified, this agent informs the other agents that the global goal is achieved, otherwise, it sends a stimulus to the CenterUpdater to start a new iteration.

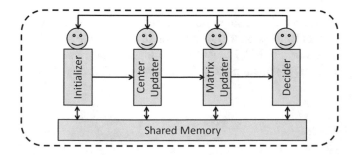

**Fig. 1** Overall system architecture

# 5  Demonstration

## 5.1  Visual Demonstration

To demonstrate the effectiveness of our algorithm, we compare it with the standard c-means algorithm. Actually, we test both algorithms upon the Iris data set [6]; a well-known data set of 150 elements spanning into three clusters (Iris Setosa, Iris Versicolor and Iris Virginica) where each datum contains four numerical attributes (SL, SW, PL and PW). To help see well the clustering results in a three-dimensional space, we eliminate one of the attributes.

From Figs. 2 and 3 we notice that both algorithms have succeeded in clustering the Iris dataset into three clusters and their results are considerably similar. In order to demonstrate which result is the closest to the pre-specified classification, we calculate the clustering accuracy (CA) as follows:

$$CA = \frac{\text{Number of Well Classified Elements}}{\text{Total Number of Elements}} \times 100\ \%. \qquad (1)$$

We found: $CA_{FCM} = 82.41\ \%$ and $CA_{FCM\_MD} = 93.62\ \%$.

$CA_{FCM}$ (resp. $CA_{FCM\_MD}$) is the clustering accuracy of the c-means result (resp. our result). From these results, we realize that our algorithm is better than the standard.

Because of the covariance matrices and their pseudo-inverses, our algorithm requires a longer running time than the standard c-means. So, this main drawback can be a bottleneck in its integration in many applications. That is why we decided to implement it as a MAS that can be easily extended or incorporated into other systems.

## 5.2  Running Time Alteration

In this subsection we demonstrate how the implemented MAS can influence the running time of an application that is based on it. As we have mentioned before, we are intending to use this MAS to extract elliptical shapes, using an ellipse fitting approach as an approximation tool, from a series of scattered two-dimensional points.

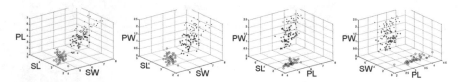

**Fig. 2** Result of the c-means algorithm

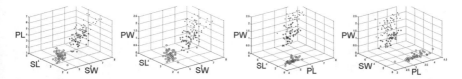

**Fig. 3** Result of our algorithm

Our MAS is implemented in a sequential way. In other words, clustering a dataset using our MAS is equivalent, in a running time standpoint, to clustering it using the sequential algorithm proposed in [2]. In contrast, if the clustering of a dataset is just a task in a long processing, then in this case our MAS can alter positively the running time of the hole application.

Assume that we have N datasets ($D_1$, $D_2$, ... and $D_N$) and each one has to go through a clustering step, to extract elliptical clusters, followed by a fitting step, to fit the extracted clusters. This application can be easily implemented by incorporating another agent, that tries to fit the extracted clusters, into our MAS. In the rest of this section, we simulate the execution of this application in order to compare the running times carried out using and without using our MAS. A simulation of the application execution is depicted in Table 2.

Let $\Delta t$(resp. $\Delta t'$) be the mean time needed to cluster a dataset (resp. to fit its extracted clusters) and $\Delta t_{max}$ be the maximum of $\Delta t$ and $\Delta t'$. Thus, the mean time needed to process these N datasets, in a sequential way, will be:

$$\Delta T_1 = N \cdot (\Delta t + \Delta t'). \tag{2}$$

According to Table 2, the mean time needed to process the given datasets using our MAS is:

$$\Delta T_2 = (N - 1) \cdot \Delta t_{max} + (\Delta t + \Delta t'). \tag{3}$$

We can easily verify that $\Delta T_2$ is much lower than $\Delta T_1$, which means that the extended MAS performs much faster than the sequential method.

**Table 2** Simulation of the application execution (using our MAS)

| Instants | Tasks | Running time |
|---|---|---|
| $T_1$ | Clustering $D_1$ | $\Delta t$ |
| $T_2$ | Clustering $D_2$ and fitting $D_1$ | $\Delta t_{max}$ |
| $T_3$ | Clustering $D_3$ and fitting $D_2$ | $\Delta t_{max}$ |
| ... | ... | ... |
| $T_N$ | Clustering $D_N$ and fitting $D_{N-1}$ | $\Delta t_{max}$ |
| $T_{N+1}$ | Fitting $D_N$ | $\Delta t'$ |

# 6  Conclusion

In the aim of keeping up with the evolution of the agent technology and get benefits from it, we developed a MAS based on a fuzzy clustering algorithm. In Sect. 5, we noticed that the proposed algorithm exceeds the standard c-means algorithm and its MAS version has the ability to reduce the running time of the applications based on it. For that reason, we are intending to use it to segment the MRI images.

# References

1. Jennings, N.R., Sycara, K., Wooldridge, M.: A roadmap of agent research and development. Auton. Agents Multi-Agent Syst. **1**(1), 7–38 (1998)
2. Barrah, H., Cherkaoui, A.: A stabilizer mahalanobis distance applied to ellipses extraction using the fuzzy clustering. In: International Conference on Multimedia Computing and Systems (ICMCS), pp. 1059–1064 (2014)
3. Woolbridge, M., Jennings, N.R.: Agent theories, architectures, and languages: a survey. In: Woolbridge, Jennings (eds.) Intelligent Agents, pp. 1–22. Springer (1995)
4. Ghosh, S., Dubey, S.K.: Comparative analysis of K-means and fuzzy C-means algorithms. IJACSA **4**, 35–38 (2013)
5. Krishnapuram, R., Kim, J.: A note on the Gustafson-Kessel and adaptive fuzzy clustering algorithms. IEEE Trans. Fuzzy Syst. **7**(4), 453–461 (1999)
6. Bache, K., Lichman, M.: UCI Machine Learning Repository. http://archive.ics.uci.edu/ml. University of California, School of Information and Computer Science, Irvine, CA (2013)

# Model Driven Architecture as an Approach for Modeling and Generating Graphical User Interface

Sarra Roubi, Mohammed Erramdani and Samir Mbarki

**Abstract** Integrating the modeling process throughout all development phases of any application is promising. On the one hand, this is the basis of Model Driven Architecture approach (MDA) which advocates the massive use of models during different steps of any application's development. On the other hand, the achievement of the User Interface (UI) of an application as Human Computer Interaction (HCI) is among the key factors of its success. Along with that, we present the approach we adopted to demonstrate the tandem of the MDA and HCI communities. Based on the MDA approach and its principles, we first elaborate the Platform Independent Model (PIM) which is actually our way to describe the interface's functionalities and not using just a simple Unified Modeling Language (UML) diagram. Second, we established the Platform Specific Model (PSM) for Swing to describe UI. Then, we focused on implementing mapping rules between the two models and also the code generation. Finally, with this method, graphical interfaces can easily be analyzed designed and generated to increase the productivity of the system.

**Keywords** Model driven architecture · User interface · Meta model · Model · Transformation · Code generation

S. Roubi (✉) · M. Erramdani (✉)
MATSI Laboratory, EST, Mohammed First University, Oujda, Morocco
e-mail: roubi.sarra@gmail.com

M. Erramdani
e-mail: m.erramdani@gmail.com

S. Mbarki (✉)
Department of Computer Science, Faculty of Science Ibn Tofail University, Kenitra, Morocco
e-mail: mbarkisamir@hotmail.com

© Springer International Publishing Switzerland 2016
A. El Oualkadi et al. (eds.), *Proceedings of the Mediterranean Conference on Information & Communication Technologies 2015*, Lecture Notes in Electrical Engineering 381, DOI 10.1007/978-3-319-30298-0_72

# 1   Introduction

Modeling a system before starting its construction allows having a generic view of what is expected from it. This is why we can say that modeling a system is a better way to manage its complexity and ensure consistency and coherence between its different parts. There are different forms and methods that help modeling software systems. The Object Management Group (OMG) has developed a number of standards for software development under its Model Driven Architecture (MDA) approach [1]. Thus, MDA promotes extensive use of models in the software development process. As defined by the OMG, MDA is a way to organize and manage enterprise architectures supported by automated tools and services to help define these models and transformations between different types of models. These transformation rules describe together how the base models are transformed into target models. It also emphasizes the separation of business logic from the technical platform.

Along with the development of new technologies and the variation of the multiplicity of interaction platforms, User Interfaces (UI) are now facing several challenges, among them, unify the design and implementation around model [2] and ensure the independence of platforms.

This is why we can say that reconciliation between the two communities MDA and HCI is obvious. The anchor is to apply the key concepts of MDA, and be able to model and represent user interfaces in a variety of platforms.

# 2   Related Works

Given the generic aspect that MDA provides, it has been applied to several system development layers. The application of these concepts has shown its efficiency when applying them to generate N-Tiers web application as defined in [3].

Recently, several works related to MDA area and its application for Human Computer Interface has emerged. The first work is based on the plasticity of UI and the application of MDA concepts for the purpose of unifying the modeling [4]. This approach evince that the two communities have a lot in common and their basic objective is to be able to introduce plasticity when it comes to User Interface. Different meta models were realized to represent Task, Domain, Concrete UI and the Final UI.

Another related work on applying MDA approach for Rich Internet Application (RIA) is found in [5]. The approach is based on XML User Interface description languages using XSLT as the transformation language between the different levels of abstraction. Besides, an MDA approach for AJAX web applications [6] was the subject of a study that propose a UML scheme for modeling AJAX user interfaces by adopting AndroMDA framework for creating an AJAX cartridge to generate the corresponding AJAX application code, in ICEFACES, with back-end integration.

To do this, a meta model of AJAX has been defined using the profiling approach on UML diagrams. At last but not least, in [7] the authors demonstrate the generation of GUI of the Amazon Integration and Cooperation Project for Modernization of Hydrological Monitoring by using the stereotyping method on UML models with AndroMDA tool as well.

In terms of our work, we focused on demonstrating the reconciliation between the two communities MDA and HCI by applying the MDA principles and come up with a new way of modeling graphical interfaces.

## 3 The Proposed Model Driven Approach

### 3.1 Overall View of the Approach and the Proposed Base Meta Model

In order to establish our Platform Independent Model (PIM) base meta model, we took as a main element the use case that describes an expected functionality of the system. This feature is logically divided into several activities that require the user's interaction with the system to accomplish it. Hence, we had the idea to develop a new model, which is neither a use case nor an activity diagram; but it is a model that describes the purpose of the use case in the form of operations and activities and each activity has a type that reflects the component that is connected to it. Figure 1 shows the proposed meta model.

Figure 1 describes the UML base PIM meta model with all its meta classes. The **UMLPackage** includes all elements of the model which are the **UseCase**,

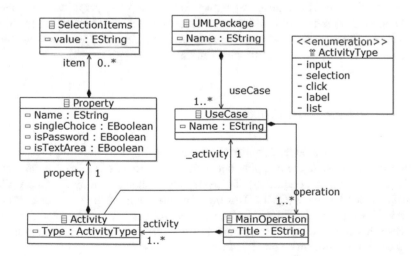

**Fig. 1** Base meta model: contains the meta classes to describe the Graphical User Interface to be generated

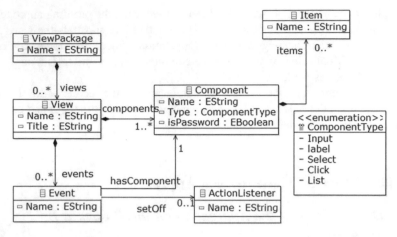

**Fig. 2** The proposed PSM describing the basic structure of a user interface in swing

describing the main functionality offered by the system, and it is composed of **MainOperation** that gathered all the **Activities** to refer to the tasks done by the user. Finally, the **Property** defines all the features of the activity to help define the required graphical component. The activity has an important attribute; **ActivityType**, to show the type that an activity could offer; which are input, select, click, labels or lists. This meta model captures the main structure of the user interface described throughout these elements. The meta model describes the graphical elements that make up this interface and the links between them. Figure 2 describes the PSM Meta model related to swing platform. Below we give a description of each element.

The meta model is composed of **views** that regroup the graphical component. Each and every **component** could have its **listener** to set off the **event** related to the user action. The component may have **selection items** depending on its type. All of these elements are gathered in the **viewPackage**.

## 3.2 Transformation Process

Once the two meta models established, comes the stage of defining the transformation rules that make the mapping between each element of the base meta model, PIM, and meta target model, PSM. These rules were then coded by the Query View Transformation language (QVT). This transformation engine gives us automatically the Java Swing model of our interface respecting the PSM meta model defined earlier in this chapter. This file contains the elements defining the desired graphical interface and is conform to our PSM meta model for Java Swing platform.

Since the code generation is the main goal of the MDA approach, we defined the templates with Acceleo to transform our target model conform to our PSM meta

**Fig. 3** The generated
Registration Form

model into Java classes with the appropriate generated swing code to generate the
final Interface. We defined several templates and each one creates a code fragment
and their combination gives us the class Java for the graphical components and the
classes for the listeners related to the equivalent components. After execution, the
result is the 'Registration View' as shown in the following figure. This Graphical
Interface is automatically generated from the base model as an instance of the PIM
meta model proposed above. Each graphical component describes the activity to be
executed by the user to achieve the final goal, which is the registration in our case.

## 3.3 Case Study: Registration Form

To validate the approach, we took the use case of Registration. The form requires a
number of operations to perform, depending on the choice of the user, in order to
accomplish its finality, which is in our case, get the registration done. To do this, the
user performs several tasks or operations that will be offered by the interface
through well-defined graphical components. This brings us to the activities that the
user should accomplish in order to fill in the form. Each activity is translated in
terms of operations that would be performed by the user. Each operation is associated with a component. For instance, "Enter Name" is closely related to the "text
field" component; the "choose gender" is described as a drop down list to choose a
single gender from the list. This model will be the input for our transformation
engine described in the next section. Figure 3 shows the generated interface handling all the user's events.

## 4 Conclusion

In this paper, we presented our Model Driven Approach to generate Graphical User
interface based on a simple model that describes the user's interactions with the
interface. Then we defined the PSM for the Java Swing platform. The result model

was introduced in the transformation engine developed by the template approach to generate the desired source code for the user interface. In the future, we aim to extend this work to allow the generation of graphical interfaces for other platform as a target taking into account Web application.

# References

1. OMG: MDA (2008). http://www.omg.org/mda
2. Sottet, J.-S., Calvary, G., Coutaz, J., Favre, J.-M.: A Model-Driven Engineering Approach for the Usability of Plastic User Interfaces; Engineering Interactive Systems, pp. 140–157. Springer, Berlin (2008)
3. Esbai, R, Erramdani, M., Mbarki, S., Arrassen. I, Meziane. A., Moussaoui. M.: Model-driven transformation with approach by modeling: from UML to N-tiers web model. Int. J. Comput. Sci. Iss. (IJCSI) **8**(3), 1694–0814 (2011). ISSN (Online)
4. Sottet, J.S., Calvary, G., Favre, J.M., Coutaz, J., Demeure, A., Balme, L.: Towards model-driven engineering of plastic user interfaces. In: Conference on Model Driven Engineering Languages and Systems (MoDELS'05) Satellite Proceedings, Springer LNCS, pp 191–200 (2005)
5. Martinez-Ruiz, F.J., Munoz Arteaga, J., Vanderdonckt, J., Gonzalez-Calleros, J.M.: A first draft of a model-driven method for designing graphical user interfaces of Rich Internet Applications. In: LA-Web '06: Proceedings of the 4th Latin American Web Congress, pp. 32–38. IEEE Computer Society (2006)
6. Gharavi, V., Mesbah, A., Deursen, A.V.: Modelling and generating AJAX applications: a model-driven approach. In: Proceeding of the 7th International Workshop on Web-Oriented Software Technologies, New York, USA, 38 pp. (2008). ISBN: 978–80-227-2899-7
7. Monte-Mor, J.A., Ferreira, E.O., Campos, H.F., da Cunha, A.M., Dias, L.A.V.: Applying MDA approach to create graphical user interfaces. In: Eighth International Conference on Information Technology: New Generations Las Vegas, NV, IEEE pp. 766–771 (2011)

# Automated Classification of Mammographic Abnormalities Using Transductive Semi Supervised Learning Algorithm

Nawel Zemmal, Nabiha Azizi, Mokhtar Sellami and Nilanjan Dey

**Abstract** Computer-aided diagnosis (CAD) of breast cancer is becoming a necessity given the exponential growth of performed. CAD are usually characterized by the large volume of acquired data that must be labeled in a specific way that leads to a major problem which is labeling operation. As a result the community of machine learning has attempted to respond to these practical needs by introducing the semi-supervised learning. The motivation of the current research is to propose a TSVM-CAD system for mammography abnormalities detection using a new Transductive TSVM with comparison of its kernel functions. The effectiveness of the system is examined on the Digital Database for Screening Mammography database DDSM using classification accuracy, sensitivity and specificity. Experimental results are very encouraging.

**Keywords** Computer aided diagnosis (CAD) · Transductive support vector machine · Semi supervised learning (SSL) · Mammographic abnormalities

## 1 Introduction

The reduction in the death rate caused by this type of cancer as well as favoring the chances of recovery is only possible if the tumor was supported in the early stages of its appearance. Faced with the increasing number of mammograms during the last decades, various researches make the effort to automatically interpret

N. Zemmal (✉) · N. Azizi · M. Sellami
Labged Laboratory, Computer Science Department, Badji Mokhtar University,
PO BOX 12, Annaba 23000, Algeria
e-mail: zemmal@labged.net; nawel.zemmal@gmail.com

N. Azizi
e-mail: azizi@labged.net

N. Dey
Department of Computer Science, Bengal College of Engineering & Technology,
Durgapur, India

© Springer International Publishing Switzerland 2016
A. El Oualkadi et al. (eds.), *Proceedings of the Mediterranean Conference on Information & Communication Technologies 2015*, Lecture Notes in Electrical Engineering 381, DOI 10.1007/978-3-319-30298-0_73

657

mammogram abnormality through (CAD) systems [1–3]. In a CAD system, there is in effect always a large volume of data which must be recognized and labeled in a specific way. However, this criterion may not always be satisfied, for reasons of cost or imperfect knowledge of the problem to solve [4]. The statistical learning community has attempted to respond to these practical needs, by formalizing more general problems that supervised learning like semi-supervised learning (SSL) [5].

As the medical images should be represented by different sources of information, it is interesting to integrate different families of characteristics. The used features in this research contain a combination of the three heterogeneous families based on texture and shape which are: co-occurrence matrix, Hu moments and central moments. In the classification step, a semi-supervised classification using Transductive Support Vector Machine (TSVM) has been opted because of its performance and it is well proven in several areas including the field of medical diagnosis [6, 7].

The remainder of the paper is organized as follows: In Sect. 2, the semi supervised learning is exposed and the mathematical concept of TSVM is shown. General scheme of proposed approach accompanying with description of each stage is presented in Sect. 3. Section 4 illustrates the experimental part and the obtained results using TSVM classifier and mixture of features. Conclusion and some perspective points for future extensions achieve the paper.

## 2   Transductive Support Vector Machine (TSVM)

The SSL classification is achieved, not only with the labeled data, but also with the unlabeled datasets. Among the semi-supervised learning techniques, Transductive Support Vector Machine (TSVM) is the most popular approach and has a strong theoretical base, which inherits from the notion of "large-margin" of supervised SVM [8]. The principle of Transductive SVM algorithm can be expressed as, given a group of independent and identically distributed labeled data sets:

$$D = \{(x1, y1), \ldots, (xl, yl)\}, \quad x \in Rn, \quad y \in \{+1, -1\} \tag{1}$$

And unlabeled data sets: $x_1$, $x_2$, ..., $x_k$.
The mathematical formula of Transductive SVM can be defined as follows:

$$\min\left(y1, y2, \ldots, yk, \omega, b, \delta1, \delta2, \ldots, \delta l, \delta_1', \delta_2', \ldots, \delta_k'\right) \frac{1}{2}||\omega||^2 + C \sum_{i=1}^{l} \delta_i + D \sum_{i=1}^{k} \delta_j'$$

$$\text{Subject to} \begin{cases} yi(\omega * xi + b) \geq 1 - \delta i; & \delta i \geq 0; \ i = 1, 2 \ldots, l \\ yj(\omega * xj + b) \geq 1 - \delta j; & \delta_j' \geq 0; \ j = 1, 2 \ldots, k \end{cases}$$

$$\tag{2}$$

where, C, D as a parameter, and D for the impact factor.

# 3　Proposed CAD System

The proposed CAD represented in Fig. 1 consists into three main steps: tumor contour extraction from the input mammogram images, Features extraction transforms the input data to characteristics and compact representation and classification in order to identify the abnormalities.

## 3.1　Tumor Contour Extraction

In the segmentation step, the outline of the shape is extracted and analyzed using a tool for image processing "Image J" [9] and the mammographic image will have the following form (See Fig. 2).

## 3.2　Features Extraction

The methods of image analysis are variable according to the types of features extracted from the image as texture features, the shape features etc.... In the proposed approach, the image is represented by three families of characteristics which are: the co-occurrence matrix [10], Hu moments [11] and central moments [12]. For more detail of these three families, the reader is referred to the previous study in [13].

## 3.3　Learning and Classification

Transductive SVM is chosen in this study because of its performance in several areas such as medical diagnosis comparing with other classification techniques.

In this work, Digital Database for Screening Mammography (DDSM) is used as test dataset [14]. It was assembled by a group of researchers from University of South Florida. The DDSM database contains 2620 cases.

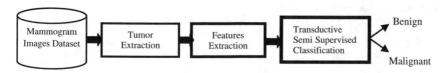

**Fig. 1**　Proposed CAD system

**Fig. 2** Extraction of the mass outline by *ImageJ* tool

# 4    Results and Discussion

The proposed system for abnormalities breast diagnosis has been trained on a sample of 400 images (benign and malignant) taken from DDSM data base. Proposed system was built using JAVA SE 1.8.0 with a simple user interface.

## 4.1    Classification Performance

Proposed TSVM-CAD for mammogram abnormalities detection and classification using a sample of 200 images (90 malignant and 110 benign) from the DDSM dataset. After contour extraction of the mass and feature extraction step (co-occurrence matrix, Hu moments and central moments). This feature vector contains over than 26 characteristics. Thereafter, a Transductive Support Vector Machine (TSVM) classifier is applied. During this phase, several empirical tests are made to keep the ones that generate the highest rate of classification. First, each family was tested independently with the Transductive SVM classifier then all features are grouped together in a single vector that will be the entrance of the classifier. Also, three best-known kernel functions are tested (Gaussian, Triangle and Linear). Table 1 summarizes the accuracy rates of each kernel function with the three families of characteristics.

In order to evaluate the classification performance, other related metrics are also calculated as (see Table 2):

$$\text{Sensitivity} = \frac{\text{TP}}{\text{TP} + \text{FN}} \tag{3}$$

**Table 1** Obtained results of TSVM classifier with the three families of characteristics and the different kernel functions

| Kernel functions | TSVM with co-occurrence matrix (%) | TSVM with Hu moments (%) | TSVM with central moments (%) | TSVM with all features (%) |
|---|---|---|---|---|
| Gaussian | *84.13* | 80.10 | 73.58 | *92.95* |
| Triangle | 77.65 | 72.25 | 69.95 | *89.44* |
| Linear | 75.58 | 70.13 | *67.13* | 82.11 |

**Table 2** Obtained results of the different metrics of the proposed CAD system

| Kernel functions | Sensitivity | Specificity | Accuracy |
|---|---|---|---|
| Gaussian | 0.89 | 0.93 | 0.929 |
| Triangle | 0.82 | 0.86 | 0.894 |
| Linear | 0.79 | 0.82 | 0.821 |

$$\textbf{Specificity} = \frac{TN}{TN + FP} \tag{4}$$

$$\textbf{Accuracy} = \frac{TP + TN}{TP + TP + FN + FP} \tag{5}$$

where TP (resp. TN) is True Positive (resp. true Negative) and FP (resp. FN) is False Positive (resp. False Negative).

From this table TSVM has again proved its effectiveness in the field of medical diagnosis and especially in indentifying the abnormalities in the mammographic images with a high accuracy rate (92, 95 %).

## 5  Conclusion

Current study was conducted to determine the right path for the future evolution of image processing in medicine and health. In this work, a new tool for the diagnosis of mammography abnormalities is proposed which is based on semi-supervised classification.

During the learning phase TSVM, several empirical tests are made to keep the ones that generate the highest rate of classification. Also, the three best-known kernel functions (Gaussian, Triangle and Linear) are tested in order to analyze the behavior of the classifier TSVM. However, Transductive SVM takes all unlabeled data without prior selection of data that allows better accuracy what can be considered as a drawback. As future study, it will be interesting to analyze how to overcome the latest inconvenient by introducing new concepts to increase the performance of Transductive SVM classifier and to select the most appropriate unlabeled images using active learning as obvious solution.

## References

1. Cedolini, C., Bertozzi, S., Londero, A., Bernardi, S., Seriau, L., Concina, S.: Type of breast cancer diagnosis, screening, and survival. Clin. Breast Cancer **14**, 235–240 (2014)
2. Zheng, B., Yoon, S., Lam, S.: Breast cancer diagnosis based on feature extraction using a hybrid of K-means and support vector machine algorithms. Expert Syst. Appl. **41**, 1476–1482 (2014)

3. Abbadi, N.E., Taee, E.A.: Breast cancer diagnosis by CAD. Int. J. Comput. Appl. **100**, 25–29 (2014)
4. Côme, E., Oukhellou, L., Denoeux, T., Aknin, P.: Learning from partially supervised data using mixture models and belief functions. Pattern Recogn. **42**, 334–348 (2009)
5. Chapelle, O., Scholkopf, B., Zien, A.: Semi-Supervised Learning. MIT Press, Cambridge (2006)
6. Filipovych, R., Davatzikos, C.: Semi-supervised pattern classification of medical images: application to mild cognitive impairment (MCI). NeuroImage **55**, 1109–1119 (2011)
7. Pang, S., Ban, T., Kadobayashi, Y., Kasabov, N.: Personalized mode transductive spanning SVM classification tree. Inf. Sci. **181**, 2071–2085 (2011)
8. Vapnik, V.N.: The Nature of Statistical Learning Theory. Springer (1996)
9. Grishagin, I.V.: Automatic cell counting with ImageJ. Anal. Biochem. (In Press)
10. Nanni, L., Brahnam, S., Ghidoni, S., Menegatti, E.: A comparison of methods for extracting information from the co-occurrence matrix for subcellular classification. Expert Syst. Appl. **41**, 7457–7467 (2013)
11. Žunić, D., Žunić, J.: Shape ellipticity from Hu moment invariants. Appl. Math. Comput. **226**, 406–414 (2014)
12. Grubbström, D., Tang, O.: The moments and central moments of a compound distribution. Eur. J. Oper. Res. **170**, 106–119 (2006)
13. Azizi, N., Zemmal, N., Tlili, Y.: Kernel Based classifiers fusion with features diversity for breast masses classification. In: Proceedings of the 8th International Workshop on Systems, Signal Processing and Their Application (WoSSPA), pp. 116–121 (2013)
14. Heath, M.D., Bowyer, K.W.: Mass detection by relative image intensity. In: 5th International Workshop on Digital Mammography, Toronto, Canada, pp. 219–225 (2000)

# A New Cryptographic Scheme Based on Cellular Automata

Said Bouchkaren and Saiida Lazaar

**Abstract** Some electronic data exchanged on internet are very critical and may be intercepted by malicious people for passive or active attacks; to protect these data, many cryptographic algorithms have been elaborated to ensure confidentiality and integrity. In this context we define and we implement a new secure and fast symmetric cryptographic scheme named Novel Encryption Scheme using Cellular Automata (NESCA) using irreversible cellular automata and a series of non-uniform cellular automata to generate cipher texts and to reconstruct plain texts. The process of encryption is applied on blocks of data and runs according to a number of rounds applying sub keys generated by cellular automata. To prove the reliability of the proposed system, a couple of tests are carried out and a comparison with AES-256 is presented.

**Keywords** AES · Cellular automaton · Confusion · Decryption · Diffusion · Encryption · Secret key

## 1 Introduction

Most internet applications send critical data that can be intercepted by malicious users, to protect those data a number of researches have been carried out in the field of cryptography and target data confidentiality and integrity.

Modern cryptography methods are divided into two types: asymmetric cryptosystems which use two keys one for encryption and other for decryption. The second type is symmetric cryptosystems which use the same key for encryption and decryption [1, 2]. However, in cryptanalysis, many attacks make these cryptography

S. Bouchkaren (✉) · S. Lazaar
Department of Mathematics and Computing, National School of Applied Sciences of Tangier/LTI, AbdelMalek Essaadi University, Tangier, Morocco
e-mail: saidbouchkaren1@hotmail.com

S. Lazaar
e-mail: s.lazaar2013@gmail.com

© Springer International Publishing Switzerland 2016
A. El Oualkadi et al. (eds.), *Proceedings of the Mediterranean Conference on Information & Communication Technologies 2015*, Lecture Notes in Electrical Engineering 381, DOI 10.1007/978-3-319-30298-0_74

663

methods sometimes vulnerable, and this vulnerability increases with the progress of new technologies. In this context, we define and implement a new symmetric cryptosystem based on reversible and irreversible cellular automata. To prove the reliability of the proposed system, a couple of tests are carried out and a comparison with AES-256 is presented.

The remainder of this paper is organized as follow: The second section presents a brief review on cellular automata, the third section describes the proposed algorithm, the fourth section explains sub keys scheduling process. To test the reliability of the algorithm, some results are presented on the fifth section, and we conclude the paper in the last section.

# 2 Brief Review on Cellular Automata in Cryptography

More details on cellular automata are due to the rich contribution of S. Wolfram, see [3, 4]. In cryptography, numerous contributions with CA were released demonstrating that CA are able to cipher texts and to generate robust secret keys starting from a chaotic and complex state. In [5], a description of a novel and fast two-dimension private key cryptosystem using reversible CA and Margolus neighbourhoods is presented, the results prove the reliability and the robustness of the proposed algorithm. More applications of CA in cryptography can be found in [6–11].

# 3 The Proposed Cryptosystem

The proposed algorithm encrypts and decrypts data in blocks with reversible automata; for each block, the process is executed according to a number of rounds and for each round a sub key is generated with irreversible automata. The algorithm uses the following parameters: The block size is equal to 128 bits; the key size corresponds to 256 bits and the number of iterations or rounds is equal to 10.

## 3.1 Encryption and Decryption Algorithm

Encryption process starts by taking a block data of 128 bits into a block $M$ of size $4 \times 4$ bytes. $M$ passes through a number of transformations named *Shift*, *OddMix*, *EvenMix* and *AddSubKey* see Algorithm 1. These transformations are new and will be explained and illustrated thereafter.

---

**Algorithm 1** Encryption algorithm

---

1: **procedure** CIPHER($M, K$)  ▷ M is the plain text message block, K is the encryption key
2:    $SK[10] \leftarrow GenerateSubKeys(K)$  ▷ SK is the sub keys array
3:    **for** i from 1 to 10 **do**
4:       M=Shift(M)
5:       M=OddMix(M)
6:       M=EvenMix(M)
7:       M=AddSubKey(M,SK[i])
8:    **end for**
9:    return M  ▷ In this step M is the encrypted message
10: **end procedure**

---

For decryption process, we apply the inverse transformations InvShift, InvOddMix and InvEvenMix; we keep the AddSubKey unchanged.

## 3.2 Shift and InvShift Transformations

The Shift transformation acts on bytes of data for each column of the block, it uses a reversible cellular automaton defined as fallow: The states are the bytes of a column $C$ and the transition rule is: byte $B[k]$ becomes $B[(k+C)\%4]$.

The InvShift is the inverse transformation of Shift. The cellular automaton used in Shift (respectively InvShift) is down (respectively up) bytes shifting of a column $C$ by 8 bits.

## 3.3 OddMix, EvenMix, InvOddMix and InvEvenMix Transformations

These transformations are executed with the entire data block of 128 bits using a two dimension reversible cellular automaton as explained: (1) Convert data block to binary and fit it into matrix $BM$ [8] [8+8]. (2) Partition $BM$ to sub matrix $B$ of 4 bits ($2 \times 2$). (3) States 0 or 1 are the bits of $B$. (4) Let $B = b_{00}b_{01}b_{11}b_{10}$. (5) Compute $Y = f(B)$ where $f$ is the transition rule. (6) $f$ is expressed as: $[15, 0, 14, 1, 13, 2, 12, 3, 11, 4, 10, 5, 9, 6, 8, 7]$ for EvenMix transformation And $[7, 8, 6, 9, 5, 10, 4, 11, 3, 12, 2, 13, 1, 14, 0, 15]$ for OddMix transformation. (7) Use periodic boundary conditions.

The OddMix transformation is applied to block $B$ starting with an odd column index, and EvenMix is applied to block $B$ starting with an even column index. InvOddMix and InvEvenMix are the reverse transformations of OddMix and EvenMix.

### *3.4   AddSubKey Transformation*

The AddSubKey transformation takes two arguments, a data block $B$ of size 128 bits and a sub key $K_i$ of size 128 bits, and calculates $B \oplus K_i$ where $\oplus$ denotes the XOR operator.

## 4   Implementation and Results

In this section, we present a comparison between our algorithm (*NESCA*) and *AES*-256 bits; the comparison holds on diffusion and confusion property and CPU time. These tests use the same keys and the same plaintext and they are executed on a PC equipped with an *intel core* i5 and 4 GB of *RAM* under *windows* 8.1 x64. The performance test measures the CPU time needed to encrypt and decrypt data using *NESCA* and *AES*-256 bits as illustrated on the Fig. 1. We remark that the system *NESCA* is much faster than *AES*-256 bits.

The diffusion property evaluates the impact of changing some bits in a plain text message on the resulting cipher text while keeping the encryption key unchanged. The Fig. 2 gives the diffusion property of *NESCA* and *AES*-256 bits. We can deduce that *NESCA* has a good diffusion compared to *AES*-256. The Confusion property evaluates the impact of changing some bits in the encryption key on the resulting cipher text while keeping the plain text unchanged. The Fig. 3 gives the confusion property of *NESCA* and *AES*-256 bits. We can deduce that *NESCA* has a good confusion compared to *AES*-256.

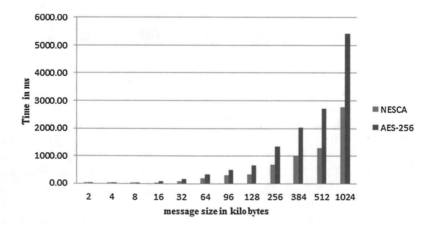

**Fig. 1** CPU comparison between NESCA and AES-256

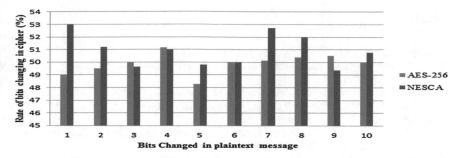

**Fig. 2** Diffusion property of NESCA and AES-256

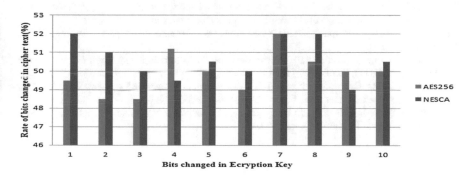

**Fig. 3** Confusion property of NESCA and AES-256

A brute force attack is not possible on our system because even one has a super machine which calculates $10^{50}$ random keys per seconds, it will needs approximately $10^{19}$ years to browse $2^{256}$ possible keys in order to find a key of 256 bits. We deduce also that according to the diffusion and confusion tests, *NESCA* is a promising cryptosystem.

## 5 Conclusion and Perspectives

In this paper, we defined and implemented a new fast and secure secret key cryptographic algorithm based on cellular automata; it encrypts and decrypts data in blocks and according to a number of rounds, each block cipher and each sub key is generated by cellular automata. According to the tests carried out on our system it has been shown that our system is fast and robust. In addition to the tests we realized for the security of our algorithm, other new tests are being carried out such as Differential attacks and NIST randomness tests.

# References

1. Schneier, B., et al.: Applied Cryptography. Wiley (1996)
2. NIST Aes: Advanced encryption standard. Federal Information Processing Standard, FIPS-197, 12 (2001)
3. Wolfram, S: A New Kind of Science, vol. 5. Wolfram Media Champaign (2002)
4. Wolfram, S.: Cryptography with cellular automata. In: Advances in CryptologyCRYPTO85 Proceedings, pp. 429–432. Springer (1986)
5. Bouchkaren, S., Lazaar, S.: A fast cryptosystem using reversible cellular automata. Int. J. Adv. Comput. Sci. Appl. **5**(5), 207–210 (2014)
6. Anghelescu, P., Ionita, S., Sofron, E.: Block encryption using hybrid additive cellular automata. In: 7th International Conference on Hybrid Intelligent Systems, 2007, HIS 2007, pp. 132–137. IEEE (2007)
7. Phani Krishna Kishore, M., Kanthi Kiran, S.: A novel encryption system using layered cellular automata. In: Proceedings of the World Congress on Engineering, vol. 1 (2011)
8. Jin, J.: An image encryption based on elementary cellular automata. Opt. Lasers Eng. **50**(12), 1836–1843 (2012)
9. Faraoun, K.M.: A genetic strategy to design cellular automata based block ciphers. Expert Syst. Appl. **41**(17), 7958–7967 (2014)
10. Faraoun, K.M.: Fast encryption of RGB color digital images using a tweakable cellular automaton based schema. Opt. Laser Technol. **64**, 145–155 (2014)
11. Ping, P., Feng, X., Wang, Z.-J.: Image encryption based on non-affine and balanced cellular automata. Signal Process. **105**, 419–429 (2014)

# A Novel Neighboring Information-Based Method for Motion Object Detection

Bingshu Wang, Xuefeng Hu, Wenqian Zhu and Yong Zhao

**Abstract** Motion Object Detection is a fundamental and crucial step in the applications of intelligent video surveillance. Various techniques including background subtraction are proposed to achieve good results of foreground extraction. This paper presents a novel background model that relies on neighboring information of pixel. The main innovation concern is that we bring life value concept to each background sample and adopt different update strategies based on different life values of background samples. Neighboring information is used to construct and update the background model. Experiment results on Change Detection 2014 dataset demonstrate that the proposed method works more robustly and outperforms some state-of-the-art methods.

**Keywords** Motion object detection · Neighboring information · Life value · Change detection

## 1 Introduction

Motion object detection, which paves the way for segmentation, tracking and other post-processings, is a very important stage. Specially, background subtraction [1] is to compare current frame with background model and to classify foreground and background.

B. Wang · X. Hu · W. Zhu · Y. Zhao (✉)
School of Electronic and Computer Engineering, Shenzhen Graduate School of Peking University, Shenzhen, China
e-mail: zhaoyong@pkusz.edu.cn

B. Wang
e-mail: wangbingshu@sz.pku.edu.cn

X. Hu
e-mail: huxuefeng@pku.edu.cn

W. Zhu
e-mail: zhuwenqian@sz.pku.edu.cn

© Springer International Publishing Switzerland 2016
A. El Oualkadi et al. (eds.), *Proceedings of the Mediterranean Conference on Information & Communication Technologies 2015*, Lecture Notes in Electrical Engineering 381, DOI 10.1007/978-3-319-30298-0_75

669

Six mainly challenges are analyzed on the basis of static and fixed camera. **Illumination**: Indoor scene is produced by artificial light while outdoor scene contains unstable lighting in morning, noon and sunset. **Shadow**: Static shadows are projected by objects such as trees, buildings with immobile shapes. Moving shadows always have the same motion property with moving objects. **Dynamic Background**: Some scenes contain moving background (i.e., waving trees, fountain). **Camouflage**: Moving objects have the similarity with background so that the difference is poorly distinguished. **Intermittent**: When a moving object which stops for some time or a background object starts moving, it will result in the false detection in the region of still object. **Small object**: The small objects are not easy to detect because of their small and ambiguous shapes and outlines.

To cope with these challenges, lots of methods or algorithms are put forward in reviews [2, 3]. Meanwhile, numerous datasets are generated to test methods [2–4]. Mahadevan et al. [5], who were motivated by the biological mechanisms of motion-based perceptual grouping, proposed a spatiotemporal saliency algorithm. This inspires us to think of the biological features and we have an assumption: each background sample has its own life value. Experiments show that the proposed method deals with shadows, illumination and the small objects well.

The rest of this paper is organized as follows. Section 2 describes related work of background subtraction. Section 3 presents our neighboring information-based method. In Sect. 4, we give experimental setup. Section 5 is the conclusion.

## 2   The Related Work

The classification criteria of background subtraction generally include:

**Pixel-level or Region-level (blocked-level)**. Pixel-level model is based on the point that each pixel is independent. Each pixel needs to model based on color space, intensity, property, or other features of pixel. The typical representation is Mixture of Gaussians (MOG [6]). Region-level [7, 8] models take spatial relations into account such as texture and histograms.

**Parametric or non-parametric**. Parametric model is required to adjust its parameters to the changing of scene while nonparametric model [9] is required to possess the optimal parameters.

**Temporal or Spatial**. Temporal data is always utilized by most of models firstly with pixel's recent history. Spatial models are attempted to use pixel's neighboring information or block data to represent the background. Spatiotemporal saliency [5] containing two aspects shows high effectiveness in dynamic scenes.

MOG2 [10] is pixel-level, parametric and temporal. ViBe is level-pixel, non-parametric and spatial. Our method is pixel-level, parametric, spatiotemporal.

## 3   The Proposed Method

**Build the background model and classify pixel**. We select the fixed neighboring pixels as background samples. We denote $S(x)$ as the background samples and $L(x)$ as the life values of the pixel model. The description of detailed formulas is the Eq. (1):

$$S(x) = \{s_1(x), s_2(x), \ldots, s_N(x)\}, \quad L(x) = \{l_1(x), l_2(x), \ldots, l_N(x)\}. \tag{1}$$

where $l_i(x)$ is the life value of corresponding background sample $s_i(x)$ with an index $i$. In addition, we denote $DT(x)$ as decide-thresh of pixel to determine whether the current pixel matches the background sample or not.

We denote $p(x)$ as the pixel value at position $x$. To classify $p(x)$, we compare it with its background samples. $p(x)$ is then classified as background if the match number, denoted by $\Phi$, is more than the cardinal number $\Phi_{min}$ [1].

$$\Phi\{p(x) \cap S(x)\} \tag{2}$$

**Update of background sample and life value**. Figure 1 shows the partition of background sample life value. Three disperse values of the life borders: *lower*, *default* and *higher* are used to ensure the update of background sample timely. The higher life value is, the more appropriately background sample describes background model and vice versa. At the initialization stage, each sample is assigned to a *default* value.

**Update of decide-thresh**. Due to the influence of complex scenes, it is necessary to adjust the decide-value to handle the dynamic environments. We denoted $ED(x)$ as Euclidean distance. The update of decide-thresh is:

$$DT(x) \leftarrow DT(x) + \varepsilon \cdot \min\{ED(x)_i\}. \tag{4}$$

Pseudo code is as the follows ($\chi$ stands for region of Fig. 1, $a, b, c$ represents the magnitude value, $s_i(x)$ is the current background sample with index $i$).

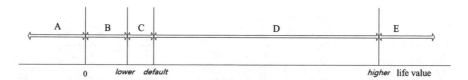

**Fig. 1**  Background sample life value is partitioned for different update mechanism

| Pseudo code of updated stage: |
|---|

If $p(x)$ is classified as background, update each background sample

    for $i = 1$ to $N$ do

        decide $l_i(x)$ which region it belongs to

$$\chi \in \{A \cup B\}, \quad l_i(x) = \begin{cases} 0 & \chi \in A \\ l_i(x) - a & \chi \in B \end{cases}$$

$$\chi \in C, \quad s_i(x) \leftarrow p(x), \quad l_i(x) \leftarrow l_i(x) + a;$$

$$\chi \in \{D \cup E\}, \quad s_i(x) = \begin{cases} (1-\alpha)\cdot s_i(x) + \alpha \cdot p(x) & \chi \in D \\ (1-\beta)\cdot s_i(x) + \beta \cdot p(x) & \chi \in E \end{cases}, \text{ where } \alpha > \beta$$

$$l_i(x) = \begin{cases} l_i(x) + b & \chi \in D \\ higher & \chi \in E \end{cases};$$

        update of decide-thresh, $DT(x) \longleftarrow DT(x) + \varepsilon \cdot \min\{ED(x)_i\}$

        select randomly one neighboring pixel with minimum life value

        $s_j(x_k), \quad j, k \in \{1, 2, ..., N\}, \quad s_j(x_k) \leftarrow p(x), l_j(x_k) \leftarrow default\ life;$

Else

    for $i = 1$ to $N$ do

        $l_i(x) \leftarrow l_i(x) - c;$

## 4  Experimental Setup

Sixteen typical videos are selected from CDnet dataset [4] to evaluate the methods (see Table 1). Without any post-processing each method is operated with fixed parameters. The default parameters derive from the original authors. Our method's parameters are: $N = 8$, $\Phi_{min} = 2$, $lower = 20$, $default = 30$, $higher = 40$, $\varepsilon = 0.125$. Evaluation metric $F$-measure is widely used [7]. Figure 2 presents the qualitative result and Table 1 the quantitative result. The time of execution is 38 ms per frame for resolution $320 \times 240$ under the conditions of Intel(R) Core(TM)

**Table 1** F-measure evaluation of three methods

| Datasets | MOG2 | ViBe | Ours | Datasets | MOG2 | ViBe | Ours |
|---|---|---|---|---|---|---|---|
| Pedestrians | 0.8366 | 0.5349 | **0.8929** | DiningRoom | 0.4321 | **0.8375** | 0.608 |
| Highway | 0.7549 | 0.6805 | **0.7916** | Park | 0.7294 | 0.6376 | **0.7669** |
| Office | 0.2995 | **0.7732** | 0.4166 | Sofa | 0.4738 | **0.6427** | 0.4808 |
| PETS2006 | **0.7559** | 0.5201 | 0.7072 | StreetLight | 0.1817 | **0.6412** | 0.3679 |
| Skating | 0.6386 | 0.5448 | **0.7021** | WinterStreet | **0.3671** | 0.3265 | 0.3234 |
| SnowFall | 0.128 | **0.6935** | 0.57 | BridgeEntry | 0.1881 | 0.1292 | **0.2017** |
| TramCross_1fps | 0.7647 | 0.1864 | **0.8079** | Fall | **0.3527** | 0.1245 | 0.1466 |
| Turnpike_0_5fps | 0.7036 | 0.5233 | **0.812** | Overpass | **0.4886** | 0.2158 | 0.2129 |

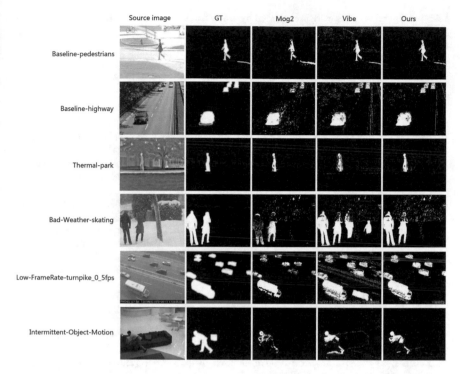

**Fig. 2** From *left* to *right columns* are original frame, ground truth, mog2 result, vibe result and our result, respectively

i7-3770 CPU, Visual Studio 2010 ultimate and Windows 7 OS. We have to address the challenge: Intermittent object motion. When a moving object stops for a long time, it will be absorbed into background. Though on the dataset our results are lower than the compared methods, *F-measure* shows that our method is more stable than the compared methods.

## 5 Conclusion

In this paper, we introduce the life value concept of background sample to propose a novel neighboring information-based method for motion object detection. By comparing with MOG2 and ViBe the method achieves the average highest value of F-measure in the experimental dataset. The life value makes the model to be updated timely and fits for complex situations. In order to meet the intermittent scenes, robust parameters needed to be obtained from sufficient experiments. In this regards, further investigation needs to be done.

**Acknowledgments** This work is supported by grants from the Program of Knowledge Innovation of Shenzhen, China. We really appreciate it for the CVPRW's contribution of providing the ground-truth results.

# References

1. Barnich, O., Van Droogenbroeck, M.: ViBe: a universal background subtraction algorithm for video sequences. J. IEEE Trans. Image Process. **20**, 1709–1724 (2011)
2. Brutzer, S., Hoferlin, B., Heidemann, G.: Evaluation of background subtraction techniques for video surveillance. In: IEEE Conference on Computer Vision and Pattern Recognition, pp. 1937–1944. IEEE Press, New York (2011)
3. Bouwmans, T.: Recent advanced statistical background modeling for foreground detection-a systematic survey. J. Recent Patents Comput. Sci. **4**, 147–176 (2011)
4. Wang, Y., Jodoin, P.M., Porikli, F., Konrad, J., Benezeth, Y., IshWar, P.: CDnet 2014: an expanded change detection benchmark dataset. In: IEEE Conference on Computer Vision and Pattern Recognition Workshops, pp. 393–400. IEEE Press, New York (2014)
5. Mahadevan, V., Vasconcelos, N.: Spatiotemporal saliency in dynamic scenes. J IEEE Trans. Pattern Anal. Mach. Intell. **32**, 171–177 (2010)
6. Stauffer, C., Grimson, W.E.L.: Adaptive background mixture models for real-time tracking. In: IEEE Computer Society Conference on Computer Vision and Pattern Recognition, vol. 2. IEEE Press, New York (1999)
7. Kim, W., Kim, C.: Background subtraction for dynamic texture scenes using fuzzy color histograms. J. IEEE Signal Process. Lett. **19**, 127–130 (2012)
8. Heikkila, M., Pietikainen, M.: A texture-based method for modeling the background and detecting moving objects. J IEEE Trans. Pattern Anal. Mach. Intell. **28**, 657–662 (2006)
9. Elgammal, A., Harwood, D., Davis, L.: Non-parametric model for background subtraction. In: Computer Vision—ECCV 2000, pp. 751–767. Springer, Berlin (2000)
10. Zivkovic, Z.: Improved adaptive Gaussian mixture model for background subtraction. In: 17th International Conference on Pattern Recognition, vol. 2, pp. 28–31. IEEE Press, New York (2004)

# A New Approach for Texture Classification: A Comparative Study

**Khalid Salhi, El Miloud Jaara and Mohammed Talibi Alaoui**

**Abstract** This paper present a comparative study of two method for unsupervised texture image classification, which is based on both Kohonen maps and mathematical morphology, using two texture features extraction methods, namely, Haralick extraction method based on Grey Level Co-occurrence Matrix (GLCM), and extraction features from the fractal dimension using differential box counting method. These features are then used to train the Kohonen Network, which will be represented by the underlying probability density function (pdf). Under the assumption that each modal region of the presentation pdf represents a homogeneous region in the texture image, segmentation of this map is made by morphological watershed transformation. Our comparative study covers the results obtained by the two methods of extraction taking into account the execution time and the error rate.

**Keywords** Image processing · Texture image · Clustering · Kohonen network · Watershed transformation · Fractal dimension · Fractal features · Co-occurrence matrix

## 1 Introduction

The purpose of the segmentation of an image is to seek the number and types of textures in the image, and what regions have which textures. The application of the segmentation generally involves two steps, the first step is to extract the texture features for each pixel in the image, and the second step is to use these features to determine the uniform regions in the image.

K. Salhi (✉) · El MiloudJaara · M. Talibi Alaoui
Faculty of Sciences, LaRi Laboratory, University of Mohammed First, Oujda, Morocco
e-mail: salhi.0.khalid@gmail.com

El MiloudJaara
e-mail: emjaara@yahoo.fr

M. Talibi Alaoui
e-mail: talibialaouim@yahoo.fr

© Springer International Publishing Switzerland 2016
A. El Oualkadi et al. (eds.), *Proceedings of the Mediterranean Conference on Information & Communication Technologies 2015*, Lecture Notes in Electrical Engineering 381, DOI 10.1007/978-3-319-30298-0_76

In this paper, we present two texture feature extraction methods. The first calculates the Haralick features extracted from Grey Level Co-occurrence Matrix (GLCM) [1, 2], and the second computes local fractal features using the differential box counting method [3, 4].

After calculating the features of one of the two methods, we use our approach for unsupervised classification of texture images based on morphological and neural concept [5–7], We first place the feature vector of each pixel into the feature space which forms a cloud of observations. To classify these vectors we make a projection of these on a self-organizing map, named Kohonen self organizing feature map, We present in the last section of our paper we the result of our comparative study of the effectiveness of two different features (Haralick and Fractal), by calculating the error rate and the execution time.

## 2 Texture Features Extraction

Every classification process begins with an acquisition step of observations which consists in determining relevant attributes that characterize better the objects. In our study we use two samples of observations, the first is constituted by features extracted from GLCM matrices of a texture image, and the second consists of Fractal features calculated by the differential box counting method.

### 2.1 Haralick Features

The gray co-occurrence matrix (GLCM) is statistical method of examining texture proposed by Haralick in the 1970s that considers the relationship between the values of neighboring pixels.

A Gray level co-occurrence matrix $P_{(dx,dy)}(i,j)$ is a square matrix, which describes the relative frequencies with which two pixels separated by a distance $d = (dx, dy)$ on the image $I$, one with graytone $i$ and the other with graytone $j$. Mathematically, The GLCM can be defined as:

$$P_{(dx,dy)}(i,j) = \sum_{p=1}^{n} \sum_{q=1}^{m} \begin{cases} 1, & \text{if } I(p,q) = i \text{ and } I(p+dx, q+dy) = j \\ 0, & \text{otherwise} \end{cases} \quad (1)$$

The large size of information obtained by GLCM matrix, makes treatment difficult. Thus, instead of directly using GLCM, we calculate some features defined by Haralick, in our work, we use only five main Haralick's coefficients [1]: Homogeneity $(f_1)$, Energy $(f_2)$, Entropy $(f_3)$, Contrast $(f_4)$ and Correlation $(f_5)$.

## 2.2  Fractal Features

During image analysis application, the fractal geometry is in most cases used through the concept of fractal dimension (FD). Many methods exist to estimates this dimension, in our study we have chosen to work with the differential box counting method, as it can be computed automatically and can be applied to patterns with or without self-similarity.

The differential box counting method [4] consist in partitioning the image space into boxes of different sizes $r$, and the probability N(r) is calculated as the difference between the maximum and minimum gray levels for each box. Then, the fractal dimension is estimated using the equation:

$$FD = \lim_{r \to 0} \frac{\ln[N(r)]}{\ln\left(\frac{1}{r}\right)} \qquad (2)$$

In this paper we use the differential box counting method to extract different features from the textural image. We use not only the original image but also derived images [3], high gray valued image ($I_2$), low gray valued Image ($I_2$), horizontally smoothed image ($I_4$) and vertically smoothed image ($I_5$), finally, we have for each pixel $(i, j)$ a vector $Xq = \{f_1, f_2, f_3, f_4, f_5\}$.

## 3  Representation of the Image Texture Information on the Kohonen Map

The unsupervised classification methods are very powerful tools for the automatic detection of relevant subgroups in a data set, one of these methods is self adaptive map proposed by Kohonen [8].

In our work we use this method to classify our cloud of observation composed from the extracted features.

### 3.1  Kohonen Map Learning Phase

Let's $\Gamma = \{X_1, X_2, X_3, \ldots, X_Q\}$ be a sample of $Q$ observations in a N-dimensional space such as $X_q = [x_{q,1}, x_{q,2}, \ldots, x_{q,N}]^T$, $q = 1, 2, \ldots, Q$. The Kohonen network is made of two layers, the first one the input layer is composed of $N$ attributes of the observation $Xq$. The output layer is composed of $M$ neural units regularly distributed on the map which elaborates prototypes of the data.

The neural units of the first layer are connected to the units of the second layer. Each interconnection from an input unit $j$ to an output unit $m$ has a weight $W_{m,j}$.

That means that each output unit m has a corresponding weight vector $W_m = [W_{m,1}, W_{m,1}, \ldots, W_{m,N}]^T$.

The steps of the learning algorithm:

a. Initialize the weights of the neurons in the Kohonen layer by giving them small random values.
b. Present an input vector $X_q$.
c. Find the winning node $m^*$ using the Euclidean distance between the vector $X_q$ and the nodes of the output layer.
d. Update the weights $Wi$ winner node, as well as those around him, using the Eq. (3).
e. Decreasing the size of the neighborhood area winners nodes.
f. Decreasing the learning coefficient $(t)$.
g. Go to Step b, or else complete learning.

$$\begin{cases} W_m(t) = W_m(t-1) + \alpha(t) \cdot [X_q - W_m(t-1)] & \text{if } m \text{ is the winning node.} \\ W_m(t) = W_m(t-1) + \alpha(t) \cdot h_m(t)[X_q - W_m(t-1)] & \text{if } m \in V(m^*, r(t)). \end{cases}$$

(3)

## 3.2 Visualization of the pdf on the Kohonen Map

Once the learning phase is processed, the determined weight vectors in the multi-dimensional data space are used to estimate the underlying probability density function (pdf). For this purpose, we use the nonparametric Parzen estimate defined by [9, 10]:

$$p(W_m) = \frac{1}{Q} \cdot \sum_{q=1}^{Q} \frac{1}{V[D(W_m)]} \Omega\left(\frac{W_m - X_q}{h_Q}\right)$$

(4)

To detect modal regions of the pdf, as a first step, we apply a numerical morphological opening on this estimation, after that, we use our watershed technique to extract modal regions of the pdf.

## 4 Comparing the Classification Results

This section presents experiments that evaluates the proposed method and compares the performance of the two texture features extraction methods (Table 1), in our textural image (cf. Fig. 1).

**Table 1** Execution time and error rate of the both features extraction methods, using for Haralick method respectively 3 and 5 features, and for Fractal method 1 and 3 features

| Features | Extraction time | Learning time | Classification time (s) | Error rate (%) |
|----------|-----------------|---------------|-------------------------|----------------|
| Haralick (3) | 1 h 24 min | 3 min 9 s | 6 | 6.27 |
| Haralick (5) | 1 h 43 min | 4 min 30 s | 11 | 6.11 |
| Fractal (1) | 7 s | 2 min 26 s | 5 | 41.32 |
| Fractal (5) | 29 s | 4 min 51 s | 10 | 44.5 |

**Fig. 1** Textural image

## 5 Conclusion and Perspective

In this paper we present classification experimental results based on the combination of Kohonen map and morphological watershed transformations.

We present two different texture feature extraction methods, the first based on GLCM matrix, while the second method is based on local fractal dimension.

The results of this comparative study show that the Haralick method is more effective despite the greater computation time than the Fractal method.

As perspective, we search in this classification approach to compare the results with other feature extraction, and we search to apply this approach on both color textural image and 3D image.

## References

1. Haralick, R.M., Shanmugam, K., Hak Dinstein, I.: Textural features for image classification. IEEE Trans. Syst. Man Cybern. **6**, 610–621 (1973)
2. Haralick, R.M.: Statistical and structural approaches to texture. Proc. IEEE **67**(5), 786–804 (1979)
3. Chaudhuri, B.B., Sarkar, N.: Texture segmentation using fractal dimension. IEEE Trans. Pattern Anal. Mach. Intell. **17**(1), 72–77 (1995)
4. Sarkar, Nirupam, Chaudhuri, B.B.: An efficient differential box-counting approach to compute fractal dimension of image. IEEE Trans. Syst. Man Cybern. **24**(1), 115–120 (1994)

5. Talibi-Alaoui, M., Sbihi, A.: Texture classification based on co-occurrence matrix and neuro-morphological approach. In: Image Analysis and Processing–ICIAP 2013, pp. 510–521. Springer, Berlin (2013)
6. Talibi-Alaoui, M., Touahni, R., Sbihi, A.: Classification des images couleurs par association des transformations morphologiques aux cartes de Kohonen. CARI, Morocco (2004)
7. Talibi-Alaoui, M., Sbihi, A.: Application of a mathematical morphological process and neural network for unsupervised texture image classification with fractal features. IAENG Int. J. Comput. Sci. **39**(3), 286–294 (2012)
8. Kohonen, T.: Self-Organisation and Associative. Memory (1984)
9. Parzen, E.: On estimation of a probability density function and mode. Ann. Math. Stat. 1065–1076 (1962)
10. Asselin De Beauville, J.-P.: L'estimation des modes d'une densité de probabilité multidimensionnelle. Statistique et analyse des données **8**(3), 16–40 (1983)

# FPGA-Based Sobel Edge Detector Implementation for Real-Time Applications

**Ismaïl El Hajjouji, Aimad El Mourabit, Abdelhak Ezzine, Zakaria Asrih and Salah Mars**

**Abstract** This paper describes an implementation of real-time edge detection algorithm based on Field Programmable Gate Arrays (FPGA). The presented block is the first step for an embedded application for automotive system. From the nature of the application, challenges were hard real time constraints, robust edges detection with optimized hardware resources. In order to respect the requirements our choice is an implementation of the Sobel detector for its better compromise noise/material resources. Furthermore, to enhance noise performances of our architecture, we implement a dual threshold instead of a one threshold in the final stage of the detector. The obtained results show that proposed edge detector can reliably detect edges even if in the presence of luminance variation, using optimal hardware resources.

**Keywords** Sobel edges detection · Field Programmable Gate Arrays (FPGA) · Real-time · Dual threshold · Video processing

I.E. Hajjouji (✉) · A.E. Mourabit · A. Ezzine · Z. Asrih · S. Mars
Laboratory of Information and Communication Technologies,
National School of Applied Sciences, Tangier, Morocco
e-mail: eismail89@gmail.com

A.E. Mourabit
e-mail: elmourabit_aimad@yahoo.fr

A. Ezzine
e-mail: ezzine.abdelhak@gmail.com

Z. Asrih
e-mail: asrih.zakariaee@gmail.com

S. Mars
e-mail: mars.ietel@yahoo.fr

© Springer International Publishing Switzerland 2016                    681
A. El Oualkadi et al. (eds.), *Proceedings of the Mediterranean Conference
on Information & Communication Technologies 2015*, Lecture Notes
in Electrical Engineering 381, DOI 10.1007/978-3-319-30298-0_77

# 1 Introduction

The real time and embedded systems place severe constraints on the architecture in terms of latency, power consumption, performance reliability and cost, especially for image and video processing applications. Because it's optimal ration the price/performance ratio price/performance, FPGA is actually the best choice for real time video and images processing [1, 2].

Even though its sensitivity to noise and the dependence of its results to the luminance condition, the Sobel operator gives a better tradeoff between complexity, cost and computing time [3]. For this reason, we decided to implement an improved version of Sobel detector, where gradient is calculated by the same kernel as traditional detector and the result is thresholded with a dual threshold to enhance noise performances.

The rest of the paper is organized as follow: The second section is reserved for theoretical and mathematical description of the algorithm. In the third section we describe the implemented architecture. At the end, we present experimental results.

# 2 Sobel Edge Detection Algorithm

The Sobel filter is based on the use of two convolution kernels for horizontal $G_x$ (1) and vertical $G_y$ (2) gradients:

$$G_x = \begin{matrix} +1 & 0 & -1 \\ +2 & 0 & -2 \\ +1 & 0 & -1 \end{matrix} \tag{1}$$

$$G_y = \begin{matrix} +1 & +2 & +1 \\ 0 & 0 & 0 \\ -1 & -2 & -1 \end{matrix} \tag{2}$$

These cores may then be combined together to find the absolute magnitude of the gradient at each point as given in Eq. (3):

$$G = \sqrt{G_x^2 + G_y^2} \tag{3}$$

Hardware implementation of Eq. (3) is expensive in terms of material resources. These operations can be fairly approximated by relation of Eq. (4):

$$|G| = |G_x| + |G_y| \tag{4}$$

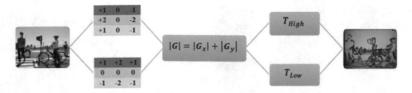

**Fig. 1** Sobel edge detector architecture with dual threshold

Conventional Sobel operator uses one static threshold to generate binary images from gradient images. In the case of detecting the edge in a road scene configuration, illumination varies during the day.

As a results, non-uniform illumination occurs. For such a situation, we propose to use of hysteresis threshold. A high and low thresholds are adopted which are compared to the gradient at each pixel. The Edge selection criterion is: (1) The pixel is rejected if its gradient is lower than the low threshold. (2) The pixel is an edge if pixel is high than high threshold. (3) If the gradient is between high and low thresholds, the pixel is an edge if it is connected to an accepted point. The implemented operation is showed in Fig. 1.

## 3 Implementation on FPGA

The hardware adopted for target system is showed in Figs. 2 and 3. The Sobel edge detector is implemented in a parallel and pipelined architecture in order to reduce the amount of data and speed up the overall operation. The operation of the implemented detector as described below:

The video input feed required by our project is generated by a real time digital camera in standard NTSC video format. The next step after image acquisition is a conversion from 24-bit RGB color standard into grayscale.

In the second stage, the pixel data are fed into a shift register which contains all the pixel information for an entire line of the VGA. We also output a pixel block $3 \times 3$ centered on the pixel in question, which is then used in the edge detection modules.

The module "detectors edges" decomposes into two separate and independent modules, although they are executed in the same way. This pipeline stage occurs after the first three scan lines stored in the buffer thus producing two successive blocks of $3 \times 3$ pixels. The input block of $3 \times 3$ pixels from the buffer is convolved

**Fig. 2** Sobel edge detector architecture

**Fig. 3** System hardware block diagram

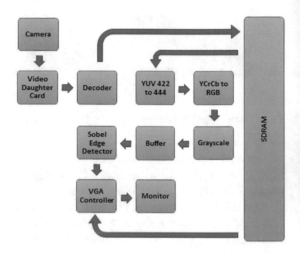

with the two components of Sobel and performs the addition and subtraction operations. The result of the convolution operations is compared with a predetermined threshold value according to the lighting. Thus, the detector saturates every pixel with a gradient value greater than its threshold value. The output of these two blocks are added up, resulting in a stream where the saturated pixels are considered part of the edges and the other pixels different than zero are considered edge candidates. Then, a sequence of operators test all pixels within the image to determine if edge candidates are connected to edge pixels for reducing the discontinuity of contours in the edge map.

## 4 Experimental Results

In order to evaluate the proposed architecture, we have implemented it in the Altera development board DE2 containing a Cyclone II EP2C35F672C FPGA device along with a digital camera Nikon COOLPIX P510 connected through NTSC connector.

Table 1 presents the overall FPGA resources used for the system. Regarding resource utilization, the FPGA Sobel detector version occupies only 5 % of the

**Table 1** Statistics of the compilation report of our final design (Image acquisition with Sobel Operator)

| Logic utilization | Used | Available | Utilization (%) |
|---|---|---|---|
| Total logic elements | 1,665 | 33,216 | 5 |
| Total combinational functions | 1,379 | 33,216 | 4 |
| Dedicated logic registers | 1,053 | 33,216 | 3 |
| Total memory bits | 72,268 | 483,840 | 15 |

| Original Image | Low Threshold | High Threshold | Dual Threshold |
|---|---|---|---|

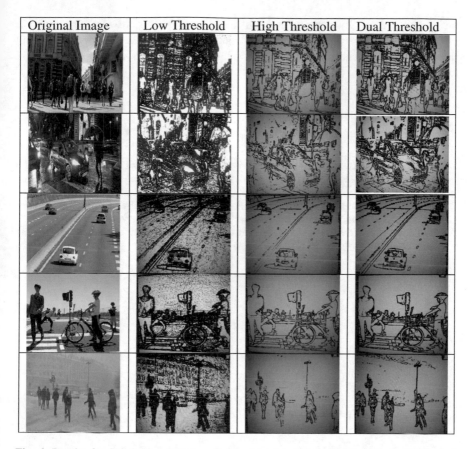

**Fig. 4** Results for Sobel edge detection. Original image, low threshold, high threshold, dual threshold

resources as described in Table 1. In terms of latency, timing analysis obtained during synthesis using the tool Quartus II, the frequency of the FPGA shows that the maximum operating frequency (Fmax) was: $\frac{1}{9.09\,\text{ns}} = 110$ MHz, is the fastest speed that the design clock can run without violating internal setup and hold time requirements. Figure 4 shows the proposed detector results on multiple images degraded by Gaussian noise and illumination variations. One can see that the proposed edge detector can reliably detect edges even if variation of luminance.

## 5  Conclusion

The proposed implementation of Sobel edge detector detection on FPGA shows the compromise that can be offered by the flexibility of the platform based on Field Programmable Gate Arrays. Thanks to the association of parallel and pipeline architecture, real time operation with more optimized hardware resource are obtained. Moreover edge detection performances is enhanced by adopting a dual threshold for the stage detection with slight modification of the conventional architecture.

## References

1. Xu, M., Xia, C., Huang, S.: A real-time noise image edge detector based on FPGA. In: 4th International Conference, ICSI (2013) Harbin, China, June 2013 Proceedings, Part II, pp. 420–427 (2013)
2. Halder, S., Bhattacharjee, D., Nasipuri, M., Basu, D.K.: A Fast FPGA Based Architecture for Sobel Edge Detection, vol. 7373, pp. 300–306 (2012)
3. Sanduja, V., Patial, R.: Sobel edge detection using parallel architecture based on FPGA. Int. J. Appl. Inf. Syst. 3(4), 20–24 (2012)

# Control Architecture for Mobile Robot Teleoperation

Mohamed Emharraf, Mohammed Saber, Mohammed Rahmoun
and Mostafa Azizi

**Abstract** Remote robots control system (telerobots) is an indispensable module for mobile robot navigation. The system based mainly on two models. The first one is used for controlling robot that work in well-defined environments. The second model concerns the monitoring and surveillance robots acting in an unknown and dynamic workspace. The disadvantages of these two models are the lack of real dynamic control of the robot tasks and limited control interfaces. The present paper proposes an implementation of an interactive control model. The model allows the robot not only to perform the operator commands, but also to run local functions, such as dealing with unexpected events. The experimental tests performed to evaluate the reliability and effectiveness of the interactive control model are promising.

**Keywords** Telerobot · Remote control · Interactive control

M. Emharraf (✉) · M. Saber · M. Rahmoun
Laboratory Electronics, Computer and Image Systems,
National School of Applied Sciences, Oujda, Morocco
e-mail: m.emharraf@gmail.com

M. Saber
e-mail: mosaber@gmail.com

M. Rahmoun
e-mail: moha1rahmoun@gmail.com

M. Azizi
Laboratory Mathmatiques appliques, traitement du signal et informatique,
First Mohammed University, Oujda, Morocco
e-mail: azizi.mos@gmail.com
URL: http://www.ump.ma

© Springer International Publishing Switzerland 2016      687
A. El Oualkadi et al. (eds.), *Proceedings of the Mediterranean Conference
on Information & Communication Technologies 2015*, Lecture Notes
in Electrical Engineering 381, DOI 10.1007/978-3-319-30298-0_78

# 1    Introduction

Nowadays, robots are not only able to perform basic motions [1], they are also capable of interacting with human operators [2–12]. Telerobots present in different areas of remote control with several potential applications such as telemedicine, distance learning, industrial automation and military. The main difficulties and limitations of remote telecontrol include the problems related to the control network such as bandwidth scarcity, transmission delays, as well as lost packets. All these limitations affect remote control telerobotics performances [13]. To find new solutions to these problems, some issues of telerobotics via network have been explored in recent years.

Currently the implementation of telerobotics are developed by several research teams (some of them are available online). The first generation of telerobots was mainly based on simple robot manipulators or mobile robots directly controlled by human operators [2, 3, 6]. These telerobots work in a structured environment with little uncertainty, and have no local reaction (intelligence). However, current research focuses on autonomous mobile robots navigating in dynamic and uncertain environments [4, 5, 14]. This generation of telerobots founded on autonomous and interactive robots; these robots can navigate in real world while dealing with uncertainties [13].

The existing telerobots systems can be classified into two categories: robot manipulators [2, 3], and mobile robots used for navigation [4–8]. For both, there is a multitude of control methods. Some robots use direct control [6, 8] which is not really suitable for performing remote operations by mobile robots. This is due to its high latency and other problems inherent to the use of network.

The paper organized as follows: Sect. 2 presents the overall framework for telecontrol; Sect. 3 discusses our telecontrol model; in Sect. 4, we present the obtained results.

# 2    The Interactive Model of Teleoperation

## 2.1    The Interactive Model Architecture

Figure 1 shows the architecture of developed telecontrol system. Is a basic interactive architecture, the operator orders issued by interactive commands with the application running on the robot system. The control interfaces allow the operator to streaming the video captured by the robot, display the information about the obstacle within a radius of 2 m and the actual speed of robot.

The telecontrol system is based on four modules: The Command Processor module (CPM) which processes and executes user commands, the transfer sensing

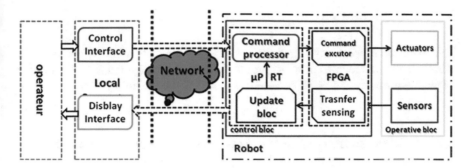

**Fig. 1** Telecontrol architecture

module that collects data from different robot sensors and prepares it for processing, the command executor module managed the feedback loop between the control system, actuators, and the sensors measurement. which allow the robot to react quickly to unexpected events derived from the real world; The update block module is responsible for transforming sensor data from low-level to high-level (e.g., the distance between robot and obstacle in a precise angle). The event detected by the robot allows CPM to make deliberation plan and respond to the encountered situation.

## 2.2 The Telecontrol System

The present telecontrol system is similar to the direct control used for vehicles driven by a human being. The operator uses the Direction command (change the direction left or right) to turn the robot, Up and down commands to control the speed of robot, to stop or even reverse the direction of robot walking.

The operator sends commands only if necessary (changing speed, direction, etc.), this greatly reducing the number of operator commands (compared to classic case). The robot decides autonomously about current situations; for example in the case of danger detection (obstacles), the robot autonomously reduces the speed to a reasonable level and sends a warning message to the operator control panel. If the danger is immediate, the robot stop. In some situations, the robot gets in the autonomous behavior mode that dominates potentially the operator control and it remains there until the danger disappeared.

The UP and DOWN commands affect the robot speed, the direction control affects the steering angle of the robot, we define the UP function in a way to allow the robot autonomously reducing its speed to a reasonable value, in the case of rotation or obstacle detection.

## 3 Experimentation

The test system is a unicycle mobile robot with four wheels and two motors, an ultrasonic sensor mounted on a DC motor. WIFI router configured as a client bridge of the main router, and an IP camera connected to the control system.

### 3.1 Implementation of Some Telecontrol System Functions

#### 3.1.1 Obstacle Avoidance

The control system calls the obstacle avoidance function in the case where the robot detects a near obstacle (10 cm), autonomously the execution of this function gets away the robot from the obstacle. Once the robot detects a distance of trust (30 cm) from the obstacle, the control system returns to handle the operator commands.

#### 3.1.2 The Motion Control

The engine control system is implemented in two parts; the first on the FPGA and the second on the real time $\mu$C. On the FPGA, we used a nonlinear control system based on a PID controller to stabilize the setpoint and to avoid the problem of engines pumping, over the control pulses management system which adjusts the motor speed.

To calculate the speed of the motors we used a module that is based on the pulses arriving from the encoder to deduce the actual speed of motor. The engine control module implemented on the microcontroller allows:

- An increase/decrease of velocity controlled by the operator with a maximum value.
- Display actual speed of each motor.
- Turn robot either left or right.

## 4 Results and Discussions

We test our telecontrol system using the path shown in Fig. 2. The human operator uses the commands UP, DOWN and DIRECTION to control the robot to move from A to D (Fig. 2). We notice that the robot moves more slowly around the points B, C, D. This is normal because the user changes the direction in these points which imply an automatic decrease of speed.

**Fig. 2** Robot test set in the path from *A* to *D*

The robot starts to increase the speed every time it receives an UP command from the operator, and the speed becomes stable once it gets the max value defined by the operator. When the robot becomes closer to an obstacle, the control system informs the operator about the obstacle and automatically reduces the speed to a reasonable value. When the robot is more close to the obstacles ($\sim 10$ cm), the speed jump to 0 rad/s and the control system uses the function of obstacle avoidance autonomously to get the robot out of danger; once the robot is brought out of danger, the control system comes back to execute the operator commands.

## 5 Conclusion

This paper propose an interactive remote control model (telecontrol). The control system is designed to perform various tasks independently and to react to expected events while the robot is dealing with unexpected events, for navigating in an unknown and dynamic world. The control interfaces allow the operator in the normal situation to control robot movements, display information about the robot, once the robot detects a danger situation the control system moves to standalone mode to get away the robot from the risk, based on local intelligence, and return to execution of the operator commands otherwise.

The experiments demonstrate that the results are promising. As perspective, we plan to go deep in this work and to add some extensions about quality and fulfillment of real-time properties.

## References

1. Nehmzow, U.: Mobile Robotics: A Practical Introduction. Springer, London (2000)
2. Taylor, K., Trevelyan, J.: Australia's telerobot on the web. In: International Symposium on Industrial Robots, pp. 39–44 (1995)
3. Goldberg, K., Gentner, S., et al.: The Mercury project: a feasibility study for Internet robots. IEEE Rob. Autom. Mag. **7**(1), 35–40 (2000)

4. Simmons, R., Fernandez, et al.: Lessons learned from Xavier. IEEE Rob. Autom. Mag. **7**(2), 33–39 (2000)
5. Thrun, S., Bennewitz, M., et al.: MINERVA: a second-generation museum tour-guide robot. IEEE Int. Conf. Rob. Autom. **3**, 1999–2005 (1999)
6. Saucy, P., Mondada, F.: KhepOnTheWeb: open access to a mobile robot on the Internet. IEEE Rob. Autom. Mag. **7**(1), 41–47 (2000)
7. Siegwart, R., Saucy, P.: Interacting mobile robots on the web. In: ICRA'99, Detroit, MI, USA, May 1999
8. Huosheng, H., Lixiang, Y., et al.: Internet-based robotic systems for teleoperation. Int. J. Assembly Autom. **21**(2), 1–10 (2001)
9. Li, P., Lu, W.: Implementation of an event-based Internet robot teleoperation system. In: Fourth World Congress on Intelligent Control and Automation, vol. 2, pp. 1296–1300 (2002)
10. Stein, M.R.: The PumaPaint project. Autonom. Rob. **15**, 255–265 (2003)
11. Cui, J., Sun, Z., Li, P.: Visual technologies in shared control mode of robot teleoperatioin system. In: Fourth World Congress on Intelligent Control and Automation, vol. 4, pp. 3088–3092 (2002)
12. Li, P., Lu, W.: Implementation of an event-based Internet robot teleoperation system. In: Fourth World Congress on Intelligent Control and Automation, vol. 2, pp. 1296–1300 (2002)
13. Sheridan, T.B.: Telerobotics, Automation and Human Supervisory Control. The MIT Press, London, England (1992)
14. Luo, R.C., Chen, T.M.: Development of a multi-behavior based mobile robot for remote supervisory control through the Internet. IEEE/ASME Trans. Mechatron. **5**(4), 376–385 (2000)

# Erratum to: A New Approach Based on PCA and CE-SVM for Hepatitis Diagnosis

Naoual Elaboudi and Laila Benhlima

## Erratum to:
"A New Approach Based on PCA and CE-SVM
for Hepatitis Diagnosis" in: A. El Oualkadi et al. (eds.),
*Proceedings of the Mediterranean Conference
on Information & Communication Technologies 2015,*
Lecture Notes in Electrical Engineering 381,
DOI 10.1007/978-3-319-30298-0_10

The original version of the Chapter 10 was inadvertently published with incorrect author name "Naoual EL Aboudi" instead of "Naoual Elaboudi". The erratum chapter and the book have been updated with the changes.

---

The updated original online version for this chapter can be found at
10.1007/978-3-319-30298-0_10

---

N. Elaboudi (✉) · L. Benhlima
Ecole Mohammadia d'ingénieurs, Mohammed V University, Rabat, Morocco
e-mail: nawal.elaboudi@gmail.com

L. Benhlima
e-mail: benhlima@emi.ac.ma